A TEXTBOOK OF BIOINFORMATICS

Information-theoretic Perspectives of Bioengineering and Biological Complexes

A TEXTBOOK OF BIOINFORMATICS

Information-theoretic Perspectives of Bioengineering and Biological Complexes

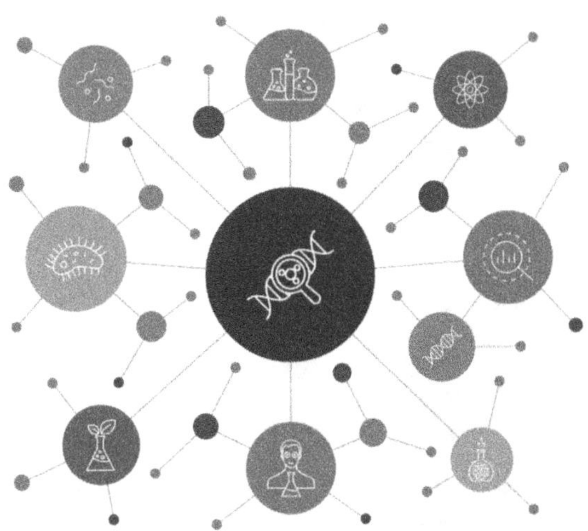

Perambur S Neelakanta

Ph.D., C. Eng., Fellow IET

Florida Atlantic University, USA

NEW JERSEY • LONDON • SINGAPORE • BEIJING • SHANGHAI • HONG KONG • TAIPEI • CHENNAI • TOKYO

Published by

World Scientific Publishing Co. Pte. Ltd.
5 Toh Tuck Link, Singapore 596224
USA office: 27 Warren Street, Suite 401-402, Hackensack, NJ 07601
UK office: 57 Shelton Street, Covent Garden, London WC2H 9HE

Library of Congress Cataloging-in-Publication Data
Names: Neelakanta, Perambur S., author.
Title: A textbook of bioinformatics : information-theoretic perspectives of bioengineering and biological complexes / Perambur S. Neelakanta, Florida Atlantic University, USA.
Description: New Jersey : World Scientific, [2021] | Includes bibliographical references and index.
Identifiers: LCCN 2020032967 | ISBN 9789811212888 (hardcover) | ISBN 9789811213847 (paperback) | ISBN 9789811212895 (ebook) | ISBN 9789811212901 (ebook other)
Subjects: LCSH: Bioinformatics. | Bioengineering.
Classification: LCC QH324.2 N433 2021 | DDC 570.285--dc23
LC record available at https://lccn.loc.gov/2020032967

British Library Cataloguing-in-Publication Data
A catalogue record for this book is available from the British Library.

For any available supplementary material, please visit
https://www.worldscientific.com/worldscibooks/10.1142/11627#t=suppl

Dedicated to...

My Grand Children, Tenzin, Maya, and Devan

Contents

Preface

Bioinformatics — It is a subject devised to address the wide span of bioengineering enclaving informatic details of biology and technical aspects of biotechnology. It refers to a potpourri of branches comprised of algorithms and computational schemes based on information-theoretic aspects of biological systems. The art of bioinformatics is conceived to include thematic details on information flow within the biological systems alluding to genomic data stored in the basic structure of deoxyribonucleic acid (DNA) strands, and the conceived scope is further extended to include intricate details of the complex profile of proteomics consistent with the suites of so-called *central dogma of microbiology.*

Beyond the classical theoretics and wet-lab details of biological sciences, fused into the perspectives of modern bioinformatics today, are *in silico* considerations that enable newer scope on state-of-the-art considerations covering topics, such as next-generation sequencing, translational/ clinical bioinformatics, viral informatics, cytogenetic informatics, and analytics of phylogeny. As such, bioinformatics today broadly includes dogmatic notions of genomics and proteomics, principles of genetics, heuristics of evolution-theoretics, concepts of information theory, and science of computing. The underlying paradigms are further expanded to address the associated molecular informatics through cellular/cytogenetic feature details culminating in descriptive notes on the whole gamut of living systems, and the associated suites of analytical modeling, instrumented and/or wet-lab inferences plus biochemical assessments (at cellular/molecular and whole-life levels) have become heavily dependent on computer-assisted analyses and data storage endeavors.

This book is specifically written on certain unique considerations in bioinformatics and presented as: (a) a set of comprehensive topics on both traditional and state-of-the-art (modern) details; (b) summary details at the end of each chapter supplemented with a collection of tutorials, exercises, and examples on the analytic perspectives and/or case studies. They are designed as end-of-chapter sections with an objective to enable readers to improve their knowledge on the subject as well as enhance their skill toward collecting (and learning) more details on the relevant topics from the archival of information deliberated in the vast literature and supercharged in the knowledge engines of multitudes of web pages; and (c) a need-based gist of detailing (*via* examples and rider problems) to portray the "bigger picture" of practical implications in assimilating the concepts of bioinformatics in the perspectives of modern bioengineering and biotechnology. Novelty of this "big picture" also stresses analytical questioning and problem-solving skills as being the key toward successful understanding of the concepts and making best use of the prevailing knowledge of the subject-matter in question. Such details are presented not just for pedagogical and classroom applications of the *academia* — they are also framed to help developing research concepts toward conceiving thesis/dissertation exercises as well as assist managerial aspects of bioinformatics.

Perambur S. Neelakanta
Boca Raton, Florida, USA
August 2019

Acknowledgments

Books are conceived, written, published, marketed as well as read, resold and possibly forgotten!! But deeply in the heart of the author, perennially lingers, all the hardships faced, patience strived, and above all, help received in making of the book to its completion!

As such, the author of this book places on record and gratefully extends special thanks primarily to all his students who participated in his research, attended his lectures (on bioinformatics), and partly enabled this book to take its shape. First to mention, are those students who were his research scholars and contributed abundantly by their untiring thesis/dissertation efforts. They shared their immense knowledge and provided constructive criticisms — all such details have become the seed in conceiving the scope of this book.

Acknowledging gratefully thereof are his Ph.D. students, Dr. Tomás Arredondo and Dr. Sharmistha Chatterjee, who inspired him to indulge into distinct avenues of bioinformatics while being his doctoral students and exercising concomitant research tasks under his guidance and gracefully shared their knowledge bilaterally with him. Their academic contributions are liberally inherited, generously borrowed, and partly adopted across all chapters of this book on *ad hoc* basis. Their efforts rendered toward solving intricate problems and their untiring help in performing computational exercises were remarkable.

Furthermore, the author's heartfelt thanks are extended to students who did their MS theses (in bioinformatics discipline) under his supervision. To mention, pertinent scholastic research outcomes are due to: Ms. Shivani B. Pandya, Ms. Alice N. DuPont, Mr. G. A. Thengum Pallil, Ms. Jagdeshwari Karri, Ms. Sandhya Sharma, Ms. Deepti Pappusetty, Ms. Poornima Shankar, and Mr. Meta Leesirinkul. Their impressive contributions, matchless tasks, and extensive studies were useful in building the scope and partly projecting the objectives in this book. Also, contributions on triage (*via* directed independent studies) by Ms. Scarlett Somarriba are thankfully acknowledged here. To all those students — saying THANK YOU is the author's reciprocally phrased gesture.

Also, the author deeply appreciates several batches of graduate students (who were enrolees in his lectures on bioinformatics) for their patience in doing assigned exercises and delivering their well-thought-out solutions with in-depth feedback. Pertinent academic contributions can be seen as exemplars across all chapters of this book. Again, the priceless phrase to praise them — a hat-tip of admiration.

Apart from the gamut of student folks and team of fellow researchers, the intellectual support and inspiring discussions received from the collegial friends, like Dr. Mirjana Pavlovic, Dr. Joseph C. Park, and Dr. Dolores De Groff are duly recognized here. Specifically, the computational help rendered by Dr. Dolores De Groff and Dr. Joseph C. Park are remarkably noteworthy and liberally adopted across the chapters of this book; further, the contribution from Dr. Kalpana R. Tirth (of Anna University, Chennai, India) on Haralick's parameters is mentionable and it is aptly used as a part of an exercise in the text.

In all, devotedly, the author thanks them all — friends who represent a vanguard in developing useful concepts, incredible debates, and dogmatic deliberations on various bioinformatic topics profusely considered and profitably assimilated in this book, and their useful suggestions plus expressed opinions have rendered proper setting of the contents in the chapters in right perspectives.

Lastly, a line-of-appreciation is gratefully extended to the Editor and the Editorial staff of World Scientific Publishing Co. — for their untiring assistance, cooperative rapport, and prompt processing of the manuscript of this book.

In end, but not in the least, the author's heartfelt thanks are due to his family — his wife Manjula Neelakanta, his two sons, their spouses and above all, his three lovely grandchildren. Furthermore, in conceiving the pedagogy of this book, the author gratefully remembers the inheritance of knowledge from his parents and pedagogical impressions gained from two of his beloved teachers, Professor Sisir Kumar Chatterjee and Professor (Mrs.) Rajeshwari Chatterjee (of I. I. Sc. Bangalore, India), who though demised, but "*... like a candle they consume his light-of-life*"

Mother (Matha) and father (Pitha) take us to the teacher (Guru), and it is the Guru, through his or her teachings, points us to the bliss of Supreme Divinity (Deivam)...

Section I
Bioinformatics: An Introduction

This section is written as a prelude consisting of a pair of chapters with an objective of learning the fundamental aspects of bioinformatics; hence, relevant introductory notions are narrated and supplemented with underlying concepts, explanatory details, and necessary elaborations. Further, the synergism of interdisciplinary subjects — namely biosciences, informatics, statistics, mathematics, biochemistry, biophysics, technology, engineering, and computer science — forms the focused deliberations outlined in the following two chapters.

Chapter 1: Introduction
Chapter 2: Information Theoretics: A Review

Chapter 1

Introduction

Outlined as the constituent theme of this chapter is a primer on biological science and heuristics of informatic concepts extended to narrate the basics of microbiology that frames the notions of bioinformatics.

Bioinformatics "... This (informatic) spud may or may not be for you; ... but, (it is) bioengineered to (near) perfection"

1.1 General

In a precise statement, *bioinformatics* refers to a uniquely defined art deliberated to view the information-theoretic aspects of biological complex. That is, bioinformatics implies an exclusive subject intended to comprehend the informatic framework of biosystems, and relevant pedagogy applies to learning synergistic details of biology, bioengineering, biotechnology, as well as the allied fields of medical science and health issues in the perspectives of information theoretics.

Thus, bioinformatics represents a potpourri of branches that covers the intricacies of biological details alluding to the informatics of genetic message inscribed across what are known as *nucleotide sequences* of the microbiological complex, as well as bioinformatic studies govern functional aspects of entities that participate in the eventual translation of the said nucleotide complex into *molecules of life* — known as the *proteins* that elegantly frame the gamut of cellular details of the whole-life systems. Hence, bioinformatics is laid on the notion of biological details pertinent to informatic aspects of *genomics* and *proteomics* of living systems at molecular level, and such omic considerations are stretched further to study the cytogenetic informatics at cellular level and expanded to include so-called phylogenetic aspects of whole life.

The term *genome* refers to genetic constituent of an organism. Genome thus includes the basic microbiological entities, namely, the so-called *deoxyribonucleic acid* (DNA) and *ribonucleic acid* (RNA), and it is constituted by a set of "genes" that carry genetic information expressed by a set of biomolecules. These are polymeric stretch of macromolecules, known as *nucleotides*, which represent four organic monomer compositions (known as, *base residues*), namely, *the adenine* (A), the *cytosine* (C), the *thymine* (T), and the *guanine* (G), and this set of {A, C, T, G} depicts the *alphabets of life*.

Typically framed as a double-helix structure, pertinent type of DNA, known as the *double-stranded DNA* (dsDNA), consists of two biopolymeric strands; each strand being made of random dispositions of nucleotides in one-dimensional (1D) sequence format. (Also exists another class of DNA known as *single-stranded DNA* (ssDNA) in certain species, for example, in viruses, and the distinguishing details of dsDNA and ssDNA are explained in pertinent sections of other chapters.) The randomly disposed {A, C, T, G} set of nucleotides in the DNA sequence provide genetic instructions encoded along the stretch of the DNA strands. Descriptive illustrations of the DNA, its transcribed version (depicting the RNA) and eventually converged translated versions of the protein complex are presented in Appendix 1A.

The RNA mentioned above is a transcribed derivative of the DNA, and it contains the mapped details of genetic information present in the DNA (or, it could be an entity all by itself possessing relevant template of genetic information in certain living systems such as in *retroviruses*). The RNA is a single-stranded stretch of nucleotides constituted again by the four nucleotide residues {A, C, U, G} with a monomer, *uracil* (U) replacing thymine (T) of the DNA. The genetic informatics of the RNA is also implicated across various microbiological roles in coding, decoding, regulation, and expression of genes.

The so-called *central dogma of microbiology* specifies the steps of the protocol on how the DNA is being transcribed into RNA, which then translates into a *protein complex*. Protein represents an entity of microbiological interest, and the entire gamut of protein complex denotes the proteome. It is comprised of, and expressed by, the associated genome housed in the cellular premise of an organism, at any specified time. That is, a proteome depicts a set of expressed proteins in a given type of cell or organism at any given time under a defined ambient and condition.

In short, topics of bioinformatics judiciously merge the constitutive details of DNA, central dogma of molecular biology, genetics of living systems, concepts of information theory, analytics of mathematical strategies, theoretics of stochastic processes, methods of enumerative statistics, heuristics of probability theory, and tools of modern computing science. Thus, bioinformatics emerges as a cohesive subject on the informatics of biology — a meaningful set of informatic profiles (or *negentropy* details in Shannon's sense[1,2] as will be described later).

Written across the chapters of this book is a compendium of *in silico* details (of bioinformatics) outlined[1,2] as computer-assisted evaluations of bioinformatic data and details. Concomitantly addressed are ways and means of disseminating and sharing the associated (massive) data using globally pervading vast information networks. Hence, particulars on exhaustive databases storing massively large sets of data (collected worldwide and conceived as repertoires of bioinformatics) are described, and the associated computational tools and dissemination strategies *via* global Internet infrastructure are described.

1.2 Scope and Objectives of Bioinformatics

If the query, "why bioinformatics?" is posed, the answer that emerges is simple — it is to set a wedlock between the explicit knowledge of information engineering and technology and the not-so-well understood or assimilated domain of biological information processing. The theme of bioinformatics also signifies another query: "Why is bioinformatics important?" What follows as a reply depicts the related details of eventual elaborations in ensuing chapters. Further posing the question, "How is bioinformatics constituted?" the answer is that it cohesively includes the scientific themes of information-theoretics on molecular, cellular and system biology, and it is based on related analytical aspects of stochastic processes, methods of mathematical biology, tools of computational methods, schemes of massive data storage and prospects of (tele)communication networks.

1.3 Computers and Bioinformatic Tasks

In view of the large-sized data structures linked to abundant details of biological domains implies the inevitable suite of computer-in-loop assistance. "Bioinformatics is grounded in molecular biology, clinical medicine, a solid information technology infrastructure and business" and the associated computational analyses form the cradle of bioinformatic tasks. As Bergeron[3] indicates, "... using computational methods to leverage the technique of molecular biology is a viable approach to increasing the rate of innovation and discovery" in bioengineering methods and bioinformatic strategies.

In posing the question: "Why bioinformatics at all?" the answer is simple — bioinformatics is conceived toward realizing a definite set of efforts in biotechnological informatics, where the art of computer science is co-mingled with the subject of biological processing. This realization duly involves human participation, meaning (human-expert)-in-the-loop. Relevant exercises involve the following: (i) Creating databases that allow storage and management of extensive biological datasets; (ii) developing algorithms and applications of statistics, which determine the relationship among the members of large datasets; (iii) evolving software and tools exclusive for bioinformatic problem-solving; and (iv) handling data sets pertinent to the entirety of biosystems, namely, the DNA/RNA sequences, the protein complex, structural and functional aspects of the proteins, *gene expression profiles, biochemical pathways*, cellular-level *cytogeny*, evolution-theoretic *phylogeny, translational bioinformatics*, whole-life biology and *next-generation sequencing* (NGS) methods.

1.4 Genomics and Proteomics

The baseline of bioinformatic studies is drawn on the plane of biological systems. Relevant studies on such systems are covered in the science of *biology*, which globally refers to the science of living systems.[4] That is, biology is a scientific framework of knowledge devoted to study life-bearing biosystems, which depict the entire set of self-replicating and persistently evolving entities of enormous complexity and exist parallel to, but distinct from, the so-called physical systems. The observable distinction seen in living systems (*versus* the physical systems) primarily conforms to the constituent, large complex molecules of cells that infuse "life" into the biosystems; in addition, the biological living systems bear reproductive ability and sustain, thereof, a genetic continuity. They can also capture energy and dissipate it in different forms to maintain life processes, sustain the associated reconstructive mechanisms, perform repairs (that involve replacing the structural parts of the biosystem as needed) and excrete waste products.

In the context of molecular biology and genetics, the implied details of biocomplex organization are concerned with the genetic material of any organism.[3–5] Such a genetic material is known as the *genome*, which refers to the gene-structure representing the entire set of biosequences containing *genes* that bear the necessary information required to build and maintain the sustenance of an organism. Parallel to the subject of genomics exists proteomics, which depicts the details on the protein-structure of the biocomplex organization. That is, generalized in terms of what is known as the *proteomes*, proteomics is a study comprised of the entire gamut of protein complex as expressed by the genome in the cell of an organism at a specified time. This study includes both physical (structural) details and functional characteristics of the proteins.

Further, in system biology, the term *cellular proteome* is exclusively used in denoting a collection of proteins seen in a particular cell type under a given ambient. A *complete proteome* is also defined to conceptualize the entire set of proteins resourced from various cellular proteomes, approximately equating the proteome as a protein-equivalent of the genome. Lastly, a collection of proteins in certain sub-cellular biological systems may also depict the proteome, for example, all of the proteins in a virus (called *viral proteome*).

Within the purview of central dogma, the nucleotide residues making the primary sequence (of DNA/RNA) in an organism, *transcribe* into a sequence of *amino acids* (AAs) made of triplets (or *codons*) of the nucleotides, and the AA-sequence eventually *translates* into a three-dimensional (3D) structure of the end product, namely, the protein complex. A part of proteomic study refers to assessing such structural details, and this is known as *protein-folding problem.* Theoretically speaking, the secondary structure of a protein can be specified from the primary sequence details (pertinent to DNA/RNA residues) using appropriate methods of statistics and other optimization strategies *via* multiple sequence alignment data.

1.5 Bioinformatic Resources

Details on bioinformatic resources are elaborated in Chapter 5. An outline of such details can be stated in terms of the following topics: (i) Comprehensive repositories (or databases) containing extensive biological information; (ii) worldwide communication networks that provide bases for global dissemination of biological data; and (iii) a host of computational tools in vogue adopted for data mining and bioinformatic analyses. Enumerated below are salient bioinformatic resources.

A. Bioinformatic Information and Database Resources

The biological information resources include a gamut of repositories hosted by a number of databases worldwide. These are exclusively conceived in two categories, namely *genome information resources* and *protein information resources,* outlined below.

Genome information resources: Biological raw data are stored in public databanks (like *EMBL*[6] and *GenBank*[7,8]) that store primary DNA sequences. The other major nucleotide database is DDBJ[9,10] which evolved as DNA sequence databases. The genomic sequence databases are widely used in the contexts of molecular biology for *homology searching.* Relevant databases assist understanding the biochemical function, chemical structure and evolutionary history of organisms. Databases, apart from being simple data repositories, are also intended to provide additional functionality, such as access to sequence homology searches and links to other databases and analysis results.

Protein information resources (PIR): These information resources on proteomics (as will be detailed in Chapter 5) represent protein sequence databanks containing verified protein sequences plus annotations describing the function of a protein. Protein-specific databases provide details on levels of data stored on primary, secondary and tertiary structures of proteins specifying the formats of primary and secondary sources. Further, these resources globally constitute the evolution of composite databases and database integration efforts. Protein sequence databanks (e.g., the *trEMBL*) are provisioned to provide the most likely translation of all coding sequences that remain as repositories in the EMBL databank.

The primary protein database is concerned with protein structure information available in the *protein databank* (PDB). The PDB is the key resource on structural biology, such as *structural genomics.* Other databases also use protein structures deposited in the PDB. For example, *SCOP* and *CATH* are databases that classify protein structures, while *PDBsum* provides a graphic overview of PDB entries using information from other sources, such as gene ontology.

Primary protein sequence databases: These include the PIR and *protein identification resource* and provide efficient on-line computer system designed for the identification and analysis of protein sequences so as to establish their corresponding coding sequences. The consortium of PIR is made of National Biomedical Research Foundation (NBRF),

International Protein Information Database of Japan (JIPID), and Martinsried Institute for Protein Sequence (MIPS). The PIR maintains *protein sequence database* (PSD), which is a comprehensively annotated protein database containing over 283,000 sequences and covering the entire taxonomic range.[11]

Bioinformatic literature databases: Commensurate with the exponential growth of biological information disseminated in the previous decades, also came into being are databases (such as, *PubMed, Medline*) that include scientific literature (with detailed references and topic annotations) providing additional functionality that supplements the art of bioinformatics.[12,13] Separate entries relevant to various literature reports and publications prevail across many sequence databases for any given protein sequence. As such, the database, such as the Swiss-Prot,[12] merges all such data, minimizing the redundancy of repeated details.

Bioinformatic database management resources: Considering the size, the number and complexity of bioinformatic databases, dynamic and efficient management on the stored data for accession and use is imperative. Also as required, the databanks need to be accessed simultaneously and correlated with each other. Regardless of co-mingled and diverse sets of biological details present across databases, unified data retrieval is implicit. Typically, this is facilitated by special languages (e.g., the *Sequence Retrieval System,* or SRS).

B. *Bioinformatic Information Networks Resources*

In the hierarchy of bioinformatic resources, as noted above, the databases constitute the core of bioinformatics. Relevantly implied objective of bioinformatics is not only storing the associated "big-data" but also facilitating the dissemination of such data effectively among the users. Hence, global bioinformatic information network has emerged with the advent of telecommunication linkages facilitated by the Internet and network options that allow interconnectivity between the entire plethora of bioinformatic databases worldwide as well as the host nodes of the users. The bioinformatic databases are designed as a structured resource of repository storing information and characterized by a hierarchy of (bioinformatic) resources with effective dissemination capability of the data among the users sharing genomic and proteomic details and facilitating information networks via telecommunication linkages with the Internet, WWW, etc.[3] as described in Chapter 5.

C. *Bioinformatic Tools*

The collection of tools adopted in computer-based efforts of bioinformatic tasks forms another essential set of bioinformatic resources. Such tools are software programs developed toward extracting meaningful information from the vast details on experimental and/or theoretical studies exercised in the realms of molecular and system biology and stored across worldwide biological databases. In essence, bioinformatic tools enable biosequence-related analyses, and they can be identified as follows: (i) Sequence analysis tools; (ii) homology and similar search tools; (iii) protein function analysis tools; (iv) protein structural analysis tools; (v) sequence comparison tools; (vi) multisequence alignment tools, and (vii) phylogenetic tree-making tools. All these tools refer to specifically written software programs.[14–21]

1.6 Beneficiaries of Bioinformatics

Described below are some details on eventual beneficiaries of the underlying science, engineering, and technology of bioinformatics.

Biotechnology: An Overview

In modern context, biotechnology (or bioengineering) refers to the art of making optimum use of natural processes or products of living systems. This strategy has been enabled as a result of discoveries and better understanding of metabolisms of natural processes in living systems in the recent part of yesteryears. In essence, as defined by The European Federation of Biotechnology, the art of biotechnology represents an "integral application of knowledge and techniques of chemistry, microbiology, genetics, and chemical engineering to draw benefits at the technological level from the properties and capacities of microorganisms and cell cultures." The subsets of areas of biotechnology are as follows.

Medical and pharmaceutical biotechnology: Underlying methods use microorganisms (such as bacteria or fungi) to make antibiotics or vaccines. In practice, among the compounds synthesized for pharmaceutical applications, less than 10% are tested on humans and only about 2% hits the market. Almost a decade or more is spent on average on preclinical and clinical trials following protocol of phases, I through IV indicated in Ref. 3: Phase I: Dosage studies. Phase II: Evaluation of effectiveness and side-effects. Phase III: Long-term surveillance; and,

Phase IV: Clinical trials in envisaging these phases, the usage of computers is obviously significant. The vocabulary on terms of bioinformatic, medical and biological significance has not been totally and comprehensively standardized as outlined in sources on medical vocabulary with its own strengths and weaknesses.[3]

Pharmacological relevance: Many aspects of bioinformatics are relevant for pharmacology. Drug targets in infectious organisms can be revealed by whole genome comparisons of infectious and non-infectious organisms. The analysis of single nucleotide polymorphisms reveals genes potentially responsible for genetic diseases. Prediction and analysis of protein 3D structure is used to develop drugs and understand drug resistance.

Industrial biotechnology: This deploys microorganisms to make enzymes (e.g., to add to biological washing powders) or to produce beer, cheese or bread. *What is industrial biotechnology*? It is a technological venture of using of biological organisms or related process in manufacturing and service industries so as to reap certain product and/or service benefits. Bacteria, yeast, algae, the cells and the tissues of higher plants or the enzymes isolated from the organisms provide active ingredients as substitutes for existing/traditional chemicals and the associated process methods leading to improved end products and better industrial processes based on microbiological concepts. Industrial biotechnological strategies also find wide applications in pollution-free leather processing, and in food technology. Another major penetration of biotechnology in the industrial side refers to managing the biomass.

Environmental biotechnology: Here microorganisms or plants are utilized to clean up land or water that is polluted with sewage and/or industrial waste. That is, environmental issues can be viewed across biotechnology efforts with reference to the following: *Adaptation and complexity, environment-related health concerns, diversity considerations* and the so-called *precautionary principle* as outlined in the following subsections.

Adaptation and complexity: The classical definition of bioinformatics is the science of solving biological problems using mathematical, statistical and computing methods; it is evolved to suite uniquely the tasks pertinent in assessing the potential aspects of environmental and natural public health risks (due to, for example, inevitably pervading presence of bacteria and viruses), as well as the hazards that arise from man-made biotechnological and other activities. Across this broad regime of interest, environment is seen as an extremely complex natural adaptive system that includes both living and non-living entities constituted by *biotic* and *abiotic* inclusions, and the associated complex systems are disturbed the way in which they respond to the disturbance depends on their makeup. It is believed that the stability of an *ecosystem* is that, which allows a rich diversity of species to survive.

Environment-related health concerns: The role of genetic engineering essentially implies genetic modification or manipulation of an organism's genome using biotechnology. A host of methods conceived thereof can change the genetic makeup of cells, including the transfer of genes within and across species boundaries to produce improved or novel organisms.[22]

Studies show that about 50% of the Earth's species may vanish within the decades of the century. In Ref. 23, it is remarked: "We're talking about nothing less than the preservation of human life-support systems. ...We neglect the issue at our peril." In this state of predicament, bioinformatics-related studies can prudently help the society to engineer ways to tackle the environmental debacles indicated above.

Bio-diversity issues: *Bio-diversity*, or *biological diversity*, is an attribute observable in an environment as indicated by numbers of different species of plants and animals. Relevant considerations are important in order for the ecosystem to properly function and maintain its stability. Eventually, the concern is not necessarily whether humanity will survive in some form or another within this declining adaptive environment, but how this declining adaptive environment will impact the growing human population under "biopollution."

The precautionary principle: This concept was proposed as a new guideline in environmental decision-making.[24] It has four central components: (i) Taking preventive action in the face of uncertainty; (ii) shifting the burden of proof to the proponents of an activity; (iii) exploring a wide range of alternatives to possibly harmful actions; and (iv) increasing public participation in decision-making. The precautionary principle as established in the Rio Declaration from the 1992 United Nations Conference on Environment and Development, also known as the *Agenda 21*, states the following: "In order to protect the environment, the precautionary approach shall be widely applied by States according to their capabilities. Where there are threats of serious or irreversible damage, lack of full scientific certainty shall not be used as a reason for postponing cost-effective measures to prevent environmental degradation."

Agriculture biotechnology: This enables producing better crops, "natural" fertilizers or feed additives. Modern biotechnology when viewed as a gene technology has the potential to contribute to sustainable agricultural practices toward abundant produces notwithstanding the prevailing public debates about its potential impact on the environment and human health.

Warfare biotechnology: This refers to (biological) warfare methods using harmful biological agents against enemy forces. Although harshly criticized, this technology has gained a worldwide repudiation and has been marked as a technology in the bad interest of humanity as a whole.

1.7 Chapter Topics: An Outline

Commensurate with the essentials topics of bioinformatics addressed in this book, the chapters that follow contain salient features deliberated across five major sections listed below, and the relevant chapters in each section are indicated in the table of contents.

Section I: Bioinformatics: An Introduction
Section II: Informatics of Biomolecular Sequences
Section III: Bioinformatic Resources and Translational/Clinical Informatics
Section IV: Bioinformatics of cytogenetic Complex and Viral Omic Landscape
Section V: Phylogenetic and Species Informatics

1.8 Review Section: Tutorials, Examples, and Exercises

Commensurate with the scope of this book (Chapter 1), this section offers some examples and exercises along with tutorials on pertinent topics and *ad hoc* heuristics of bioinformatics. Thus, gracefully merging the aspired goals of bioengineering feats and biotechnological efforts, the following subsections are presented with necessary examples and self-study exercises as review efforts. Appendix 1A concurrently provides some basics of biology/microbiology as needed. Additionally, relevant literature annotations and website details (for example as in Refs. 25–40) can be availed as necessary in comprehending the topics narrated.

Tutorial T-1.1

DNA: An overview: As mentioned earlier, the *deoxyribonucleic acid* (DNA) depicts a sequenced set of four, large monomers (called *nucleotide bases*): *Adenine* (A), *Cytosine* (C), *Guanine* (G) and *Thymine* (T). These four molecules form the so-called "alphabets of life" and constitute basic building blocks of the DNA sequence, and the associated statistics of these alphabets construct the genetic information supported by the DNA. The chemistry of polymeric nucleotides frames the backbone structure of the DNA chain. It consists of alternating sugar and phosphate residues, and the sugar constituent is a deoxyribose having five carbons, and the sugar constituents are linked to each other. Further, attached to each sugar is a nitrogenous base at carbon atom 1′ (one prime) and the sugar with its attached base depicts a *nucleoside.* A nucleoside plus an attached phosphate group to its 5′ or 3′ carbon represents a *nucleotide* (which as stated above, forms the basic building block of the DNA strand).

Typically, the DNA is structured as anti-parallel double-helix strands* (dsDNA), and each strand is constituted by a sequence of nucleotide bases {A, C, T, G}, as indicated above. (*There are also *single-stranded DNA* sequences (ssDNA), for example, as in certain viruses outlined in Chapter 8). The opposing (anti-parallel) strands of the double-helix have opposite directions linking 5′ to 3′ (in forward direction); and, the stretch 3′ to 5′ (making a *complementary strand*) in reverse direction. (As stated above, a phosphate group attached to its 5′ or 3′ carbon on the sugars in the nucleoside.) Across the twisted pair of double-helix strands, the base A pairs with T, and G pairs with C (or *vice versa* in both cases). They are known as *Watson–Crick* (WC) *pairs.*

Considering the genetic details associated with DNA/RNA sequence, it refers to the inscribed information by the set {A, C, T, G}, and the associated protocol, known as the *central dogma of microbiology,* augments the gene information, so as to get the DNA transcribed into RNA, which eventually is translated into the chemistry of a protein complex. The underlying steps involved (in central dogma of microbiology) are illustrated in Figs. 1.1 and 1.2.

With reference to a dsDNA, the first step corresponds to transcription process of the cell's machinery copying the gene information (in the unwound complementary strand) of DNA sequence into a *pre-messenger ribonucleic acid* (pre-mRNA) sequence. The molecules of (pre-mRNA sequence are the same four nucleotides; however, the *Thymine* (T) is

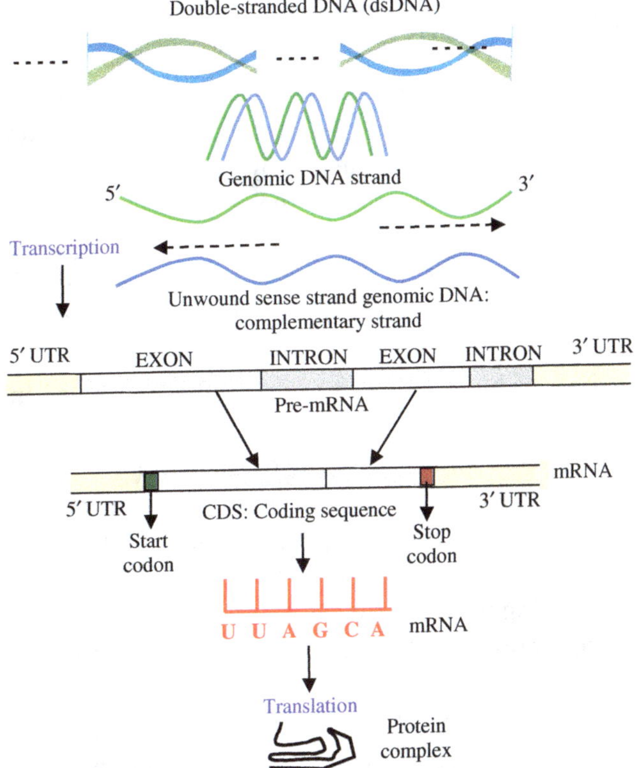

Fig. 1.1 Illustration of transcription through translation steps involved in a dsDNA as per the *central dogma of microbiology* with explicit indications of *untranslated regions* (UTR) and *coding sequence* (CDS) parts of the sequence, and the associated pre-mRNA and mRNA stages are also marked.

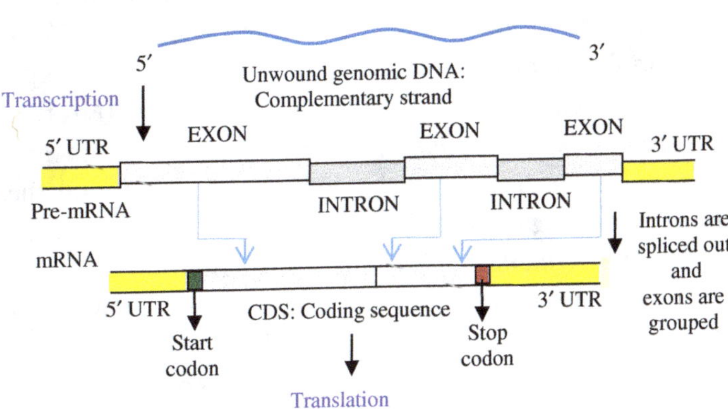

Fig. 1.2 UTR and CDS parts of the pre-mRNA and mRNA stages: The unspliced mRNA precursor and the resulting mRNA (after introns are removed *via* splicing). This mature mRNA sequence is ready for translation.

replaced by another base, *Uracil* (U) subsequently; and the pre-mRNA now is designated as the *messenger RNA* (mRNA). It is now arranged in triplets of nucleotides, called *codons*, that is, there are 64 codons possible and the pre-mRNA depicts a random sequence of the codons. Some segments of such codon arrangement may contain the gene information carried forward *via* transcription, and they are known as *exons* (implying entities with expressed genetic details).

There are also segments of codons that bear no gene-implied informatic messages. They are called *introns* (or *junk codons* depicting a set of "introduced" codons possessing *nil* or junk details). Further, the set of 64 codons (framed by triplets of {A, C, T, G} are grouped into 20(+2) *amino acids* (AAs) as listed in Table 1.1, where the AAs are indicated with their three-letter and single-letter designations.[36]

As indicated in Fig. 1.1, the transcription step involves forming a *precursor messenger RNA* (pre-mRNA) sequence using the reverse strand unwound from the DNA helix. Next *RNA splicing* takes place. It is a RNA processing in which the pre-mRNA transcript is transformed into a mature, *messenger RNA* (mRNA). During splicing, introns are removed (spliced out) and exons are joined together. For eukaryotic genes that contain introns, this splicing is usually required in order to create an mRNA molecule that can be translated into protein as per the gene encoding. Largely, for such eukaryotic introns, splicing is carried out in a series of reactions that are catalyzed by the *spliceosome*: a complex of small nuclear *ribonucleo-proteins* (snRNPs). The coding region in Fig. 1.1 of a gene refers to the *coding sequence* (CDS) depicting the portion of a gene's DNA (or RNA) that codes for protein. The region usually begins at the 5′-end by a start-codon and ends at the 3′-end with a stop-codon. The sum total of the coding regions of an organism's *genome* is called *exome*. The CDS is that portion of an mRNA transcript that is translated by a *ribosome*. The coding region in an mRNA is flanked by the five prime *untranslated regions* (5′-UTR) and the three prime *untranslated regions* (3′-UTR). The CDS is also synonymous with the terms, *coding sequence* and *coding region*. In all, the CDS is a key feature used to denote the "protein-coding sequence" in a gene feature table by the major sequence databases of International Nucleotide Sequence Database Collaboration (INSDC). It is a collaborative initiative that operates between DDBJ, EMBL-EBI, and NCBI.

Table 1.1 The sixty-four (64) codon triplets constituted from the base set: {A, C, T, G} and grouping of triplets into amino acids (AAs). Each AA is shown by its single-letter (1D) and triple-letter (3D) designations.[36]

1st and 2nd nucleotides	3rd nucleotide added	Amino acids (1D) and (3D)	1st and 2nd nucleotides	3rd nucleotide added	Amino acids (1D) and (3D)
TT	TTT	(F) Phe	**TC**	TCT	(S) Ser
	TTC	(F) Phe		TCC	(S) Ser
	TTA	(L) Leu		TCA	(S) Ser
	TTG	(L) Leu i		TCG	(S) Ser
CT	CTT	(L) Leu	**CC**	CCT	(P) Pro
	CTC	(L) Leu		CCC	(P) Pro
	CTA	(L) Leu		CCA	(P) Pro
	CTG	(L) Leu i		CCG	(P) Pro
AT	ATT	(I) Ile	**AC**	ACT	(T) Thr
	ATC	(I) Ile		ACC	(T) Thr
	ATA	(I) Ile		ACA	(T) Thr
	ATG	(M) Met i		ACG	(T) Thr
GT	GTT	(V) Val	**GC**	GCT	(T) Ala
	GTC	(V) Val		GCC	(T) Ala
	GTA	(V) Val		GCA	(T) Ala
	GTG	(V) Val		GCG	(T) Ala
TA	TAT	(Y) Tyr	**TG**	TGT	(C) Cys
	TAC	(Y) Tyr		TGC	(C) Cys
	TAA	(*) Ter		TGA	(*) Ter
	TAG	(*) Ter		TGG	(W) Trp
CA	CAT	(H) His	**CG**	CGT	(R) Arg
	CAC	(H) His		CGC	(R) Arg
	CAA	(Q) Gln		CGA	(R) Arg
	CAG	(Q) Gln		CGG	(R) Arg
AA	AAT	(N) Asn	**AG**	AGT	(S) Ser
	AAC	(N) Asn		AGC	(S) Ser
	AAA	(K) Lys		AGA	(R) Arg
	AAG	(K) Lys		AGG	(R) Arg

(*Continued*)

Table 1.1 (*Continued*)

1st and 2nd nucleotides	3rd nucleotide added	Amino acids (1D) and (3D)	1st and 2nd nucleotides	3rd nucleotide added	Amino acids (1D) and (3D)
GA	GAT	(D) Asp	GG	GGT	(G) Gly
	GAC	(D) Asp		GGC	(G) Gly
	GAA	(E) Glu		GGA	(G) Gly
	GAG	(E) Glu		GGG	(G) Gly

Note:

1. ATG ⇒ (M) Met i: Start codon

The codon ATG (or AUG in mRNA) provides the "start" (initiation) message for a ribosome that signals the initiation of protein translation from mRNA. Hence, methionine appears in the *N*-terminal position of all proteins in eukaryotes and archaea during translation (although it is usually removed by post-translational modification).

2. TAA ⇒ (*) Ter: Stop codon
3. TAG ⇒ (*) Ter: Stop codon
4. TGA ⇒ (*) Ter: Stop codon

The three codons, TAG, TAA and TGA as above (also referred to as amber, ochre and opal codons, corresponding to UAG, UAA, and UGA in mRNA, respectively) perform "terminate the translation (stop)" function.

Non-standard/non-canonical amino acids

A set of non-standard and non-coded amino acids, known as *non-proteinogenic* or "unnatural" amino acids, also exist. They are not naturally encoded or found in the genetic code of any organisms. Apart from the tabulated 23 amino acids (21 in eukaryotes) or the proteinogenic amino acids used by the translational machinery in constructing the proteins, there are over 140 natural amino acids are known, and thousands of more combinations are indicated as being feasible.[31] The non-proteinogenic amino acids are considered in practice in view of the following: (i) They are intermediates in biosynthesis and post-translationally incorporated into protein. (ii) They possess certain physiological roles (e.g., as components of bacterial cell walls, neurotransmitters and toxins). (iii) They could be natural and man-made pharmacological compounds, and (iv) they are present in meteorites and in certain prebiotic experiments. There are three extra proteinogenic amino acids, namely, *selenocysteine* (Sec/U), *pyrrolysine (Pyl/O)*, and *N-formylmethionine*. The formylmethionine is an amino acid encoded by the start codon AUG in bacteria, mitochondria and chloroplasts, but removed post-translationally.

Example Ex-1.1

Statement of the Exercise

Relevant to details on DNA/RNA outlined in the tutorial above, assuming the query sequence given below represents a hypothetical DNA message strand and performs the following exercises.

(A) Write down the following for the query sequence: (i) The complementary DNA strand, (ii) pre-RNA strand, and (iii) possible mRNA.
(B) Decode the query DNA sequence into its associated amino acid (AA) message.

Query DNA sequence: 5′... CAC **GCA TCG AAT** CGG TAT AAA GCT CCC **TTA ATG**...3′

Solution

Relevant to the example exercise given above, the following pursuit is indicated toward solution sought:

(A) (i) The complementary strand corresponds to: 3′–5′ Watson–Crick pair-matched sequence of the query. Resulting complementary strand:

3′... GTG **CGT AGC TTA** GCC ATA TTT CGA GGG **AAT TAC**...5′

(ii) The Uracil replaces Tysine in the complementary strand in framing the pre-mRNA strand. Further, in the precursor mRNA, the introns are spliced out in the pre-RNA strand with U replacing, T. The result is:

3′... GUG **CGU AGC UUA** GCC AUA UUU CGA GGG **AAU UAC**...5′

(iii) mRNA strand is obtained (with introns spliced out from mRNA sequence).

3′... GUG GCC AUA UUU CGA GGG ...5′

(B) Availing the details on triplet/amino-acids from Table 1.1, the query DNA sequence is decoded into its AA sequence. That is, each triplet message in mRNA is decoded and expressed in terms of the designated amino acids. Resulting AA Sequence:

5′... G S F I P V...3′

Problem P-1.1

Repeat Example Ex-1.1 to deduce solutions for the associated exercises (A) and (B) in the case of following hypothetical DNA sequence (assume the triplets indicated in bold as intron regimes).

5′... GGG CCG ACA **CAC ATC** CTA CTC CTC **GAG CTT** GAG ...3′

Problem P-1.2

Statement of the Problem

(a) A hypothetical forward strand (X) of a double-strand DNA (dsDNA) query sequence, X is as follows:

X: 5′... CAC GTA TCG **ATT GCT CCC** GTA **ATC** AAA CCT AAC TCT... 3′

Write down the complementary strand of the query sequence

(b) With the codon triplets shown in bold along the strand X are introns. Find the relevant RNA to decide the designated amino acid sequence expressed in its: (i) Single-letter and (ii) three-letter formats (as indicated in Table 1.1)

Tutorial T-1.2

Codons and Amino Acid Designations ... More Details[31,36]

Referring to the genetic code listed in Table 1.1, the following additional details can be noted: Each amino acid is coded for, by a 3-nucleotide sequence on the mRNA. As mentioned earlier, such triplets depict codons formed by 64 triplets, each constituted by three bases, and there are 22 amino acids. (20 plus 2 recently discovered 21st and 22nd amino acids, selenocysteine and pyrrolysine.)

Of the 64 possible triplets formed by four nucleotides {A, C, T, G}, 60 are specified as coding residues of the amino acids. The amino acids on the mRNA strand are protein-making participants. The other four codons, as stated earlier, signal the translation stop and start points of protein-making process envisaged by enzymatic ribosome.

- The *start* codon is AUG. It is the Methionine and it is the only amino acid specified by just a single codon, AUG.
- The *stop* codons are UAA, UAG and UGA. They encode no amino acid. The stretch of codons between AUG and a stop codon is called an *open reading frame* (ORF). (An analysis of DNA sequence can predict the existence of genes based on ORFs as detailed in Chapter 3.)
- Other amino acids are specified by more than one codon, usually differing at only the third position. This is due to *Wobble effect* in codon triplets. The wobble effect is caused by the degeneracy, that is, the redundancy found in the genetic code. Each amino acid can be coded for, by more than one codon and for any amino acid, the

first 2 nucleotides in the codon are always identical. Its third nucleotide can change and it depicts the wobble position. The "wobble hypothesis" proposed by Francis Crick can be stated as follows: Rules of base pairing are relaxed at the third position, so that a base can pair with more than one complementary base.

As seen above, the DNA carries the genetic instructions on how to make proteins. In the stepped procedure implied by central dogma, the DNA does not leave the nucleus, and the cells make a copy of the messenger RNA (mRNA) — the *transfer RNA**. (The tRNA* is thus a member of a nucleic acid family, and it represents an adaptor molecule made of RNA [typically 76–90 nucleotides in length].) Francis Crick (one of the discoverers of DNA) suggested the existence of tRNA. It serves as the physical link between the mRNA and the amino acid sequence of proteins. That is, the tRNA carries an amino acid to the protein synthetic machinery of a cell (ribosome) as directed by a 3-nucleotide sequence (of codon triplets) in mRNA. That is, the tRNA brings the protein subunits, namely, the amino acids to the ribosome where proteins are constructed.

The involvement of tRNA thereof is as follows: The so-called wobble hypothesis correctly predicts the minimal number of tRNAs required to determine fully the genetic code (and also accounts for the observed degeneracy in the code). If wobble is permitted in the third position — the codons UUA and UUG, for example — can both be recognized by a single tRNA containing a UAA *anticodon**. That is, either the standard A: U or the wobble G: U could occur, and these codons, therefore, cannot specify two different amino acids, and indeed, they both would specify leucine. *Anticodon**: The anticodon region of a transfer RNA is a sequence of three bases that are complementary to a codon in the messenger RNA. During translation, the bases of the anticodon form complementary base pairs with the bases of the codon by forming the appropriate hydrogen bonds.

Example Ex-1.2

Statement of the Exercise

The following hypothetical strand of amino acids denotes a subsection of a tRNA. Identify the associated RNA strand with possible set of codons: 5′... Ile Met Glu Leu Leu Ser Gly Stop ... 3′.

Solution
Back translating protein to RNA (or DNA) is not normally advocated inasmuch as it is difficult to inverse-translate amino acids into RNA as there are only 20 amino acids and 64 different variations of the bases T, C, A, and G. This *degeneracy* implies that different variations of bases (in the third wobbling position) could result in the same amino acids. Therefore, the given query sequence: [5′... Ile Met Glu Leu Leu Ser Gly Stop ... 3′] translates to the following:

5′ ... [AU(U,C,A)] [AUG] [GA(A,G)] [CU(U,C,A,G), UU(A,G)], [CU(U,C,A,G), UU(A,G)], [UC(U,C,A,G), AG(U,C)], [GG(U,C,A,G)], [UA(A,G), UG(A)] ... 3′

(Note the redundant degeneracy seen with reference to each AA, and the resulting translation is not unique.)

Problem P-1.3

Statement of the Problem

Each of the following hypothetical strands of amino acids denote a subsection of a tRNA. In each case, identify its associated RNA strand.

(i) 5′... R O U T A W H O M R T ... 3′
(ii) 5′... Phe Glu Met Pro Leu Ala Gly Stop Asp ... 3′
(iii) 5′... Y O U H Q R C P U X Z ... 3′

Example Ex-1.3

Statement of the Exercise

The following is a pair of hypothetical triplet sequences. Check in the context of central dogma, whether the sequences in the pair represent identical messages encoded (mRNA). If so, justify.

5′... TTA CTT ATT GTT TCT GCT TTA ATC...3′
5′... TTA CTG ATT GTG TCG GCG TTA ATG...3′

Solution
Use Table 1.1 on triplet/amino acids and write down the amino acids (AAs) of the triplets given in each sequence. Hence, infer the required conclusion. The given pair translates to:

5′...L L I V C A L I...3′
5′...L L I V S A L M...3′

They do not represent identical messages, because they both have differences within their primary amino acid sequence.

Problem P-1.4

Statement of the Problem

Repeat the exercise indicated in Example Ex-1.3 with reference to the following pair of hypothetical triplet/codon sequences:

5′... TTG CTT ATT GTT TAT GCT TTA ATC...3′
5′... ATT TCT ATT GTG CAT CGG TTA GGG...3′

Tutorial T-1.3

Coding Sequence (CDS): A Revisit

Considering transcription through translation steps of central dogma of microbiology indicated earlier, the coding region, or the *coding sequence* (CDS), shown in Fig. 1.1 depicts the portion of a gene's DNA (or RNA) that codes for protein. This region usually begins at the 5′-end (by a start-codon) and ends at the 3′-end (with a stop-codon). The CDS is also synonymous with the terms, *coding sequence* and *coding region.* The sum total of coding regions of an organism's *genome* is called its *exome.* The CDS portion of an mRNA transcript is translated by a *ribosome.* The coding region in mRNA is flanked by 5′ *untranslated region* (5′-UTR) and 3′ *untranslated region* (3′-UTR) as illustrated in Figs. 1.1 and 1.2. In all, CDS is the key feature used to denote eventual "protein-coding sequence."

Example Ex-1.4

Statement of the Exercise

Based on details pertinent to the central dogma of molecular biology, construct the possible upstream coding DNA (indicating *untranslated region* [UTR] and *coding sequence* [CDS] parts) of the following hypothetical DNA double string. Justify your answer. (The intron segment (IS) is marked in the hypothetical substrand.) UTR and CDS details are explained in the tutorial above.

. T G	A T A T G G C G G T A C C T T
. A C	T A T A C C G C C A T G G A A

3′

T T C C A T A C A	IS	G C .
A A G G T A T A G	IS	C G .

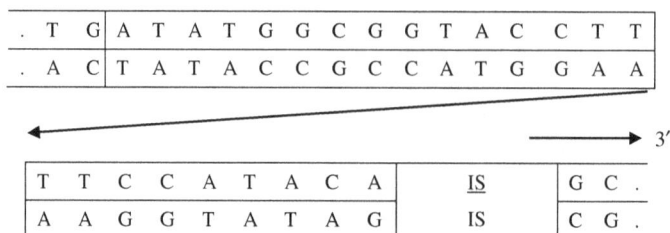

Solution method

Considering only the unwound complementary strand, and splicing out the segment indicated as an intron segments (IS), (i.e., after the intron is edited out), one can group the residues (in the complementary strand) into a pre-mRNA sequence with triplet codons explicitly written as:

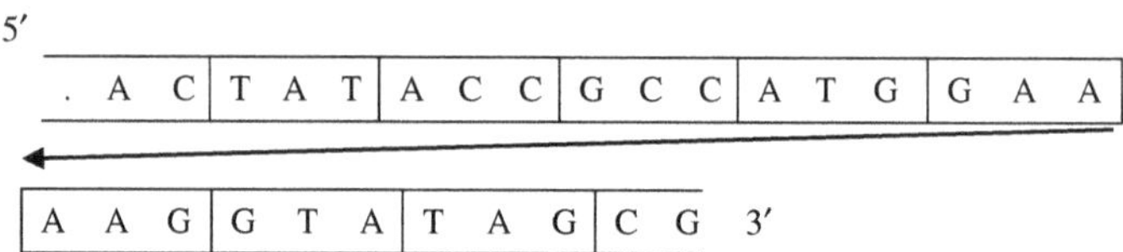

The corresponding coding strand of DNA (with T replaced by U) depicts the following mRNA:

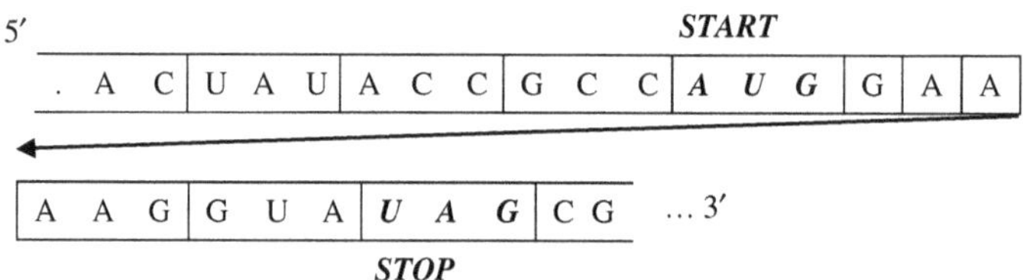

In terms of amino acids, the codons can be represented by the following sequence. (Note, however, this representation is not *unique* inasmuch as there could other possible amino acids that may represent the codons involved; that is, degeneracy is accounted for.)

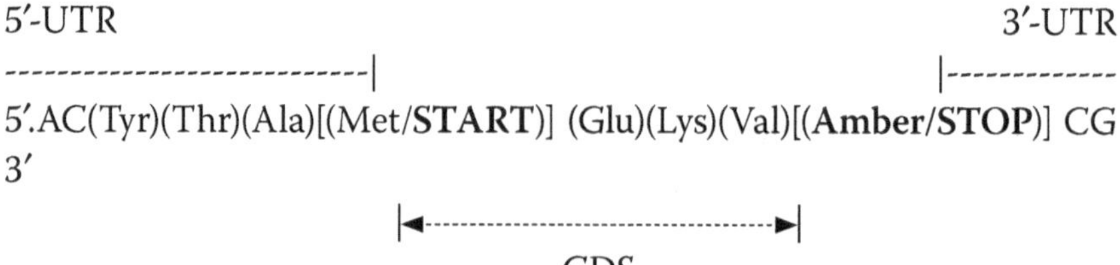

Problem P-1.5

Statement of the Problem

Given an upstream coding DNA of the following hypothetical DNA double string and assuming the bases marked in bold correspond to two intron segments, construct the corresponding, possible upstream coding DNA (indicating the UTR and CDS parts).

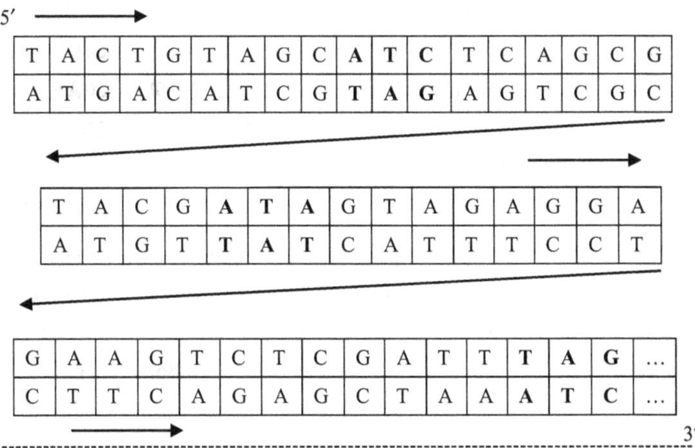

Tutorial T-1.4

Relative Frequency of Occurrences of Codons: An Example

Each species is characterized by its DNA/RNA sequence structure specified by random occurrence of codons. The underlying occurrence statistics of 64 triplets can be specified by relative frequency of occurrence of 64 codons in the sequence. For example, by visiting the website indicated in Ref. 41, the codon usage on relative frequency of occurrences of codons of *Drosophila melanogaster* (fruit-fly) as listed in Table 1.2 can be noted.

By virtue of this relative frequency of occurrences (or corresponding probability of occurrence) of codons, an entropy attribute can be inferred implying the associated (gene-related) informatics as described in Chapter 2. [Probability of occurrence of the triplet (p_i) = (Frequency of occurrence of individual triplet, f_i)/(Total of all occurrence frequencies 64 triplets = $\Sigma_{i=1}^{64} f_i$).]

Table 1.2 Relative frequency of occurrence of 64 codons in the DNA/RNA sequence is observed across 42417 CDS's in 21945319 codons.

fields: [triplet] [frequency (f_i): per thousand] ([number])

UUU 13.2 (289916)	**UCU 7.0 (154186)**	UAU 10.8 (236811)	UGU 5.4 (118088)
UUC 21.8 (479372)	**UCC 19.6 (429341)**	UAC 18.4 (403675)	UGC 13.2 (288853)
UUA 4.5 (97715)	**UCA 7.8 (171695)**	UAA 0.8 (17807)	UGA 0.5 (10767)
UUG 16.1 (353621)	**UCG 16.6 (365159)**	UAG 0.7 (14362)	UGG 9.9 (217518)
CUU 9.0 (196787)	CCU 6.9 (151856)	CAU 10.8 (236061)	CGU 8.8 (192276)
CUC 13.8 (303153)	CCC 18.1 (396168)	CAC 16.2 (354699)	CGC 18.0 (395106)
CUA 8.2 (180360)	CCA 13.5 (297071)	CAA 15.6 (342415)	CGA 8.4 (185119)
CUG 38.2 (839127)	CCG 15.8 (347206)	CAG 36.1 (792657)	CGG 8.2 (180473)
AUU 16.6 (363497)	ACU 9.5 (208889)	AAU 21.0 (460669)	AGU 11.5 (252555)
AUC 22.9 (502821)	ACC 21.3 (467509)	AAC 26.2 (575297)	AGC 20.4 (447808)
AUA 9.5 (208315)	ACA 11.0 (241893)	AAA 17.0 (372524)	AGA 5.1 (112784)
AUG 23.6 (518200)	ACG 14.4 (315479)	AAG 39.5 (866960)	AGG 6.3 (137902)
GUU 11.0 (240735)	GCU 14.4 (315879)	GAU 27.6 (604730)	GGU 13.3 (291161)
GUC 13.9 (304893)	GCC 33.6 (736394)	GAC 24.6 (540386)	GGC 26.7 (587016)
GUA 6.4 (139476)	GCA 12.8 (280181)	GAA 21.1 (462468)	GGA 18.0 (395377)
GUG 27.8 (609794)	GCG 14.0 (307977)	GAG 42.5 (933622)	GGG 4.7 (102708)

Example Ex-1.5

Statement of the Exercise

With reference to Table 1.2, the following example set of frequencies of occurrence (per thousand) of four triplets is shown in bold: {UCU 7.0; UCC 19.6; UCA 7.8; UCG 16.6}, and from Table 1.2, all these four triplets as above depict a single amino acid Ser/S. This exemplifies how degeneracy of codons could depict redundancy in the genetic code exhibited as multiplicity of three-base pair codon combinations that specify single amino acid. The degeneracy of the genetic code is what accounts for the existence of *synonymous mutations*, which denote point mutations, depicting a miscopied DNA nucleotide that only changes one *base pair* (bp) in the RNA copy of the DNA. The codons in the query RNA depict a set each of three nucleotides that encode a specific amino acid. As indicated earlier, more than one codon (triple) can translate into more than one amino acid. This *degeneracy* implies that in spite of possible variations of bases (in the third wobbling position), it could result in same amino acids. As such, if the third nucleotide suffers mutation, it could result in coding for different amino acids or the same amino acid.

Problem P-1.6

Statement of the Problem

Visit the website referenced as Ref. 42: to access the codon usage table of *Homo sapiens* [gbpri]: 93487 CDS's (40662582 codons) fields: [triplet] [frequency: per thousand] ([number]).

(i) Find the frequencies of occurrence (per 1000) of the following triplet set: {CGU CGC CGA CGG}.
(ii) What single amino acid (AA) is represented by these four triplets?
(iii) Similarly, decide on a degeneracy of another example set of codons denoting a single AA.

Example Ex-1.6

Statement of the Exercise

When a hypothetical DNA strand is analyzed, segments (windows) of unequal residual lengths depicting codon and non-codons can be randomly observed as mentioned before. Typically, the non-informative, non-codon sections would correspond to those, where each of the four residues {A, C, T, and G} could "equally-likely" be present. That is, relevant probabilities (p) of occurrence will be such that $\{p(\text{A}) = p(\text{C}) = p(\text{T}) = p(\text{G}) = 0.25\}$. Such "equally-likely" entities would constitute a posentropic statistics bearing "no information" (in Shannon's sense) as will be discussed in Chapter 2.

Problem P-1.7

Statement of the Problem

Justify (with reasons) the segments along a DNA strand having the following probabilities of occurrence {as regard to each base of the set: $\{x\} \equiv \{\text{A, C, T, G}\}$: $\{p(\text{A}) = 0.247059$; $p(\text{C}) = 0.242157$; $p(\text{T}) = 0.245098$; and $p(\text{G}) = 0.265686\}$ may most likely depict non-codons.

Tutorial T-1.5

Purine and Pyrimidine

Purine and *pyrimidine* are nitrogenous bases that make up the two different kinds of nucleotide bases in DNA and RNA. The two carbon nitrogen ring bases (adenine and guanine) are purines, while the one carbon nitrogen ring bases (thymine and cytosine) are pyrimidines. That is, {A, G}: Purines and {T, C}: Pyrimidines, and they are compared in Table 1.3.

Tutorial T-1.6

Mutations

Mutation implies a change observed in a DNA, usually in its sequence, the number of copies of the sequence that are present, how the DNA is arranged, or in the location of the DNA (which is the chromosome), and such a mutation on a nucleotide may correspond to changes conforming to the following cases of residues: (i) Purines-to-purines; (ii) purines-to-pyrimidines; (iii) purines-to-pyrimidines; and (iv) pyrimidines-to-purines. One or more of the following processes may mutate a DNA sequence.

- *Point mutation*: Substitution of an individual base residue with another. Some common substitutions: A for C; A for G; C for T; G for T, A for T; or G for C.
- *Deletion*: Deletion of a segment of residues in the sequence.
- *Inversion*: Inversion of a segment of base residues in a sequence keeping its place unaltered in the overall order (of occurrence).
- *Transposition*: Moving a segment of the sequence from one place to the other in the overall order.

Table 1.3 Comparing purines and pyrimidines.

	Purines {A, G}	Pyrimidines {T, C}
Structure	Double carbon nitrogen ring with four nitrogen atoms	Single carbon nitrogen ring with two nitrogen atoms
Size	Bigger	Smaller
Source	A and G exist in both DNA and RNA	C exists in both DNA and RNA T is seen only in DNA T/U prevails only in RNA

- *Duplication*: Repetition of a section of residues along the sequence, one or more times.
- *Insertion*: Insertion of another sequence pattern into the query sequence.

Considering the purines and pyrimidines, the possible versions of mutations associated with them are as follows.

Transition and *transversion*: Transversions are interchanges of purine for pyrimidine bases, which therefore involve exchange of one-ring and two-ring structures. Although there are twice as many possible transversions, because of the molecular mechanisms by which they are generated, transition mutations are generated at higher frequency than transversions. That is, transition mutations are more common than transversion mutations and are less likely to produce a difference in the amino acid sequence compared to transversions. This is the basic difference between transition and transversion. In all, transition mutations (purine ↔ purine and pyrimidine ↔ pyrimidine, that is, A ↔ G and T ↔ C) are more common than transversion type mutations (purine ↔ pyrimidine, that is, [A, G] ↔ [T, C]).

Problem P-1.8

Statement of the Problem

Construct a matrix to illustrate the characteristic of mutations with reference to the nucleotide bases {A, C, T, G}, by scoring 100% to depict the element of the matrix pertinent to no mutation as shown in the matrix below.

	A	C	T	G
A	100%			
C		100%		
T			100%	
G				100%

A prorated set of percentages thereof can be used to represent other elements illustrating the changes such as [A, G] ↔ [T, C]). Gather and cite the pertinent references showing annotated details on the possible, relative percentage scoring.

Problem P-1.9

Statement of the Problem

Suppose certain mutational influences occur on four segments of a query sequence, X:

X:	5′ ... CAC GTA	TCG ATT	GCT CCC	GTA ATC ... 3′
	Segment A	Segment B	Segment C	Segment D

The mutational changes on the query sequence X conform to following cases: (i) Purines-to-urines on any two residues in Segment A; (ii) purines-to-pyrimidines on any one residue in Segment B; (iii) purines-to-pyrimidines on any two residues in Segment C; and (iv) pyrimidines-to-purines on any three residues in Segment D. (As indicated earlier, purines and pyrimidines depict, respectively, the sets {A, G} and {T, C}.) Find the resulting complementary RNA strand of X and decode each triplet message expressed in terms of the designated amino acids.

Problem P-1.10

Statement of the Problem

Suppose mutations of the following types are (a) point mutation, (b) transition, and (c) transversion. A query sequence (X) of codons is shown below with its corresponding AA sequence:

Triplet #		1	2	3	4	5	6	7	8	9	10	
X:	5′...	CAC	GUG	UCG	AUU	GCU	CCC	GUC	AUC	AAU	CCU	... 3′
AA		H	V	A	I	T	P	V	V	N	P	

Suppose each version of mutational change (listed above) can possibly occur at one of the three positions of the triplets as specified. Construct three resulting AA sequences (with associated degeneracy) in the following cases.

(i) The wobble position in each codon suffers point mutation of changing to A.
(ii) The first position in each codon suffers a point mutation of transition.

(iii) The first position in each codon suffers a point mutation of transversion.

[Answer hint: Case (i) This refers to substituting the base residue say, A at the wobble position of each triplet that takes place. Hence, the resulting modified codon sequence X* is:

Triplet #		1	2	3	4	5	6	7	8	9	10	
X:	5′ …	CAA	GUA	UCA	AUA	GCA	CCA	GUA	AUA	AAA	CCA	… 3′
AA		Q	V	A	I	T	P	V	V	K	P]	

Example Ex-1.7

Perform a visual comparison of a pair of sequences X and Y shown below and indicate in your answer the resulting identities by: 1's on matches and 0's on mismatches. Also indicate the mutational change in Y with reference to X, by scoring indices, S for transition and V for transversion. (Note: {A, G}: Purines and {T, C}: Pyrimidines.)

Solution

```
X ...   C A C    G T A    T C G ...
          |      |   |    | |
Y ...   G A T    G A A    T C A ...
        0 1 0    1 0 1    1 1 0
        V   S    V        S
```

Scores (on 9 residues): 5 Matches — (ID 1); 4 Mismatches — (ID 0); 2 Transitions — (ID S); and 2 Transversions — (ID V).

Problem P-1.11

Statement of the Problem

Perform visual comparison of the following two putatively related* nucleic acid sequences u and v. As in the example above, indicate in your answer the identities, by 1s on matches and 0s on mismatches. Also show mutations by their scoring indices, S for transition and V for transversion:

```
u:  A A G C T T A C G C A A A C C G -
v   - G C T C A C G G T T G C C A C T
```

(The term "putative" is used to describe an entity or a concept that is based on what is generally accepted or inferred even without a direct proof of it. *A putative gene*, for instance, may refer to a nucleotide sequence believed to be a gene, on the basis of its *open reading frame* (ORF) although its function and the protein it codes have not been (fully) identified. For example, *putative gene 57* is the (temporary) name suggested for the gene coding for a protein produced by *Bacteriophage SP01* that infects the bacterium, *Bacillus subtilis*.)

Tutorial T-1.7

Conservative and Non-conservative Mutations

With reference to 20 (+2) amino acids (in Table 1.1), some of these share similar biochemical properties and physico-chemical characteristics. For example, both leucine (K) and isoleucine (I) are aliphatic and both have branched hydrophobes and as such, are substitutable.

In general, conservative substitutions happen between amino acids that are similar in size, polarity, hydrophobicity, charge, and share other physiochemical properties. As such, in spite of substitution mutation having taken place, the function and the properties of the participant AAs are not radically changed. There are two possible types of substitution mutations.

(i) *Conservative replacement* (also known as *conservative mutation* or a *conservative substitution*): This version of mutation refers to a replacement of a given amino acid by a different amino acid, which has similar physico-chemical properties (such as charge, hydrophobicity, and size).
(ii) Conversely, a *radical replacement*, or *radical substitution*, is a type of mutation wherein an exchange of one amino acid takes place with another amino acid having different physico-chemical properties. Tables 1A-5 and 1A-6 (of Appendix 1A) contain the list of amino acids and show their common properties.

A cursory survey of the AAs by considering their properties indicated, the following examples of comparative details can be assessed in terms of the associated similar properties leading to most probable or possible conservative substitutions as identified in Table 1.4.

Table 1.4 Examples of conservative substitutions.

Glycine (Gly/G)	↔	Alanine (Ala/A)	Both AAs are hydrophobic because they are alkanes.
Glycine (Gly/G)	↔	Valine (Val/V)	Both AAs are similar due to same characteristics of the molecule.
Glycine (Gly/G)	↔	Serine (Ser/S)	Serine is neutral, polar and hydrophobic, and glycine is aliphatic and neutral.
Alanine (Ala/A)	↔	Proline (Pro/P)	Alanine is aliphatic, hydrophobic and neutral, and proline is hydrophobic and neutral.
Glutamine (Glu/E)	↔	Valine (Val/V)	Both AAs have similar biochemical properties, pertinent to electric charge, hydrophobicity, and size.
Aspartate (Asn/N)	↔	Glutamate (Glu/E)	Both AAs are small and negatively charged residues.

Typically, the associated biochemical properties of the AAs enable classifying the amino acids, into six classes on the basis of their structure and the general chemical characteristics of their R groups. The following are those identifiable groups.

(i) Serine, cysteine, selenocysteine, threonine, methionine-aliphatic class.
(ii) Serine, cysteine, selenocysteine, threonine, methionine-hydroxyl class.
(iii) Proline-cyclic class.
(iv) Phenylalanine, tyrosine, tryptophan-aromatic class.
(v) Histidine, lysine, arginine-basic class.
(vi) Aspartate, glutamate, asparagine, glutamine acidic and their amides class.

Conservative replacements may not cause adverse effects on the functions of a protein complex as compared to the influences due to nonconservative replacements. Such reduced effect of conservative replacements on protein function can be seen as occurrences of different replacements in nature, and nonconservative replacements between proteins could mean deleterious effects; as such, they are mostly removed by natural selection.

Mutations thus imply a permanent, heritable change that may occur in the nucleotide sequence in a gene or a chromosome. In such mutational instances of evolutionary biology, a substitution matrix is constructed in bioinformatic contexts to show the rate at which one character in a sequence changes to another character state with the passage of time. Such matrices arise in amino acid and/or DNA sequence alignments where the similarity between sequences depends on their divergence time and the substitution rates as represented in the matrix.

The matrix shown in Table 1.5 is an example where two sets of amino acids are compared in terms of the number of substitutions that may encounter. In practice, an amino acid substitution matrix is constructed with the assigned score to a pair of aligned amino acids, *j* and *k*. Correspondingly, a substitution matrix can be conceived in view of the following considerations.

(a) With reference to biophysical properties of amino acid (AA) residues that differ in size and charge: Some AAs are acidic, some are basic and some have aromatic side chains as outlined earlier. Further, as mentioned earlier, replacement of one amino acid with another amino acid having similar properties is less likely to affect the protein to pose considerable changes in its function. In contrast, if the replacement refers to an amino acid having distinct properties, the observable effects on the translated product, namely the protein, could be very adverse. A substitution matrix is designed to reflect such differing features.
(b) Evolutionary divergence should be reflected in the matrix conceived. In highly diverged protein families, observing identical or functionally conservative amino acids at the same site could be rare than in families characterized by sparse or nil sequence divergence. As exemplified with Table 1.4, a substitution matrix could be based on the statistics of amino acid replacements consistent with typical degree of evolutionary divergence of the proteins being considered. Hence, it is preferred to frame a family of matrices that is parametrized by sequence divergence.
(c) The possibility of multiple substitutions should not be ignored. The score assigned to an amino acid pair (*j* and *k*) should reflect the

Table 1.5 Example of [20 × 20] matrix of amino acid substitutions. Number shown in each cell denotes the statistics on enumerated substitution replacement encountered in the comparison exercise of AAs.

AAs in		1	2	3	4	5	6	7	8	9	10
1-letter code		A	R	N	D	C	Q	E	G	H	I
1	A										
2	R	58									
3	N	10	37								
4	D	2	10	30							
5	C		7	66	25						
6	Q	1	3	8	21	6					
7	E	1	3	3		2					
8	G	45	77	4	3	2	2	12			
9	H	5	59	19	5	13	3	1	70		
10	I	16	11	1	4	4			43	17	
11	L	3	9	3	8	1	2		5	4	5
12	K	16	15	2		1			10	6	53
13	M	11	27	4	2	4	1		9	3	9
14	F	6	6	2	4	4	9		17	20	32
15	P	1	3	2	2	3	2	1	14	2	2
16	S	1	2	3	4			1	3	1	23
17	T	2	2	1	17	9	2		4	1	1
18	W		2	2	2	1		3	2	2	4
19	Y				1				2		
20	V	5	35	5	4	1		1	27	7	3

AAs in		11	12	13	14	15	16	17	18	19	20
1-letter code		L	K	M	F	P	S	T	W	Y	V
1	A										
2	R										
3	N										
4	D										
5	C										
6	Q										
7	E										
8	G										
9	H										
10	I										
11	L										
12	K	8									
13	M	42	83								
14	F	15		10							
15	P	12	9		48						
16	S	24	4	2	2	10					
17	T					1	2				
18	W			1	1		4	26			
19	Y					3		1	1		
20	V	9	1	4	4	7	5	1			

probability of observing *j* aligned with *k*, and the possibility of multiple replacements at the same site should be duly considered.

Problem P-1.12

Statement of the Problem

With reference to Table 1.5, identify all the most probable conservative substitutions (i.e., those substitutions that do not affect the property, or function, of the protein) and *via* literature search, justify with reasons why such conservative substitutions could have taken place. (For example, glycine is commonly replaced by alanine or *vice versa* because these two amino acids bear smallest side chains.)

Problem P-1.13

Statement of the Problem

Plot the associated *hydropath* profile[43] along the following amino acid sequences (1) and (2). Learn from the literature in more detail on *hydrophilic* and *hydropathic* properties of AAs and different versions of indices indicated thereof (Table 1.6).[6]

(1) 5′... Phe Met Ile Leu Thr Ser Ser His Gly Stop ... 3′
(2) 5′... A R Q U T K W H O P R T Z ... 3′

More Exercises, Projects, and Open Questions

Consistent with the text of this chapter, additional details are presented below with queries and exercises (with solution hints as needed) to promote the self-study pedagogy of the readers.

Table 1.6 Twenty amino acids are sorted as per their hydropathy indices (based on the data due to Kyte and Doolittle).[43]

R	K	N	D	Q	E	H	P	Y	W
−4.5	−3.9	−3.5	−3.5	−3.5	−3.5	−3.2	−1.6	−1.3	−0.9

S	T	G	A	M	C	F	L	V	I
−0.8	−0.7	−0.4	+1.8	+1.9	+2.5	+2.8	+3.8	+4.2	+4.5

Problem P-1.14

What are XenoNAs? Identify their roles in conceiving synthetic cells and new forms of life. Could they replace the real DNA in gene therapy?

Xeno nucleic acid (XNA) refers to a synthetic form of DNA considered as an alternative to real/natural form of nucleic acids, namely, DNA and RNA. The genetic information stored in XNA is "invisible" and therefore useless to natural DNA-based organisms. Research on creating synthetic polymerases to transform XNA, and its production toward application potentials have led to the new science of *xenobiology*.[27–29]

Problem P-1.15

What are *Mendel's laws*? Identify the roles of gene factor *versus* offspring considerations. Explain (i) law of segregation, (ii) law of independent assortment, and (iii) law of dominance specific to Mendel's theory.

Problem P-1.16

Statement of the Problem

Define the following terms: (i) phenotype and (ii) genotype. Explain with examples: The outwardly expressed end results of the genes that are inherited as phenotypical characterization and genetic make-up of an organism deciding genotypical features.

Learn and write an explanatory note on: Allele depicting one of the several possible versions of a gene.

Distinguish a genome and a gene pool. Describe the underlying considerations with examples.

Discuss how to cluster efficiently a set of gene expression data.

Project Pr-1.1

Statement of the Project Exercise

(a) Learn about exclusive segments such as CG-islands (CpG), TATA box in a genome. Explain a possible method of detecting a CpG island in a genome.

(b) Learn the heuristics of the *Burrows–Wheeler transform** and its algorithmic complexity as adopted in genome-assembly efforts. The Burrows–Wheeler transform* (BWT) implies a block-sorting compression that rearranges a character string into runs of similar characters.

(c) What are *k*-mers pertinent to a genome assembly? Elucidate the merits and demerits in having long *versus* short *k*-mers* in a genome assembly. (*k-mer** refers to all the possible substrings of length *k* that are contained in a given string. Such *k*-mers of various lengths are used to explain variability in mutation rates in human genomes.)

(d) Learn the difference between local and global alignment. Discuss the complexity of the algorithm used in finding a global alignment between two DNA sequences that have a high degree of similarity.

(e) Indicate an example of multiple alignment. *Why does a relevant implementation, using the software Clustal, need a guide tree*?

(f) Define phylogeny (Chapters 9 and 10) and learn the principle of the UPGMA algorithm. What does the ultrametric property of a phylogenetic tree inform about the evolutionary process?

(g) Describe the significance of a "sequence logo" in bioinformatics.

(h) In modeling a metabolic process, describe the advantages and disadvantages of using a stochastic approach (e.g., agents) as opposed to using a set of deterministic differential equations.

(i) Present the aim of BLAST software (Chapter 5): (i) Describe the principle behind its algorithm. Learn the output format of a BLAST search.

(j) Describe the purpose of microarray data analysis and explain the format of a microarray data.

(l) The simple members of eukaryota domain of life are Baker's yeast, fruit fly, and nematode worm. Using the search engine Entrez, find the number of DNA molecules and base pairs in each case. Find the open reading frames (ORFs) in this sequence using Genscan with "Vertebrate" as the query organism.

(m) Using Entrez search tool, find the number of base pairs in *human chromosome 1*.

(n) Visit NCBI > Entrez > nucleotide and obtain the nucleotide sequences for the *enzyme protein kinase*. (The organism or sub-family of enzyme is not of interest.)

1.9 Closure

Contents of this chapter are written to project introductory details on the heuristics of bioinformatics with necessary foundation topics grounded in molecular biology and underlying central dogmatic details. Hence, the scope of bioinformatics is narrated in terms of a graceful merger of system biology and aspired feats of omic informatics pertinent to biosequences.

Appendix 1A

Concepts in System Biology and Basics of Microbiology

Reference to the realm of studies concerning bioinformatics, there are a numbers of concepts and terms that have their origin from life sciences (specifically, from microbiology).[3,27–35] Such biological details as needed are adopted in bioinformatics along with the fusion of engineering and technology perspectives, principles of computing methods and heuristics of information theory.[1,2] Relevant concepts in system biology and heuristics of microbiology are outlined in the following subsections.

Biology and Its Organization

Stated in a nutshell, *biology* is a study on living systems depicting the science of living organisms. The embodiment of living entities represents a self-replicating and evolving universe of entities of life of enormous complexity. They pose genetic continuity and contain extensive amount of large (and complex) molecular units, termed as the "cells."[37–40]

An associated degree of complexity of hierarchy inherently resides in the organization of living systems set forth by (i) a molecular *stratum* of DNA residing in the cells wherein the chemistry of life begins; (ii) an *organelle* level of organized cellular activities with each cell representing the smallest embodiment of life; that is, the primary activity of a living system begins at the cell. Finally, (iii) the sum of cellular functions culminate into surviving details of full-life complex at the organism level,[41–47] and relevantly, all organisms possess common molecular patterns and exhibit similar *expressions of life* in spite of their diverse characteristics (or *diversity*).

Further, the biological systems need energy for all their functions and survival. They store it in the form of chemical energy in the chemical bond of a compound called *adenosine triphosphate* (ATP), which represents a reservoir and is a carrier of energy in the cellular body. Deriving chemical energy from the molecular details of the ATP is as follows: ATP, in essence depicts a triphosphate group attached to the ribose (denoting the sugar backbone of the DNA), and it conforms to a set of three phosphates linked *via* high-energy chemical bonds. These phosphates are negatively charged and considerable energy is involved in binding them together. In this bonded structure, if the end-most phosphate is broken off, then a substantial extent of energy is released, and the ATP reduces to *adenosine diphosphate* (ADP). By removing another phosphate, the ADP will reduce to *adenosine monophosphate* (AMP) releasing more energy, if needed. All in all, the ATP serves as the resource of energy and it enables availing it by the entire cellular domains.

The Hierarchical Structure of the Biological System

The living systems, in general, exist in stratified levels of hierarchy commencing from non-living morsels of matter (namely, atoms and molecules) toward constituting the primitive part of the life system — the "cells." Thus, the hierarchical structure of living systems can be outlined as a stream of [*Non-living entities of matter at microscopic level: Atoms and molecules*] → [*Living cells: 90% made of: H, C, N, O, P, S*] → [*Living system of families*] → [*Multicellular organisms*] → [*Multifamilies or living society*].

In essence, *what is a living system*? It is a biosystem: an embodiment of thousands of macromolecules that interact with each other and undergo (or catalyze) chemical changes as necessary. The macromolecules of living systems fall under three groups, namely: (i) proteins (in most part), (ii) nucleic acids, and (iii) carbohydrates. The living system is constituted by multitudes of a basic biological entity termed as: The *cell*. That is, the premise of cell theory (proposed in 1800s) suggests that all living systems are constituted by one or more cells, a cell is the basic unit of life, and all cells are derived from pre-existing ones. The biological cell denotes the lowest level of the living structure. It is capable of performing the required activities of life. The exclusive study of cells is called *cell biology*.[30]

Organisms can be classified as *unicellular* (consisting of a single cell) or *multicellular*. The bacteria, for example, are single-cellular system, and plants and animals are multicellular systems. Essentially, the cell is confined to, and surrounded by, a *plasma membrane*, an outer covering that defines the cell boundary and protects the cell from its immediate environment. This membrane mediates the flow of nutrients and waste products across it. The core of the cell is known as the *nucleus*, which houses the genetic material namely, the *deoxyribonucleic acid* (DNA), and the cell also includes other contents such as the protein complex. During non-divisional cell-cycle, this DNA plus protein complex prevails in an uncoiled dispersed state (known as *chromatin*). During the other phases, known as *mitosis* and *meiosis*, the chromatin coils up and condenses into structures called *chromosomes*. Mitosis refers to a form of cell division resulting in the production of two cells, each with the same chromosome and genetic complement as the parent cell, and meiosis is the process wherein one replication of the chromosomes is followed by two nuclear divisions to produce four *haploid** cells. (*A *haploid* is a cell or organism having a single set of unpaired chromosomes.)[26]

When the nucleus is absent altogether, the cells correspond to being *prokaryotic*. That is, a prokaryotic cell has a simple membrane, contains one chromosome, and has no organelles. As mentioned before, bacteria and blue–green algae are typical prokaryotic systems. The chromosome in prokaryotes is an intact DNA molecule containing the genome; in eukaryotes, a DNA molecule exits with another genetic material, namely, *RNA (ribonucleic acid)* and proteins to form a thread-like structure. This complex has the chemistry and a structure (or form) decided by genetic information in the omic details arranged in a linear sequence.

The cell chemistry can be specified by a set of macromolecules that, in turn are made of smaller units, are known as *monomers* (which need not be identical). Thus, the bulk of a living system is a biosystem with a constituent of thousands of macromolecules that interact with each other and undergo or catalyze chemical changes. (The *catalysts* are substances present in small amounts [relative to main reactants] and accelerate chemical reactions without being consumed in the process. The biological catalysts are known as *enzymes*.)

The cell also includes other contents, such as the *protein complex*.[30] Hereditary information flows from DNA to RNA eventually in making of the protein. That is, protein recipes, encoded in the DNA, first get transcribed into a messenger RNA (mRNA). Then these transcripts are used as templates to synthesize the proteins. The translated products depicting the proteins are building blocks of tissues and form the basis for complex chemical reactions. They act as sensors, transducers, and energy transformers. Also, as enzymes (catalysts), they form substrates for appropriately configured chemical reactions. The number of proteins in various functions would differ. They contain C, H, O, N, and S and may temporarily associate with P toward shape changes while losing or gaining an enzymatic activity. The molecules that store energy to drive cellular processes are *carbohydrates* that contain equal amount of C atoms and H_2O.

Cell Structure

Considering the two versions of the biological cells, namely the *eukaryotic* and *prokaryotic* types mentioned above, the former contains a nucleus and exist either as single-celled or multicellular organisms, and the prokaryotes are single-celled organisms with the cell not having the nucleus. Further, the eukaryotic cells could be about 10–100 μm in size, and the size of a typical prokaryote cell is approximately 1–5 μm; as such, comparatively, the eukaryotic cell can be thousand times greater in volume.

Prokaryotic cells are the most primitive form of life on the Earth, possessing the vital biological processes including cell signaling and being self-sustaining. The structural detail of a prokaryotic cell implies that it lacks nucleus, and it is simple with extensive sections with membrane separators. It is shaped as rod-like, spherical, or spinal chain structure. The constituent parts of prokaryotic cell structure and the associated structural details (plus relevant definitions and functional considerations) can be seen in books on general and cell biology. The DNA in prokaryotic life is in a condensed form known as the *nucleoid*, and the term nucleus refers to a "kernel" which is absent in prokaryotic cell.

In contrast, in eukaryotic cells, there are compartments within the cell, wherein designated, specific metabolic activities take place. Many eukaryotic cells are ciliated with *primary cilia*, which denote minute hair-like organelles. Cilia can be "viewed as sensory cellular antennae that coordinates a large number of cellular signaling pathways, sometimes coupling the signaling to ciliary motility or alternatively to cell division and differentiation." As such, the primary cilia play vital roles in chemosensation, mechanosensation, and thermosensation. Further, in the eukaryotic cell structure, there is a nucleus containing chromosomes. It is a system made of interior membranes (*endoplasmic reticulum*) and membrane-bounded organelles, which are functional compartments of the cell and specific organelles participate in energy-related efforts as in mitochondria, which is a site for oxidative metabolism and chloroplast, which is a site of photosynthesis. Illustrative structural details of a eukaryotic cell along with relevant definitions of constituent parts, descriptions, and functional considerations of cellular constituents can be seen in books on general and cell biology.

Cellular and Phospholipid Structure

The biological cell structure is an embodiment of a unit that has the interior filled with organelles plus a semi-fluid matrix of cytoplasm, and the membrane covering the cell is an encasing made of a delicate sheet (a few molecules thick), depicting a plasma membrane as described above. The plasma membrane has a phospholipid structure as illustrated in Fig. 1A-1.

Functionally, the lipid bilayer controls the transport of materials across the cell membrane. Such transported materials include food, water, etc. required for the survival as well as toward excretion of processed waste materials out of the cell. The transit of materials is controlled by the plasma membrane as follows: Essentially, there are two regions, Region I and II as shown in Fig. 1A-1. Region I is a polar or *hydrophilic* region. It is hydrophilic because it has phosphate chemical group attached,

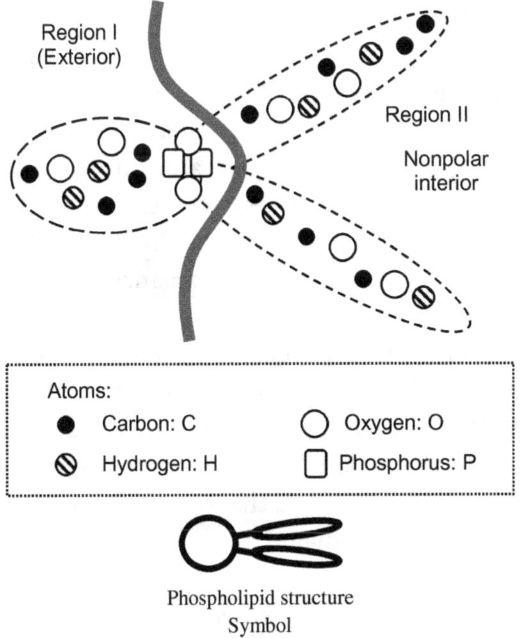

Fig. 1A-1 Phospholipid structure of a biological membrane.

which is water-soluble and, hence, hydrophilic. Region II is the nonpolar interior and it is *hydrophobic,* because it consists of fatty-acid chains that are highly nonpolar and hydrophobic. The interior of lipid bilayer comprising of completely nonpolar chemical composition is and hydrophobic, it repels any water-soluble molecule that attempts to pass through it. The modes of transport of materials across the cell membrane are as follows: First, material transport may occur due to structural imperfections such as water leaking across the membrane; second, material transport is regulated by proteins within the membranes, which act as gate-keepers and admit the flow of the molecules selectively.

Such a diffusion is a passive transport process; there is also passive diffusion process involving restricted transport that allows only selective substances to pass through as decided by the semipermeable property of the membrane. The semipermeable diffusion across plasma membrane allows certain types of molecules such as H_2O, urea, glycerol, CO_2; oil-soluble hydrophobic molecule like O_2, N_2 and nonpolar benzene. Molecules that do not easily pass through the phospholipid bilayer are charged ions H^+, Na^+, HCO_3, K^+, Ca^{2+}, Cl^-, Mg^{2+}; large, uncharged glucose molecules and polar sucrose molecules. Thus, the (semipermeable) diffusion across the cell membrane is decided by polarity (hydrophobicity *versus* hydrophilicity), electric charge (being charged/uncharged), and dimension (being small or big). Apart from simple diffusion process, a facilitated diffusion also exists. This also refers to diffusion across a membrane, but the rate of diffusion is enhanced by certain membrane proteins that act as carrier molecules aiding the diffusion process. These "carrier proteins" are called *permeases.* Another version of facilitated diffusion across the cell membrane is based on *channel protein-aided transport.* Here, the diffusion is facilitated to the solute to flow through a tunnel-like transmembrane protein (called *channel protein*).

The transmembrane transports take place *via* active modes with the expense of energy derived by the cells in the form of chemical energy stored in the molecular body of ATP as indicated earlier. In such active transport processes, the materials are transported across the membrane against the concentration gradient, that is, from lower to higher concentration regions. This is mediated by a class of proteins known as *membrane carrier molecules.*

The lipid bilayer is a coarse structure with a variety of proteins attached to its surface and embedded in the membrane. Each of such

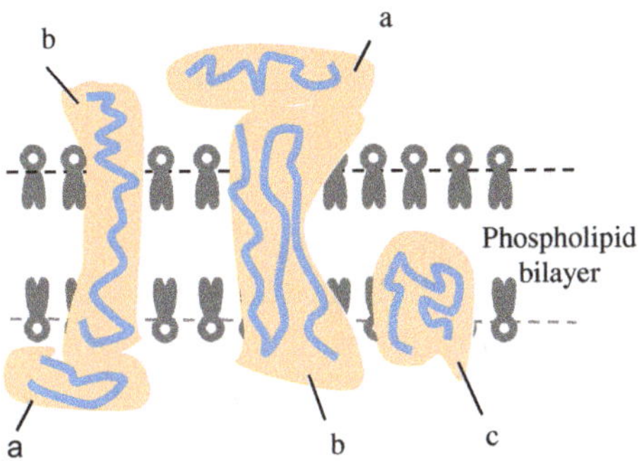

Fig. 1A-2 Membrane proteins: (a) Peripheral type, (b) transmembrane type, and (c) integral type.

proteins has a specific purpose. Examples of membrane proteins include ion channels, receptor proteins, and proteins that allow cells to connect to each other, and so on. Further, the membrane proteins are of two types: peripheral proteins and integral proteins.

Thus, integral membrane proteins are polypeptide chains that exist in the form of five "heavy chains" framed as three extracellular domains, one transmembrane domain, and one cytoplasmic (intracellular) domain. The transmembrane, peripheral membrane, and integral membrane proteins are illustrated in Fig. 1A-2. The transmembrane proteins may exist as a single-pass heavy chain across the bilayer or as a set of a definite number of multiple passes of polypeptide chains. They may also prevail as a barrel-shaped spun.

Gene and the Genetic Code

Gene represents the smallest physical and functional unit of heredity in biological systems that encodes for a specific biological structure or function. In other words, a gene contains hereditary information encoded in the form of DNA and is located at a specific position on a chromosome in the nucleus of a cell. Genes determine many aspects of anatomy and physiology of living systems by controlling the production of proteins. Each individual form of life has a unique sequence of genes known as *genetic code.*

The functional and physical unit of heredity is passed from parent to offspring. Genes are pieces of DNA, and most genes contain the information for making a specific protein. Though gene was originally described as a "unit of inheritance," in more general terms, it can be defined as an abstraction, which is useful for the purposes of nomenclature and for the assignment of a symbol. Structurally, gene is a basic unit of hereditary material and denotes an ordered sequence of nucleotide bases that encodes one polypeptide chain (*via messenger RNA* or *mRNA*). That is, physically, genes denote structures of the blueprints (called proteins) that control the infinite variety of life forms.

Each individual form of life has its own unique sequence of genes characterized by a genetic code. As indicated earlier, structurally, a gene specifies a continuous stretch of a genomic chain constituted by the stretch of DNA strand consisting of thousands of genes, and each gene serves as a recipe on how to build a protein molecule. The proteins perform important tasks for the cell functions and serve as building blocks of living systems. A protocol of flow of genetic information from the genes culminates in determining the protein composition and thereby the functions of the cell.

The cells of the living systems denote the macromolecules of life. The living cell, 70% by weight, is made of H_2O. It is a macromolecule made of chains of smaller units called *monomers*, which in general need not be identical. For example, as will be described later, the DNA molecule is made of four distinct monomers called nucleotides, and a protein complex is constituted with 20 different monomers called *amino acids*.

The gene includes regions preceding and following the coding region (leader and trailer), as well as (in eukaryotes) intervening sequences (introns) between individual coding segments (exons). Functionally, the gene is defined by the *cis-trans* test that determines whether independent mutations of the same phenotype occur within a single gene or in several genes involved in the same function. As will be indicated later, a comprehensive definition of a gene specifies a "continuous stretch of a genomic DNA molecule, from which a complex molecular machinery can read information" (encoded as a string of A, T, G, and C) and make a particular type of a protein or a few different proteins. Specific sequences of nucleotides along a molecule of DNA (or, in the case of some viruses, RNA) represent the functional units of heredity. The majority of eukaryotic genes contain coding regions (*codons*) that are interrupted by non-coding regions (*introns*) and are therefore labeled as split genes.

A gene is a DNA segment that contributes to *phenotype/function*. In the absence of a demonstrated function, a gene may be characterized by *sequence, transcription,* or *homology.*[31] A gene is a set of connected *transcripts*. A transcript is a set of *exons via* transcription followed (optionally) by *pre-mRNA splicing*. Two transcripts are connected if they share at least part of one exon in the genomic coordinates. At least one transcript must be expressed outside of the nucleus and one transcript must encode a protein. In all, the gene represents the smallest physical and functional unit of hereditary, which encodes for a specific biological structure or function. It is a term often used to denote a unit of genetic information sufficient to decipher an observable trait in living systems. Such traits are the genetic trends passed on from parent organisms to the offspring and so on.

Functional and structural aspects of genomics: The goal of genomics as narrated earlier is to determine the complete DNA sequence for all the genetic material contained in an organism's complete genome. *Functional genomics* (sometimes referred to as *functional proteomics*) aims at determining the function of the proteome (the protein complement encoded by the entire genome of an organism). It expands the scope of biological investigation from studying single genes or proteins to studying all genes or proteins together in a systematic fashion, using large-scale experimental methodologies combined with statistical analysis of the results.

Structural genomics: This refers to the systematic effort to gain complete structural description of a defined set of molecules, ultimately for an organism's entire proteome. Structural genomics projects apply X-ray crystallography and NMR spectroscopy in a high throughput manner.

In all, genomics is subset pedagogy of bioinformatics with a comprehensive base in the complex biological studies applied to modern biomedical systems enclaving medical discipline, public health, agricultural, as well as veterinary sciences. The modern art of genomics enables basics of high throughput, *next-generation sequencing* (NGS), and advances in biomolecular imaging. It has also bears the extended scope of translational medicine bringing the facts of base pairs to the bedside of the patients. It offers directions towards realizing new drug designs and implementing personalized healthcare.

Genomic efforts in modern sense warrant supercomputing exercises supported by advanced statistical learning and intricacies of *artificial intelligence* (AI). Relevant computational approaches include handling massive bioinformation and healthcare data profoundly deployed as bioinformatics strategies in improving the human, plant and animal health. The associated functional informatics span across microbiology, medical/health sciences, computational biology, technocentric arts of smart drug designs, and interdisciplinary prospects of biology plus allied life, as well as physical sciences.[32] In essence, genomics plays a crucial role in understanding the concepts of biological systems useful in diagnostic and therapeutic medicine seen in terms of all levels of organization in the living systems — commencing from the rudiments of molecular biology, walking through the pavement of cellular structures and culminating in the highways of whole-life complexity. Related details of life sciences are laid in the foundations of bioengineering and computer sciences, bourgeoning out across the studies on biosequences. Relevant efforts have necessitated development of appropriate algorithms, computational methods based on dynamic programming, stochastic concepts, statistical techniques, and theoretical frameworks of formal and practical biomedical problems. Specific topics include a broad range of subject areas as follows: (i) Microarray data analysis, (ii) genome and sequence analysis, (iii) protein structure prediction and classification, (iv) gene regulation elements analyses, (v) disease classification using machine-learning techniques, (vi) biological network construction, and (vii) genome and database search tools.

Genome: Inner Details of the DNA

The genomic details can be comprehended by learning the inner anatomy of the DNA complex. As mentioned earlier, the DNA is typically a double-stranded helix bearing chemical instructions on the genetic information of a cell. (*Single-stranded DNA* or ssDNA also exists as in the case of viral genes and pertinent details will be exclusively presented in Chapter 8). Essentially, the DNA consists of two such polynucleotide chains wound helically around a long central axis

Specific disposition of bases randomly along DNA sequence (or, in the case of some viruses, along a molecule of RNA) represent the functional informatic units of *heredity*, and genetic information is encoded (in majority of eukaryotic genes) in specific coding regions or *codons* (called *exons*) that are interrupted by non-coding regions (*introns*). Such a structure of DNA sequence conforms to and is labeled as *split genes*.

The gene that facilitates holding of genetic information is an embodiment of subunits, as outlined below: Chemically, as shown in Fig. 1A-3, the DNA refers to a double-stranded molecule twisted as a helix (to form a spiral form, except in certain viruses like ϕX174). The genetic message associated with the DNA complex controls the chemistry of

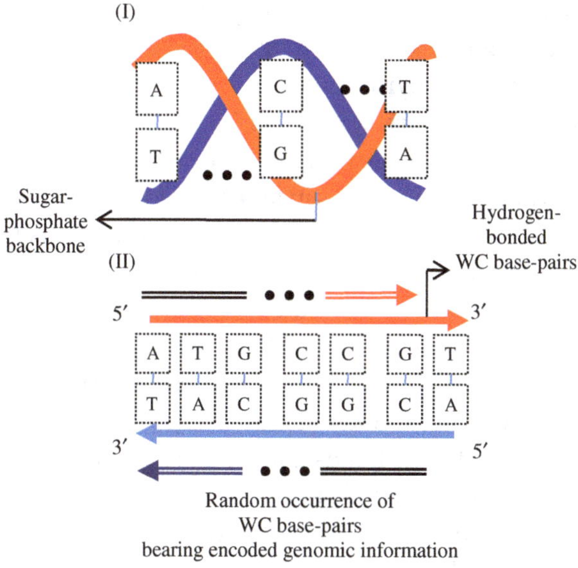

Fig. 1A-3 A double-stranded DNA (dsDNA) structure: (I) actual helical structure and (II) a hypothetical parallel format of the dual strands.

every cell of the body acting in a specific way. Each spiraling strand of the DNA is comprised of a sugar-phosphate backbone and attached *bases*. It is further connected to a complementary strand by non-covalent hydrogen bonding between paired bases. These bases, as indicated earlier are A, T, C, and G, and they are classified into two versions of bases: the purine base (A and G) and the pyrimidine base (T and C).

In Fig. 1A-3, the DNA molecular chain depicts a double-stranded DNA (dsDNA). It is formed by two twisted linear-chains of molecules conforming to a helix. The DNA macromolecule is made of four nucleotides {A, C, T, G} which are linked covalently, framing into a polynucleotide chain (or a *strand*) with a backbone chemistry of sugar-phosphate (from which the bases extend). The two DNA strands are held together by hydrogen bonds between paired nucleotide bases, namely, (A ↔ T), (T ↔ A), (G ↔ C), or (C ↔ G). These pairs are known as *Watson–Crick (WC) base-pairs*. Arrowheads are placed at the ends of the DNA strands to indicate the polarities of the two strands that run anti-parallel to each other in the DNA molecule.

As illustrated in Fig. 1A-4, the constituents of a nucleotide in a DNA (as well as, in the RNA) are framed by five-ringed carbon sugar either as a *ribose* or as a *deoxyribose* compound and the sugar group bonds across the nitrogen base and the phosphate group. The genetic message is written in the language of these four nitrogenous bases of the DNA, namely, the purines and pyrimidines.

Both DNA and RNA are *nucleic acids.* The backbone of the strands of the nucleic acid is made of phosphodiester bonds (that correspond to exactly two of the hydroxyl groups in phosphoric acid react with hydroxyl groups on other molecules forming two ester bonds). That is, in DNA (and in RNA), the phosphodiester bond forms the linkage between 3′ carbon atom of one sugar molecule and 5′ carbon atom of another, (deoxyribose in DNA and ribose in RNA). Strong covalent bonds form between the phosphate group and two 5-carbon ring carbohydrates (pentose) over two *ester* bonds. Thus, a chain of nucleotides joined together by phosphodiester bonds constitutes a *nucleic acid.*

The essential chemical bonds in the DNA linkage of macromolecules are as follows: (i) Ester bond: This refers to the oxygen–carbon linkage between the triphosphate group and the 5′ carbon of the ribose sugar group in a single DNA or RNA nucleotide; (ii) glycosidic bond is the nitrogen–carbon linkage between the 9′ nitrogen of purine bases or 1′ nitrogen of pyrimidine bases and the 1′ carbon of the sugar group; and (iii) hydrogen bonding: This is a weak, non-covalent linkage between a donor and an acceptor which, when lined up next to each other, have favorable electrostatic interactions. It provides a small extent of stability to DNA and RNA helices and decides the specificity of interactions between polynucleotide strands

The associated hydrogen bond acceptor depicts a group with at least one free lone pair of electrons (In DNA and RNA, the common *acceptor* groups include carbonyls, hydroxyls, and tertiary amines), and the hydrogen bond *donor* denotes a free-hydrogen group. (In DNA and RNA, common donors include secondary amines and hydroxyl groups.)

The phosphodiester linkage indicated above refers to the bond between the 3′ hydroxyl of a sugar group in a nucleotide and a phosphate group attached to the 5′ carbon of another sugar group. The nucleotide in a DNA is framed with the sugar phosphate plus the base as illustrated in Figs. 1A-4 and 1A-5. The DNA is thus a polymer made up of nucleotide units. The linear sequence of base-pairs in a polynucleotide chain is called a *primary structure* where, as described earlier a nucleotide unit consists of a chemical base, a deoxyribose sugar and a phosphate and the chemical bases involved correspond to A, T, C, and G. Each base is connected to a sugar *via* a *β-glycosyl* linkage. The nucleotide units are connected *via* O3′ and O5′ atoms forming phosphodiester linkages.

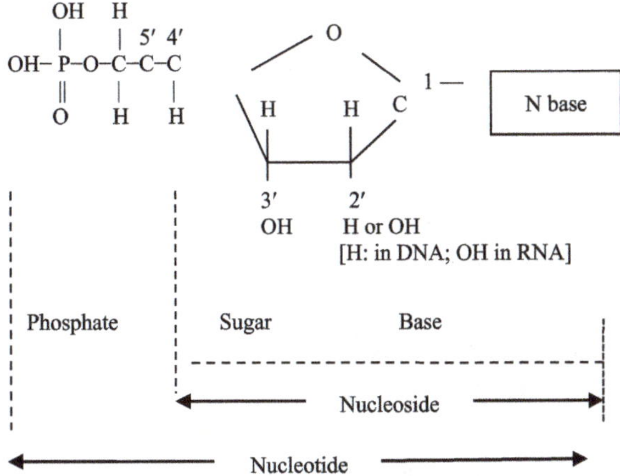

Fig. 1A-4 Constituents of a nucleotide in a DNA/RNA. The form of five-ringed carbon sugar shown is either a *ribose* or a *deoxyribose,* and the sugar group bonds across the nitrogen base and the phosphate group. These bases are A, T, C, and G.

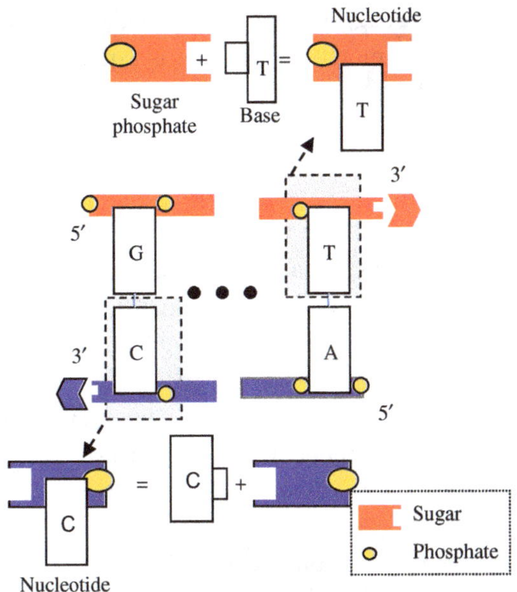

Fig. 1A-5 Formation of the nucleotide in a DNA with sugar phosphate plus the base.

In summary, the double-stranded DNA being a pair of polynucleotide chains (wound helically around a long central axis), chemically it is made of four deoxyribonucleotide monomers or deoxyribotides. Each deoxyribonucleotide consists of pentose sugar (deoxyribose), a phosphate group, and a nitrogenous base (either purine or pyrimidine). The purine bases (A and G) are heterocyclic and two-ringed bases, and the pyrimidines (T and C) are one-ringed bases. A nucleoside is formed by association of the nitrogen containing the base (A, T, C, or G) with the pentose.

For purine bases, the bond is between atom 9 of the base and the carbon 1′ of the pentose. For pyrimidine bases, the bond is between atom 1 of the base and the carbon 1′ of the pentose. The bases A and T are connected by two hydrogen bonds, and G and C are connected by three hydrogen bonds. Watson–Crick base-pair model implies that the bases occur in the DNA structure in the most plausible and stable forms with relevant pairing of adenine (purine) with thymine (pyrimidine) and guanine (purine) with cytosine (pyrimidine). In other words, if an adenine forms one member of a pair on either chain, then the other member must be the thymine; similarly, for guanine, the pair is cytosine. Suggested in 1948 by Erwin Chargaff, a rule that indicates that regularity prevails in the base composition of the DNA molecules isolated from several organisms. Explicitly, this rule suggests that $A/T = G/C$ or $(A + T)/(G + C) \rightarrow 1$, meaning that the total amount of purines is equal to total amount of pyrimidines of the DNA.

Ribonucleic Acid

Ribonucleic acid (RNA) has the same primary structure as DNA — a polymer of four nucleotides, each containing D-ribose, phosphoric acid, and nitrogenous base, namely, the purines (A, G) and the pyrimidines (C, U), where the uracil (U) substitutes for thymine (T). The RNA is usually single-stranded except in some viruses, like TMV, yellow mosaic virus, influenza, virus, etc. It consists of a backbone made of sugar phosphate with nucleotides attached to the 1′ carbon of the sugar. The differences between DNA and RNA are as follows.

- RNA has a hydroxyl group on the 2′ carbon of the sugar.
- RNA uses uracil instead of thymine nucleotide.
- Because of the extra hydroxyl group on the sugar, RNA exists as a single-stranded molecule. The RNA molecule with its hairpin loops is also said to have secondary structure.
- Inasmuch as the RNA molecule is not restricted to a rigid double helix, it can assume many different tertiary structures, as well.

There are different kinds of RNA made by the cell. They are as follows: mRNA-messenger RNA is a copy of the gene; tRNA-transfer RNA is a small RNA that has a very specific secondary and tertiary structure; rRNA-ribosomal RNA is one of the structural components of the ribosome; *small nuclear RNA* (snRNA) processes RNAs as they travel between the nucleus and the cytoplasm. A summary on salient features of DNA *versus* RNA is presented in Table 1A-1.

The nucleotides of the DNA make triplets. There are 64 codons, each formed by a set of any three nucleotide bases namely, A, T, C, and G and eventually are classified into a set of 20 amino acids. The underlying encoding or the genetic code determines eventually the insertion of a specific amino acid in a polypeptide chain during protein synthesis or the signal-to-stop protein synthesis.

Table 1A-1 A comparison of the genomic entities — DNA *versus* RNA.

Deoxyribose nucleic acid (DNA) is the basic genetic material seen in almost all living systems at the molecular level.

The DNA is a very long typically two-stranded giant molecule composed of four different types of nucleotides or base chemicals, Adenine, Guanine, Cytosine, and thymine (abbreviated as A, T, G, and C). Single-stranded DNA also exists in some viral species.

Ribonucleic acid (RNA) refers to any class of single-stranded molecules transcribed from DNA in the cell nucleus or in the mitochondrion or chloroplast. It contains along its strand depicting a linear sequence of nucleotide bases, and this strand is complementary to the DNA strand from which it is transcribed: The composition of the RNA molecule is identical with that of DNA except for the substitution of the sugar ribose for deoxyribose and the substitution of the nucleotide base uracil for thymine. That is, the nucleotides of the RNA are {A, T, G, U}. The RNA is the genetic material that prevails restrictedly in some plants, animals, and bacterial viruses.

DNA *versus* RNA: A summary of frequently asked questions	
By origination, which precedes the other?	
DNA originated after RNA	RNA is more primitive than DNA
What is the physical shape?	
Most DNA are double-stranded helix except in a few viruses (e.g., ϕX174)	Most RNA are single-stranded except in some viruses (e.g., retrovirus)

(*Continued*)

Table 1A-1 (*Continued*)

What is the version of pentose sugar constituent?	
Deoxyribose in the DNA	Ribose in the RNA
What are the constituent nucleotide bases?	
{A, G, C, T} in the DNA	{A, G, C, U} in the RNA
How is Watson–Crick base-pairing done?	
[A-T] and [G-C] in the DNA	[A-U] and [G-C] in the RNA
Where does the base-pairing occur?	
Across the entire length of DNA molecule	Only at the hair-pin structure and in the helical region (in the RNA)
What is the extent of nucleotide contents?	
About 4 million in the DNA	Lesser extent (about 12,000) in the RNA
What are the types of molecules involved?	
Single type in the DNA	rRNA, mRNA, and tRNA in the RNA
Where it is found?	
DNA is in chromosomes	rRNA in cytoplasm, mRNA in nucleolus, tRNA in cytoplasm
What is the function?	
DNA: Encoding the genes into transcripts	RNA: Translating the transcripts of DNA into protein complex
Does it need an enzyme vis-à-vis its functions?	
Not needed in the DNA	Genetic RNA needs the enzyme reverse transcriptase during replication process

The codon-specified AAs are depicted by three-letter codes (as in Table 1.1) and furnished in Table 1A-2 is the inverse table of standard genetic codes. Functionally, the coding statistics can be defined as the statistics that assays the likelihood that a sequence is coded for a protein. Hypotheses prevail addressing the abundance and locations of introns within the protein — *coding DNA sequence* (CDS) of transcripts. The intron positions are largely random within transcripts or occur at "proto-splice sites" that carry short sequences similar to conserved exon sequences flanking introns that are otherwise context free.[33,34] Intron positions may thus be purely fortuitous or involve selection for features that influence transcription or translation.

Table 1A-2 Inverse table for the standard genetic code (compressed using IUPAC notation).

Amino acid	Codons	Compressed	Amino acid	Codons	Compressed
Ala/A	GCU, GCC, GCA, GCG	GCN	Leu/L	UUA, UUG, CUU, CUC, CUA, CUG	YUR, CUN
Arg/R	CGU, CGC, CGA, CGG, AGA, AGG	CGN, MGR	Lys/K	AAA, AAG	AAR
Asn/N	AAU, AAC	AAY	Met/M	AUG	—
Asp/D	GAU, GAC	GAY	Phe/F	UUU, UUC	UUY
Cys/C	UGU, UGC	UGY	Pro/P	CCU, CCC, CCA, CCG	CCN
Gln/Q	CAA, CAG	CAR	Ser/S	UCU, UCC, UCA, UCG, AGU, AGC	UCN, AGY
Glu/E	GAA, GAG	GAR	Thr/T	ACU, ACC, ACA, ACG	ACN
Gly/G	GGU, GGC, GGA, GGG	GGN	Trp/W	UGG	
His H	CAU, CAC	CAY	Tyr/Y	UAU, UAC	UAY
Ile/I	AUU, AUC, AUA	AUH	Val/V	GUU, GUC, GUA, GUG	GUN
START	AUG	—	STOP	UAA, UGA, UAG	UAR, URA

The genetic code is said to be universal because nearly all living organisms on Earth use the same four-letter code to direct their metabolic functions and build bodies. A single sequence of nucleotides is able to synthesize a single protein, regardless of the species it is in. Further, the genetic code or coding statistics is a *universal code* because all known organisms use the same four nucleotide bases; organism differ according to the arrangement of the nucleotide bases. In summary, the exons are parts of the DNA within a gene that form the RNA transcript. It contains code sequence for protein to be synthesized, and the introns correspond to fill-in spaces between the exons in the gene locus. These are transcribed but spliced out during mRNA processing just before translation (or production of protein). In addition, certain *promoters* also exist on the DNA, which represent regulatory sequences controlling the rate of RNA transcription.

Amino Acid (AA)

The 64 triplets indicated above can be grouped into 20 types known as amino acids, which are commonly referred to as the "building blocks" of proteins. Figure 1A-6 shows the structure of an amino acid. As mentioned earlier, the sequence of nucleotides in a DNA molecule depicts the genetic information.

The nucleotide sequence formulates the set of 20 amino acids, and a sequence made of selective amino acid triplets may constitute a protein. Such a sequence representing a protein is encoded within the DNA sequence in question. The relation between the sequence of nucleotides and the sequence of the corresponding amino acid sequence depicting the protein is called *genetic code*. The characteristics of each amino acid are dependent on their side chain, and they can be divided into several classes. These classifications are denoted either as polar, nonpolar, acidic, or as basic.

Transcription and Translation Processes: Illustrations

The transcription process relevant to a double-stranded DNA sequence and subsequent translation to protein of amino acid sequence are illustrated in Figs. 1A-7a and 1A-7b. The mRNA is then transported out of the nucleus and into the cytoplasm of eukaryotes where proteins are formed through a process called *translation*. Messenger RNA carries coded information to ribosome, and the ribosomes "read" this

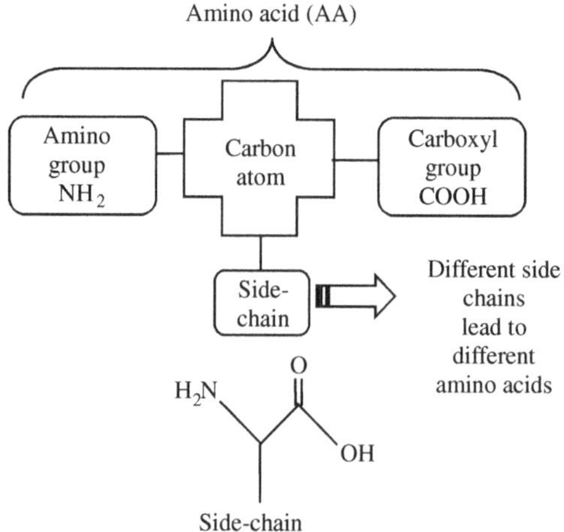

Fig. 1A-6 Amino acid structure: a simplified representation.

Table 1A-3 Unusual codon usage in nuclear and mitochondrial genes.[36]

Codon (triplet)	Standard universal code	Unusual code 3-letter format	*Occurrence in: Species
UGA	Stop	Trp	*Mycoplasma*, *Spiroplasma*, Mitochondria of several species
CUG	Leu	Thr	Mitochondria in yeasts
UAA, UAG	Stop	Gln	*Acetabularia*, *Tetrahymena*, Paramecium, etc.
UGA	Stop	Cys	*Euplotes*

"Unusual code" is used in nuclear genes of the listed organisms and in mitochondrial genes as indicated.

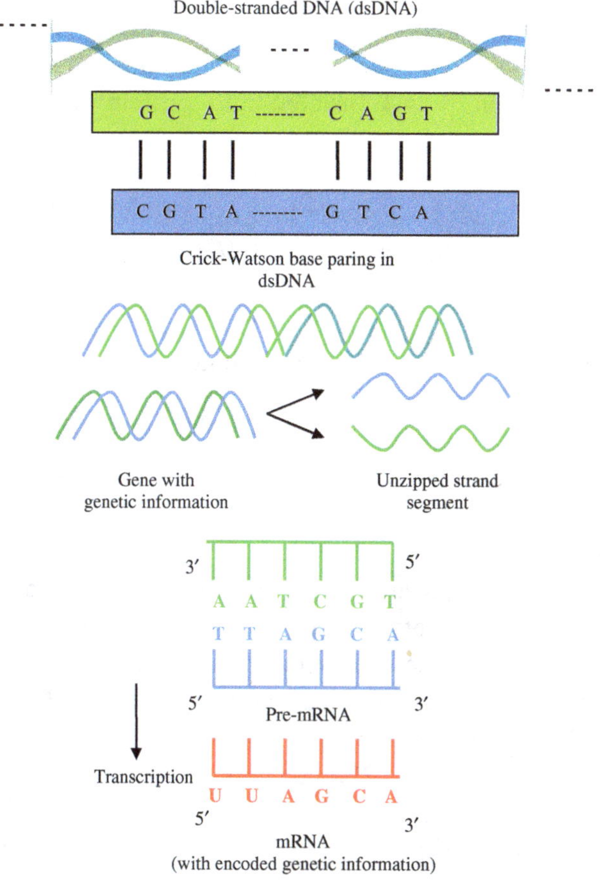

Fig. 1A-7a Transcription process relevant to a double-stranded DNA sequence.

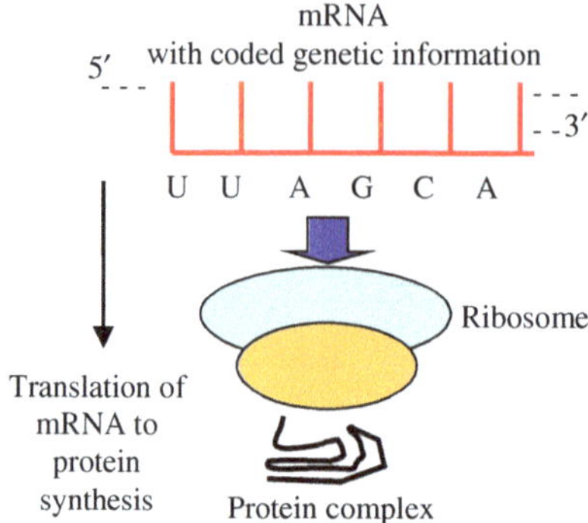

Fig. 1A-7b Translational process on the transcribed genetic information shown in Fig. 1A-14.

information and uses it for protein synthesis. In summary, the translation process refers to:

(a) mRNA, when transported through the nucleus membrane to the cytoplasm, then it is translated into protein with the help of ribosomes.
(b) The ribosomes contain a variety of different proteins and assortment of RNA molecules collectively known as ribosomal RNA (rRNA). These short-lived, but abundant, rRNAs are involved in the binding of mRNA to ribosome during protein-making translation process.

In summary, transcription implies the following.

- RNA synthesis within cell nucleus.
- DNA converted to a single-cell RNA (nRNA), which is then processed to a matured RNA (mRNA) known as messenger RNA.

The process of protein-making through the central dogma of molecular biology, illustrated in Figs. 1A-7a and 1A-7b, involves first the transcribing of information in sections from the DNA strand into an intermediate polymer called *messenger RNA* (or mRNA), which is similar to DNA except that the sugar residue R is replaced with a slightly different one, namely ribose R′, and a base called uracil (U) replaces the thymine (T) of the DNA.

The stages of making proteins *via* translation involve step-by-step addition of amino acids to a growing protein chain by a ribozyme (called a *ribosome*). That is, when mRNA is made available at the ribosome acting as message centers situated throughout the cell, the coded information in the mRNA gets translated into a biochemical polymeric complex as decided by a protocol of operations dictating, which building blocks are needed and in what order so that eventually an assembled composition, namely, the protein is synthesized.

The above operation happens with the help of *ribosomal RNA*. This *ribosomal ribonucleic acid* (rRNA) is the RNA component of the ribosome. It is the predominant material within the ribosome with two subunits, namely, the *large subunit* (LSU) and *small subunit* (SSU). The LSU rRNA acts as a ribozyme enabling the formation of catalyzing peptide bond in synthesizing the protein.

Thus, the culmination of translation yields proteins made up of several or many polypeptides. As is evident from the stages involved in protein-making, such polypeptide building blocks are chains of amino acids constituted by the triplets of the set {A, C, U, G}, and in essence, the protein is the culminating result of an one-dimensional sequential arrangement of amino acids (residues) bearing information consistent with the genetic code, as read from an mRNA template, and the RNA itself is a template copy of one of the organism's genes derived *via* the steps of central dogma as described above. Thus, the agenda of how the information in DNA is turned into protein is the tale of central dogma in microbiology. The amino acid participating as residues in deciding ultimately the protein composition (as per genetic code) are triplets or codons formed by the alphabet set of the DNA, namely {A, C, T, G}. Taken in triplets of bases, it results in set of codons with a cardinality of 64. In terms of these triplets (or codons) grouped appropriately, a set of 20 distinct amino acids are prescribed. These 20 amino acids are encoded by the standard genetic code and are called *proteinogenic* or *standard amino acids* as tabulated earlier.

Post-Transcription Process ... Toward Translation

Subsequent to transcription, the translation process involved conforms to the mRNA being transported into the cytoplasm where the encoded information in the gene-template is "de-coded" or "translated," so as to produce the correct order of amino acids in a protein. This translation is enabled by the intervention of certain enzymes listed as follows.

- Ribosomal RNA (rRNA): These are RNA molecules representing certain proteins to form the ribosomes, and each ribosome at a time can accept two *transfer RNAs* (tRNAs) and one mRNA.

- The tRNAs are small RNA molecules that carry a specific amino acid, at one end, and an *anticodon* region that recognizes and binds mRNA, at the other end. The tRNA that binds to that mRNA codon determines what amino acid is added to a protein chain.

Thus, the three RNAs (namely, mRNA, tRNA, and rRNA) participate collectively to turn the genetic information in the DNA into an eventual 3D protein. In all, the translation is the process by which the nucleotide sequence of mRNA is converted to the amino acid sequence of a polypeptide. (In prokaryotic single-celled organisms, such as the bacteria, the translation process takes place in the cytoplasm.)

The genetic information contained originally in the DNA is carried forward when the bases are set in triplet forms. With the four {A, C, T, G} bases, the possible permuted triplets, as stated earlier are 64 and they are grouped into 20 amino acids, and each amino acid being the triplet of bases bears the mapped genetic code of the DNA. The transcription occurs at a specific site on one strand of DNA known as *transcription initiation site*, marked by a characteristic base sequence. The transcription proceeds through a specific chemical pairing, namely, the WC-paring (A ⇔ T/U) and (G ⇔ C) mentioned earlier. The transcription process, in essence, is an information retrieval technique from the original memory units of the DNA. In the subsequent process of translation, the information contained in amino acids constraints the cell in the order of the amino acids strung together in making of the protein constituents. The eventual (correct) translation of eukaryotic genomic data into a protein complex is, however, subject to the effects of mutations on the evolutionary conservation. Any underlying corruptions may manifest at the so-called *splice junctions* that separate/delineate two subsequences in a DNA sequence, namely, the (genetic) information-bearing codon segment (called an *exon*) and the non-informative "junk" codon, also known as noncodon or *intron*. Exons bear necessary information toward protein-making, whereas non-codons are non-informative and their genetic role has not been fully elucidated. Exons and introns appear randomly along the DNA sequence. Exons tend to be typically no more than 200 characters long, while non-codons (introns) could be tens of thousands of characters in length. Thus in majority, introns prevail mostly in a typical eukaryotic gene. Toward the process of protein-making, introns are first scissored out (in the transcription stage) from the sequence and the remaining exons are spliced together

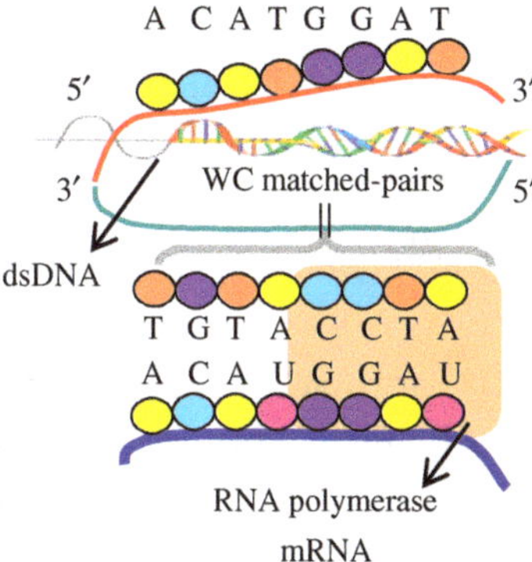

Fig. 1A-8 Pre-mRNA synthesis through transcription.

constituting the mRNA as in Fig. 1A-8, which is rendered ready for translation into a protein complex (at the cell interior). Should any error occur (due to mutations), they would give room to the possibility of evolving wrong or cryptic splice junctions and lead to (imperfect) translations. That is, aberrant splice junctions may result from mutational spectrum and would hamper the making of correct proteins.

In short, the concept of central dogma adopted resulting in the culmination of protein-making can be summarized as follows. Transcription: Relevant to a dsDNA helix, a portion of which corresponds to a gene is first unzipped to form a messenger RNA (mRNA). This unzipped part thus belongs to the one-side of the DNA, namely, the 3′–5′ reverse strand. In the unzipped part, the nucleotide base T is replaced by uracil (U). The resulting product is the pre-mRNA. Of its contents, namely, the exons and introns, the introns are removed and the exons are spliced together to constitute a mature mRNA, which leaves the nucleus to be transcribed by the ribosome. The sequence of DNA, which encodes the sequence of the amino acids in a protein, is thus copied into a mRNA chain.

Protein Structure

A framed protein structure resulting in the translation step described above is known as a *protein fold* (or *conformation*). It refers to the way the associated secondary structure elements are arranged relative to each other in space. The functions of proteins depend on the configured 3D shape resulting from the fold. Two proteins with related folds but unrelated sequences are called *analogous proteins*. During evolution, analogous proteins independently developed the same fold.

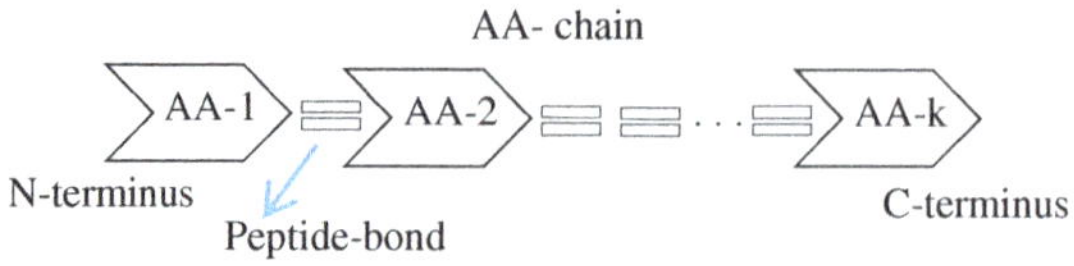

AA- chain

AA-1 AA-2 ... AA-k

N-terminus Peptide-bond C-terminus

Fig. 1A-9 Peptide-bonded amino acid (AA) chain with loose ends of N- and C-*termini.*

The mechanism of fold formation is consistent with the underlying chemistry of amino acids being used toward protein synthesis. As indicated earlier, the four nucleotide bases, A, C, T, and G, are organized in triplets to form 64 trinucleotides or codons and are grouped into 20 amino acids. The chemistry of bonding the amino acids in constituting an AA-chain acid is illustrated in Fig. 1A-9. The basic AA-chain forms into folds as a result of interaction between the chained amino acids as decided by underlying electrostatic interaction, Van der Waal forces, volume constraints, chemistry of hydrogen and disulfide bonds involved, and interaction between amino acids and water. Once formed, the folds conform to 3D structures with required rigidity of backbone. The levels of such protein folds are as follows: primary structure, secondary structure, tertiary structure, and quaternary structure.

The primary structure is the basic AA-chain as illustrated in Fig. 1A-9. The secondary structure known as α-*helix* refers to a helical twist of the primary structure. Further coiling of folded individual peptides causes the tertiary unit and aggregation of two or more peptides to conform to a bundled up quaternary structure.

Two common versions of secondary structures are α-helices and β-strands, and another type is a coil or loop category. Apart from not-so-well-defined random structural possibility, such coils may assume formats such as beta-twins, omega-loops, etc. In all, a *folding class* can be specified as decided by the abundance and disposition of secondary structure elements in a protein. Relevant seven classes can be specified in terms of the associated percentages of the contents as tabulated in Table 1A-4.

Some folds could be super-secondary structures (or protein motifs). For example, the so-called 4-helix bundle and the *TIM barrel* refers to a conserved protein-fold consisting of eight α-helices and eight parallel β-strands that alternate along the peptide backbone. The term TIM stands for *triosephosphate isomerase,* a conserved metabolic enzyme. The protein structures are determined in most cases *via* experimental studies based on X-ray crystallography and/or nuclear

Table 1A-4 Classes of protein folds.

	Contents		
Class	**α-helix**	**β-sheet**	**Cysteine**
α-class	>50%	<10%	<5%
β-class	>10%	<10%	<5%
α/β-class	>10%	<10%	<5%
Random class	<20%		—
Cys-rich α-class	>50%	<10%	>5%
Cys-rich β-class	>40%	<10%	>5%
Cys-rich α/β-class	>10%	>10%	>5%

magnetic resonance (NMR). However, such studies are largely time-consuming and expensive.

Relevant computational models and fold predictions are specified by: (i) Class (based on secondary structure composition), (ii) architecture (defining the overall shape of the domain structures), and (iii) fold (topology specifying the overall shape and connectivity of the domain structures). Such prediction methodology is pertinent to (a) 1D cases done *via* secondary structure details, solvent accessibility considerations, and transmembrane helical characteristics; (b) 2D cases performed with inter-residue strand contact details; and (c) 3D cases done *via* homology modeling, fold recognition technique (called *threading* described below), and by *ab initio* predictions (carried out through molecular dynamics).

Exclusively, the secondary prediction relevant to protein structure recognition refers to the following: Given an amino acid sequence, suppose it is required to predict a secondary structure (α, β or coil) for each residue in the structure, (e.g., as in the AA sequence, K E L V L A L Y D Y Q ..., the pursued approaches are based on two methods: (1) Prediction for a given residue is done by considering a short window (of 13–21 residues) in the neighborhood (2) by the notion of *homology modeling,* where the proteins with similar sequences are presumed to have the tendency to form folds of similar structures, that is, the homology modeling aligns sequence-to-sequence and ascertains the similarity between them and *via protein threading method,* a process that attempts to map a constructed, partial secondary structure to a (compatible) completely known structure.

Threading process: The strategy of threading indicated above adopted in predicting the protein structures needs a library of core fold templates. The protein threading is also known as *fold recognition*, involved in modeling those proteins that have the same fold as proteins of some known structures, but may not have homologous proteins with known structure. That is, the protein threading is used for proteins that do not have their homologous protein structures deposited in the Protein Databank (PDB), whereas homology modeling is used for those proteins that can be recognized against protein in the PDB. The prediction by "threading" implies proper placing, aligning of each amino acid in the target sequence to a position in the template structure, and evaluating how well the target matches the template.[44] The threading technique is illustrated in Fig. 1A-10.

In the realization of protein secondary structures, understanding the concept of *amphipathic helices* is necessary. The *amphipathic* and *amphiphilic* considerations apply to the two-sided disposition of hydrophobic or hydrophilic amino acids, and correspondingly, an *amphipathic* or *amphiphilic helix* is evolved, and it has a hydrophobic face, on one side, and a hydrophilic face, on the other. That is, a commonly encountered motif in various proteins and peptides is known as a *membrane-binding amphipathic helix* (AH) where the underlying *amphipathicity* refers to the segregation of hydrophobic and polar residues between the two opposite faces of the α-helix, a distribution that well conforms to membrane binding. An AH model observed *via* X-ray diffraction has an orientation parallel to the membrane plane with its central axis positioned at the level of the lipid glycerol group,

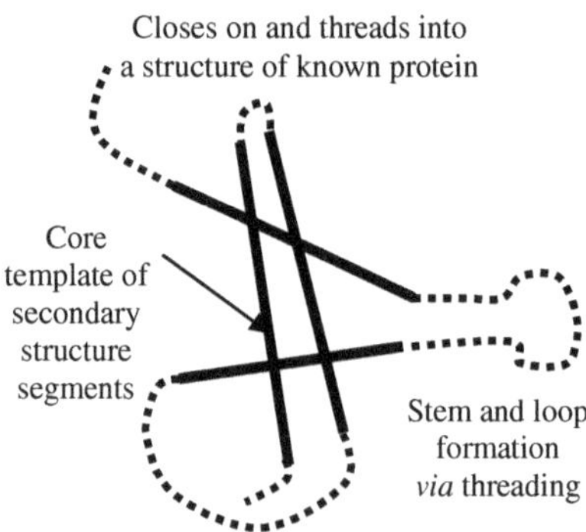

Fig. 1A-10 Protein-threading process.

which are hydrophobic residues insert between fatty acyl chains (unlike the polar residues that face lipid polar heads).

Mechanistic studies suggest that AH binds to membrane in three steps as follows: (i) The unfolded peptide sequence accumulates at the vicinity of negatively charged membranes through long-range electrostatic interactions. (ii) Then it is transferred to the membrane through an entropy-driven process (the *hydrophobic effect*) corresponding to the insertion of hydrophobic side-chains between the lipid acyl-chains and the release of water molecules. (iii) A random coil to α-helix transition occurs so as to reduce the energy penalty of having exposed peptide bonds in a hydrophobic environment and it accounts for about 50–60% of the free energy of binding.[45–47]

Relevant to regular secondary structures, a "periodicity" is prescribed. It refers to the hydrophobic periodicity, a pattern (up or down side-to-side alternating) exhibited as a repetition of hydrophobic amino acid side chains (with a frequency of repetition one for every 3–4 residues) in a given peptide sequence. This is shown in Fig. 1A-11. The periodically patterned residues of the AA sequence, when viewed down the helical axis show grouping of similar kinds of amino acids placed on the periphery of a helix, and such periodicity pattern (of polar and apolar residues) is indicative of alpha helix in the secondary structure. For β-sheets, if completely buried within a protein, uniform occurrence of apolar sections will be seen, and such sheets remain on the surface, alternating polar and apolar residues will occur. A view down the helical axis shows the groupings of similar kinds of amino acids as above, that is, looking down the axis of an α-helix gives the *helical wheel* projections of polar surface residues, on one side, and nonpolar interior residues, on the other.

Correspondingly, the *hydrophobic periodicity* can be measured in terms of *hydrophobic moment*. It is a quantitative measure of the

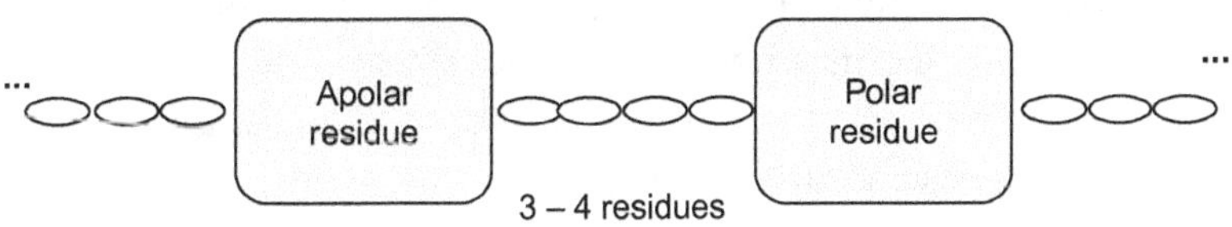

Fig. 1A-11 Periodicity pattern of polar and apolar residues indicative of alpha helix in the secondary structure.

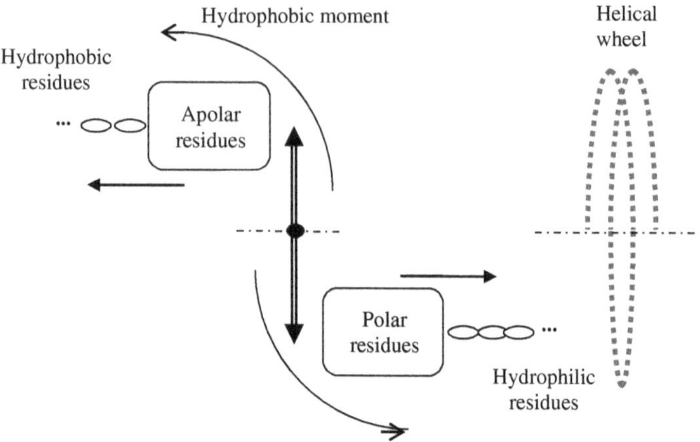

Fig. 1A-12 Pattern of polar and apolar residues leading to hydrophobic moment and helical wheel representation of AA sequence.

distribution and periodicity of hydrophobic amino acids over a defined sequence length as indicated in Fig. 1A-12 showing how proteins structurally contain amphipathic helix made of hydrophobic (nonpolar) residues, on one side, of the helical cylinder and hydrophilic (polar) residues, on the other side, resulting into hydropathic moment[48] seen as a helically wound diagram.

The helical wheel formed with amino acid side chains when projected down the axis (denoting the axis of an α-helix, orthogonal to plane of the paper) is an ideal α-helix consisting of 3.6 residues per complete turn, and the angle between two residues is chosen to be 100° and thus there exists a periodicity after five turns and 18 residues. When coiled-coil situation exists, it leads to the so-called *super helices*. Considering an AA sequence of 18 residues, for example, MLQSM ..., the corresponding helical wheel can be visualized as illustrated in Fig. 1A-13.

In summary, a helical wheel denotes the plot or visual representation that illustrates the characteristics of alpha helices in proteins. The AA sequence that forms a helical region of the secondary structure of a protein when plotted in a rotating manner with the angle of rotation between consecutive amino acids is 100°. The helical wheel indicates whether hydrophobic amino acids are concentrated on one side of the helix, usually with polar or hydrophilic amino acids on the other. Such an arrangement is commonly seen in alpha helices within globular proteins, where one face of the helix is oriented toward the hydrophobic core and one face is oriented toward the solvent-exposed surface.[40]

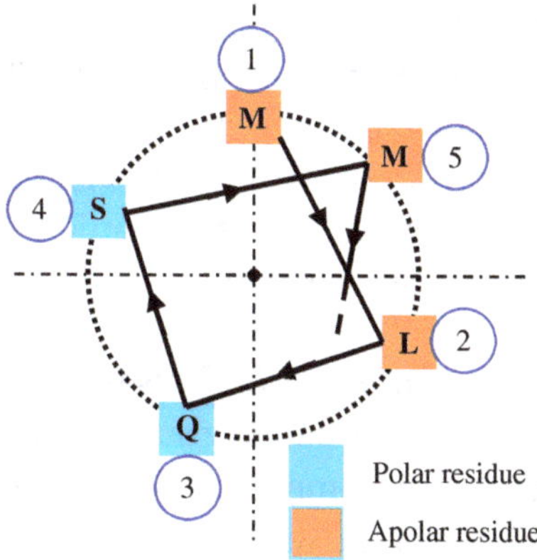

Fig. 1A-13 Making of a helical wheel.[49]

Homologous Proteins

Two proteins with related folds and related sequences are called homologous. Commonly, homologous proteins are further divided into orthologous and paralogous proteins. While orthologous proteins have evolved from a common ancestral gene, the paralogous proteins were created by gene duplication.

Additional Gene- and Chromosome-Related Details

Further to the genomic, proteomic, and cellular details outlined above, the following additional definitions, notions, and explanations are useful in bioinformatic contexts.

Motif: This refers to a consecutive string of amino acids in a protein sequence whose general character is repeated or conserved.

Transposons: In the biological cells, *transposons* are mobile genetic elements containing additional genes unrelated to transposition functions, and they represent genes that move from one location to another on a chromosome. Transposons in essence are *mutagens*.

Plasmids: These are small, circular, extra chromosomal DNA molecules. They can replicate independently of the genome and are found in numbers ranging from one to hundreds per cell. This is called "copy number." Plasmids frequently carry genes for antibiotic resistance. While antibiotic resistance signifies a problem in medical science, it is a useful marker in the so-called *recombinant DNA* technology.

Bioremediation: This is a treatment process that uses naturally occurring microorganism (such as yeast, fungi, or bacteria) to breakdown, or degrade, hazardous substances into less toxic or nontoxic substances. Microorganisms, like humans, eat and digest organic substances for nutrition and energy. In chemical terms, "organic" compounds are those that contain carbon and hydrogen atoms.

Fluorescent in situ hybridization (Fish): This refers to a genetic-marking technique and is a method used to visualize the presence of DNA sequences or whole chromosome. In the relevant technique, DNA sequences from the region of interest are labeled with a chemical that is fluorescent when test sequence parts of interest are labeled with a chemical that is fluorescent when exposed to ultraviolet (UV) light. This labeled molecule is called a "probe." Chromosomes are prepared in such a way that allows the probe to interact and form base pairs with homologous DNA sequences on the chromosomes. This process of probe binding to chromosomes is called *hybridization*. When the labeled DNA probe bound to the chromosomes is exposed to UV light, it fluoresces. This shows the presence and location of DNA sequences homologous to the probe. This is an ideal technique for both visual observation and automatic image analysis systems to be applied to genetic research, screening, and diagnosis. The first step in this method is to break apart (*denature*) the double strands of DNA in both the probe DNA and the chromosome DNA. Next, the probe is placed on the slide and glass cover slips placed on top. The slide is then placed in a 37°C incubator overnight for the probe to hybridize with the target chromosome. Overnight, the probe DNA seeks out its target sequence on the specific chromosome and binds to it. The strands slowly rejoin (or go through *re-annealing*).

Hybridoma: It is a method used to produce *monoclonal antibodies*. Large amounts of homogenous antibodies with tailor-made specificities can be readily prepared by this technique. Mammalian cells are fused with cultivated tumor cells to form hybridomas, which produce monoclonal antibodies.

Ramachandran plot (RP): This plot also known as Ramachandran diagram or [φ, ψ] *plot* developed in 1963 by Ramachandran *et al.*[50] allows visualizing the backbone dihedral angles φ against ψ in the structures of amino acid residues in a protein complex, and it enables identifying sterically permitted regions for these angles. (The angles φ and ψ were originally designated as φ and φ' by Ramachandran in Ref. 50.) While in geometry, a dihedral angle, in general, represents the angle between two intersecting planes; in chemistry, however, it depicts the angle between planes through two sets of three atoms, having two atoms in common.

Ramachandran *et al.*[50] used computer models of small polypeptides to systematically vary the angles phi and psi with the objective of finding stable conformations. For each conformation, the structure was examined for close contacts between atoms. Atoms were treated as hard spheres with dimensions corresponding to their van der Waals radii. Therefore, phi and psi angles that cause spheres to collide correspond to sterically disallowed conformations of the polypeptide backbone. The RP can be used in two distinct ways: (i) To show theoretically, which values or conformations of the angles (ψ and φ) are possible for an amino acid residue to assume in a protein and (ii) to show the empirical distribution of data points observed in a single structure (or in a database of many structures) useful in applications such as structure validation. In either of the two cases, the plot outlines for the theoretically favored regions.

Physico-Chemical Properties of Amino Acids

With reference to Tables 1.1 and 1A-2 of standard 20 amino acids,[36] it is essential to know their physico-chemical properties useful in bioinformatic tasks. Hence, listed in Tables 1A-5 and 1A-6 are relevant physico-chemical considerations *versus* standard amino acids.

Apart from the four main physico-chemical classifications indicated above for amino acids, there are other characteristics that can be used to describe them. This is best illustrated in the properties diagram in Fig. 1A-14. In the Venn diagram shown, the word "tiny" is used to describe very short side chains, while the term "small" denotes small side chains. Further, the terms aliphatic, aromatic, and hydrophobic indicated commonly refer to the chemical composition of the side chain of an amino acid. These side chains are typically composed of only carbon and hydrogen atoms. The additional terms mentioned as charged, negative, positive, and polar designate the electronic characteristics of a side chain.

With reference to Fig. 1A-14 denoting the set of overlapping physico-chemical properties of amino acids, the summary of such properties can also be listed as given in Table 1A-6. Another physico chemical parameter of interest in bioinformatic studies refers to a set of numerical

Table 1A-5 List of amino acids and their physico-chemical properties.

Amino acid	3-letter code	1-letter code	Properties
Alanine	Ala	A	Aliphatic; hydrophobic; neutral
Arginine	Arg	R	Polar; hydrophilic; charged (+)
Asparagine	Asn	N	Polar; hydrophilic; neutral
Aspartate	Asp	D	Polar; hydrophilic; charged (–)
Cysteine	Cys	C	Polar; hydrophobic; neutral
Glutamine	Gln	Q	Polar; hydrophilic; neutral
Glutamate	Glu	E	Polar; hydrophilic; charged (–)
Glycine	Gly	G	Aliphatic; neutral
Histidine	His	H	Aromatic; polar; hydrophilic; charged (+)
Isoleucine	Ile	I	Aliphatic; hydrophobic; neutral
Leucine	Leu	L	Aliphatic ;hydrophobic; neutral
Lysine	Lys	K	Polar; hydrophilic; charged (+)
Methionine	Met	M	Hydrophobic; neutral
Phenylalanine	Phe	F	Aromatic; hydrophobic; neutral
Proline	Pro	P	Hydrophobic; neutral
Serine	Ser	S	Polar; hydrophilic; neutral
Threonine	Thr	T	Polar; hydrophilic; neutral
Tryptophan	Trp	W	Aromatic; hydrophobic; neutral
Tyrosine	Tyr	Y	Aromatic; polar; hydrophobic
Valine	Val	V	Aliphatic; hydrophobic; neutral

representation of the values that conform to the so-called *electron–ion interaction-pseudopotentials* (EIIP) assigned to the nucleotides in a DNA sequence. The EIIP of nucleotide bases are identified and listed in Refs. 51 and 52 as follows: A: 0.1260; T: 0.1335; G: 0.0806; and C: 0.1340.

Exemplars of Codon Occurrence Statistics

Considering the 64 triplets (codons) specified from the nucleotide set {A, C, T, G}, their occurrence statistics expressed in terms of their probabilities of occurrence are unique for a given organism. This occurrence statistic is useful in deciding the information-theoretic details of the genes of the species. Furnished in Table 1A-7 are exemplar sets of *codon usage frequency* (CUF) of: (a) *Homo sapiens*; (b) *Escherichia coli*; (c) *Methanococcus jannaschii; and* (d) *Rickettsia prowazekii.*[41,42,53] These

Table 1A-6 Properties of amino acids: A summary.

Property	Amino acid	Property	Amino acid
Hydrophobic-aliphatic	Ala (A), Val (V), Leu (L), Ile (I), Met (M)	Basic amino acids	His (H), Lys (K), Arg (R)
Hydrophobic side chain	Val (V), Leu (L), Ile (I), Met (M), Phe (F)	Conformationally important: Special structural properties	Gly (G), Pro (P)
Hydrophobic-aromatic	Phe (F), Tyr (Y), Trp (W)	Disulphide bond former	Cys (C)
Hydrophilic side chain	N (Asn), E (Glu), Q (Gln), H (His), K (Lys), R (Arg), D (Asp)	Cyclic amino acid	Pro (P)
Neither hydrophobic nor hydrophilic side chain	G (Gly), A (Ala), S (Ser), T (Thr), Y (Tyr), W (Trp), C (Cys), P (Pro)	Hydroxyl	Ser (S), Thr (T), Tyr (Y)
Neutral-polar side chains	Ser (S), Thr (T), Asn (N), Gln (Q), C (Cys), M (Met)	Sulfur-containing amino acids	Cys (C), Met (M)
Acidic amino acids	Asp (D), Glu (E)	Smallest side chains	Ala (A), Gly (G)

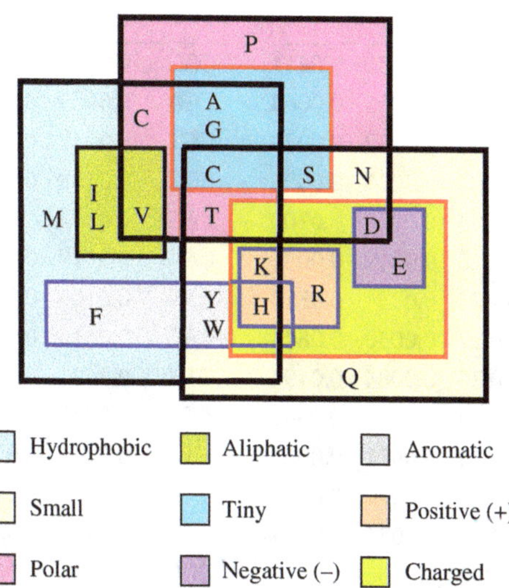

Fig. 1A-14 Venn diagram representation of the set of overlapping physico-chemical properties of amino acids (indicated via 1-letter codes).

Table 1A-7 Relative *codon usage frequency* (CUF) (normalized and rounded off values) of the 64 triplets: (a) *Homo sapiens*; (b) *Escherichia coli*; (c) *Methanococcus jannaschii*; and (d) *Rickettsia prowazekii*.[41,42,53,54]

Codon triplet	(a) CUF	(b) CUF	(c) CUF	(d) CUF	Codon triplet	(a) CUF	(b) CUF	(c) CUF	(d) CUF
GGG	0.0165	0.0113	0.0057	0.0514	AGG	0.0116	0.0021	0.0034	0.0238
GGA	0.0164	0.0096	0.0148	0.0088	AGA	0.0117	0.0037	0.0152	0.0033
GGT	0.0108	0.0254	0.0286	0.0676	AGT	0.0120	0.0099	0.0190	0.0208
GGC	0.0226	0.0270	0.0067	0.0104	AGC	0.0194	0.0151	0.0067	0.0041
GAG	0.0403	0.0187	0.0132	0.0254	AAG	0.0326	0.0125	0.0155	0.0198
GAA	0.0290	0.0390	0.0443	0.0038	AAA	0.0240	0.0353	0.0672	0.0010
GAT	0.0221	0.0327	0.0433	0.0145	AAT	0.0167	0.0207	0.0576	0.0133
GAC	0.0257	0.0191	0.0051	0.0560	AAC	0.0193	0.0214	0.0101	0.0233
GTG	0.0287	0.0243	0.0061	0.0265	ATG	0.0223	0.0264	0.0213	0.0046
GTA	0.0070	0.0116	0.0258	0.0258	ATA	0.0072	0.0070	0.0446	0.0339
GTT	0.0109	0.0198	0.0212	0.0212	ATT	0.0158	0.0298	0.0516	0.0041
GCC	0.0146	0.0143	0.0038	0.0038	ATC	0.0213	0.0236	0.0112	0.0332
GCG	0.0076	0.0300	0.0043	0.0043	ACG	0.0062	0.0137	0.0038	0.0055
GCA	0.0160	0.0212	0.0266	0.0266	ACA	0.0149	0.0094	0.0186	0.0238
GCT	0.0187	0.0171	0.0273	0.0273	ACT	0.0129	0.0105	0.0260	0.0033
GCC	0.0284	0.0242	0.0034	0.0034	ACC	0.0191	0.0219	0.0047	0.0208

Codon triplet	(a) CUF	(b) CUF	(c) CUF	(d) CUF	Codon triplet	(a) CUF	(b) CUF	(c) CUF	(d) CUF
TGG	0.0130	0.0139	0.0070	0.0417	CGG	0.0118	0.0059	0.0008	0.0103
TGA	0.0014	0.0010	0.0005	0.0064	CGA	0.0063	0.0038	0.0028	0.0031
TGT	0.0101	0.0053	0.0082	0.0024	CGT	0.0047	0.0199	0.0095	0.0006
TGC	0.0124	0.0061	0.0025	0.0006	CGC	0.0108	0.0196	0.0019	0.0088
TAG	0.0007	0.0003	0.0005	0.0200	CAG	0.0344	0.0284	0.0068	0.0128
TAA	0.0008	0.0020	0.0019	0.0038	CAA	0.0119	0.0146	0.0246	0.0024
TAT	0.0121	0.0175	0.033	0.0313	CAT	0.0106	0.0126	0.0160	0.0037
TAC	0.0155	0.0122	0.0050	0.0844	CAC	0.0150	0.0093	0.0030	0.0189
TTG	0.0127	0.0130	0.0084	0.0194	CTG	0.0401	0.0482	0.0028	0.0044
TTA	0.0073	00.014	0.0538	0.0540	CTA	0.0070	0.0042	0.0115	0.0353
TTT	0.0171	0.0222	0.0411	0.0065	CTT	0.0129	0.0119	0.0202	0.0083
TTC	0.0204	0.0160	0.0070	0.0553	CTC	0.0195	0.0102	0.0029	0.0186
TCG	0.0045	0.0085	0.0034	0.0164	CCG	0.0070	0.0208	0.0040	0.0070
TCA	0.0119	0.0090	0.0166	0.0417	CCA	0.0168	0.0086	0.0106	0.0103
TCT	0.0148	0.0104	0.0189	0.0064	CCT	0.0173	0.0075	0.0158	0.0031
TCC	0.0175	0.0091	0.0026	0.0024	CCC	0.0200	0.0054	0.0008	0.0006

CUF details (normalized and rounded off values) as listed in Table 1A-7 are used in other chapters on *ad hoc* basis.

References

[1] C. E. Shannon: A mathematical theory of communication. *Bell System Technical Journal,* 1948, vol. 27, 379–423 and 623–656.

[2] R. V. L. Hartley: Transmission of information. *Bell System Technical Journal,* 1928, vol. 7(3), 535–563.

[3] B. Bergeron: *Bioinformatics Computing.* Pearson Publication Inc., Upper Saddle River, NJ, USA: 2003.

[4] C. L. Stone: *The Basics of Biology.* Greenwood Press Inc., Santa Barbara, CA, USA: 2004.

[5] T. K. Attwood and D. J. Parry-Smith: *Introduction to Bioinformatics,* Pearson Education Ltd., Essex, England: 1999.

[6] T. Kulikova, P. Aldebert, N. Althorpe, W. Baker, K. Bates, *et al.*: The EMBL nucleotide sequence database. *Nucleic Acids Research,* 2004, vol. 32 (database issue), D27–D30.

[7] D. Benson, I. Karsch-Mizrachi, D. J. Lipman, J. Ostell, D. L. Wheeler, *et al.*: GenBank. *Nucleic Acids Research,* 2008, vol. 36 (Database), D25–D30.

[8] D. Benson, I. Karsch-Mizrachi, D. J. Lipman, J. Ostell, E. W. Sayers, *et al.*: GenBank. *Nucleic Acids Research,* 2009, vol. 37 (Database), D26–D31.

[9] Y. Tateno, T. Imanishi, S. Miyazaki, K. Fukami-Kobayashi, N. Saitou, H. Sugawara, *et al*:. DNA data bank of Japan (DDBJ) for genome scale research in life science. *Nucleic Acids Research,* vol. 30(1), 27–30.

[10] E. Kaminuma, T. Kosuge, Y. Kodama, H. Aono, J. Mashima, *et al.*: DDBJ progress report. *Nucleic Acids Research,* 2011, vol. 39 (Database issue), D22–D27.

[11] R. Apweiler, A. Bairoch and C. H. Wu: Protein sequence databases. *Current Opinion in Chemical Biology,* 2004, vol. 8, 76–80.

[12] A. Bairoch and R. Apweiler: The SWISS-PROT protein sequence databank and its new supplement TREMBL. *Nucleic Acids Research,* 1996, vol. 24(1), 21–25.

[13] Z. Ghosh and B. Mallick: *Bioinformatics — Principles and Applications.* Oxford University Press, New Delhi, India: 2008.

[14] A. J. Gibbs and G. A McIntyre: The diagram, a method for comparing sequences. Its use with amino acid and nucleotide sequences. *European Journal of Biochemistry,* 1970, vol. 16, 1–11.

[15] S. F. Altschul: Global and local sequence alignment. Available online at: *https://www.cs.umd.edu/class/fall2011/cmsc858s/Alignment.pdf.*

[16] S. B. Needleman and C. D. Wunsch: A general method applicable to the search for similarities in the amino acid sequence of two proteins. *Journal of Molecular Biology*, 1970, vol. 48(3), 443–453.
[17] P. H. Sellers: On the theory and computation of evolutionary distances. *SIAM Journal on Applied Mathematics*, 1974, vol. 26(4), 787–793.
[18] T. F. Smith and M. S. Waterman: Identification of common molecular subsequences. *Journal of Molecular Biology*, 1981, vol. 147, 195–197.
[19] O. Gotoh: An improved algorithm for matching biological sequences. *Journal of Molecular Biology*, 1982, vol. 162, 705.
[20] S. F. Altschul and B. W. Erickson: Optimal sequence alignment using affine gap costs. *Bulletin of Mathematical Biology*, 1986, vol. 48, 603–616.
[21] P. Hogeweg and B. Hesper: The alignment of sets of sequences and the construction of phyletic trees: an integrated method. *Journal of Molecular Evolution*, 1984, vol. 20, 175–186.
[22] M. L. Pearson and D. Soll: The human genome project: a paradigm for information management in the life sciences. *Federation of American Societies for Experimental Biology Journal*, 1991, vol. 5(1), 35–39.
[23] P. R. Ehrlich and A. H. Ehrlich: How humans cause mass extinctions. Available online at: https://www.project-syndicate.org/commentary/mass-extinction-human-cause-by-paul-r--ehrlich-and-anne-h--ehrlich-2015-08?barrier=accesspaylog.
[24] D. Knebel, J. Tickner, J. Lemons, R. Levins, E. L. Loechler, M. Quinn, R. Rudel, T. Schettler and M. Stoto: The precautionary principle in environmental science. *Environmental Health Perspectives*, 2001, vol. 109(9), 871–876.
[25] J. D. Watson and F. H. C. Crick: Molecular structure of nucleic acids: A structure for deoxyribose nucleic acid. *Nature*, 1953, vol. 171, 737–738.
[26] W. S. Klug and M. R. Cummings: *Concepts of Genetics*. Macmillan Publishing Co., New York, NY, USA: 1986.
[27] M. Schmidt (Ed.): *Synthetic Biology: Industrial and Environmental Applications*. Wiley-VCH Verlag GmbH & Co., Weinheim, Germany: 2012, p. 151.
[28] M. Schmidt: Xenobiology: A new form of life as the ultimate biosafety tool. *BioEssays*, 2010, vol. 32(4), 322–331.
[29] R. Gonzales: XNA is synthetic DNA that's stronger than the real thing. Available online at: *http://io9.com/5903221/meet-xna-the-first-synthetic-dna-that-evolves-like-the-realthing*.
[30] A. Maton: *Cells Building Blocks of Life*. Prentice-Hall, Upper Saddle River, NJ, USA: 1997.
[31] H. M. Wain, E. A. Bruford, R. C. Lovering, M. J. Lush, M. W. Wright and S. Povey: Guidelines for human gene nomenclature. *Genomics*, 2002, vol. 79(4), 464–470.

[32] J. Y. Yang, M. Q. Yang, M. M. Zhu, H. R. Arabnia and Y. Deng: Promoting synergistic research and education in genomics and bioinformatics. *BMC Genomics*, 2008, vol. 9(Suppl 1), Article number: I1 (2008).

[33] T. Cavalier-Smith: Selfish DNA and the origin of introns. *Nature*, 1985, vol. 315, 283–284.

[34] W. F. Doolittle: Genes in pieces: Were they ever together? *Nature*, 1978, vol. 272, 581–582.

[35] A. Ambrogelly, S. Palioura and D. Söll: Natural expansion of the genetic code. *Nature Chemical Biology*, 2007, vol. 3(1), 29–35.

[36] IUPAC-IUB Joint Commission on Biochemical Nomenclature: Nomenclature and symbolism for amino acids and peptides. *European Journal of Biochemistry*, 1984, vol. 138, 9–37.

[37] S. Osawa, T. H. Jukes, K. Watanabe and A. Muto: Recent evidence for evolution of the genetic code. *Microbiology and Molecular Biology Reviews*, 1992, vol. 56(1), 229–264.

[38] F. H. C. Crick: Central dogma of molecular biology. *Nature*, 1970, vol. 227, 561–563.

[39] J. D. Watson: *The Molecular Biology of the Gene*. W. A. Benjamin. Inc., New York, NY, USA: 1965.

[40] F. H. C. Crick: On protein synthesis, The Biological Replication of Macro Molecules. In Symposia of the Society of Experimental Biology, Number XII (1958), pp. 138–163, Cambridge University Press, Cambridge, England.

[41] Codon Usage Database: *Drosophila melanogaster* [gbinv]: 42417 CDS's (21945319 codons). Available on-line at: https://www.kazusa.or.jp/codon/cgi-bin/showcodon.cgi?species=7227.

[42] Codon Usage Database: *Homo sapiens* [gbpri]: 93487 CDS's (40662582 codons). Available on-line at: https://www.kazusa.or.jp/codon/cgi-bin/showcodon.cgi?species=9606.

[43] J. Kyte and R. F. Doolittle: A simple method for displaying the hydropathic character of a protein. *Journal of Molecular Biology*, 1982, vol. 157(1), 105–132.

[44] D. T. Jones and C. Hadley: Threading methods for protein structure prediction. In D. Higgins and W. R. Taylor: *Bioinformatics: Sequence, Structure and Databanks*. Springer-Verlag, Heidelberg, Germany: 2000, pp. 1–13.

[45] K. Hristova, W. C. Wimley, V. K. Mishra, G. M. Anantharamiah, J. P. Segrest and S. H. White: An amphipathic alpha-helix at a membrane interface. A structural study using a novel X-ray diffraction method. *Journal of Molecular Biology*, 1999, vol. 290, 99–117.

[46] J. Seelig: Thermodynamics of lipid–peptide interactions. *Biochimica et Biophysica Acta*, 2004, vol. 1666, 40–50.

[47] D. Eisenberg, R. M. Weiss and T. C. Terwilliger: The helical hydrophobic moment: A measure of the amphiphilicity of a helix. *Nature*, 1984, vol. 299, 371–374.

[48] G. Drin and B. Antonny: Amphipathic helices and membrane curvature. *Frontiers in Membrane Biochemistry*, 2010, vol. 584(9), 1840–1847.

[49] M. Schiffer and A. B. Edmundson: Use of helical wheels to represent the structures of proteins and to identify segments with helical potential. *Biophysical Journal*, 1967, vol. 7(2): 121–135.

[50] G. N. Ramachandran, C. Ramakrishnan and V. Sasisekharan: Stereochemistry of polypeptide chain configurations. *Journal of Molecular Biology*, 1963, vol. 7, 95–99.

[51] A. S. Nair and S. P. Sreenadhan: A coding measure scheme employing electron–ion interaction pseudopotential (EIIP), *Bioinformation*, 2006, vol. 1(6), 197–202.

[52] K. D. Rao and M. N. S. Swamy: Analysis of genomics and proteomics using DSP techniques. *IEEE Transactions on Circuits and Systems. I. Regular Papers*, 2008, vol. 55(1), 370–378.

[53] J. Athey, A. Alexaki, E. Osipova, A. Rostovtsev, L. V. Santana-Quintero, U. Katneni, V. Simonyan and C. Kimchi-Sarfaty: A new and updated resource for codon usage tables. *BMC Bioinformatics*, 2017, vol. 18, 391–401.

[54] Codon Usage Database: *Escherichia coli, Saccharomyces cerevisiae, Rickettsia prowazekii, Methanococcus jannaschii etc.* Available on-line at: https://www.kazusa.or.jp/codon/.

Chapter 2

Information-Theoretics: A Review

Deliberated in this chapter are concepts of entropy vis-à-vis their relation to information-theoretic considerations. Hence, identified are their bioinformatic significance relevant to molecular, cellular, and/or at whole-life systems. Appropriate metrics and measures of information-theory (in Shannon sense) are defined thereof and explained in the contexts of bioinformatics.

"... decrease of entropy ... by the intervention of intelligent beings ..."

Leo Szilard

2.1 Biological Information: A Prelude

Biology being the science of knowledge specified across the entire gamut of living systems, it primarily includes and exhaustively elaborates the diverse aspects of the phenomena associated with origin, growth, reproduction, structure, and behavior of organisms. The associated spatio-temporal characteristics of living entities (as well as their diverse activities) refer to a complex representation of underlying *organized* (or *disorganized?*) (micro)/macro states, and such states comprehensively depict both deterministic and stochastic aspects of the portfolio detailing the information on perception, cognition, and signaling inherently present in all organisms. Hence, the bioinformatic heuristics of entropy and concepts of information theory implicitly describe core details of biological sciences and facilitate inferring pertinent considerations that portray a plethora of information on the perspectives of living organisms, and such informative details manifest as "expression(s) of information, execution of programs, and interpretation of (genetic) codes"[1] of the biological media. Commencing at biomolecular level across the chains of DNA/RNA sequences and implied by structural aspects of protein complex, the aforesaid manifestation of details is concerned with functional/structural aspects of cytogenetic cellular/organelle units. Pertinent information, in addition, formulates the phenetic details and phylogenetic characteristics of whole species.

In the perspectives as above, any living system can be regarded as being inherently information rich. Processing the gamut of such information ("expressed as a message" in genetic, cellular, and whole-life contexts) and ascertaining the associated details across complex domains of various biological structures (of diverse species) specify, in general, the scope of biological informatics.[2]

The principle of information theory (counting from the days of Claude Shannon[3]) describes the methods and means of assessing the embodiment of details in a given system, and such details implicitly depict the so-called *entropy* (in Boltzmann sense), which corresponds to the uncertainty state (of a system) and denoted as, S. Entropy becoming less, means that the system attains more ordered and predictable micro- and/or macro states.[3–5] Originally, Boltzmann's heuristics of entropy was indicated in the context of thermodynamics, where it is specified by the following *Boltzmann's entropy relation*: $S = \{k_B \times \log_b(W)\}$; here, the Boltzmann constant, k_B links the thermodynamic entropy (S) associated with the (thermal) energy (or implicitly, by the *temperature*, T °K) *versus* plausible number of (micro)/macrostates, W existing in the system, and the observed stochastical probability of microstates (or uncertainty of the states) is implied by the associated (conserved) energy and mass. That is, the entropy or uncertainty in a system is estimated by considering statistical probabilities of the features (or states) in the system, and correspondingly, Shannon defined *information* (H) associated in a stochastic system comprised of observable states (having probabilistic and hence entropic attributes) as follows: Information depicts the associated *negentropy*; that is, H (information) ≡ $-S$, with S denoting corresponding *posentropy*. Functionally, H depends on the probabilities of the state leading to the so-called *Shannon's law*.

The base of the logarithm in the above expression for S or H decides the unit for entropy. If the base is taken as the Euler number, "$e = \approx 2.71828$" (corresponding to the natural logarithm), then the unit of S is called "nats." If $b = 2$, then S or H is expressed in "bits" meaning *binary digits*. (Note also, in rarely used contexts, if $b = 10$, the unit for H is "hartley," and the unit "bits" was originally indicated as the unit "Shannon," but relevant usage is defunct now). The base-changing formula of logarithm, namely $H_b(X) = [\log_b(a) \times H_a(X)]$ or $[I_b(X) = \log_b(a) \times I_a(X)]$, enables converting the units of $H(X)$ or $I(X)$ as needed.

The link between aforesaid thermodynamic entropy (due to Boltzmann) and information entropy (in Shannon's sense) can be established on the basis of following considerations: Suppose W denotes the total number of microstates (depicting a set of random variables [RVs] that constitute the macrostate) of a stochastic system, and relevant occurrence statistics of the microstates corresponds to a set of *a priori* probabilities $\{p_i\}$, with i denoting the ith state, then, the corresponding Shannon entropy denoting the information (or known details *sans* uncertainty in a message-bearing system) is given by $H = -\sum_{i=1}^{W}[p_i] \times [\log_e(p_i)]$ nats. Identically, referring to Boltzmann entropy formula, $S \equiv [k_B \times \log_e(W)]$ nats, the function $\log_e(W)$ can be interpreted as *thermodynamic entropy* per nat. In all, entropy measures the amount of disorder (or uncertainty) in a physical system, or more precisely, "entropy measures the lack of information about the actual structure of the system" prevailing spatio-temporally, and this lack of information would correspond to "actual disorder in the hidden degrees of freedom" prevalent at any time and space.

2.2 Information Theory: Basic Concepts

The thermodynamic entropy *versus* information theory heuristics are detailed in Chapter 8 of Neelakanta[6,8] by the author. In essence, Boltzmann implicitly specified the energy ($\mathfrak{E}$) of a thermodynamic system by the well-known (Boltzmann) equation, $\mathfrak{E} = (k_B \times T)$ with T denoting the system temperature and k_B representing the Boltzmann constant. Considering the macro states (W) set forth by thermodynamic energy in a system, the Boltzmann entropy relation, $S \equiv [k_B \times \log_e(W)]$, nats was commensurately formulated, and the associated stochasticity of W was interpreted in terms of $\{p_i\}$, depicting probabilistic attributes of micro-states. Hence, the axiomatic definition of entropy as a "probability measure" was indicated by Gibbs in 1878[7]) as: $S = [k_B \sum_i p_i \times \log_e(p_i)]$ nats. That is, entropy was interpreted as a measure of disorder of subsets of microstates residing within a set of macroscopic universe.

Consistent with the details as above, Léon Nicolas Brillouin (in 1889), elucidated the close connection between information theory, entropy, and thermodynamics, that is, by expressing as a negated value of thermodynamic entropy, S,[6–13] the Shannon entropy (denoted by H [= $-S$]) depicts the amount of information observed in messages. And, John von Neumann (the "information philosopher") demonstrated in 1932 the heuristics of *irreversibility of observation* in which, the information about the state of a system could become statistically inaccurate,[13] and this inaccuracy, in a negative sense, would correspond to H, the measure of information (in Shannon sense). In summary, the relatedness between the concepts of S, $\mathfrak{E}$, and H aggressively swept through the heuristics of Neelakanta,[14] and Neelakanta and Baeza,[15] and it was judiciously enunciated by Hartley (in 1928)[16] and pursued by Claude Shannon as the notions of information-theory.

Considering the informatics of biological systems, it is ascertained (*via* Shannon's perspectives) that a mammalian cell could embrace a diverse set of choices of (bioinformatic) microstates yielding about (2×10^{10}) bits of information,[6] and the information associated with a protein molecule, for example, is about 100 to 200 bits (which is typically equivalent to a written paragraph fully packed with messages). Further, information-theoretic entropy (H) of Ralph Hartley[3,4] measured by Claude Shannon in terms of uncertainty is just the complement of entropy[17,18] in the context of statistical thermodynamic entropy (S) envisioned by Ludwig Boltzmann and Willard Gibbs. As stated by Norbert Wiener[19] in 1949, the "amount of information is the negative of the quantity usually defined as entropy in similar situations." Relevantly, the term "*negentropy*" for the negative entropy ("a characteristic of free or available energy, as opposed to heat energy in equilibrium") came into being due to Leon Brillouin.[20,21] Further, Jaynes[22] indicated that the maximum entropy depicts a "least-biased estimate" of the surprise and related theorems of information *versus* entropy could be seen in the works due to Alekseev,[23] and Bennett and Landauer.[24] Further, exclusive to complex system representation of biological systems having inevitable realizations of different random attributes and corresponding information (and entropy) features can be seen in Neelakanta *et al.*[25] The underlying details are built on heuristics of statistics and related axioms of probability.[4,5] Thus, the probability of a discrete or continuous variable X is denoted by, $p(X)$ and its three versions are: *Self* (marginal), *conditional,* and *joint probabilities*[25,26] are adopted in formulating the entropy and/or information measures as regard to random events, ($X \in \mathcal{X}$) ($Y \in \mathcal{Y}$), ($Z \in \mathcal{Z}$), etc. represented in the Venn space.[25–27]

Further, a redundancy feature can also be specified across inevitable realizations of different random attributes (depicting a wasted "information space") that may prevail, for example, in bioinformatic contexts considering long sequences of residues (such as, nucleotides and/or trinucleotide codons in biological sequences.

2.3 Entropy *Versus* Information: A Summary

Additional information/entropy related definitions, hypotheses, and formulations can be summarized as follows: Denoting *unpredictability* (or uncertainty features) associated with random outcomes, by the buried information, I, it is related to entropy, H *via Boltzmann's H theorem* as follows: $H(X) = E[I(X)\} = E[-\log_b\{p(X)\}]$, where $E[\cdot]$ denotes the expectation operator.

Entropy is defined in four perspectives as follows: (i) *Posentropy — a measure of (implied) uncertainty* denoting a *measure of unpredictability* of underlying events encountered in a subspace of stochastic variables. (ii) *Negentropy:* Since the entropy (or the *Shannon entropy* in the information-theoretic plane) denoted by: $H(X) > 0$ pertinent to the discrete random variable (X) having a mass (density) function $p(X)$, it concurrently represents a *measure of certainty.* (iii) *Conditional or cross-entropy*: Given two events, ($X \in \mathcal{X}$) and ($Y \in \mathcal{Y}$), with corresponding *a priori* probabilities $p(X)$ and $p(Y)$, respectively, as well as the associated conditional (*a posteriori*) probabilities are: $p(Y|X)$ and $p(X|Y)$, then one may define the conditional entropy of the two event sets, $\{X_i\}$ and $\{Y_j\}$, as follows: $H(X|Y) = \Sigma_{i,j}[p(X_i,Y_j)] \times \log_b[\{p(Y_j)\}/\{p(X_i \cdot Y_j)\}]$ where $p(X_i, Y_j)$ denotes the joint probability of X_i and Y_j, and the entity $H(X|Y)$ refers to the amount of randomness associated with the random variable X, given the condition that the random event Y prevails. Likewise, $H(Y|X)$ denoting the extent of randomness associated with the random variable Y given the condition that the random event X prevails can be specified as follows: $H(Y|X) = \Sigma_{i,j}[p(X_i,Y_j)] \times \log_b[\{p(X_j)\}/\{p(X_i \cdot Y_j)\}]$. (Here the conditional entropy is also measured in the units of bits, nats, or hartleys.) Whenever a transition probability is encountered (implying an associated mutual entropy), it implies a random process wherein the random variable (RV) undergoes a sequence of transitions from one state to another on a state-space implying a *Markov process and Markov chain* (due to Andrey Markov).[28–31]

2.4 Statistical Divergence

As discussed above, relative entropy refers to a measure of how one probability distribution diverges from a second, expected probability distribution. It characterizes the mutual information (in Shannon's sense) due to randomness seen in stochastic processes. In information-theoretics, it also implies the *statistical divergence* (D) or a *contrast function* that establishes the "distance" of one probability distribution (p) with respect to the other (q) on a statistical manifold of *information geometry.*[32] The statistical divergence establishes a metric to assess the *similarity* or *dissimilarity* details between two probability measures in *Kullback–Leibler* (KL) sense. That is, it provides a value on the *information gain* as regard to comparing statistical models of inference. It is in contrast to the so-called *variation of information* and as such, it does not represent a distribution specific *asymmetric* measure; hence, D is not a statistical *metric of spread*. Thus, $D = 0$ indicates that one can expect similarity as regard to the behavior of two different statistical distributions, and $D = 1$ implies that two given distributions are such that the expectation given in the first distribution approaches zero. Further, D denotes a *measure of surprise* and it is adopted in diverse applications of statics and stochastic processes.

The statistical divergence can be interpreted as the expected additional information required to convey a message when a given (incorrect) distribution q is used instead of the true distribution, p. Classically, the notions of information entropy due to Shannon led to the concept of *maximum entropy*[33] to represent the most unbiased representation of one's ignorance (uncertainty) of the state of the system.

Classes of Divergence Measures

In view of the entwined notions of entropy and information, an entropy-based metric can be stipulated to elucidate the expected value of the *likelihood ratio* of the statistics pertinent to the informatic-data being contrasted. Such expected values of likelihood ratio are general representations of the cross-entropy (or mutual information) between the statistics of the two regions under consideration. Hence, a perceivable statistical divergence that denotes a discriminant or a contrast measure is the cross-entropy that determines the mutual information between the statistical frameworks of information-bearing domains. It is popularly known as *f-divergences* as described below. (As an alternative to such entropy-based *f-divergences* metrics, a set of *statistical distance measures* can also be defined toward contrasting information across two message frames.)

f-Divergence Measures

The *f-divergence* can be defined as a function $D_f(p||q)$ that measures the difference or divergence between two probability distributions, p and q. Intuitively, it portrays the divergence as an average, weighted by the function ϕ of the *odds ratio* given by p and q. (This odds ratio refers to a measure of association between an exposure and an outcome. It depicts the odds that an outcome will occur given a particular exposure, compared to the odds of the outcome occurring in the absence of that exposure.)

The *f-divergences* measures were introduced and studied independently by Csiszár in 1963,[34] Morimoto in 1963,[35] and by Ali and Silvey in 1966.[36] As such, they are known as *Csiszár f-divergence measures, Csiszár–Morimoto divergence measures* or *Ali–Silvey distance measures, respectively.* Mathematically, a *f-divergences* measure can be stated as follows: Suppose p and q are two probability distributions over a space Ω such that p is absolutely continuous with respect to q, then for a *convex function** ϕ such that $\phi(1) = 0$, the *f-divergences* of q from p is defined as follows: $D_f(p||q) \equiv \int_\Omega \phi(dp/dq)dq$. Further, suppose both p and q are absolutely continuous with respect to a reference distribution γ on Ω, then their probability densities p and q satisfy the relation, $dp = (p \times d\gamma)$ and $dq = (q \times d\gamma)$; hence, the corresponding *f-divergences* can be specified as follows: $D_f(p||q) \equiv \int_\Omega \phi[p(X)/q(X)] \times q(X)d\gamma(X)$. In the above expressions, $\phi(\cdot)$ is a convex function defined in the domain of ϕ or $dom(\phi)$ such that $\phi(1) = 0$, and ϕ: $(0, \infty) \subseteq dom(\phi) \rightarrow [0, \infty]$ following Jensen's inequality, also $\phi'(1) = 0$, and the scale of divergence is set by $\phi''(1) = 1$ with the p. The *f-divergences* are pertinent to a corresponding set of *generators*, and the ϕ-functions, for example, can be specified as a generator, $\phi(u) = -\log_b(u)$ that would lead to the so-called Kullback–Leibler (KL) version of *f-divergences* with $f(u = 1) = 0$.[37] For other *f*-divergences measures, the associated ϕ-functions are tabulated later in this chapter. In information theory, the *f-divergences* belong to unique family of convex separable divergences that satisfies the so-called *information monotonicity* property. This simple concept has unusual implications in information theory, pertinent to, for example, information hiding, security, quantum information. (*A *convex function* (Φ) refers to a real-valued function defined on an interval when the line segment between any two points on the graph of the function lies above or on the graph, in a Euclidean (vector) space.[38,39])

To bypass the integral evaluation of $D(p||q)$, which is often mathematically intractable, stochastic certain approximations of *f-divergences* are indicated in Csiszár,[34] Morimoto[35] and Ali and Silvey,[36] so as to carry out a summation instead, as follows:

$$\hat{D}_f(p||q) \approx \frac{1}{2n}\sum_{n=1}^{n}[\phi(q/p) + p/q \times \phi(q/p)] \qquad \text{Eq. (2.1)}$$

with $(s_1, \ldots, s_n)$ and $(t_1, \ldots, t_n)$ denote *independent and identically distributed* (IID) samples of the statistics of p and q, respectively. The *f-divergences* can be expressed using Taylor series and rewritten using a weighted sum of chi-type distances.[40–49]

Kullback–Leibler (KL) Notion of f-Divergence

A conceived distance measure of dissimilarity between the mappings of a pair of *probability density functions* (pdfs) in the domain of information-theoretics elucidated from the interpretations in probabilistic terms refers to the relative information (mutual information) concept of divergence between the pdfs mentioned before. Suppose a basic set of elements or random parameters, $\{\mathbf{y}_k\}$ that deviates from a target set of (random) values $\{y_\ell\}$ has probability measures, $\{p_k\}$, and the probabilities associated with the random parameters $\{y_\ell\}$ are $\{q_\ell\}$ *via* an *affine transformation* of a random parameter space (Ω_P) into a corresponding entropy space (Ω_H).

The extent of randomness associated with a subspace in a stochastic system relative to an (organized) reference (or target) subspace (O_R) depicts the *disorganization,* and the corresponding entropy functional are dictated by the relative probabilistic attributes (or randomness) of the kth realization y_k against the ℓth realization y_ℓ (or *vice versa*) (summed over all values of $k = 1, 2, 3, \ldots, n$ and $\ell = 1, 2, 3, \ldots, m$), $D(k\colon \ell, y) = \Sigma_k[p_k] \times [\log_b(p_k/q_\ell)]$.

Hence, considering the disorganization attributes of the subspaces $\{\mathbf{y}_k\}$ and $\{y_\ell\}$, the corresponding pair of entropy functionals depicting the relative distance measures, KL_1 and KL_2 between the subspaces are as follows: $\mathrm{KL}_1 = D_{\mathrm{KL}_1}(p\colon q) = \Sigma_i[p_i] \times [\log_b(p_i/q_i)]$ and $\mathrm{KL}_2 = D_{\mathrm{KL}_2}(p\colon q) = \Sigma_i[q_i] \times [\log_b(q_i/p_i)]$. (Here, the subscript KL in the conventional designation of D_{KL} refers to as the same Kullback–Leibler measure[33,37] mentioned earlier.) In essence, the KL measure constitutes a metric that measures the distance between the stochastic patterns of $\{\mathbf{y}_k\}$ and $\{\mathbf{y}_\ell\}$ of the subspaces mentioned earlier. Concurrently, it measures the *average information* involved in the discrimination in favor of $\mathcal{H}_{y_k}$ *vis-à-vis* $\mathcal{H}_{y_\ell}$; further, a corresponding *symmetrical measure of divergence* can also be specified as follows:[33,37] $D_J[(k\colon \ell, y_\ell)\colon (\ell\colon k, y_k)] = D(k\colon \ell, y_\ell) + D(\ell\colon k, y)$, where D_J is known as *Jensen measure* or *J-divergence.*[33,37,50] Explicitly, $\mathrm{D}_J = \Sigma_{k,\ell}(p_k - p_\ell) \times [\log_b(p_k/p_\ell)]$ depicting the statistical divergence between the set $\{p_i, q_i\}$.

Further, $D_J(p_i{:}q_i) = \Sigma\, p_i \times [\log_b(p_i/q_i)] + \Sigma\, q_i \times [\log_b(q_i/p_i)\,]$. In addition, corresponding to each of the random realizations pertinent to a stochastic space, a weighted divergence can be stipulated as follows: Suppose π_k and $\pi_\ell(\pi_k, \pi_\ell) \geq 0$ and $(\pi_k + \pi_\ell) = 1$ are the weights of the two probabilities p_k and p_ℓ, respectively. Then, a generalized divergence measure (known as *Jensen–Shannon* (JS) *measure*) can be stipulated in terms of the entropies associated with the random or probabilistic attributes of p_k and p_ℓ as follows:[50–52]

$$D_{\mathrm{JS}}(p_k\colon p_\ell) = H(\pi_k p_\ell + \pi_j p_\ell) - \pi_k H(p_\ell) - \pi_j H(p_\ell). \qquad \text{Eq. (2.2)}$$

Here, the entropy function $H(z)$ refers to Shannon's information measure specified by the (negative) entropy, namely, $-k\sum_i[p(z_i)] \times [\log_b\{p(z_i)\}]$, corresponding to a set of random variables $\{z_i\}$ with probabilities $\{p(z_i)\}$ and k is a constant as decided by the base of the logarithm involved. The measure given by $D_{\mathrm{JS}}(\cdot)$ is non-negative and equal to zero only when $p_k = p_\ell$. It also provides the upper and lower bounds for the *Bayes' probability of estimated divergence.* The JS divergence is ideal to describe the variations between the subspaces of random realizations as in an entropy space, and it measures the distances between the random-graph depictions of such realizations pertinent to the entropy plane $\Omega_{\mathcal{H}}$. Written in terms of p_i and q_i, the divergence measure, D_{JS} is given by $D_{\mathrm{JS}}(p\colon q) = \varepsilon_{\mathrm{JS}} = [\pi_1 \times \sum_i[p_i] \times \log_b(p_i/q_i)] + [\pi_2 \times \sum_i[q_i] \times \log_b(q_i/p_i)]$ with $(\pi_1, \pi_2 < 1)$ and $(\pi_1 + \pi_2) = 1$. A special case of KL measure of directed divergence with a weighting parameter $0 \leq \mu < 1$ can be written in a weighted symmetrized form as follows,[33,53] $D_{\mathrm{WKL}}(\,)$ given by:

$$\begin{aligned} D_{\mathrm{WKL}}(p\colon q) = & \sum[p_i] \times \log_b\{[p_i/[\mu p_i + (1-\mu)q_i]\} \\ & + \sum[q_i] \times \log_b\{[q_i/[\mu q_i + (1-\mu)p_i]\}, \end{aligned} \qquad \text{Eq. (2.3)}$$

where $(0 \leq \mu < 1)$, and this divergence-measure can be designated as *weighted and symmetrized Kullback–Leibler measure* (specified with the subscript, WKL), constituted by weighted fractions of the probabilities. Apart from the symmetrized details on the KL measure indicated above, a framework of summarization pertinent to an exhaustive set of divergence measures and the associated implications are described in Chapter 4 of Neelakanta[6] by the author. Further, using the definition of KL-measure another measure known as the *K-measure* can be defined as follows: $K(p_1, p_2) = \mathrm{KL}[p_1, (p_1 + p_2)/2] = \sum p_1 \log_b[p_1/((p_1 + p_2)/2)$. Also, in terms of the *K*-measure, another *L-measure* can be defined as follows: $L(p_1, p_2) = K(p_1, p_2) + K(p_2, p_1) = 2 \times H[(p_1 + p_2)] - H(p_1) - H(p_2)$, where $H(\cdot)$ denotes Shannon's entropy.

Jeffrey-Divergence (J-Divergence)

The *J*-divergence is the symmetric version of KL-divergence specified with respect to both $p(x)$ and $q(x)$, and it is given by: $D_J(p||q) = \int[p(x) - q(x)] \times \{p(x)\} - \log_b\{q(x)\}]dx$. Inasmuch as the cross-entropy or mutual information functions between the statistics stipulated *via* $p(\cdot)$ and $q(\cdot)$ are given by the KL-measures, namely ($KL_1 = I_1$ and $KL_2 = I_2$), the corresponding Jeffery's *symmetrized KL measure* (*J*-measure or distance) is given by: $J = (I_1 + I_2)$, and the Jensen–Shannon (JS) measure indicated earlier represents a weighted sum of I_1 and I_2; that is, $JS = [n_1/(n_1 + n_2)] \times I_1 + [n_1/(n_1 + n_2)] \times I_2$ where n_1 and n_2 depict the prorated number of random variables of the pdfs, $p(\cdot)$ and $q(\cdot)$, respectively, totaling, $(n_1 + n_2) = N$.

Csiszár's Family of Minimum-Directed Divergence Measures

The concept of minimum-directed divergence measure indicated in terms of Kullback–Leibler measure described above can be further generalized to represent a family of measures[33] $D(p{:}\ q)$ and $D(q{:}\ p)$ given explicitly as: $D(p{:}\ q) = \sum_{i=1}^{n} q_i \times [\Phi(p_i/q_i)]$ and $D(q{:}\ p) = \sum_{i=1}^{n} p_i \times [\Phi(q_i/p_i)]$, where $\Phi(u) = f(u)$ denotes the generator function (mentioned earlier), and it is twice differentiable convex function for which $\Phi(u = 1) = 0$. Again, $D(\cdot)$ satisfies the essential and desirable conditions relevant to the axioms indicated earlier.

In *f-divergences* measures (introduced and characterized by Csiszár[43]), "*f*" represents the function, Φ, and *f*-measures can be regarded as generalized set of information measures, which when applied to a specific class of problems may indicate certain interesting properties akin to Shannon-type information measures. As Aczél[54] points out that specific to those problems, "if the properties are indeed intuitive and significant, then there is a good chance the measures thus obtained may have future application." The *f-divergences* measures are numerical measures of relative "informativity" of two stochastic experiments specified in terms of the associated probability distributions.

The relative information measure of statistical divergence can be specified uniquely by a family of functions representing Φ, which concurrently assures the non-negativity of the probabilities, namely $(p)_i = (1, 2, 3, \ldots, n \geq 0)$. Such specific generator functions are explicitly listed in Kapur and Kesavan[33] across the following set of divergence measures: (a) *Kullback–Leibler (KL) measure;* (b) *Kapur (KP) type-1 measure;*

(c) *Havrda and Charvát (HC) measure*;[55] (d) *Sharma and Mittal (SM) measure*;[56,57] (e) *Rényi (R) measure*;[56] and (f) *Kapur (KP) type-2 measure*.[58]

Apart from the six cases of Csiszár's measures as indicated above, another set of generalized Csiszár's measures are obtained with necessary modifications to represent certain general forms[33] obtained (similar to the Kullback–Leibler format of the function, $\Phi(x) = [(x) \times \log_b(x)]$. Such general forms conform to a set of *generalized Csiszar's measures* (GCZ1 through GCZ5) that are explicitly given by the author in Neelakanta[6] aud indicated in Kapur and Kesavan[33] with relevant descriptions of their properties. The family of *f-divergences* can be generalized through functions $f(u)$, convex on $u > 0$, such that $f(1) = 0$ as follows: $D_f(p\,||\,q) = \int p(x) \times \phi[p(x)/q(x)]dx$. Hence, the members of typical *f*-divergences family can be explicitly identified as listed below in Table 2.1. The ϕ-function or generating function $\phi(u)$ indicated earlier can be explicitly written in formalizing certain *f*-divergences measures as shown in Table 2.2.[59]

Table 2.1 Some salient examples of *f-divergences* family members.

f-Divergence family member	$D(p\|\|q)$
Kullback–Leibler divergence	$\int p(x) \times \log_b[p(x)/q(x)]dx$
Exponential divergence	$\int p(x) \times \{\log_b[p(x)/q(x)]\}^2 dx$
Kagan's divergence	$(1/2) \times \int\{[p(x) - q(x)]^2 / p(x)\}dx$
(α, β)-product divergence	$[2/(1-\alpha) \times (1-\beta)] \times \{\int[1-\{p(x)/q(x)\}^{(1-\alpha)/2}]$ $\times [1-\{q(x)/p(x)\}^{(1-\beta)/2}] \times p(x)dx\}$
Chernoff α-divergence	$[4/(1-\alpha^2)] \times [[1-\int\{p(x)\}^{(1-\alpha)/2} \times \{q(x)\}^{(1+\alpha)/2}dx]$
Squared Hellinger divergence	$[D(p\,\|\|\,q)]^2 = 2 \times \int\left[\sqrt{p(x)} - \sqrt{q(x)}\right]^2 dx$

Table 2.2 Examples of ϕ-functions or generating functions $\phi(u)$ of typical *f*-divergences family members.

f-Divergences family member	Generating ϕ-function
Kullback–Leibler divergence	$u \times log_b(u), -log_b(u)$
Hellinger divergence	$[\sqrt{u} - 1]^2, 2 \times (1 - \sqrt{u})$
χ^2-divergence	$[u - 1]^2, \quad (u^2 - 1)$
Chernoff α-divergence	$\frac{4}{(1-\alpha^2)}\left[1-u^{(1+\alpha)/2}\right]$ if $\alpha \neq \pm 1$ $(u \times \log_b(u)$ if $\alpha = +1$ $-\log_b(u)$ if $\alpha = -1$

Non-Parametric Distance Measures

The non-parametric models used in the assessment of statistical distances[60] are relevant to discrete set of random variables expressed in terms of histograms. That is, a *non-parametric distance* refers to a distance measure deduced in the contexts of so-called *non-parametric statistics*, which are not based on parametrized families of probability distributions and they include both descriptive and inferential statistics. The following are typical of such non-parametric dissimilarity measures indicated to specify the statistical distance *via* histogram features. (a) *Minkowski distance (MkD) measure:* It is a L^{p_0} *distance* of *order* p_0 between two points, $\mathscr{X} = (x_1, x_2, \ldots, x_n) \in \mathscr{R}^n$ and $\mathscr{Y} = (y_1, y_2, \ldots, y_n) \in \mathscr{R}^n$ is defined as:

$$\text{MkD} = \sum_{i}^{n}\left[|x_i - y_i|^{p_0}\right]^{1/p_0} = \left[\sum_{n} | K_1(x_n; \mathbf{T}_1) - K_2(y_n; \mathbf{T}_2)|^{p_0}\right]^{(1/p_0)} \qquad \text{Eq. (2.4)}$$

where $\mathbf{K}_1(x_n; \mathbf{T}_1)$ and $\mathbf{K}_2(y_n; \mathbf{T}_2)$ correspond to two distributions, specified in the bounded convex Euclidean *n*-space, $\mathscr{R}^n$; and $\mathbf{T}_1$ and $\mathbf{T}_2$ denote the subsets with $(x, y) \in \{\mathbf{T}_1, \mathbf{T}_2\}$ for $(p \geq 1)$. (b) *Chebyshev distance* (CD): In the limiting case of p_o (indicated above) reaching infinity, yet another distance can be specified as follows: $\text{CD}|_{\max} = \text{Limit}|_{p_0 \to \infty} \left[\sum_{i=1}^{n} | x_i - y_i |^{p_0}\right]^{(1/p_0)} = \max|_{i=1}^{n} \left(| x_i - y_i |\right)$. Likewise, when $p_o \to (-\infty)$, it follows that $\text{CD}|_{\min} = \text{Limit}|_{p_0 \to \infty} [\sum_{i=1}^{n} | x_i - y_i |^{p_0}]^{(1/p_0)} = \min|_{i=1}^{n} (| x_i - y_i |)$. (c) *Euclidean and Manhattan distances:* Now, considering the length of a vector $\mathbf{x} = \{\mathbf{x}_1, \mathbf{x}_2, \ldots, \mathbf{x}_n\}$ in the *n*-dimensional real vector-space, $\mathbf{R}^n$, the Euclidean norm defined as above refers to: $||\mathbf{x}||_2 = (\mathbf{x}_1^2 + \mathbf{x}_2^2 + \cdots, \mathbf{x}_n^2)^{1/2}$; that is, the Euclidean distance between two points x and y is simply, the length $||x - y||_2$ of the straight-line between the two points. In contrast, the Manhattan distance accounts for the paths either orthogonal or parallel to each other capturing the notion of actual distances in a given space (as decided by Manhattan taxi-cab drivers who duly specify the measure distance, not in terms of just the length of the straight-line to their destination).

In other words, the Euclidean distance function estimates the passage length covered "as-the-crow-flies" between two locales. That is, considering a point in the vector-space **X**: (x_1, x_2, etc.) and another point in the vector-space **Y**: (y_1, y_2, etc.), the Euclidean distance is explicitly equal to $\sqrt{\sum_{i=1}^{n}[x_i - y_i]^2}$. It is a shear manifested version of the famous Pythagoras theorem. Relevantly, furnished in Fig. 2.1 are illustrations to portray the difference between *Euclidean* and *Manhattan distances.*

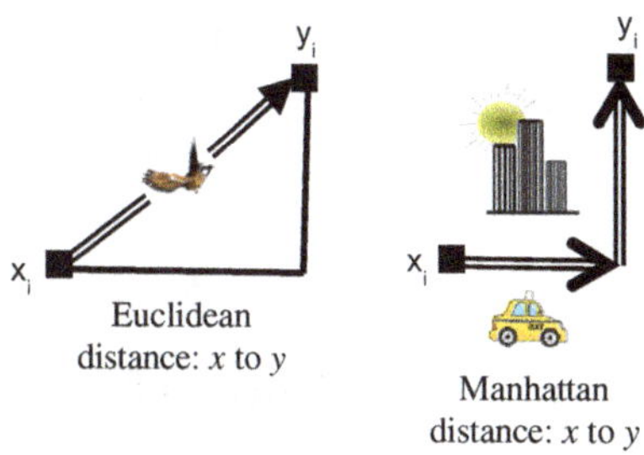

Fig. 2.1 The Euclidean and Manhattan distances.

(d) *Bhattacharyya distance measure:* Originally introduced by Anil Kumar Bhattacharya in 1943,[61] the *Bhattacharyya distance* (B) also known as the *Hellinger distance* (HE),[62–64] denotes another non-parametric distance measure. It quantifies the similarity between two probability distributions. It is a type of *f*-divergences and defined in terms of the *Hellinger integral* as follows: $\mathrm{HE}^2(p,q) = (1/2)\times\int\left[\sqrt{p(x)} - \sqrt{q(x)}\right]^2 dx$. The squared Hellinger distance indicated above can also be expressed as $\mathrm{HE}^2(p,q) = 1 - \int\left[\sqrt{p(x)\times q(x)}\right]dx$, and the Hellinger distance $H(p, q)$ satisfies the property, $(0 \le H(p, q) \le 1)$, derivable from the Cauchy–Schwarz inequality concept. Corresponding to a pair of discrete sets of random variables with distributions, $\{p_1, p_2, \ldots, p_k\}$ and $\{q_1, q_2, \ldots, q_k\}$, the Bhattacharyya distance measure (B) can be derived as follows: $B(p,q) = (1/\sqrt{2})\times\sqrt{\Sigma_{i=1}^{k}\left[\sqrt{p_i} - \sqrt{q_i}\right]^2}$.

Alternatively, in terms of $\mathbf{K}_1(n; \mathbf{T}_1)$ and $\mathbf{K}_2(n; \mathbf{T}_2)$ defined earlier, the corresponding, geometrically implied Bhattacharyya distance measure between two discrete set of points can be specified as follows: $B = D(\mathbf{K}_1,\mathbf{K}_2) = \left[1 - \Sigma_n |\mathbf{K}_1(n;\mathbf{T}_1)\times\mathbf{K}_2(n;\mathbf{T}_2)|^{1/2}\right]^{(1/2)}$. Further, with reference to distributions (p_1 and q_2), B is given by $B_{i=1,2} = -\log_b(\rho_i)$, $0 < B < \infty$, where ρ_i is known as the *Bhattacharyya coefficient* given by $\rho_{i=1,2} = \Sigma[\log_b(p_1\times q_2)]^{1/2}$, $(0 < \rho_{i=1,2} < 1)$.

(e) *Kolmogorov variational distance:* Another measure closely related to ρ_i is the so-called *Kolmogorov variational distance* (KV) defined as follows: $\mathrm{KV}_{i=1,2} = 1/2\Sigma|\pi_1 p_1(x) - \pi_2 q_2(x)|$, where π_1 and π_2 denote the *a priori* probabilities of the hypotheses, $\mathcal{H}_1$ and $\mathcal{H}_2$ pertinent to the statistics of p_1 and q_2. Further, a total *variation distance* (TD) for probability distributions p and q specified on a sigma-algebra $\mathcal{F}$ of subsets* of the sample space Ω can be defined as follows:[64] $\mathrm{TD}(p,q) = \sup_{x\in\mathbf{F}}|p(x) - q(x)|$. The *Kolmogorov(–Smirnov) statistics* specifies the maximum vertical distance between the cumulative distribution functions of the two datasets. (*The *sigma-algebra* [also known as *σ-algebra, σ-field* or *sigma-field*] on a set **X** is a collection Σ of subsets of **X** (including the null subset.)

Distance Measures on Statistical Similarity

There exists yet a plethora of other statistical distance measures evolved to determine the "similarities" between two sets of stochastic entities identified with subscripts *i* and *j*. Such estimates of similarity features are also regarded as "distances." Described below are some well-known similarity-based distance measures.

(a) *Mahalanobis measure* (*M*)
A commonly used measure of such distances or similarities is known as *Mahalanobis measure* (*M*).[65] It evaluates the distance of a point *P* with respect to a statistical distribution, *D*. In essence, it is a multi-dimensional generalization of a concept outlaid to measure the extent of standard deviations away that a point *P* is offset from the mean of the statistics specified by *D*. Basically, it is based on the concept of Euclidean distance. As indicated earlier, the Euclidean distance between a pair of two vectors $\mathbf{x}_i$ and $\mathbf{x}_j$ is given by $\mathrm{d}(\mathbf{x}_i, \mathbf{x}_j) = \|\mathbf{x}_i - \mathbf{x}_j\|$, as such, it implies greater similarities between the vectors with smaller distance assessed.

When the vectors $\mathbf{x}_i$ and $\mathbf{x}_j$ are from two different populations (pools) of data, the Euclidean distance measure can be specified in terms of $\mu_i = E[\mathbf{x}_i]$ and $\mu_j = E[\mathbf{x}_j]$ depicting the mean values of the vectors $\mathbf{x}_i$ and $\mathbf{x}_j$, respectively, with $E[.]$ representing the expectation operator. Now considering a set of observations specified by $\{\mathbf{x}_i\}_{i=1,2,\ldots,N}$ from a set of observations with mean $\{\mu_j\}_{j=1,2,\ldots,N}$, the squared value of the distance from $\mathbf{x}_i$ and $\mathbf{x}_j$ is given by $d_{ij}^{\ 2} = (x_i - \mu_j)^T \Sigma^{-1}(x_i - \mu_j)$, where *T* is the transpose operation and Σ^{-1} is the inverse of the covariance matrix Σ given by $\Sigma = E[(x_i - \mu_i)(x_i - \mu_i)^T] = E[(x_j - \mu_j)(x_j - \mu_j)^T]$. Hence, the Mahalanobis measure (*M*) or *Mahalanobis distance* (MD) is given by $M(x_i, x_j) = \sqrt{(x_i - \mu_j)^T \Sigma^{-1}(x_i - \mu_j)}$.

The Mahalanobis distance is preserved under full-rank linear transformations of the space spanned by the data. As such, if the data has a nontrivial null-space then $M(\cdot)$ can be computed after projecting the data (non-degenerately) down onto any space of the appropriate dimension for the data. In summary, the Mahalanobis distance (or "*generalized squared inter-point distance*" for its squared value) is a dissimilarity measure between two sets of random vectors, $\mathbf{x}_i$ and $\mathbf{x}_j$ having the same distribution with the covariance matrix, Σ. If the covariance matrix is the identity matrix, then $M(\cdot)$ reduces to the Euclidean distance.

(b) *Normalized Euclidean distance*
If the covariance matrix is diagonal, then the resulting distance measure between $\mathbf{x}_i$ and $\mathbf{y}_i$ is known as the *normalized Euclidean distance* given by: $d(x_i, y_i)\,|^{\text{Euclidean}}_{\text{Normalized}} = [\sum_{i=1}^{N}\{(x_i - y_i)^2 / \sigma_i^2\}]^{1/2}$, where σ_i is the standard deviation of the x_i and y_i over the sample set. The Mahalanobis distance thus denotes the distance of a point of observation from the center-of-mass divided by the width of the ellipsoid in the direction of the observation point. In the case of normal distribution (specified in any number of dimensions), the probability of an observation is uniquely determined by the Mahalanobis distance, and M^2 is *chi-squared* (χ^2-squared) distributed.

(c) *Dot or inner product distance measure*
Another Euclidean measure of distance is based on the *concept of the dot or inner product.* That is, given a pair of vectors $\mathbf{x}_i$ and $\mathbf{x}_j$ of the same dimension, their inner product is $\mathbf{x}_i^T\mathbf{x}_j$, which can be written in an expanded form as follows: $(\mathbf{x}_i, \mathbf{x}_j) = \mathbf{x}_i^T\mathbf{x}_j = \sum_{k=1}^{m} x_{ik}x_{jk}$, where k refers to the number of elements of the vector space, $\mathcal{R}$. From the definition of dot product $[\mathbf{x}_i, \mathbf{x}_j]/[||\mathbf{x}_i||\ ||\mathbf{x}_j||]$ is the cosine of the angle subtended between the vectors $\mathbf{x}_i$ and $\mathbf{x}_j$. In short, the Euclidean distance and inner product measures of distance specify the similarities between vectors. More similar are the vectors $\mathbf{x}_i$ and $\mathbf{x}_j$, larger will be the inner product $\mathbf{x}_i^T\mathbf{x}_j$.

Measures of "Distance" in a Random Sampled-Data Space

These denote histogram-based distances relevant to discrete sampled-data random entities as described below.

Euclidean distance of sampled-data of standardized values: As indicated earlier, the Euclidean distance function estimates the passage length between two locales, and as such, it denotes the Pythagorean distance. It can be extended in the multivariate contexts of sampled-data as follows: Typically, the random sample-data entree depicts "locally" fluctuating entities due to the ambient of exogenous and endogenous influences of the statistics involved. So a standardization is advocated to balance out such contributions by resorting to a normalized random variable defined as follows: Considering the random n entities in each dataset of a vector-space $\mathbf{X}$: $(x_1, x_2, \ldots, x_n)$ and in another vector-space $\mathbf{Y}$: $(y_1, y_2, \ldots, y_n)$, the corresponding normalized sets can be defined as follows: $\tilde{\mathbf{X}} = \{(\mathbf{x}_i - \mu_i)/\sigma_i\}_{i=1,2,\ldots,n}$ and $\tilde{\mathbf{Y}} = \{(\mathbf{y}_j - \mu_j/\sigma_j\}_{j=1,2,\ldots,n}$, where (μ_i, σ_i) and (μ_j, σ_j) denote the expected mean and standard deviation pairs of $(x_1, x_2, \ldots, x_n)$ and $(y_1, y_2, \ldots, y_n)$, respectively. Hence, the standardized Euclidean distance is computed *via* the relation, $\sqrt{\Sigma_{i,j=1}^{n}[\tilde{\mathbf{x}}_i - \tilde{\mathbf{y}}_j]^2}$ using the normalized (standardized) data value. Further, in terms of weighting coefficients, $w_{i,j} = 1/(\sigma_{ij})^2$, a weighted form of standardization can be conceived by defining new sets as: $\tilde{\mathbf{X}}_{\mathbf{w}} = \{\mathbf{x}_i/\sigma_i\}_{i=1,2,\ldots,n}$ and $\tilde{\mathbf{Y}}_{\mathbf{w}} = \{\mathbf{y}_j/\sigma_j\}_{j=1,2,\ldots,n}$, and corresponding weighted (standardized) Euclidean distance is computed *via* the relation, $\sqrt{\Sigma_{i,j=1}^{n}[(\tilde{\mathbf{x}}_w)_i - (\tilde{\mathbf{y}}_w)_j]^2}$ using the weighted (standardized) data values deduced.

χ^2*-distance of sampled-data:* This refers to a specific class of metrics envisaged in assessing the distance measures of random entities of χ^2-statistics of multivariate analysis. That is, it measures the likeliness of one distribution being drawn from another under χ^2-statistics specified *via* histograms.

The chi-squared distance can be specified as follows: By denoting this measure as $D(h_1, h_2)|_{\chi^2}$ corresponding to the vector sets being compared, it is given by: $D(h_1, h_2)|_{\chi^2} = [\Sigma_n[h_1(n) - h_2(n)]^2/h_n]^{1/2}$, where $h_n = [h_1(n) + h_2(n)]/2$ depicts the mean histogram.

Histogram-specific KL divergence measures

Relevant to two KL measures as indicated earlier, namely $\text{KL}_1(p_i, q_i) = \Sigma_i(p_i) \times \log_b(p_i/q_i)$ and $\text{KL}_2(p_i, q_i) = \Sigma_i(q_i) \times \log_b(q_i/p_i)$, the histogram-based quadratic distance can be defined to measure the weighted similarity between histograms. It provides better results than the "like-bin"-only-based similarity comparisons. However, such quadratic measures would impose computational burdens. The histogram quadratic distance metric essentially measures the weighted similarity between histograms that provides more desirable results than "like-bin"-only comparisons. Typically, the quadratic distance between histograms specified as $D_{\text{QD}}(h_1, h_2) = [h_1(i) - h_2(j)]^T \times \mathbf{A}[h_1(\text{i}) - h_2(\text{j})]$, where $\mathbf{A} = [\mathbf{a}_{i,j}]$ is a $[N \times N]$ matrix and $\mathbf{a}_{i,j}$ denotes the similarity coefficient between the dataset with indices i and j being compared. It is equal to $1 - [d_{i,j}/d_{\text{max}}]$, where $d_{i,j} = |h_1(i) - h_2(j)|$ depicts the similarity between the bins i and j.

Other Statistical Divergence and/or Distance Measures

There are identifiable statistical distances that are different from those described above. Relevant set includes the following.

(a) *Total variation distance*: This is an example of a statistical distance metric, and hence called "the" statistical distance. It is denoted as $V(p,q)$ and defined as follows: It corresponds to the largest possible difference between the probabilities that two probability distributions can assign to the same event. This *V-distance measure* satisfies relevant metric properties. In the finite sample-space, the *V*-distance measure is the 1-norm of the difference of the two probability distributions namely: $D_V(p,q) = (1/2)||p - q||_1 = (1/2)\Sigma_x|p(x) - q(x)|$.

(b) *Earth mover's distance*: The *earth mover's distance* (EMD) (also known as *Wasserstein* or *Kantorovich metric*) refers to a distance function defined between probability distributions on a given metric space **M** in the following intuitive context.[66] Suppose "each distribution is viewed as a unit amount of 'dirt' piled on **M**, the metric is the minimum 'cost' of turning one pile into the other, which is assumed to be the amount of dirt that needs to be moved times the distance it has to be moved."

(c) *Lévy and Lévy–Prokhorov metrics*: *Lévy metric* is a special case of the so-called *Lévy–Prokhorov metric* specified on the collection of probability measures on a given metric space.[67] Considering two cumulative distribution functions F,G: $\mathbf{R} \rightarrow [0,1]$ of 1D random variables, the *Lévy distance* between them is defined as follows: $D_L(F,G) = \inf\{\varepsilon > 0 \,|\, [F(x - \varepsilon) - \varepsilon) - \varepsilon] \leq G(x) \leq [F(x + \varepsilon) + \varepsilon]\}$ for all $x \in \mathbf{R}$.

This $D_L(\cdot)$ measure corresponds to the following: Suppose between the graphs of *F* and *G*, one inscribes squares with sides parallel to the coordinate axes (and at points of discontinuity of a graph vertical segments are added), then the side-length of the largest such square is equal to $D_L(F, G)$.

(d) *Energy distance*: The *energy distance* corresponds to a statistical measure of distance between probability distributions of two independent random vectors **X** and **Y** specified in the space **R** with cumulative distribution functions *F* and *G*, respectively, and the energy distance D_{EN} between the distributions *F* and *G* is given by the following expression: $D_{EN}(F, G) = [2 \times E||X - Y|| - E||X - X'|| - E||Y - Y'||]^{1/2} \geq 0$, where

$\{X, X'\}$ and $\{Y, Y'\}$ are independent and identically distributed (iid) sets of random variables, E is expected value, and $(||.||)$ denotes the length of a vector. The energy distance satisfies all axioms of a distance metric, and as such, characterizes the equality of distributions: $D_{EN}(F, G) = 0$ *iff*, $F = G$. This distance measure is useful in relevant statistical applications. It was introduced in 1985 by Székely.[68,69] It has been proved that for real-valued random variables, energy distance is exactly equal to twice the so-called *Cramér's distance*, (D_{CR}).[70] It is given by $D_{CR}(F,G) = \int_{-\infty}^{+\infty}[F(X) - G(X)]^2 dX$. In higher dimensions, the two distances D_{EN} and D_{CR} will be different inasmuch as the energy distance is rotation invariant, whereas the Cramér distance is not.

(e) *Kolmogorov–Smirnov distance*: Among many ways of measuring the similarity of two probability distributions, the *Kolmogorov distance* depicts the sub-distance between the distribution functions, and the Kantorovich–Rubinstein distance refers to the maximum difference between the expectations values of two distributions of functions with Lipschitz constant 1, that is, it denotes *bounded-Lipschitz distance* of absolute value at most equal to 1.

The *Kolmogorov–Smirnov statistics* quantifies a distance between the empirical distribution function of the sample and the cumulative distribution function of the reference distribution or between the empirical distribution functions of two samples. That is, it is the statistics that represents a distance between two probability distributions defined on a single real variable. Corresponding maximum mean discrepancy can be defined in terms of the kernel embedding of distributions.

(f) *Socratic ratio measure (S)*: This is an *a priori* measure of the expected benefits (of admitting the uncertainty) relative to the expected benefits of perfect knowledge (a wisdom proposed by Socrates). The measure can be used in the context of specifying the distance with reference to two statistics with probability distributions p_1 and p_2 as follows $S = J/J_P$, with $J_p = (I_{1,p} + I_{2,p})$, where $I_{1,p} = \sum p_1 \times \log_b(p_1/u_1)$ and $I_{2,p} = \sum p_2 \times \log_b(p_2/u_2)$. Here, $u_1 = u_2 = (1/n)$ depict equally-likely occurrence probabilities of each element in the set of n entities of two random sets. The term J_p implicitly measures the perfect information (of certainty) posed by equally-likely occurrence of the n elements.

(g) *Log-quadratic distance*: With reference to the sets of random variables, $\{x_i\}$ and $\{x_j\}$, a *log-quadric (or mean-square) measure* can be defined as follows: $\mathrm{LQ}(x_i, x_j) = -\sum_x \log_b[1 - (x_i - x_j)^2]$.

Simple-ratio measure: A simple way of depicting the similar or dissimilar aspects of two sets random entities, in terms a fraction of such similarity or dissimilarity of total vector set subjected to contrasting. In terms of $\{x_i\}$ and $\{x_j\}$, the following ratio measures can be specified to indicate such fractions $r = [\sum_x |x_j - x_i| / \sum_x x_{j \text{ or } i}], \overline{r} = (1 - r)$. Such ratios can also be expressed as percentages.

(h) *Correlation measure*: This is a standard measure popularly used in statistics. It is normally expressed as a correlation coefficient (such as *Pearson correlation coefficient*) defined as follows: $C_p(x_i, x_j) = \sum_x (x_i - \overline{x}_i)(x_j - \overline{x}_j) / \sigma_{xi}\sigma_{xj}$, where $\overline{x}_i$ and $\overline{x}_j$ are averages and σ_{xi} and σ_{xj} are the corresponding standard deviations pertinent to the sets $\{x_i\}$ and $\{x_j\}$.

There is also another correlation measure defined by the so-called *Malthen correlation coefficient* defined for numbering values over the range -1 to $+1$. It is a measure depicting how the normalized variables $(x_i - \overline{x}_i) / \sigma_{xi}$ and $(x_j - \overline{x}_j)/\sigma_{xj}$ tend to have the same sign and magnitude. A value of -1 indicates total disagreement and $+1$ depicts total agreement, and the value 0 denotes completely random predictions. Thus, the Malthen correlation coefficient measures the similar or dissimilar features between the sets contrasted with respect to a random base line. The *Malthen coefficient* can be explicitly written as follows: $C_M(x_i, x_j) = [\sum x_i \sum x_j] - N \sum \overline{x}_i \sum \overline{x}_j / \{[(\sum x_i)^2 - N(\overline{x}_i)^2][(\sum x_j)^2 - N(\sum x_j)^2]\}^{1/2}$, where N is the vector dimension involved. This correlation coefficient is akin to L^2-distance except that it is allowed to take negative values (≤ -1).

(i) *Hamming distance*: Suppose the vector sets $\{x_i\}$ and $\{x_j\}$ correspond to (or are represented by) binary sets of blocks $\{\xi_i\}_z$ and $\{\xi_j\}_z$, respectively, where z is the binary number 0 or 1. The statistical distance between $\{\xi_i\}_z$ and $\{\xi_j\}_z$ can be specified by the Hamming distance. It represents the number of bit positions at which the two vectors have differing values. The *Hamming distance* (HD) is obtained by performing modulo-2 (XOR) operation between the sets being compared. In non-binary form, this Hamming distance corresponds to a simple Euclidean distance (difference) between x_i and x_j, that is, $\text{HD} = \sum_x |x_i - x_j|$.

An example of estimating HD in bioinformatic contexts is as follows: A DNA sequence can be binary formatted by representing each element of the nucleotide set {A, C, T, G} say, by assigning decimal 1 or binary 001 to A; decimal 2 or binary 010 for C; decimal 3 or binary 011 for T and decimal 4 or binary 100 for G. That is, a tri-bit representation of {A, C, T, G} as follows: A → 001; C → 010; T → 011 and G → 100.

Alternatively, a DNA sequence can also be represented in terms of its base constituents *via* dibits, for example, as follows: {A → 00, C → 10, T → 11, G → 01}. That is, in order to analyze the symbolic DNA sequences, assignment of binary digits to nucleotide set {A, C, G, T} is necessary. Further, inasmuch as the bases can be divided in classes of two elements, namely purine (R = A, G) pyrimidine (Y = C, T), with strong H-bonds (S = G, C) and weak H-bonds (W = A, T), appropriate scheme toward such binary representation can be devised. Relevant methods of depicting a symbolic DNA sequence, by assigning binary digits to nucleotide set {A, C, G, T} is indicated by the author (and others) in Neelakanta *et al.*,[70,71] also Pandya has elaborated relevant scheme.[72] Another example of such binary representation in the context of protein sequencing is suggested in Nemzer,[73] where the author introduces a binary representation of the canonical genetic code based on both the structural similarities of the nucleotides, as well as in terms of the physico-chemical properties of the encoded amino acids. Each of the four mRNA bases is assigned a unique 2-bit identifier, so that the 64 triplet codons are each indexed by a 6-bit label. The ordering of the bits reflects the hierarchical organization manifested by the DNA replication/repair and tRNA translation systems.

(j) *Fisher linear discriminant based distance measure*: The *Fisher linear discriminant* is another technique used in defining the statistical distance between two distributions.[74–76] It involves finding a linear function X_F of a set of measurements $\{x_i\}$ on the statistical attributes of a test population that will maximize the ratio of the difference between the specific means (μ_1 and μ_2) to the standard deviations (σ_1 and σ_2) of the associated statistics. This is called the *Fisher criterion*. In essence, this criterion requires finding a set of coefficients λ_1 through λ_n pertinent to the function $X_F = \sum_i^n \lambda_i x_i$ that would maximize the distinguishability between the two statistics of the populations domains subjected to n measurements.

Stated explicitly, the Fisher criterion, namely $F_C = |\mu_1 - \mu_2|^2/(\sigma_1^2 + \sigma_2^2)$, refers to a relevant measure that indicates how far apart are the two distributions being contrasted, and at the same time, it is used to assess how tightly the two statistics are distributed. A greater separation between the two populations and a tighter distribution for each will produce a higher value for the Fisher criterion. Thus, the *Fisher discriminant measure* (*F*-measure)[75,76] denotes the statistical separation between two random sets, as well as it can sieve out details pertinent to subsets of the shared similarity in a given domain of a statistical framework consisting of large-sized constituents having stochastic attributes.

The concept of Fisher discriminant was originally developed by R. A. Fisher in 1936[75] with reference to taxonomic studies toward finding out the extent to which given two sets of data are statistically similar or dissimilar. In other words, the Fisher linear discriminant with optimized coefficients applied to a set of measurements would enable an algorithm by which the subsets of the population "are best discriminated".[75,76]

The Fisher measure[75] can also be cast in the information-theoretic plane[76] so as to assess it in terms of negentropy considerations (or information content in Shannon sense) consistent with the stochasticity of a given test population. In classical sense as conceived by Fisher, the Fisher measure is specific to parametric statistical space depicting the extent of random details in a given (statistical) sample space, and such details are specified in terms of appropriate set of parameters $\{\theta\}$ in question. These entities parametrize the underlying statistics *via* a probability density function (pdf).

In fact, the term "information" pertinent to such parameters in a technically defined sense was first introduced in the realm of statistics by Fisher. Subsequently, however, Shannon and Weaver developed in their published works of 1948,[19] the logarithmic measures of information pertinent to communication theory as detailed earlier. The relation between Fisher's and Shannon's entropy[51] is though "intimate," yet both concepts are completely autonomous as explained in Cedilnik and Košmelj.[77]

Cedilnik and Košmelj[77] have suggested the following with reference to Fisher measure: For each set of measurements (pertinent to an "inherent" variable/parameter), the associated *a posteriori* distribution (depicting actual statistics of the parameter [variable] in question) can be compared against a set of *a priori*, uniformly distributed variable dataset. The underlying heuristics of such comparison leads to rewriting the Fisher information appropriately for a multivariate case. The heuristic details derived in Cedilnik and Košmelj[77] specify the following relation on Fisher information content *versus* KL distance for a univariate case: $?^{0.26} = [(B_F - A_F) \times \sigma_{KL}/(B_{KL} - A_{KL}) \times \sigma_F]$, where $\{B_F, A_F, \sigma_F\}$ and $\{B_{KL}, A_{KL}, \sigma_{KL}\}$ denote, respectively, the set {*supremum*, *infimum*; standard deviation} of the univariate in question observed, respectively, through Fisher and Kullback–Leibler (KL) statistics in information-theoretic perspective.

2.5 Information-Theoretics of Biomedia

In the biological complex, the triumvirate structure of bioentropy, thermobiodynamics, and bioinformatics remains always sustained, and it is poised robustly with bioenergetics as illustrated in Fig. 2.2.

For example, even at the microscopic level of cellular structures, the other facet of bioentropy, namely informatic biology, embarks at the most primitive level of nucleotide sequence — breaking, transcribing, and translating of the associated genetic codes; at this level, in spite of any contra or mutated information (or posentropy) that may prevail, a successful endeavor is mostly sustained *via* the streamlined protocols of central dogma in search of, and deliberating, information-content pertinent to framing the giant molecules, namely, the proteins that perform a diverse array of functions within living organisms, such as catalyzing metabolic reactions, DNA replication, responding to stimuli, and transporting molecules from one location to another as indicated in Chapter 1.

Information-theoretics at the cellular level of cytogenetic details and across phylogenetics of the whole species depict a multi-dimensional perception of bioinformation. It can be conceived as the negentropy or information ($H = -S$) viewed in the sense of improbability relevant to a set of specified ordered states in the complex biological system with co-featured details of bioenergetics — either at the nucleotide level or at the macroscopic, whole-life level. Correspondingly, as illustrated in Fig. 2.2, a hypersphere of universe (Ω) of informational biology can be framed

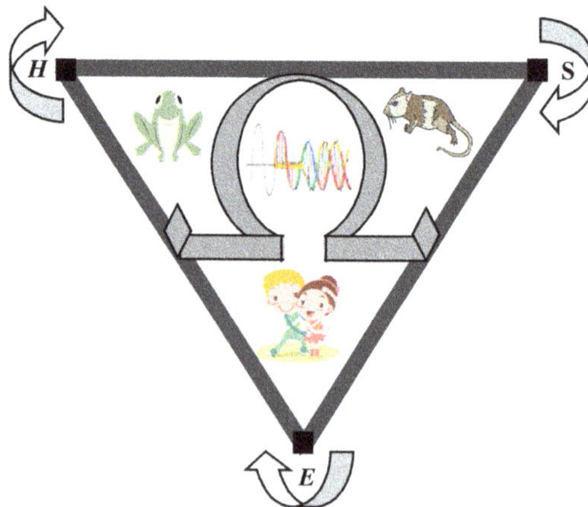

Fig. 2.2 The hyperspherical universe (Ω) of informational biology — a complex domain framed by three equilateral frames of information (*H*), entropy (*S*), and energy (*E*).

with crucial vertices of information (H), entropy (S), and energy (E) depicting the complexity of biosystems. In the perspectives of thermobiodynamics, the first law of thermodynamics can be reiterated to confirm the cent-percent transformable aspect of (mass ↔ energy) while retaining their conservative feature intact.

Relevant perspectives of entropy (S), the cybernetic attributes of bioenergetics, and the notions of information theory (due to Shannon[17,19] and Khinchin)[78,79] are the three equilateral planes (Fig. 2.2) that enclave the hypersphere of informational biology.[6]

In physical systems, an "arrow of time describes the irreversibility of physical events as entropy gives time a direction — toward more products and arrangements." Organisms "want" to organize, grow, and live or make negative entropy. If disorder increases (implying positive entropy), it would end the negative entropy trend. Possibly, this amounts to a thermobiodynamic description of death.[6,23] Any divergent trend in negentropy constrained by the constancy of energy in the universe amounts to the available energy shrinking to zero, suggesting a *catastrophe*. The internal differential entropy drops monotonically after a certain stage of adult-life indicating a progress toward the death process in conformance to a devolutionary action illustrated in Fig. 2.3. It is a universal law of biology, according to Bauer (as quoted in Alekseev[23]), that "the balanced state of the inanimate systems is stable"; (surprisingly!), the unbalanced state of animate systems is also stable; Could be

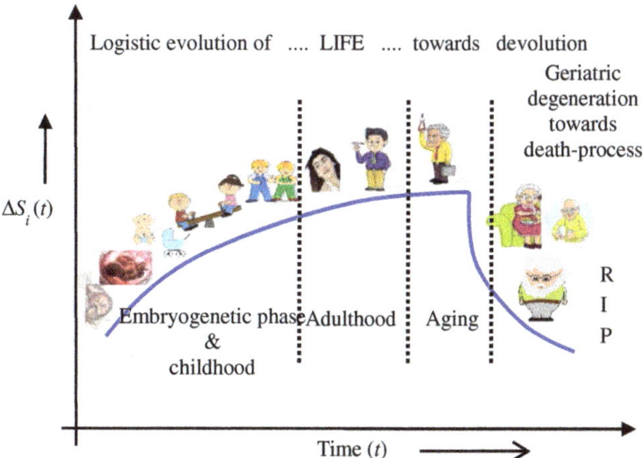

Fig. 2.3 Logistic evolution and devolution of differential entropy $\Delta S_i(t)$ in differentiating organisms.

that the animate systems carry *free-energy* which may be liberated under certain conditions so as to support the unbalanced state. "Turning back the hands of time" appears as "what is life from the point of view of physics." The second law of thermodynamics seems to pass a "death sentence" and enforces it mercilessly on the "inanimate world which is dead in advance." Life, as Alekseev[23] points out, "suspends this sentence and takes advantage of the fact that the verdict is passed without any fixed term of execution."

The fundamental concepts of thermodynamic states specified in biological systems (as per the equations of state and the extents of energy ultimately convertible into mass) constitute the entropic explanations in alternative ways that keep the bioinformatic "bits together with their energetics", and relevant processes depend on "informing" and energizing the actions perceived, and such processes are both *semantic* and *pragmatic*. They refer to meaningful, message-bearing actions executed at the molecular level. The associated information governs the matrix of molecular receptors and excitation energy in reducing the uncertainty among the degrees of freedom relevant to the expected modes of iterations of a process (e.g., photosynthesis) cycle.

Specifically, the biosystem is an ensemble of information-processing units wherein the core biological entities, namely *genes*, function as per a set of definitive, algorithmic trends with an associated informatic profile, and the cellular units which hold these genes could be presumed to follow similar informational trends in their self-organizing endeavors as well. The brain represents the information center of biological systems with the capability of information storage and management, databanking, and the related processing functions. The gross features and functions of biological entities denote an *extensive* representation and the corresponding *localized* characteristics (at least intuitively and even at the microscopic DNA molecular stage) can be analyzed by the considerations of physical sciences as governed by mass, momentum, energy, entropy, and action.

Regardless of biomolecular, cellular, or whole-life domains of living systems, *order* is a feature that presents a negentropy as it is generated, stored, transcribed, and copied (retrieved). It "informs" the system to organize toward an objective function. Hence, it constitutes an implicit entity of message-bearing information. In contrast, possibly coexists is the *disorder* feature that would try to off-set the system's objective and therefore constitutes a posentropy. The order–disorder conflict is a thermodynamic *tug-of-war* as explained by the author in Chapter 8 of Neelakanta.[6]

Bioinformation-theoretics also frames the ladder of evolution with the associated natural selection perpetually shaping the genetic information prevailing at the biomolecular state of the DNA. In such contexts of information-theoretics, pertinent heuristics of evolution are specified by the so-called *Baldwin effect*,[80] which describes the effect of "learned behavior" in the thematic of modern evolutionary synthesis, that is, it explains organism's ability to learn new behaviors (e.g., to acclimatize to a new stressor) influencing its reproductive success and consequently decide the genetic makeup of its species through natural selection. Unlike Lamarckian evolution, living things *inherited* their ancestral acquired characteristics. The Baldwin effect thus stipulates that individual biospecies explore a set of phenotypes so as to find a good match with their environment rather than encode all options genetically. That is, the "phenotypes result from learning rather than from phenotypic plasticity." In acquiring more information about their environment, the phenotypic plasticity warrants just a good cue facilitating learning by acquiring more information, and this cue-specific learning allows potentially an appropriate fitness associated with phenotype variates.

2.6 Semantic, Semiotic, and Pragmatic Aspects of Bioinformation

In studying information-theoretics, defining three essential terms, namely *semantics*, *semiotics*, and *pragmatics*, is appropriate.[6,8] Semantics refers to the study of meaning conveyed by the relationship between *signifiers* of information-bearing messages. For example, in a language, the words, phrases, signs, and symbols stand for their denotation of understanding the underlying expression of linguistic communication. In essence, semantics examines the meaning that is conventionally residing or "encoded" in a given message-bearing communication.

Semiotics is the study of meaning-making with sign processes in message-bearing, meaningfully informatic communications. Typically, the signs and related processes (*semiosis*) of indications, designations, and symbolist format the underlying signification in the contexts of communications. Semiotics includes the semantics relation between the signs and the entities that they refer *vis-à-vis* the underlying signified *denotata*. Further, the relational attributes of signs or symbols in the formal structures of communication structures are specified as syntactics.

Pragmatics is a subset of semiotics that specifies the ways in which context contributes to meaning. For example, pragmatics encompasses the implicative aspects of conversational communication. Pragmatics refers to the transmission of meaning and it depends not only on the structural aspects of the knowledge (such as grammar in a language) but it also implies the context of delivery (e.g., the *verbatim* utterance in a linguistic communication) and any *a priori* knowledge as regard to such delivery viewed as *a posterior* effort. As such, pragmatics overrides any fuzzy or ambiguous implications of the delivered information, and the pragmatics denotes the *via media* between semiotic signs, syntactic usage of signs, and the semantics conveyed by the agents or interpreters involved. The role of entire semantic, semiotic, and pragmatic aspects of the organization of life is abridged into a nutshell by the science of so-called *biosemiotics* that integrates the knowledge areas of the biocomplex into "a coherent theoretical structure studying semiosis as life and life as semiosis."[80]

Further, with reference to the semiosis depicting the performance aspects of the symbols of life, an interesting query is posed on the nature of "biological information" at its most basic genetic level. When viewed *via* relevant biosemiotics aspects of the central dogma of molecular biology, the genetic information presumably flows only in one direction, from the genome to the protein complex involving biochemical activities in the cell. This consideration presupposes "a specific referential concept of information of some kind (or perhaps only a primitive one, specifying sequences of chemical monomers)." This concept, however, is not precisely defined in the context of molecular biology, as such, the so-called "*Sarkar challenge*" has been posed to interject the heuristics of semantic information at the molecular level of biology (in central dogma sense) or *per se*, any other possible paradigm of information that exists thereof.[80]

Biological systems are intrinsically information-rich.[81,82] This can be inferred from the expression on the information-content, I ($\equiv H$), which specifies whether a particular biological entity (say, a cell) is alive or not under the isothermal condition of Boltzmann ambient of ($k_B \times T$). The richness of ordered state of a biological system would, however, decay "to the most disordered possible state unless work is performed constantly to restore order in the system." This work is done in accordance with the second law of thermodynamics following the principle of the heat pump. The source and sink of heat is provided by exterior ambiance of the biosphere, namely, earth, sun, etc. In all, biological systems represent a complex information system. It is a blend of complex domains comprised of ontogeny aspects of embryonic developments, phylogeny details of evolutions, and genetic heuristics of heredity. As Baldwin[83] rightly points out "...the problems involved in a theory of organic development may be gathered up under three great heads: Ontogeny, Phylogeny, Heredity."

2.7 Informatics of System Biology

The state-of-the-art *systems biology*[84–92] provides a collective knowledge on analytics of complex biological systems with adjunct supports of computational and mathematical modeling. Emerging as an engineering discipline applied to biological studies, systems biology efforts can be trimmed to address the analytical frameworks of molecular bioinformation. Relevant biology-based field of studies include informatic details pertinent to microbiology, cell-biology, and the whole living system.[84–92] The primary task in bioinformation studies refers to assaying the information contents of biological sequences in measurable units as outlined as follows.

Informatics of Molecular Biology

In a living system, the DNA sequences represent very long strings of the order of 10^4–10^9 base pairs. Therefore, they are stochastically viable for representation in the information-theoretic plane by virtue of their extremely large number. Mathematically, the DNA represents a linear sequence of symbols and exactly duplicates this sequence each time the molecule reproduces itself. This confirms the role of DNA as the site of information storage on the genetics and hereditary characteristics of the biological species.

In order to quantify the negentropy characteristics of the DNA/RNA, the following questions should be answered.

- *Are the bases in the DNA chain independent events?*
- *Does the occurrence of any one base along the chain after the probability of occurrence of the base next to it?*
- *What is the conditional probability that when a base A occurs, it will be followed by A or any other base designated at T, C, or G?*

In all, the salient aspects of DNA information can be summarized as follows.[92]

- DNA information is inherently redundant.
- It represents at least a *first-order Markov source* output.
- DNA information can be regarded as the output of an ergodic source.
- The total divergence of DNA information from the maximum entropy state is composed of D_1 and D_2. These indices (known as *D-indices*) also specify the contributions of D_1 and D_2 relative to the total divergence defined as follows: $RD_1 = D_1/(D_1 + D_2)$ and $RD_2 = D_2/(D_1 + D_2)$ with $(R, D_1) \Leftrightarrow (R, RD_1)$ and $(R, D_2) \Leftrightarrow (R, RD_2)$ be regarded as a set of coordinates defining a vector, and R denotes the Shannon's *redundancy factor* given by $R(\cdot) = [1 - H_m/\log_b(\alpha)]$. Here, α denotes the number of states (or events) of the statistics involved, and the factor $[\log_b(\alpha)]$ in $R(\cdot)$ corresponds to the maximum entropy of the sequence of equiprobable, independent elementary events $i = 1, 2, \ldots, \alpha$, and H_m is the entropy of the first-order Markov chain. The ratio $H_m/[\log_b(\alpha)]$ denotes the *relative entropy*. Further, D_2 is an *evolutionary index* that separates the higher organisms (such as vertebrates from lower organisms). Vertebrates have achieved higher R-values by holding D_1 relatively constant and increasing D_2. Lower organisms, on the other hand, achieve higher values of R, primarily by increasing D_2, that is, by forging divergence in the base compositions from the uniformly distributed characteristics and enabling a march toward a central tendency area that prevailed at their primitive state of evolution.

Shannon's Second Theorem in Genetic Information-Theoretics

Consistent with a typical communication system consisting of a source of information, a noisy channel, and a receptacle (that receives the information), one can model the relevant genetic information transmission following the sequence of central dogma as follows.

- *Source output*: Base sequence of the DNA in encoded form.
- *Received message*: Amino acid sequence in protein.
- *Noisy channel*: Mechanics of protein synthesis involves transcription of a base sequence denoting the DNA into a corresponding mRNA and eventual translation of mRNA into the proteins on the ribosome. Considering this chain of events, the *Shannon's second*

theorem can be applied. Accordingly, under certain conditions, it is possible to transmit the message without loss of message rate despite the presence of noise, if the message is encoded appropriately at the source. In the perspectives of a biologist, Gatlin[82] translates the application of Shannon's second theorem in genomic contexts as follows: "It is possible, within limits, to increase the fidelity of the genetic message without loss of potential message namely, provided that the entropy variables change in the proper way, namely, by increasing D_2 at relatively constant D_1." Vertebrates can accomplish such a source encoding. That is the reason for them to be "higher" organisms.

Information Space in Biological Sense

Pertinent to genetic information, or the domain of "DNA space," it can be specified as a spotty region at DNA/RNA level in microbiological sense with its information-theoretics set by the genetic code and the gene expression, and relevant concept can be extended to the larger frameworks at cellular (or cytogenetic) space. Or, more elaborately, the bioinformatic space may conform to the *compacta*[92] of whole-life of the living systems. These considerations are summarized below.

Stratum of information space at microbiological level: The bioinformation begins with the genetic code. As described in Chapter 1, there are specific regions of information involved in the DNA that enable making of proteins.

Stratum of information space at cellular level: Biological organization is built on cellular architecture, and such biological cells bear the genetic mapping conveyed to them *via* translation process of gene regulation indicated above. Hence, the bioinformatic analyses can be exclusively specified to this cellular stratum of the biological architecture.

Stratum of information space at whole-body level: Viewed in a large-scale, the cellular entities lead to conceiving the whole-body level of living systems. As such, the inherent details of genes and the associated information-theoretics can be seen mapped implicitly at this macro-level of organisms. Hence, an exclusive study can be prescribed to address the bioinformatic considerations to the organismic level. It includes the topics of viewing the bioinformation in terms of evolutionary-theoretics addressing the bioinformatic avenues in *phylogeny* based on classical and modern views on evolution theory pertinent to the origin and diversity of species and perusals of Baldwin theory.[83] Relevant aspects of information-theoretics tailored to phylogenetic concepts are presented in Chapters 9 and 10.

Information-Theoretics of "DNA Language"

As described above, the string of bioinformation extends from the genes at the DNA level and manifests as information details at the whole-body *stratum* of organisms. In all, the underlying considerations conform to studying the information-bearing message inscribed as the *DNA language* — a "language of communication between cells." This language, therefore, can be analyzed by similar methods employed in information-theoretic exercises pertinent to written/spoken languages, and it is also possible that short- and long-range statistics may prevail in the DNA language as in linguistic studies.

DNA is a message-bearer in the cellular communication system. DNA message (in Shannon's perspectives) possesses characteristics like entropy, relative entropy, mutual information, equivocation, and redundancy, and such characteristics are important in determining the functioning of the replicative and reproductive cellular system. For example, a high degree of redundancy in DNA information helps avoiding individual errors in transcription or mutation from causing fatal conditions to the cellular functions. An analysis of information-theoretic details of DNA could help in providing some insights into the relationship of the DNA to the characteristics of the cells and organisms.

Transmission of Genetic Information

Commensurate with a typical information communication system, analogously the transmission of biomolecular or genetic details can be studied by associating the concepts of *channel capacity* with the protocols of central dogma and heuristics of Shannon's second theorem. In general, the concept of entropy constitutes the basis in elucidating the *information capacity* of a system. The "channel" in an information communication system refers to an embodiment of physical units that facilitates the passage of a message. That is, the channel is constituted by a runner bearing a "scroll on which the message is inscribed", and the capacity in question is a quantification of the number of messages transacted per unit time. This transmission capacity can be hampered (lowered in value) when the semantic value of the message corrupted by entities of interference (either inherent to the system or from external influences), typically known as the *noise.*

Information-processing paradigms depend on the probabilistic attributes of the underlying complex system, and this information-theoretic approach is advocated to ascertain the *multiple input-single output* (MISO) transfer function of information. Further, the information capacity of a biocomplex can be deduced on the basis of the complex patterns viewed as a collective system. This capacity can be represented as a memory (that stores information) and is quantified *via* Hartley–Shannon's law as the logarithm of the number of strings of address lines of distinguishable functions (state-transitional process) involved. Concurrent to elucidating the underlying optimal capacity of the system, the following are the basic queries posed: (i) *What is the maximum number of pattern examples that the biocomplex can store?* (ii) *For a given set of patterns (less than the maximum value), what are the different functions that could relate the network inputs to the output? How do the statistical properties of the patterns affect the estimation of the information capacity?*

Coding and Decoding of Bioinformation

Information, in fact, "is data of value in decision-making." Thus, the biocomplex activities support a data of value in making a decision to achieve self-organization. For example, the DNA genes support a valued data on the required protein leading to subsequent central dogma-dictated, self-organized steps toward the decision on the desired protein complex. The existing body of information-theoretics as applied to biocomplex does not *per se* specify explicitly the pragmatic and semantic utility of bioinformation with regard to self-organizing (cybernetic) control goals or decision-making strategies outlined earlier. This is due to the fact that the techniques pursued by the statistical theory of information do not match *per se* the analytics of control problems. Classical statistical theory of information describes only the processes involved in the transmission and storage of information. It is not comprehensible to treat information *vis-à-vis* control strategies or the extent of contraction of the parameter subspace as a result of previous training.

In the context of biology, the physical carrier of the information changes and the base content accrues meaning or semantics. The amino acid sequence, for example, is structured in such a way that meaning is accrued, not only as a three-dimensional structure but as a functional enzyme or structural element, able to interact with other molecules.[93] Coding implies signal transduction and is ubiquitous across the entire cell as well as between cells. The state of a cell can be in an encoded format as a particular concentration level of a small molecule (referred to as the metabolic code).[94,95]

2.8 Review Section: Tutorial Notes, Examples, and Exercises

This section offers a set of notes that supplement the details furnished in the text *via* tutorials reviews, as needed on the subject matter of this chapter. Such reviews will be useful in problem-solving exercises of the next section.

Tutorial T-2.1

Binary-Formatting a String of Symbols

A string of symbols, for example, the residues in a DNA sequence (or a part of it) can be transformed into a binary format, and informatics-related analyses can be performed on the binary-formatted sequences. For example, each residue of the set {A, C, T, G} that appears in a DNA sequence can be assigned a *dibit* with their complementary (Watson–Crick pairing) relation preserved. That is, suppose A corresponds to "01," then T is depicted by "10." Likewise, assigning "00" to C, correspondingly, then G is given by "11." Note the choice of the *dibit* for any residue could be arbitrary, but the complementary relation should be maintained, hence, {A, T, C, G} ↔ {01, 10, 00, and 11}. Relevant use of binary formatted strings are described in Neelakanta *et al.*[71] and Pandya.[72]

Tutorial T-2.2

Biosequence construction and analyses: The feasibility of binary representation of DNA indicated in the tutorial above has been sparsely addressed in the archives of bioinformatic literature in sequence comparison exercises. Nevertheless, the following are possible efforts that can be conceived in framing the binary-formatted biosequences and relevant bioinformatic applications.[96–100]

- Representation of a DNA sequence in binary format representations of the associated residues.
- Construction of a DNA sequence (in binary format) using the statistics of its constituents (i.e., as per codon triplets and/or base composition statistics).
- Estimating the statistical distance, entropic divergence similarity/dissimilarity correlation metrics and power spectral measures (*via* periodogram analysis), etc. with binary-formatted biosequences for bioinformatic applications such as codon–non-codon ("junk" codon) delineation toward finding the boundary of separation between codon and non-codon regions and ascertaining symmetry/asymmetry features between a DNA sequences.

Tutorial T-2.3

Edit distance or Levenshtein distance: A measure of *edit distance* (known as *Levenshtein distance, LD*) can be adopted to compare a pair of sequences and establish the "distance" between them. In bioinformatic contexts, the LD specifies minimal number of operations (indels and substitutions) exercised (between two biosequences) so as to transform one sequence to another. That is, the LD denotes a string-metric envisaged to measure the "difference between two sequences." For example, the LD between two words is the minimum number of single-character edits (i.e., insertions, deletions, or substitutions) required to change one word into the other. (It may be noted that edit distance implies that all operations are exercised on only one sequence.)

Tutorial T-2.4

Fourier domain representation and analyses of biosequences: This tutorial refers to a brief note on power-spectrum-based comparison of biosequences. Typically, the concept of power-spectrum estimation is used in predicting certain feature details of a signal.[101] In random signal analysis, the power spectrum can be obtained by *periodogram method.* For a given window of observation of random occurrences (as in a random event sequence), the *power-spectrum density* (PSD), $S(\omega)$ can be ascertained from a standard expression based on *Wiener–Khinchin theorem* on PSD *versus* autocorrelation function available in Oppenheim and Schafer,[101] Arredondo *et al.*,[102] Leitão *et al.*,[103] Benson,[104] Tiwari *et al.*,[105] and Lathi.[106] In order to evaluate $S(\omega)$, a *window function* (also known as *apodization function* or *tapering function*) is introduced in relevant signal-processing, and it represents a mathematical function, which is zero-valued outside of some chosen interval of random data being processed. To name a few, (i) *Bartlett's or Fejér (triangular) window*; *Hanning window*, and *Hamming* or *Kaiser* window are popular in signal-processing techniques. In Chapter 8, use of $S(\omega)$ in analyzing biosequences of viral serovars is described.

Tutorial T-2.5

Evaluation of subsequence characteristics: In bioinformatic contexts, codon–non-codon or noncodon–codon demarcation can be ascertained from the coding statistics associated with a DNA sequence as indicated by the author and others in Arredondo *et al.*[102] Pertinent strategy is described in Chapter 3. In short, it implies the following: Assuming that the codon and non-codon regions along a DNA sequence depict parts of a total sequence length N and designating the codon region by subscript 1 and noncodon region by the subscript 2, relevant fractional regions being compared can be specified, respectively, as (n_1/N) and (n_2/N) so that $(n_1 + n_2) = N$. The triplets constituted by the nucleotide set $\ell \in \{A, T, C, G\}$ occur with a specific probability distribution in a given genome type. Suppose these probabilities are denoted for the regions 1 and 2 by the sets $\{P_{1,\,\ell}\}$ and $\{P_{2,\,\ell}\}$, respectively. With reference to human or bacterial genes, for example, the associated set of probabilities $\{P_{1,\,\ell j}\}$ are listed in Human codon frequency usage.[100] In non-coding parts, it can well be assumed that all the 64 triplet constituents occur with the same frequency so that $P_{2,\,\ell}$ values refer to equally-likely occurrence (or uniformly distribution), each corresponding to 1/64. This distinguishing feature (of probabilities of occurrences) of triplets in codon and non-codon parts leads to identifying the demarcation between (codon and non-codon) regions in a DNA sequence in terms of their distinguishable entropy features. A typical metric used is the log-likelihood ratio (of probabilities of occurrence of codons and non-codons), which distinctly signifies regions codons and non-codons along the DNA sequence and allows their delineations.

Alternatively, entropy-based considerations can be applied to a query segment (using the associated codon statistics) to decide whether that segment belongs to codon or non-codon category. The entropy-based scoring technique suggested thereof refers to, for example, the

Jensen–Shannon (JS) measure. Likewise, the KL measure (and hence, the JS measure or its variants) can essentially compare two vector spaces corresponding to codon and non-codon regions, respectively, by elucidating the divergence in the associated stochastic characteristics.

In the entropic segmentation method suggested in Bell and Forsdyke,[99] a DNA sequence with a heterogeneous set of subsequences (or compositional domains), the associated entropy profiles of the domains are distinctly decided by probability distributions of triplets in those regions, and it is shown in Bell and Forsdyke[99] that relevant approach could lead to accurately predicting the borders between coding and non-coding regions without any *a priori* training details on known sets. Notwithstanding the existence of aforesaid methods, yet another set of methods to establish the delineation in question using binary formatted DNA sequence and Hamming distance (HD) measure is outlined in Neelakanta *et al.*[70,71] and Pandya.[72]

Example Ex-2.1

Statement of the Problem

Pursuing the example analysis indicated in Balado,[107] this project exercise is to estimate the embedded information in DNA sequences. Essentially, the study refers to determining the maximum amount of information that can be theoretically embedded in DNA sequences implying the Shannon capacity of such sequences. That is, deducing the upper limit on the information content that can be reliably embedded within a DNA under a given error rate is the theme of Balado.[107] With reference to such embedded details in the DNA, it is of interest to know: (i) whether the DNA has additional arbitrary data, (ii) whether the embedded details within the DNA travel alongside each replication, and (iii) whether the information passage happens inside or outside the living system.

The study in Balado[107] focuses on analyzing the DNA-embedded information pertinent to: (a) the junk or non-codons (introns) in the DNA sequence that contribute nothing toward eventual protein-making and (b) the coding elements (exons) that bear genetic details toward protein-translation. The first set, namely, the non-codon elements randomly seen across a DNA sequence can be regarded analogous to arbitrary digital data of the sequence, and the information-bearing exon details conform to embedded information hiding within the genetic host. This description refers to the classic *data-hiding problem*[108] with the host sequence representing the side information at the encoder.

Through transcription toward translation, the information-carrying DNA is assumed to be functional, with its exons enabling the eventual translation (to proteins); this should, however, be true even before the embedding operation that incorporates the non-coding introns. Although when such non-coding parts are excised out during transcription (and they are regarded as junk with no contribution in the translation of the residues into proteins), studies Gibbs[109] also suggest that the non-coding part of the DNA may not be discarded as junk. Their presence may alter regulatory regions whose biological task is yet unknown. Therefore, a full understanding of coding and non-coding segments of the DNA is useful, in assessing the functional effectiveness and unobtrusiveness of these constituents. In Gibbs,[109] the goal of DNA data-embedding is viewed as twofold: (i) Tagging genetic material for tracking purposes. Reliable DNA embedding may allow new forms of genetic finger-printing by attaching unique tags to differentiate among functionally identical genetic material. Such tracking would allow spatial and temporal evolution of different instances of genes with identical protein translation, and possibly mutations can be detected *via* the embedded information, as in cases where no single host genome is used as a reference, such as in the studies regarding viral quasi-species. (ii) The other goal of DNA data-embedding is concerned with using genetic material as a massive and compact storage media. This implies that long-term storage of data in the DNA of living organisms (such as bacteria) can be actually implemented with real organisms as described in Wong *et al.*[110]

The randomly intermixed state of codons and non-codons lead to ascertain the embedded (hidden) information (in Shannon's sense) of a DNA complex and relevant analytical pursuit in Balado[107] is attempted as an open question in order to determine the maximum amount of information that can theoretically be so embedded, that is, the Shannon capacity of a DNA complex. Hence, the goal toward ascertaining the genetic meaning of C-DNA data embedded along with exons is studied in Wong *et al.*,[110] and the C-DNA capacity analysis is indicated with the constraint of nonzero inequality placed on the average Hamming distance, consistent with the indications in Pradhan *et al.*[111] and Barron.[112] Treating the DNA sequence as a message-bearing communication entity, the non-coding and coding segments are regarded as two distinct payloads. That is, the non-coding (NC) DNA is regarded as DNA bases constituting equally-likely occurring (uniformly distributed with $p_{NC} = 1/64$), 4-ary alphabets for which the Shannon information (I) is a trivial solution leading to: $I|_{NC} = \log_2(2^2) = 2$ bits/base-residue. However, for the coding exons, the average embeddable payload information is given by:

$P|_C = (1/N) \times \log_2 |M| = (1/N) \times \sum_{i=1}^{N} \log_2[\mu\{\alpha(x_i)\}$ bits/coding exon. Here, M denotes the set with cardinality, $|M|$ from which the embedded message of exons is evaluated. Further, $\{x_i\}_{i=1,2,...,N}$ are exons (coding segments) derivable from the 64-ary trinucleotides. Relevant to the set $\{x_i\}_{i=1,2,...,N}$, its translated version, namely, $\alpha\{x_i\}_{i=1,2,...,N}$ would correspond to the probability values of the codons $\{x_i\}$ involved specified for a given species. (for example, as in Table 1A.7 of Chapter 1 depicting sets of *codon usage frequency* (CUF) of: (i) *Homo sapiens*, (ii) *Escherichia coli*, (iii) *Methanococcus jannaschii, and* (iv) *Rickettsia prowazekii.*

Considering the trinucleotide set of 64 codons, they are grouped into 20 amino acids (as indicated in Chapter 1) with a given codon representing more than one amino acid. Inasmuch as there are multiples of trinucleotides that depict the same amino acid, explicit indication of this multiplicity associated with an amino acid requires a unique exon-sequence representation constructed with N residues is profiled in terms of $\mu[\alpha\{x_i\}_{i=1,2,...,N}]$. The average message of exon payload is thus given by $[P|_C]_{Ave} = E[\log_2[\mu\{\alpha(x_i)\}]]$ bits per exon.

Pertinent to the embedded information of non-coding and coding segments of a DNA indicated above, the corresponding Shannon capacities can be specified as follows: For the non-coding segments with equally-likely presence of base residues, the Shannon capacity is that of a standard M-ary symmetric channel equal to, $C|_{64}$. However, for the coding segments, the distribution of exon residues is non-uniform. As such, considering the resulting achievable information rate corresponds to the uniform input that maximizes the mutual information over a symmetric channel.[113,114] It is given by the following expression: $R = C|_{64} - H(X')$ bits per exon, where $H(X')$= denotes the information content of the set, $X' = \{\alpha(X)\}$ with X' denoting the set, $[\alpha\{x_i\}_{i=1,2,...,N}]$.[115] Summing up, the available methods implement information rates *vis-à-vis* biological sequences such as the DNA in a limited sense; the deduced results seem to be far from the realistic capacity and/or operate in the far low-error end of the achievable region.

Problem P-2.1

Statement of the Problem

Project exercise: With reference to the heuristics of deducing channel capacity of a DNA discussed in the above example, the following can be noted: The codon equivalence is not evenly spread over the amino acids ensemble, and the embedding limits for C-DNA hosts are not immediately obvious due to the possible hidden details of regulatory functions of such introns. As well-known, an intron is any nucleotide sequence within a gene that is removed by RNA splicing during maturation of the final RNA product.

The term "intron" refers to both the DNA sequence within a gene and the corresponding sequence in RNA transcripts. (Sequences that are joined together in the final mature RNA after RNA splicing are exons.) The word intron is derived from the term intragenic region, that is, a region inside or within a gene. The introns could be grossly lost or gained along the temporal evolution. Such extensively persistent events of intron loss and/or gain, pertinent to eukaryotes, correspond to episodes of "intron invasion." Relevant aspects of genomic deletions and/or mechanisms of intron gain remain elusive and controversial. Typically, it appears that over 90% of DNA base residues consist of repetitive noncoding regions, namely, the introns that have no explicitly known function. However, contraindications that introns may have other functions also, exist.[116–120] For example, it is argued that "the cell puts a huge amount of its energy into the creation of these introns, then discards them ... Nature would not go to all that trouble without a reason." On the tale of introns, Patrusky[118] comments as follows: "Nature, for reasons as yet unknown, created the intron, and evolution has chosen to keep it ... and ultimately has found new ways to use it." Additional discussions can be found in Bergman,[119]and Wu and Lindsay.[120]

Problem Statement

Consistent with the previous exercise and discussions on the role of introns indicated above, devise a method to subdue the junk or non-informative aspect of introns in a DNA by considering a fraction of 64 trinucleotide constituents that form the introns as being non-uniform. Explicitly, out of 64 residues, suppose a set of 10 residues $\{r_1, r_2, \ldots, r_{10}\}$ with corresponding occurrence probabilities, $\{p(r_1), p(r_2), p(r_3), \ldots, p(r_{10})\}$ pertinent to a species, participate in making introns that presumably play functional roles, say, in the regulatory segments of the DNA. Then, the rest of 54 non-coding introns will have uniform probability distribution, $p_{NC} = (1 - p^*)/54$, where $p^* = [p(r_1) + p(r_2) + p(r_3) \cdots + p(r_{10})]$.

With this modified p_{NC} value, determine the corresponding modified rate, $R^* = [C|_{p_{NC}} - \{H(X') + H|_{p_{NC}}(Y')\}]$, where the set Y' corresponds to $\{r_j\}_{j=1,2,\ldots,J}$ (e.g., $J = 10$ in the above case) translated into a new set of probabilities, namely, $\beta\{r_j\}_{j=1,2,\ldots,J}$ as specified for a given species (as, for example, of human genome), and again toward explicit indication of the multiplicity governing 64 trinucleotides *versus* 20 amino acids, the set $\{r_j\}_{j=1,2,...,J}$ is profiled and written in terms of $\nu[\beta\{r_j\}_{j=1,2,...,J}]$ as indicated earlier for X'. Hence, the associated average message is $H|_{p_{NC}}(Y') = E[\log_2\{\nu\{\beta(r_j)\}\}]$ bits per residue. Hence, determine and plot R^* as a function of the chosen residue cardinality (J) of the set $\{r_j\}$ being 5, 10, 15, etc. The choice of trinucleotides can be arbitrary except that start/stop codons need to be avoided. The codon statistics may conform to, for example, the human genome data discussed in Chapter 1.

Tutorial T-2.6

Complex system representation of the DNA: The complex functional attributes of a DNA structure are exhibited through their structural units or domains and lead to the consensus of classifying such structures as adaptive complex systems as portrayed in Nagl.[121] That is, considering the genome statistics of the DNA (consisting of randomly disposed codon and non-codon entities), a complex system model that describes the stochastic and informatics aspects of a DNA structure can be specified as follows: since a DNA sequence is a large-sized, binary mixture of codons and non-codons, its each constituent subset (of the mixture) depicts a conglomeration of elements occurring in a proportion as decided by some probabilistic distribution. Relevant to such a complex mixture set, an algorithm can be specified as a metric that quantitatively evaluates the associated complexity (in terms of underlying information redundancy).

Inspired by the notions in Nagl[121] and aforesaid binary mixture characteristics of codons and non-codons, the statistical mixture theory is applied to a DNA complex[25]; and a *complexity metric* compatible for bioinformatic analyses is specified. That is, a measure of global complexity is specified in terms of the maximum entropy associated with the disordered (mixture) constituent entities, namely exons and introns, in the DNA complex. Relevant heuristics are as follows.

It is assumed that the set of exons $\{x(\mu)\}_i$ and the set of introns $\{y(\nu)\}_j$ coexist, respectively, in large populations with $i = 1, 2, \ldots, \ell, \ldots, n_1 \to \infty\}$ and $j = 1, 2, \ldots, \ell, \ldots, n_2 \to \infty\}$ and each type is mixed in a specified proportion of as θ and $(1 - \theta)$, respectively, by $\{x(\mu)\}$ and $\{y(\nu)\}$. Hence, following the concept of statistical mixture theory (due to Lichtenecker and Rother[122]), the underlying heuristics can be specified in terms of a weighted probability r that describes the effective statistical attribute of a mixture proportioned. This weighted probability r can be indicated within statistical upper- and lower-bounds, such that ($r_{min} \le r \le r_{max}$). Explicitly, r, r_{min}, and r_{max} are given by: $r(\theta) = [P_{1\ell}^{\theta} P_{2\ell}^{1-\theta}]$, $r_{max} = [\theta P_{1\ell} + (1-\theta) P_{2\ell}]$, $r_{min} = [\theta / P_{1\ell} + (1-\theta) P_{2\ell}]^{-1}$ where $\{P_{1\ell}\}_{\ell = i}$ and $\{P_{2\ell}\}_{\ell = j}$ are occurrence probabilities of the sets $\{x(\mu)\}$ and $\{y(\nu)\}$, respectively, and corresponding Shannon measures of entropy (negentropy) can be written as follows: $H(r) = [-r \times \log_b(r)]$, $H(r_{max}) = [-r_{max} \times \log_b(r_{max})]$.

Eventually, it is possible to define an *information efficiency* (η) factor of the encoded structure of a DNA using the classical concepts of information theory. It refers to the ratio of the average information, $[\bar{C}]$, (per codon) of the encoded ensemble to the maximum possible (average) information (per codon), that is, $\eta = H(x)/[\bar{C} \times \{H_{max} = \log_b(N = 64)\}]$; here $H(\cdot)$ denotes an entropy functional, namely a statistical divergence (SD) measure, such as the Jensen–Shannon (JS) metric described earlier in this chapter. Concurrently, ($1 - \eta$) can be regarded as the *redundancy factor* (R). It represents the reduction in information content of an ensemble from the maximum possible. Without any loss of generality, any divergence or distance measure can be adopted for $H(\cdot)$ *in lieu* of the JS measure as indicated by the author and others in Neelakanta.[25]

Problem P-2.2

Statement of the Problem

Project exercise: With reference to discussions and heuristics as above, obtain the locations of delineation between randomly occurring exon and intron segments disposed across the DNA of *Drosophila melanogaster*. The codon occurrence statistics of this species is available at www.kazusa.or.jp[123] detailing the occurrence frequency of codon triplets of *D. melanogaster*, that is, probability of occurrence (P) of the triplet codon statistics of *D. melanogaster* [gbinv]: 42417 CDS's (21945319 codons); Coding GC 53.86%, 1st letter GC 55.79%, 2nd letter GC 41.48%, 3rd letter GC 64.32%.

Solution hints: The distinguishing feature (of probabilities of occurrence) leads to identifying codon and non-codon regions in a DNA sequence, and the entropic segmentation method due to Galvan *et al.*[99] in essence prescribes a strategy of delineating the underlying codon and non-codon regions based on the aforesaid statistical distinction expressed in terms of associated entropy considerations. As such, the entropy segmentation efforts as mentioned before essentially conform to divergence measures,

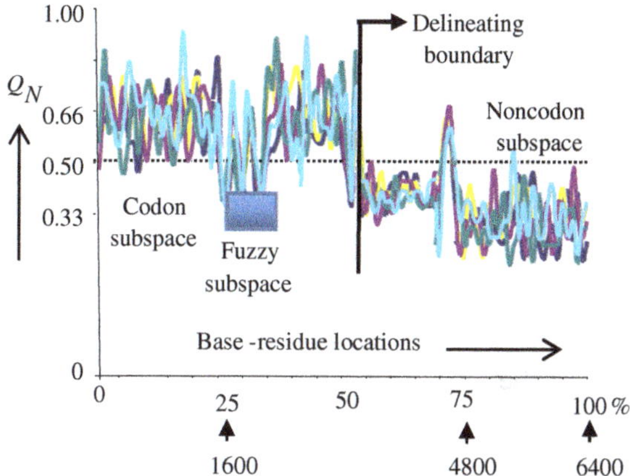

Fig. 2.4 Simulated ensemble of results on normalized contrast scores (Θ_N) *versus* segmented windows across the DNA of human codon sequence. (The test DNA chain is regarded as an admixture of codons (X) and non-codons (Y) divided into K windows of equal-length W.) The score values obtained for any Kth window correspond to KL-measure deduced with the probability sets: $\{p_A, p_C, p_T, p_G\}_{Wk}$ and $\{q_A, q_C, q_T, q_G\}_{Wk}$ that are, respectively, pertinent to the contrasting pairs of sequences, namely, TS and JS

such as the JS-measure. In essence, the relevant method partitions a heterogeneous DNA sequence into homogenous subsequences (or compositional domains), and it is shown that the corresponding approach could lead to predicting accurately the borders between coding and non-coding regions without any *a priori* training details on known sets.

Alternative to entropy segmentation algorithm (narrated in the tutorial-example above), proposed here is a method to discern the codon/non-codon borders *via* the *R-measure*, which also as stated earlier is specific to the informative profile of the constituents of a DNA structure. Such constituents are vast when viewed in terms of their attributes, variety, stochasticity, and interactions between the associated units, and hence they belong to the gamut of informative complex systems.

The results shown in Fig. 2.4 denote the simulated ensemble of data on the contrast scores (Θ_N) evaluated *versus* the base-residue locations along the segmented windows of the test human codon sequence. The results reveal subspaces that show the contrasting high-low values of the estimated scores, thus the codon–non-codon subspaces delineated as shown. However, there is a short segment of overlapping segment, which is fuzzy for a clear distinction as codon or non-codon. Associated defuzzification is described in Arredondo *et al.*[102]

Problem P-2.3

Problem Statement

Project exercise: With reference to the above illustrative exercise, obtain the locations of delineation between randomly occurring exon and intron segments disposed across the DNA of *D. melanogaster* (having codon occurrence statistics as available in www.katusa.or.jp123) using: (a) JS-divergence measure; (b) *R*-metric; and (c) Bhattacharyya distance measure. Compare the results and discuss.

Problem P-2.4

Problem Statement

Stationarity and ergodicity aspects of biosequences: A bioinformatic problem of interest is to find an adequate model that would allow for assessment of certain characteristics of a "*biological text*," while using a relatively small number of parameters. Relevant approach is to describe biosequences by stochastic processes. For example, sparsely addressed in the archival of literature is a topic on diagnosing the stationarity and ergodicity attributes of the stochastic framework pertinent to microbiological details of biosequences. Such studies govern the variant or invariant statistical details of the residues randomly disposed along the stretches of test biosequences and across the ensemble of such sequences. One can consider the underlying details as "biological texts" and the analytics of the associated stochastic complexity are indicated in Ryabko *et al.*[124] as being amenable *via* the so-called *Kolmogorov complexity.* Further, inspired by Fisher's classic works[125] on statistical testing of biological hypotheses, forming the directives suites to pursue quantitative analysis of biological data in terms of stochastic perspectives,[126,127] a study can be undertaken to elucidate the stationarity and ergodicity details of biosequence texts.

Project exercise: With reference to the descriptive note on stationarity and ergodicity aspects of stochastic systems, this project is proposed to address a sparsely studied topic of ascertaining the stationarity and/or ergodic features of bio-texts. Relevantly the problems indicated described below can be undertaken as term-paper exercises pertinent to directed-independent studies, and the studies can be extended to broader suites of thesis or dissertation.

Problem P-2.5

Problem Statement

Project exercise: This project refers to *single nucleotide polymorphisms* (SNPs or "snips"), a common type of genetic variation seen among people.[128] Each SNP represents a difference in a single DNA building block of a nucleotide. For example, a replacement of the base residue *C* with *T* in a certain stretch of DNA is a SNP. Such SNPs may occur throughout a human DNA. Typically, they occur once in every 300 nucleotides on average, which means there could be about 10 million SNPs in the human genome. Most commonly, these variations are found in the DNA between genes and they act as *biological markers* to locate genes that are associated with disease. That is, whenever SNPs occur within a gene (or in a regulatory region near a gene), they affect the gene's function and is regarded as a causative factor toward disease. However, most SNPs may not pose to be such factors with deleterious effect on health and/or development issues. But genetic differences perceived due to SNPs could help predicting "individual's response to certain drugs, susceptibility to environmental factors such as toxins and risk of developing particular diseases." Also in tracking the inheritance of disease genes within families is made feasible *via* SNPs.

The sequence of alleles seen contiguously (connected throughout in an unbroken stretch) of SNP positions along a chromosomal region is called a *haplotype*. Correspondingly, sets of alleles or DNA sequences can be clustered so that a single SNP can identify many linked SNPs, and this is called *haplotype mapping*. Knowledge of haplotype structure can possibly enable cost-effective genome-wide association studies, since haplotype tagging SNPs can be chosen to capture most of the genetic information in a region that has a block structure of similar average percentage of common haplotype in blocks. Substantial evidence has already accumulated to show that the genome can be parsed into such haplotype blocks of variable length.[129]

Problem P-2.6

Problem Statement

In view of the evidence on the possible existence of haplotype blocks along DNA sequences, perform a statistical analysis to find out the expected number such haplotype blocks in segmented windows taken along the sequence length, so as to decide on the stationarity profile of the SNP.

Approach/Solution Hint

With reference to the context SNPs outlined above, a stochastic framework can be conceived to address the stationarity profile of blocks and related subpopulations in DNA sequences. Described in Zhu *et al.*[129] is the definition of haplotype block and its application. Further indicated in the study is that the linkage disequilibrium between a given SNP and the haplotype block is a monotonic function of distance. This correlation is found to be essentially independent of the minor allele frequency of the putative causal SNP when it fell outside of the block; however, it is strongly dependent on the minor allele frequency when the SNP was internal to the block.

The genome is organized into blocks of haplotypes. Finding a genome-wide haplotype map of SNPs is of interest to current researchers. Relative performance details on diverse, algorithmically defined haplotype blocks in terms of real data and/or in terms of the impact of allele frequencies and parameter choices on the detection of haplotype blocks and the markers that tag them are sparse. Schwartz *et al.*[130] assessed the overlap of block boundaries assigned by different algorithms. They found the existence of a "generally poor agreement between block boundaries derived from different" algorithms, which was more pronounced in small samples. Also, the marker-spacing would affect the predicted length of haplotype blocks in an evolutionary modeling analysis.

A formal comparison of two major algorithms based on *linkage disequilibrium* (LD)-based method and a *dynamic programming algorithm* (DPA), in three chromosomal regions differing in gene content and recombination rate, is presented in Schulze *et al.*[131] The number of haplotype blocks and tag SNPs can be identified by each method and such variables are critical in association mapping.

The LD-based method is due to Gabriel *et al.*[132] and the DPA was developed by Zhang *et al.*[133] In the studies due to Schulze *et al.*,[131] three fully sequenced chromosomal regions that differed in their average recombination rates and gene content. Relevant results show that the two major methods as above for detecting haplotype maps and tag SNPs can produce different results for the same data and that such results are also sensitive to marker allele frequencies and parameter choices. Further, the DPA identifies fewer and larger haplotype blocks as well as a smaller set of tag SNPs than the LD method. A relevant research thereof may refer to elucidating the stationarity profile depicting the average statistical performance of a biosequence in terms of the percentage of common haplotypes in blocks. That is, for a chosen species, of a set of human genomes, the proposed study can be advocated.[134,135] As a guideline, prior study data can be looked into. For example, addressed in Zhu *et al.*[129] and Zhang *et al.*[133] is the analysis relevant to the data obtained from the public access website of the Whitehead Institute. It includes details on genotype data obtained from four population samples: (a) 30 parent–offspring trios (90 individuals) from Nigeria; (b) 93 individuals from 12 multi-generational CEPH pedigrees of European ancestry; (c) 42 unrelated individuals of Japanese and Chinese origin; and (d) 50 unrelated African Americans. Correspondingly, a total of 3,738 SNPs in 54 autosomal regions are genotyped in all four groups with the average size of a region being 250 kb.

Problem P-2.7

Problem Statement

Project exercise: This project refers to verifying the possible stationarity and ergodic aspects of certain characteristics seen across the taxa set of species with a common ancestral root.

Problem Statement: Verify whether the evolution might have altered or conserved such features seen along the sequence length as well across the ensemble of species.

Solution-hint: Two possible features that can be probed into are: (i) splice junctions and (ii) telomeres. Each chromosome is capped in both ends with a telomere that is made of a repetitive DNA sequence and a set of proteins. The telomeres prevent degradation of the ends of the chromosomes during replication. (The 2009 Nobel Prize in Physiology or Medicine was given to Elizabeth H. Blackburn, Carol W. Greider, and Jack W. Szostak for the discovery of how chromosomes are protected by telomeres and the enzyme telomerase.)

Example Ex-2.2

Problem Statement

Project exercise: This project is an exercise on statistical profiling of biomolecular sequences. A review thereof is presented first to illustrate a relevant example, and a problem is then indicated subsequently as a rider exercise on the example. The associated effort refers to an exclusive attempt to characterize the DNA sequences statistically. Particularly, focus has been placed to study the underlying "long-range correlations" that exist in DNA sequence, and mutual information considerations are used to elucidate the pertinent biosequence characteristics.

In general, informatics-based assessment of sequence characteristics alone may not be comprehensive to decide the complexity and stochastic aspects of a given sequence. As such, Herzel and Groβe[136] developed a model that applies mutual information concepts to the global length of the query sequence and also decides the statistical distribution profile of the constituents in order to generate pertinent stochastic models. That is, by considering the statistics of a given DNA sequence, such stochastic models illustrate the existence of oscillatory variations of the mutual information as well as showing decay in the level of mutual information along the sequence length. Corresponding algorithmic method developed by Herzel and Groβe[136] enables generating a set of pseudochromosomes, and in parallel, by considering the statistical description of human chromosomes.

The 1D sequences (such as DNA sequences) implies the appearance of the residues (like nucleotide bases in a DNA structure or base density) that tend to co-vary at regions separated by a "long-range or distance." A typical, academic directed study on statistical characterization could aim at finding long-range correlations in DNA sequences. Such correlations in DNA sequences largely stem from mutational events known as *string rewriting rules.*[137] These rules include repetition of promoter sites in DNA sequences where long-range correlations exist. Stochastic techniques like Levy-walks, Fourier transforms, and wavelet-theoretics[138] are prescribed to study such correlations, and these techniques facilitate establishing typically power-law correlations in order to quantify the associated critical exponent of the correlation.[138,139] For example, explored in Mansilla *et al.*[139] a method that uses the mutual information function as a sensitive measure to decide the *autocorrelation function* (ACF) in the study of long-range correlations in DNA sequences. The mutual information formula can also be used to elucidate the period of three possible oscillations prevalent in DNA sequences.

Hence, toward characterizing DNA sequences statistically, a conceivable method is to combine the concept of mutual information with some other technique(s) as indicated by Herzel and Groβe[140] and toward finding correlation features in DNA sequences relevantly, exon-length distributions, as well as non-uniform codon tables have been adopted to generate pertinent stochastic models. The details on correlations in DNA sequences and the role of protein-coding segments are deliberated in Herzel and Groβe.[140]

Tutorial T-2.7

Long-Range Stochastic Features in the DNA

To elucidate long-range stochastic features in the DNA, Herzel and Groβe[140] developed models that combine the heuristics of mutual information function and statistical attributes of the exon-length and the coding usage. That is, a measure of statistical dependence along the length of a DNA sequence is formulated in terms of mutual information or cross-entropy details of sub-segments in the DNA structure.

Relevant step indicated in developing the statistical model as above involves defining the associated mutual information function with reference to the statistics of events depicting the occurrence probabilities of nucleotide residues {A, C, T, G} in a DNA sequence. Suppose A = {A_1, A_2, A_3, A_4} represents these four nucleotide letters (≡ {A, C, T, G}) and the associated prior probability of A_i is denoted by p_i (where i = 1, 2, 3, 4). Further, the conditional probability of finding A_j, given A_i at a distance of k letters counting from A_i is $p_{ij}(k)$, and if A_i and A_j are statistically independent, then $p_{ij}(k) = (p_i \times p_j)$. Hence, the mutual information function $I(k)$ refers to the associated cross-entropy (in Kullback–Leibler sense). It corresponds to a measure of information assessed on A_j based on the knowledge of A_i. It is specified as a *correlation function* $C(k)$ as follows: $C(k) = \Sigma_{i,j=1}^{4}[p_{ij}(k) - (p_i \times p_j)] \times (a_i \times a_j)$, where a_i and a_j denote, respectively, the coefficient (numeric) values of the residue symbols, A_i and A_j (further, it is implicitly assumed that the underlying statistics is ergodic and stationarity). The correlation functions, in general, measures the linear dependencies of the entities involved, and relevantly, the characterization of long-range correlations can be described by a "degree of distribution" *via* a power-law relation of $C(k)$, such as $C(k) \propto k^{-\gamma}$, presumably entails a power-law-based decay in the mutual information function with a scaling exponent of 2γ and $I(k) \propto k^{-2\gamma}$. Using the relations as above, a second-order mutual information equation can be formulated illustrating the cumulative mutual information of all paired correlations in the distance k[140] (in terms of the associated deviations from statistical dependence.)

It turns out that most of the long-range statistical dependences can explain quantitatively the *degeneracy* of codons, which implies the redundancy of the genetic code exhibited due to the multiplicity of three-base pair codon (triplet) combinations that specify an amino acid.[141] Hence, a *codon usage bias* (CUB) exists (and it varies between organisms) as detailed in Chapter 1. The CUB represents a unique feature of an organism. Pertinent to the heuristics of the CUB, a "nonuniform codon table" can be specified, and the mutual information formulation can be applied to derive the underlying stochastic models and long-range dependence features in a DNA sequence can be assessed. Further, protein-coding segments (exons) exhibit persistent correlations between their nucleotides with a pronounced period three. It is shown that this periodicity induced by the nonuniform codon usage implies long-range correlation over hundreds of base pairs, if the length distribution of exons is duly taken into account.

Apart from mutual entropy considerations in depicting the statistical perspectives of biosequences, analytics of complexity of long-range dependence of such sequences can also be invoked in differentiating protein-coding and non-coding RNA segments. That is, by comparing the entropy of the original sequence with that of its shuffled one, one can identify the source of the difference between the two segments and their relative contributions to the sequence. To demonstrate the method, the DNA sequences of the bacterium *Clostridium difficile 630* (with G + C = 29.1%) and *Bdellovibrio bacterivorous* (with G + C = 50.6%) could be representatives of bacteria with unbalanced and balanced nucleotide content, respectively. It is shown in Lagerkvist,[141] Komar,[142] Hockenberry *et al.*,[143] and Mazaheri *et al.*[144] that regardless of nucleotide content, in both bacteria, the relative difference of the two entropies is significantly greater in protein-coding regions, when compared with non-coding RNA segments.

Problem P-2.8

Problem Statement

Project exercise: Following the example indicated above, develop analytical schedules and perform necessary computations to deduce the following.

(a) Shannon (neg)entropy profiles along the stretches of DNA sequence using the statistics of codon triplet usages in: (i) *Methanococcus jannaschii*; (ii) *Rickettsia prowazekii*; (iii) *Saccharomyces cerevisiae;* (iv) *Escherichia coli*; and (v) *Human.*

(b) Indicate in each case, the Shannon (neg)entropy is significantly greater in protein-coding regions, when compared with non-coding RNA segments.[143,144]

Tutorial T-2.8

This tutorial reviews some salient information-theoretics related topics useful in their applications in bioinformatics. Each topic is outlined with pertinent literature details for gaining more information as needed.

Functional Equations in Information Theory

A set of measurable solutions of functional equations connected with Shannon's measure of entropy, directed divergence, or *information gain*, and related *inaccuracy* considerations (along with relevant definitions and theorems) are discussed in Kannappan,[145] and Kannappan and Rathie.[146] Such details include topics on *maximum entropy formalism*, *Jayne's formalism*, *minimum discrimination information principle*, *Kullback's minimum discrimination information principle*, *direct problem*, *first*, *second*, and *third inverse problems*. Further considerations, such as *uniqueness of the solutions* and *entropy measure for maximum likelihood principle*, are explained. Such details can be viably useful in solving bioinformatic problems.

Generalized Information Functions

Daróczy has furnished in Daróczy[147] the concept of the so-called *information functions*. In terms of such information functions, the so-called *entropies of type* β are defined. Hence, it is indicated that these entropies portray a number of interesting algebraic and analytic properties similar to Shannon's entropy. Application of relevant concepts in bioinformatics is an open-question showing possible avenues of useful research projects.

Measures of Information vis-à-vis Their Characteristics

This is an *addendum* to the comprehensive presentations on information theoretics described in this chapter on conceptual details pertinent to measures of information.[148] These additional details refer to *f-entropies*, *probability of error*, and *feature selection*, and they supplement the heuristics of entropies involving a number of more interesting algebraic and analytic properties. Further, a summary of pertinent details on probability of error in feature selection, etc., can be seen in Bassat,[149] and underlying heuristics are indicated as useful in statistical pattern recognition strategies[150–152] where considerations on information and distance measures, error-bounds, and feature selection are invoked in bio-feature analyses. Similarly, bioinformatics-specifics toward deducing the feature details in 2D-biological patterns by comparing sequences of the pixels of the test pattern *versus* those of a reference pattern are indicated by the author (and others) as described in Neelakanta and Pappusetty,[153] and Neelakanta *et al.*[154]

Applications of Contingency Table in Bioinformatics

A *contingency table* (otherwise known as a *cross-tabulation* or *crosstab*) refers to a table in a matrix format that displays (multivariate) frequency distribution of the variables. Relevant matrix provides details on the interrelation between two variables leading to find the interactions between them. *Via* contingency table, every observation is cross-classified by two category variables instead of just one. Pertinent goal is to test whether true (population) relative numbers of individuals falling into the different classes for one variable is the same regardless of individual values for the second variable. An example may refer to the number of survivors and non-survivors in two classes of population in biological observations. An information-theoretic approach to the evaluation of contingency tables is proposed in Gokhale and Kullback.[155]

Axiomatic Characterization of Directed Divergence in Biological Contexts

Elaborations on comparative assessment of various measures of directed divergence,[156] descriptive aspects of divergence, and distance measures (in communication-related signal selection, etc.) are described in Kailath.[157] Further, the underlying axiomatic characterization of directed divergences and their linear combinations are furnished in Johnson.[158]

Relevantly, a plethora of novel details on cross-entropy, dissimilarity measures, diversity and dissimilarity coefficients, associated measurements, decomposition, apportionment and analysis, as well as approximation of discrete probability distributions (defining new classes of divergence measures) are explained in Rao and Nayak,[159] Rao,[160,161] and Lin and Wong,[162] and deciding associated statistical bounds (such as Chernoff bound, sharper lower bounds prescribed for discrimination information, upper- and lower limits on probability-of-error and equivocation), in terms of observable variations as in the contexts of biology and bioinformation *via* heuristics as in Hellman and Raviv,[163] Kullback,[164] and Toussaint[165] could form viably strategic pedagogical efforts in bioinformatic studies.

Decision-Theoretics and Discrimination Statistics in Bioinformatics

Implications of decision-theoretic approach, discriminating statistical sets of variables, and deducing discrimination information on the variations observed in statistical inferences are implicit in the art of bioinformatics as could be evinced from the literature.[166–169] Further, prescribing an invariant form of prior probability in the estimation problems of informatics is necessary, as observed in Jeffreys.[170] Decision-theoretics in bioinformation contexts are inevitably based on, and notionally tied to, the heuristics of entropy and profiles of distance measures, and pertinent analyses can be applied to recognizable structural and/or functional random patterns.[171–173]

Informatics of Microarray Data Classification

Ramakrishnan and Neelakanta have addressed[174] an information-theoretics inspired entropy method describing a (feature) co-occurrence detection approach toward extracting subset details in a DNA microarray data. Considered in Ramakrishnan and Neelakanta[174] by the author (and Ramakrishnan) is the so-called *Haralick feature-finding* algorithm, and the underlying Haralick feature scores are estimated in terms of the entropy co-occurrence information in an image with reference to a threshold profile. The scores obtained thereof from the intensity gradations of colored spots in a microarray are correlated to the relative extent of the color status existing in the observed spots. Hence, corresponding gene expression (depicting tumor, non-tumor and/or mixed conditions, etc., as implied by proportional abundance of color pixels in the microarray) is quantitatively inferred. Relevant strategy in using Haralick algorithms in microarray contexts is a novel approach and the details on simulation experiments using synthetic and real microarray patterns presented in Ramakrishnan and Neelakanta[174] illustrate the efficacy of the underlying method.

More Examples and Exercises

This section is written to provide illustrative examples and problems on the overall aspects of the subject matter narrated in this chapter. The exemplar exercises are indicated on *ad hoc* basis with solutions and/or solution hints consistent with the heuristics of details introduced in this chapter and supplementing tutorial reviews of the previous section.

Example Ex-2.3

Statement of the Exercise

Hamming distance (HD) based exercises: As defined earlier, the HD between two strings of equal length is the number of positions at which the corresponding symbols are different. That is, it measures the minimum number of *substitutions* required to change one string into the other or the minimum number of *errors* that could have transformed one string into the other.

(a) Find the HD between the following pairs of strings.

(i) A G T C and C G T A

Solution

A G T **C**

C G T **A** HD = 2 (A/C, C/A)

(ii) KENTUCKY and TENTURKI

Solution

K E N T U **C** K **Y**

T E N T U **R** K **I** HD = 3 (K/T, C/R, Y/I)

(iii) CUMBERLAND and TIMBERLAND

Solution

C U M B E R L A N D

T I M B E R L A N D HD = 2 (C/T, U/I)

(*Note*: Sometimes, the HD can be indicated in terms of its complement, namely, the number of positions at which the corresponding symbols being the same. For example, in the above exercise (iii), CUMBERLAND and TIMBERLAND, other than the first two symbols, the others are the same. As such, an alternative solution on HD can be stated in terms of its complement, namely, $\overline{\text{HD}}$, which in the present case is equal to 8.)

Example Ex-2.4

Statement of the Exercise

The HD measurements between locally aligned pair of nucleotide sequences (*s* and *t*) are as shown:

	I	II	III
s	GGU	AGCAA	AGCACACA
t	UGG	ACAUA	ACACACUA
HD(*s*, *t*)	2	3	6

Considering the III-group of sequences (i.e., s_{III}: AGCACACA and t_{III}: ACACACUA), align the associated eight nucleotides by introducing two gaps in each sequence randomly so that the HD score (or *cost*) is optimally improved. (The gap denotes a deletion in a sequence or an insertion in the sequence being compared.)

Solution

A random aligning of the eight nucleotides of III group by introducing two gaps in each sequence leads to:

A **G** C A C A C – A
A – C A C A C **U** A

Therefore, the new value of HD after the introduction of the gaps between s_{III} and t_{III} is 2, (G/– and –/U).

Problem P-2.9

Statement of the Problem

As in the previous example, the HD assessed between locally aligned pair of nucleotide sequences (x and y) are as shown:

	I	II	III
x	CCU	UCGUU	UCGUGUGU
y	UCC	UGUAU	UGUGUGAU
HD(s, t)	2	3	6

If the III group of the sequences x and y (i.e., x_{III}: UCGUGUGU and y_{III}: UGUGUGAU) are aligned by introducing two gaps in each sequence as needed so that the HD score is optimally improved. Determine this HD score (or cost) of the aligned subset III. (*Answer: New value of HD is 2.*)

Problem P-2.10

Statement of the Problem

Assuming 0-insertion and 2-deletion(s) performed on the sequence pairs: s: A G A C C A and t: C A C A C A, determine the resulting HD between s and t: (*Answer*: HD(s, t) = 4.)

Problem P-2.11

Statement of the Problem

Find the HD between two sequences (*s* and *t*) after applying two random insertions or deletions on the original pair. The two sequences (*s*, *t*) are *s*: T G C A C A C C and *t*: T C A C A C T C.
(*Answer*: HD(*s*, *t*) is 0.)

Example Ex-2.5

Problem Statement and Approach

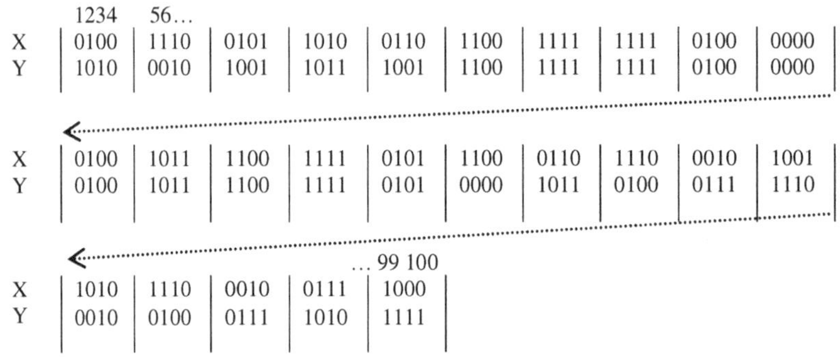

	1234	56…								
X	0100	1110	0101	1010	0110	1100	1111	1111	0100	0000
Y	1010	0010	1001	1011	1001	1100	1111	1111	0100	0000

X	0100	1011	1100	1111	0101	1100	0110	1110	0010	1001
Y	0100	1011	1100	1111	0101	0000	1011	0100	0111	1110

				…99	100
X	1010	1110	0010	0111	1000
Y	0010	0100	0111	1010	1111

Given a pair of sequences *X* and *Y* specified in their formats, calculate and plot the HD values between them as a function of binary digit locations along the stretch of 0 to (about) 100 binary residues listed below. Hence, indicate the most common substring locations between them.

The query pair of binary sequences (*X* and *Y*) contains 25 windows and each window has four residues as illustrated in Fig. 2.5. The Hamming distance, as specified before, refers to a number used to denote the difference between two binary strings; and, a logical connective depicts the "exclusive or" or exclusive disjunction. It yields true if exactly one (but not both) of two conditions is true. The XOR operation between *A* and *B* is denoted as (*A* XOR *B*) or (*A* direct sum *B*) or ($A \oplus B$).

In the above example, *X* and *Y* entities are represented as binary dibits of the symbols {A, C, T, and G}. Considering the entire binary strings, they are divided into short segments of equal-sized windows. Selecting a window of size four digits, the sequence *Y* is placed below *X*. For each window, the HD is assessed between *X* and *Y*. That is, across the binary

Solution

Window	W1	W2	W3	W4	W5	W6	W7	W8	W9	W11
X	0100	1110	0101	1010	0110	1100	1111	1111	0100	0000
Y	1010	0010	1001	1011	1001	1100	1111	1111	0100	0000
XOR score	0	2	0	1	3	0	0	0	0	0

Window	W12	W13	W14	W15	W16	W17	W18	W19	W20	W21
X	0100	1011	1100	1111	0101	1100	0110	1110	0010	1001
Y	0100	1011	1100	1111	0101	0000	1011	0100	0111	1110
XOR score	0	0	0	0	0	2	1	2	2	1

Window	W22	W23	W24	W25	W26
X	1010	1110	0010	0111	1000
Y	0010	0100	0111	1010	1111
XOR score	1	2	2	0	3

Fig. 2.5 Window-by-window representation of the two test sequences *X* and *Y* enabling the XOR operation between them leading to assess the HD value between them.

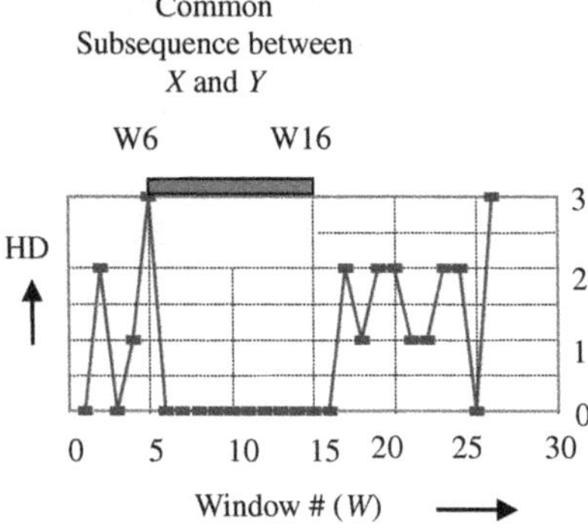

Fig. 2.6 Estimated window-by-window HD values between the two test sequences *X* and *Y* *versus* window subsequence number. The common subsequence between *X* and *Y* as illustrated is at: *W* #5 through *W* #16.

strings *X* and *Y*, the HD in each window is decided by XOR operation with reference to the two residues one below the other, and the number of 1's is counted in each window in the resulting XOR output string, depicting the HD value for that window. Then window # *versus* HD value is plotted as shown in Fig. 2.6. (Observed result: the common subsequence between *X* and *Y* as illustrated is at *W* #5 through *W* #16.)

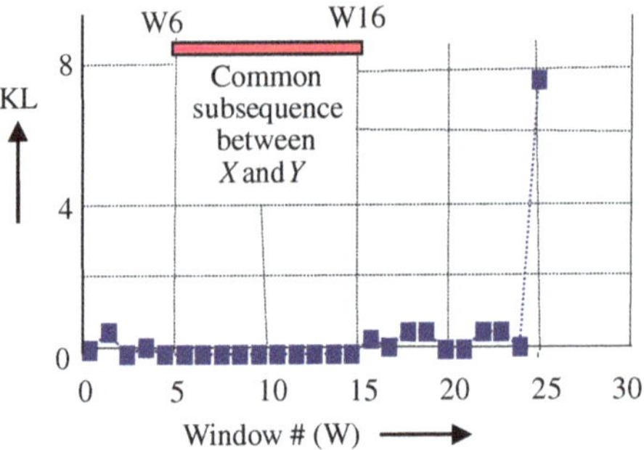

Fig. 2.7 Estimated KL values (in nats) between the two test sequences *X* and *Y versus* window subsequences: (The common subsequence between *X* and *Y* can be seen [as illustrated] at: *W* #5 through *W* #16, which is same as that decided *via* HD computations of Example Ex. 2.3.)

Problem P-2.7

Statement of the Problem

For the same two binary sequences *X* and *Y*, indicated above (in Example Ex. 2.5), plot the Kullback–Leibler (KL) measure between the two strings, window-by-window. Hence, confirm how the most common substring locations between test sequences are the same as decided *via* HD measure.

Example Ex-2.6

Statement of the Problem

Determine the Levenshtein distance (LD) between the following word sequence segments.

(a) BIOINFORMATICS and BIOINFORMATION.
(b) TELE-INFORMATICS and HAIL-FORMATION.
(c) TELEINFORMATICS and C-O-NFORMATION.

Solution

BIOINFORMATICS and BIOINFORMATION: There are two letter changes. So, LD = 2.

TELE-INFORMATICS and HAIL — FORMATION: There are two gap insertions and six letter changes. So, LD = 8.

TELEINFORMATICS and C-O — NFORMATION: There is one gap insertion and there are six letter changes. So, LD = 7.

Example Ex-2.7

Statement of the Problem

For the two sequence segments indicated below, determine the minimum number of edit operations required to transform one into the other.

X: A C C U G A
Y: A G C U – A

Solution
The minimum number of edit operations required to transform one sequence segment into the other corresponds to the LD between the segments. The steps involved are as follows.

Step (i): Substitution of C by G in *X*: → A **G** C U G A.
Step (ii): Indel operation (insertion of G) in *Y*: → A G C U **G** A.

Thus, there are (a minimum) of two edit operations required for the transformation. Therefore, the Levenshtein score is 2.

Example Ex-2.8

Statement of the Problem

Determine the edit distance in transforming one into the other of the following sequence segment pair:

X: T A C C A G T
Y: C C C G U A A

Solution
The edit distance is 5, with the following changes done.
In *X*, TACC → CCAT, next, in *X*, ATAGT → TGATA. Then in *Y*, the third C → T, in *X*, the first A → U, in *Y* the first A → T.

Problem P-2.13

Statement of the Problem

Find the edit distance between: A G – T C C and C G C T C A.
(Answer: Levenshtein distance (LD) = 3.)

Problem P-2.14

Statement of the Problem

Given that:

Position	1	2	3	4	5	6	7	8
X	A	G	C	A	A	C	C	A
Y	A	C	A	C	A	C	A	T

Reference to the pair of sequence segments as above, obtain solutions for the resulting LD values in the following two cases of arbitrary insertions and/or deletions (*indels*).

(a) Assume: 1-insertion and 2-deletions.
(b) Assume: 2-insertions and 1-deletion.

Problem P-2.15

Statement of the Problem

Consider the test two sequence segments, namely:

X: A G T G G G C A T T C C T T T
Y: T C T A G A A T. T T C T G T T

The following alignment is done with minimal editing by permitting identity matches and transitions [namely, purines (A ↔ G) or pyrimidines (C ↔ T)] to stay eventually. Determine the associated Levenshtein score.

Solution hint: The resulting sequences are:

*X**: A G T G G G C A T T C C – T T
*Y**: T C T A G A – – T T C T G T T

Is there a possibility of better scoring feasible in aligning *X* and *Y*? If so indicate it. *(Answer hint: There is no possibility for a better score.)*

Example Ex-2.9

Statement of the Problem

Perform a visual comparison of the following putatively related pair of nucleic acid sequence segments, *x* and *y*. Indicate the identities by 1's on match and 0's on mismatches. Also show mutations by the scoring indices (notations) S for transition (A ↔ G; C ↔ T) and V for transversion (A ↔ C; G ↔ T):

```
x:  T C {G} [C T] [G] G {C} G [C] {A} [A] {A C C G} –
      1 0  0 0  0 1 0 1 0  0  0  0 0 0 0
Y: – C {C} [T C] [A] G {G} G [T] {T} [G] {C A A C} A
```

Solution

Bases indicated as square-bracketed entities in x and y are transitions, and the transversions across the residues are denoted within curly brackets.

Example Ex-2.10

Statement of the Problem

As in previous example, perform a visual comparison of the following putatively related pair of nucleic acid sequences α and β. Indicate in your answer the identities by 1s on match and 0s on mismatches. Also indicate mutations by the scoring indices (notations) S for transition and V for transversion:

```
α:  A A {G C A G T C} T [C A A] T [A C] G C
    – 1  0 0 0 0 0 0  1  0 0 0  1  0 0  1 1 –
β     A {C A C T G A} T [T G G] T [G T] G C C
          V V V V V V     S S S       S S
```

Solution

Note: Transition (A ↔ G; C ↔ T) and transversion (A ↔ C; G ↔ T).

The set of bases shown in square brackets in sequences α and β corresponds to transitions (S) and those in curly brackets depict transversions (V).

Project-Level Examples and Exercises

The following examples and exercises are topics compatible as term papers and for research projects. The scope of these can be expanded to a desired level of exercise with the annotated details in the literature.

Example Ex-2.11

Statement of the Problem

This example refers to verifying the symmetry/asymmetry aspects of the forward DNA strand (5′–3′) and its reverse complement,

(3′–5′). The *complementary DNA* strand refers to DNA strand that has the opposite nucleotide sequence and chemical polarity of a DNA strand. According to the so-called *Chargaff's first parity rule*,[92] the number of A's in a single DNA strand is equal to the number of Ts and number of C's is equal to number of G's. Similarly, it can be stated that for a complementary DNA, the same base pattern is followed in opposite/reverse direction, as such following the *Chargaff's second parity rule*, the number of AG or CT will be same in both DNA and complementary DNAs. This symmetry feature between the DNA and its complementary version is of interest in bioinformatic studies.[96] The aforesaid symmetry/asymmetry property can be assessed by a *skew parameter*. A skew parameter is a metric, which is dependent on the direction and replication or gene orientation and it is independent of the count of AT. Significant work has been done on the first-order symmetry of DNA and complementary DNA sequences,[96,97] but there has not been much studies done concerning higher order symmetry.

The skew characteristics between a DNA strand and its complementary version can be determined *via* statistical models, and relevant measures can be specified thereof to verify the symmetry considerations under discussion. To compute the symmetry/asymmetry profile between the complementary DNA strands, the concept of "contrast function" is introduced in Baisnée and Hampson.[96] This contrast function is a compensative function that would react to extreme maximum or minimum values when the complementary DNA strands are compared. The terms "contrast," "comparativeness," or dissimilarity between the strands can be synonymously viewed as the "distortion" or "skewness" that depicts the statistical difference between the sequences and, therefore, can be specified in terms of the known stochastic attributes of the sequence characteristics. Suppose a sequence $\{S^+\}$ and its reverse complement $\{S^-\}$ are statistically large ($N \to \infty$), then the element features of $\{S^+\}$ and $\{S^-\}$ should bear uncertainties by virtue of their probabilities as $[q_1; (1 - q_1)] \in \{S^+\}$ and $[q_2; (1 - q_2)] \in \{S\ \}$ with q_1 and q_2 representing the probabilities of occurrence of 1's, and $(1 - q_1)$ and $(1 - q_2)$ are probabilities of occurrences of 0's.

Illustrative analysis: This illustrative case study refers to simulation experiments performed on a test DNA and its complementary reverse DNA sequences using the algorithmic concepts discussed above. The statistical data on nucleotide bases of *S. cerevisiae* available in GenBank.[98]

Table 2.3 Typical percentages of occurrence of A, C, T, and G constituents in mitochondrial chromosomes of *S. cerevisiae*.

Nucleotide	Percentage of occurrence
A	30%
C	40%
T	30%
G	40%

The probabilities (percentage) of occurrence of relevant nucleotide bases shown in Table 2.3 correspond to the test species *S. cerevisiae* (in terms of percentage). Using the percentage of occurrence data of the set {A, C, T, G} as above, first the DNA sequence of the test species and its reverse complement are constructed as explained in Pandya.[72]

Using the binary DNA strand and its complement simulated, they can be contrasted *via* one of the distance or divergence measures. For example, using the KL metric, the deduced KL-measure statistics of a test species, namely, mitochondrial chromosomes (*S. cerevisiae*) are illustrated in Fig. 2.8. The KL measure is computed in terms of A, C, T, and G occurrence probabilities evaluated window-by-window across the test DNA strands (forward and reverse). That is, considering the simulated test sequences of forward and reverse strands (denoted as FS and RS, respectively), they are divided into a convenient number of windows (W) and in each sub-segment of residues represented by the window, the KL-measure in nats is determined as follows.

The histograms of Fig. 2.8 provide a visual comparison of the symmetry/asymmetry profiles of the forward and reverse strands. However, a more accurate statistical comparison of the profiles can be done in terms of mean, variance, and standard deviation details ascertained from the data of the histograms. Hence, the following details are ascertained.

- For KL1, the hypothesis that forward strand DNA and reverse strand C-DNA are symmetric is acceptable to an extent of 99.9% of confidence-level defined *via* *t*-values* (t – 0.1%) for one-tailed sequences. (The *t*-value* measures the size of the difference relative to the variation in your sample data. That is, *t*-value refers to the calculated difference represented in units of standard error. The greater the magnitude of *t*-value, the greater the evidence against the null hypothesis.)

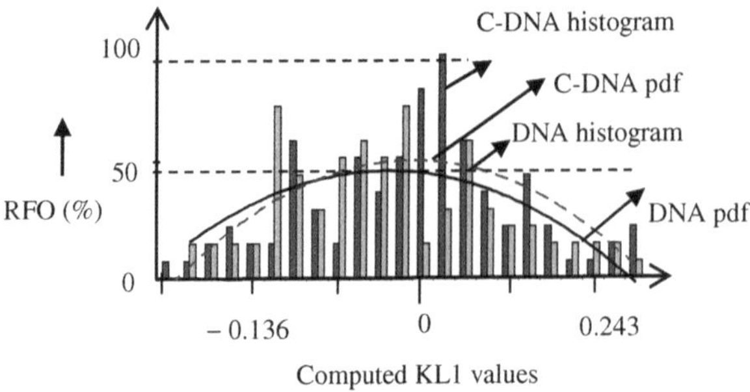

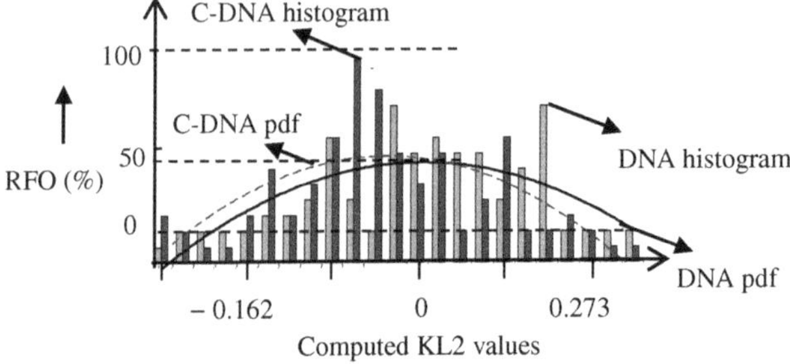

Fig. 2.8 Histograms of computed KL1 and KL2 values of the forward (DNA) and reverse (C-DNA) strands of the test species *Saccharomyces cerevisiae* evaluated window-by-window across (simulated) forward and reverse test strings, respectively. The sets $\{p_A, p_C, p_T, p_G\}_{FS}$ and $\{p_A, p_C, p_T, p_G\}_{RS}$ depicting occurrence probabilities of the residues {A, C, T, G} are used in the computations indicated (RFO: Relative frequency of occurrence).

- KL2 results exhibit a symmetry to an extent of 99% between forward-strand DNA and reverse-strand C-DNA.

Problem P-2.16

Statement of the Problem

Project exercise: Pursuing the illustrative case study relevant to statistical data on nucleotide base composition of *Saccharomyces cerevisiae* as above, repeat the study *via* pertinent simulation experiments performed on a test DNA and its complementary reverse DNA sequences of human genome. Further, instead of KL measures use another distance or divergence measure with appropriate algorithmic concepts. Hence, justify the hypothesis of the symmetry between forward-strand DNA and reverse-strand C-DNA of human genome.

Problem P-2.17

Statement of the Problem

Project exercise: In bioinformatic literature, binary representation of DNA has been sparsely studied. An example has been suggested for protein classification in Nemzer.[73] As mentioned earlier, the present study is done to elucidate methods that use exclusively the binary representation of a DNA sequence for the purpose of delineating the boundary between codon and non-codon sections within the sequence, and the binary depiction of the DNA chain is also adopted to compare the chain for similarity against a complementary chain. Relevant heuristics and test studies are as follows.

Delineation or locating transitions between coding and non-coding regions in a heterogeneous DNA sequence involves defining a contrast function that estimates the difference in composition in question between the two regions. Such codon–non-codon demarcation can be ascertained from the coding statistics associated with a DNA.[99] This coding statistics is specified by probabilities occurrence of nucleotide triplets in codon and non-codon segments of a genomic DNA (considered in a large scale of sequencing bases in the DNA of a given species). Relevant models developed to distinguish coding parts in a DNA sequence are essentially based on stochastic considerations of coding statistics as indicated earlier in the pertinent tutorial.

Notwithstanding the existence of entropy-segmentation method,[99] yet another approach is to establish the delineation in question by representing the random structure of DNA-base composition as a binary string indicated earlier. The scoring strategy with binary strings proposed here corresponds to sequences compared by subjecting them to a modulo-2 (XOR) operation and the dissimilarity between the compared sections is assessed in terms of Hamming distance or the count of 1's in the resulting string of XOR-operation. Simulated results on HD scores are obtained across differential blocks of a DNA sequence (constructed in binary form). The test sequence used in the simulations corresponds to a given species (human, bacteria, etc.) having distinct domains of codons and non-codons with coding statistics specified by random base constituent distribution for the species in question and uniform distribution of the bases, respectively.

Project exercise: With reference to discussions and heuristics as above, obtain the locations of delineation between randomly occurring exon and intron segments disposed across the DNA of *D. melanogaster.* The codon occurrence statistics of this species is available at www.kazusa.or.jp[123] detailing the occurrence frequency of codon triplets of *D. melanogaster,* that is, probability of occurrence (*P*) of the triplet codon statistics of *D. melanogaster* [gbinv]: 42417 CDS's (21945319 codons); Coding GC 53.86%, 1st letter GC 55.79%, 2nd letter GC 41.48%, 3rd letter GC 64.32%.

Solution hints: The distinguishing feature (of probabilities of occurrence) leads to identifying codon and noncodon regions in a DNA sequence, and the entropic segmentation method due to Galvan *et al.*[99] in essence prescribes a strategy of delineating the underlying codon and non-codon regions based on the aforesaid statistical distinction expressed in terms of associated entropy considerations. As such, the entropy segmentation efforts as mentioned before essentially conform to divergence measures, such as the JS measure. Shown in Fig. 2.9 are results obtained with ensemble of runs for human DNA sequence of 6,400 triplets long. It depicts the plot of relative Hamming distance (HD) *versus* the length of the test binary sequence. The required delineation or boundary of separation between codon and codon domains is distinctly seen in Fig. 2.9.

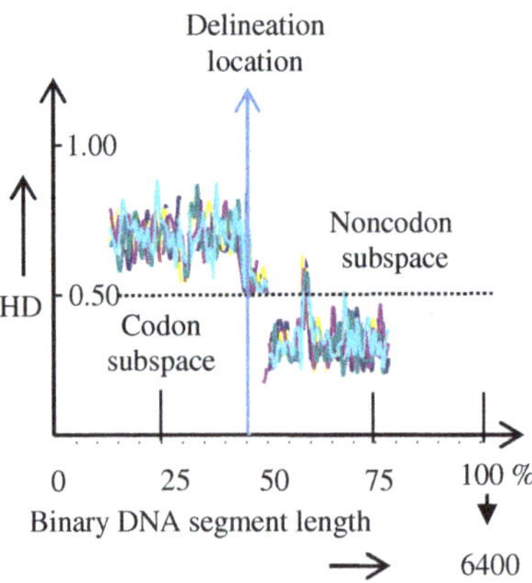

Fig. 2.9 Computed (ensemble) results on relative Hamming distance (HD) along the test binary sequence of 6,400 residues (scaled to 100%) containing codon and non-codon subspaces.

Problem P-2.18

Statement of the Problem

Project exercise: This exercise refers to using various information-theoretic measures for entropic segmentation toward finding the borders between coding and non-coding DNA regions. Relevant measures correspond to various Csiszár *f*-divergences family of metrics discussed earlier. The logic behind using different metrics (from the Csiszár *f*-divergences family) toward bioinformatic sequence comparison analysis is that such alternatives would offer: (i) flexibility of selecting an appropriate measure; (ii) options toward selecting robust evaluation methods from the many; and (iii) alternative formulations for fast algorithmic searches. Overall, these new measures provide valid alternatives and options to the JS measure previously available.

The entropic segmentation method has been shown to be capable of performing this task. This algorithm was introduced by Galván *et al.*[99] using the Jensen–Shannon (JS) contrast measure for the detection of coding *versus* noncoding regions in DNA. Necessary codon-usage frequency data can be availed from www.kazusa.or.jp.[100]

Example Ex-2.12

Statement of the Problem

This exercise refers to the use of Fourier power-spectrum method in comparing biosequences. Relevant approach can be understood with the following example: considering two biosequences, such as the forward-strand DNA and the reverse-strand complement C-DNA sequences, suppose the constituents of each sequence occur with the probabilities p_i and q_i, respectively, where $i = 1, 2, 3, \ldots, 64$ depicting the 64 trinucleotide constituents (codons) formed by four nucleotide bases, namely A, C, T, and G. The sets $\{p_i\}$ and $\{q_i\}$ for a given biological species (such as the human) conform to GenBank data available in Human codon frequency usage[100] in terms of probabilities of occurrence of $\{i\} = \{p_i\}$ trinucleotides: $\{i = 1, 2, \ldots, 64\}$ and listed in Chapter 1 (Table 1A.7) depicting the set of *codon usage frequency* (CUF) of *Homo sapiens*. Note that $\sum_{64} p_i = 1$ of Table 1A.7 (Chapter 1) guarantees *normality condition* on the probabilities involved. Likewise, for the corresponding C-DNA sequence, $\sum_{64} q_i = 1$. In terms of the occurrence statistics of the trinucleotides in the test sequences specified by the sets $\{p_i\}$ and $\{q_i\}$, the sequences can be cross-compared using a correlation measure *via* the associated periodogram (Fourier transform) of the randomly occurring suites of the codons across the test sequences. As is well known, if the randomness related with a computed entity is a white, Gaussian noise, then the power-spectrum density would exhibit the white or "flat" spectral pattern. The computed result pertinent to human DNA/C-DNA is indicated in Fig. 2.10.

Relevantly, the test sequences namely, the DNA and the C-DNA of the human species, the computed result in Fig. 2.10, illustrates the white noise attributes. It means that the DNA and C-DNA are very much similar to each other so that their difference is only noise conforming to flat spectrum characteristics.[106] Thus, the spectral domain analysis of random sequences is a viable method that can be pursued to study and compare biological sequences and genomic signal-processing. (In the above example, a simple rectangular window is used, however, without any loss of generality, the other types of windows can also be used.) A more rigorous example of using power-spectral method is indicated in Chapter 8 in the context of comparing the biosequences of viral strains.

Problem P-2.19

Statement of the Problem

This is a project exercise on performing a spectral domain analysis to compare two random biological sequences, namely, the forward strand and reverse complement of the DNA genomics pertinent to *Escherichia coli* (abbreviated as *E. coli*), which are large and diverse group of bacteria. The statistical framework of *E. coli* DNA is available in GenBank. Analyze the symmetric features of the test sequences via periodogram method. Use a simple rectangular window or other types of windows. (Similar exercise can also be pursued to study the genomic structures of the species, such as (i) *Methanococcus jannaschii*, (ii) *Rickettsia prowazekii*, and (iii) *Saccharomyces cerevisiae.*)

Example Ex-2.13

Statement of the Problem

This is a project exercise on estimating the information content of a DNA sequence leading to find homologs in a DNA bank.[103]

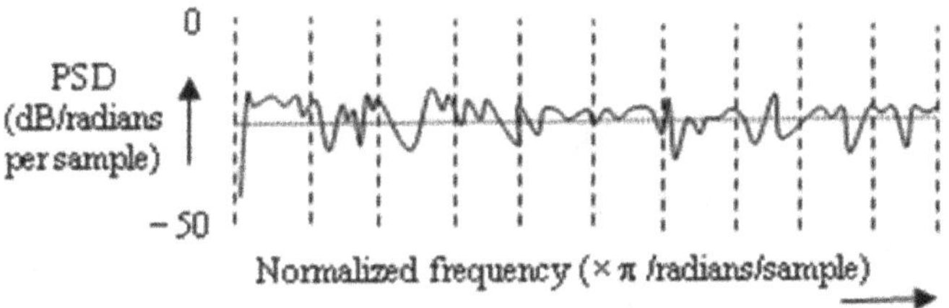

Fig. 2.10 Flat power spectral density (PSD) characteristics of the difference entity of the test sequences determined from the DNA and C-DNA statistical patterns.

Solution-Hint

Two moderately long homologous DNA sequences are presumably descendents from a common ancestor (see Chapters 9 and 10). Suppose for a given ancestral sequences, A and B are two homologs observed each having distinct noisy features due to the associated bases either mutated and/or lost.

Suppose A and B have constituting entities $\{a_1, a_2, \ldots, a_i \ldots, a_k\}$ and $\{b_1, b_2, \ldots, b_j \ldots, b_\ell\}$, respectively, and the random occurrence features of these constituent elements lead to defining *a priori* probabilities of A and B, respectively, as $p_i = \text{Prob}_A(a)_i$ and $q_j = \text{Prob}_B(b)_j$. Relevantly, $H(A)$ and $H(B)$ denote the uncertainty features of the sets, $\{a_1, a_2, \ldots, a_i \ldots, a_p\}$ and $\{b_1, b_2, \ldots, b_j \ldots, b_q\}$, which depict, respectively, the associated information features (known *a priori*); they can be specified by negentropy functions (in Shannon's sense).

Now, an information comparison exercise suite can be pursued. It refers to comparing A and B so as to know how much information A gives about B (or *vice versa*) on average. Relevantly, given the knowledge (or prior information) of A expressed in terms of the negentropy function $H(A)$, the mutual information that one can gain about B with the available statistics across A and B can be specified by the conditional probability distribution, $\text{Prob}[B: b|A = a_i]$.

With reference to A and B, the associated stochastic attributes can be modeled as symmetrically distributed Gaussian variables expressed as $(U + m)$ and $(U + n)$, where U depicts a deterministic component common to both A and B; correspondingly, a mean value of μ can be presumed to reflect this common (central) tendency in the genetic statistics of A and B stemming from their ancestral homologous characteristics. Hence, the average (expected) value of mutual information content specified by cross-entropy attributes of A about B is equal to $I(B/A) = H(B) - H(B/A)$. Similarly, $I(A/B)$ can also be specified.

Example Ex-2.14

Statement of the Problem

This is indicated as a project exercise on estimating the Shannon information content of a DNA sequence (with possible random mutational attributes of its residues) *via* Fourier coefficients, when the sequence is transformed and represented in the spatial Fourier domain. Random changes are statistically imposed on the evolution of the DNA structure of base composition as a result of mutations, etc. For example, considering a DNA marked as A, the underlying statistics of variations can be denoted by a RV, m superimposed on a (unperturbed) deterministic profile, U. The underlying stochastic can be presumed as of Gaussian probability distribution with relevant central-limit considerations. This implies that a mean tendency of variations toward a value μ exists and the corresponding variance of the RV is $(\sigma_m)^2$. Similarly, another DNA marked as B, which is homologous to A, can be considered with random changes denoted by an RV, n with Gaussian parameters, μ and $(\sigma_n)^2$.

Solution-Hint

As per the given description, the random processes involved are $(U+m)$, $(U+n)$, m and n. They all conform to independent random variates. Suppose the parametric values of 1D strings representing the sequences of A and B are subjected to Fourier transformation, the logistic behind such a transformation is as follows: it is expressed in Leitão *et al.*[103] that an obstacle may be faced in extracting useful information content in a genomic sequence represented by its parametric residues. Considering still sparsely known dependence between nearby bases and their occurrence statistics across the genomic sequence (in Markov's sense), it is indicated[103] that extraction of useful information in a genomic sequence may not be fully done with the basic statistics of the constituents. As such, it is argued that the Fourier transform may be adopted considering the fact that real and imaginary parts of Fourier coefficients are all independent random variables, and as such, they may yield two distinct sets of fortifying details on the associated statistics with augmented information. Simulation experiments using some DNA strings extracted from the GenBank database and are shown to confirm the said assertion of Leitão *et al.*[103] and Benson.[104]

Relevant to the DNA sequences A and B with stochastic variants, $(U+m)$, $(U+n)$, m and n, suppose the corresponding (rth) power spectral components deduced are $\{|U(\omega)|^2 + |M(\omega)|^2\}_r$, $\{|U(\omega)|^2 + |N(\omega)|^2\}_r$, $\{|M(\omega)|^2\}_r$ and $\{|N(\omega)|^2\}_r$. Then, the information given by each coefficient associated with A about the corresponding coefficient of B (or *vice versa*) is given by the compatible expressions $I(B_k/A)$ and $I(A_k/B)$ as in Leitão *et al.*,[103] Benson,[104] Tiwari *et al.*,[105] and Lathi.[106]

Problem P-2.20

Statement of the Problem

This project exercise follows the details in Leitão *et al.*[103] and is indicated on the following heuristics: Suppose the (rth) power-spectral Fourier component relevant to an original, unperturbed DNA sequence X of an ancestral genome is $\{|U(\omega)|^2$. Assuming that a homolog of this genome Y has a statistically perturbed stochastic variant, $(U + m)$ with the corresponding (rth) power-spectral components deduced as $\{|U(\omega)|^2 + |M(\omega)|^2\}_r$ (where $\{|M(\omega)|^2\}_r$ corresponds to the variance of the perturbations seen in Y) and a space D is specified to depict the difference between the stochastic frameworks of the ancestral genome (X) and the homolog, Y. Derive an expression for the information content of Y *versus* X expressed in terms of $\{|M(\omega)|^2\}_r$ and $\{|\Delta(\omega)|^2\}_r$, where $\{|\Delta(\omega)|^2\}_r$ is the variance of $D = (Y - X)$.

Problem P-2.21

Problem Statement

Project exercise: Deduce the relative entropy (mutual information) profiles along the stretches of DNA sequence using the statistics of codon triplet usages in: (i) *Methanococcus jannaschii*, (ii) *Rickettsia prowazekii*, (iii) *Saccharomyces cerevisiae*, (iv) *escherichia coli*, and in (v) *human*. Use in each case, a hypothetical junk codon sequence constructed with equiprobable, random occurrences of the 64 triplets as indicated earlier.

Problem P-2.22

Problem Statement

Project exercise: Expanding the scope of the problem exercises indicated above, a thesis/dissertation problem can be defined with the following objectives: (a) using various distance/discrimination measures in lieu of Kullback–Leibler measure; (b) finding relative efficacy of such measures in portraying the correlation trends of the constituent residues; and (c) verifying the existence of fuzzy attributes. Again use of a hypothetical junk-codon sequence constructed with equi-probable, random occurrences of the 64 triplets can be pursued.

2.9 Closure

This chapter provides a comprehensive foundation on salient features of entropic-theoretics and information theory enabling their extension toward knowing the intricacies of bioinformation. The crucial aspects of biological complex rely on various information-theoretics at gene level (manifesting as genomic features and functional plus conformal aspects of proteomics). Further, informatics of cytogenetic details (of cellular functions and structures) and phylogenetic characterizations (showing up as phenotypical features and/or traits of an organism along its evolutionary scale) viably dictate the vast scope of bioinformatics.

Relevantly, this chapter is an embodiment of the theme emphasizing omnipresence of *Information in Nature* ... as an endless combinations and repetitions of a very few *Laws of the Universe* ... dictating *Dogmas on Coexistent Entities* ... *Energy and Entropy* (or *Information*?) ... across the diverse span of living systems.

> *" ... far the Energy (represented by the primeval sound of Divine Illumination (or* 'Om'*) constitutes and pervades the entire Universe; it is Omnipresent — across everything, in everyone, and everywhere* ('Bhu Loka' *— the physical plane,* 'Bhuvar Loka' *— the intermediate space and* 'Swar Loka' *— the divine plane). By rightly channeling or through awakening utterance ('prachayodat'), all this Energy (the 'Savitur' or Source of all Life manifesting as Divine Illumination) would imply gaining Negentropy (... the pure form of enlightened Intelligence or* 'devasya dhlmahi'*)"... ... (that is, 'negentropy' or sustenance of purity implicitly amounts to destruction of impure or unwanted 'posentropy') ...*

Om Bhuur-Bhuvah Svah
Tat-Savitur-Varennyam!
Bhargo Devasya Dhiimahi
Dhiyo Yo Nah Pracodayaat!!
... Gayatri Mantra
(Rig Veda RV 3.62.10 in Sanskrit)

---o---

References

[1] P. Godfrey-Smith and K. Sterelny: Biological information. *Stanford Encyclopedia of Philosophy*, 2007. Available online at: *http://plato.stanford.edu/entries/information-biological/.*

[2] P. Godfrey-Smith: *Philosophy of Biology*. Princeton University Press, Princeton, NJ, USA: 2014.

[3] C. E. Shannon: A mathematical theory of communication. *The Bell System Technical Journal*, 1948, vol. 27, 379–423 and 623–656.

[4] T. M. Cover and J. A. Thomas: *Elements of Information Theory*. John Wiley and Sons, Hoboken, NJ, USA: 2006.

[5] J. R. Pierce: *An Introduction to Information Theory — Symbols, Signals and Noise*. Dover Publications, Mineola, NY, USA: 1961.

[6] P. S. Neelakanta: *Information-theoretic Aspects of Neural Networks*. CRC Press, Boca Raton, FL, USA: 1999.

[7] J. W. Gibbs: *Elementary Principles in Statistical Mechanics*. Longmans Green and Co., New York, NY, USA: 1928.

[8] P. S. Neelakanta and D. De Groff: *Neural Network Modeling: Statistical Mechanics and Cybernetic Perspectives*. CRC Press, Inc., Boca Raton, FL, USA: 1994.

[9] M. Tribus and E. C. McIrvine: Energy and information. *Scientific American*, 1971, vol. 224, 178–184.

[10] L. Szilárd: Über die entropieverminderung in einem thermodynamischen System bei eingriffen intelligenter wesen. *Zeitschrift für Physik*, 1929, vol. 53, 840–856.

[11] J. C. Maxwell: *Theory of Heat*. 1871 (Reprinted), Dover Publications, New York, NY, USA: 2001.

[12] C. H. Bennett: Demons, engines and the second law. *Scientific American*, 1987, vol. 257(5), 108–116.

[13] J. von Neumann and R. T. Beyer (Translator): *Mathematical Foundations of Quantum Mechanics*. Princeton University Press, Princeton, NJ, USA: 1955.

[14] P. S. Neelakanta: *A Textbook on ATM Telecommunication Engineering*. CRC Press, Inc., Boca Raton, FL, USA: 2000.

[15] P. S. Neelakanta and D. Baeza: *Next Generation Telecommunications and Internet Engineering*. Linus Publications Inc., Dear Park, NY, USA: 2009.

[16] R. V. L. Hartley: Transmission of information. *Bell System Technical Journal*, 1928, vol. 7(3), 535–563.

[17] C. E. Shannon: Communication in the presence of noise. *Proceedings of the Institute of Radio Engineers*, 1949, vol. 37(1), 10–21.

[18] D. Middleton: *An Introduction to Statistical Communication Theory*. McGraw-Hill Book Co., New York, NY, USA: 1960.

[19] C. E. Shannon and W. Weaver: *The Mathematical Theory of Communication*. University of Illinois Press, Chicago, IL, USA: 1949.
[20] L. Brillouin: *Science and Information Theory*. Dover Publications, Inc., Mineola, NY, USA: 2004.
[21] L. Brillouin: Life, thermodynamics, and cybernetics. *American Scientist*, 1949, vol. 37(4), 554–568.
[22] E. T. Jaynes: Information theory and statistical mechanics. *Physical Review*, 1957, vol. 106(4), 620–630.
[23] G. N. Alekseev: *Energy and Entropy*. Mir Publishers, Moscow, USSR: 1986.
[24] C. H. Bennett and R. Landauer: The fundamental physical limits of computations. *Scientific American*, 1985, vol. 253(1), 48–56.
[25] P. S. Neelakanta, T. V. Arredondo and D. De Groff: Redundancy attributes of a complex system: Application to bioinformatics. *Complex Systems*, 2003, vol. 14, 215–233.
[26] R. P. Feynman and A. Hey: *Feynman Lectures on Computation*. Addison-Wesley Longman Publishing Co., Inc., Boston, MA, USA: 1998.
[27] C. K. Chow and C. N. Liu: Approximating discrete probability distributions with dependence trees. *IEEE Transactions on Information Theory*, 1968, vol. 14(3), 462–467.
[28] A. A. Markov: Rasprostranenie zakona bol'shih chisel na velichiny, zavisyaschie drug ot druga. *Izvestiya Fiziko-matematicheskogo obschestva pri Kazanskom universitete*, 1906, 2-ya seriya, tom 15, 135–156.
[29] W. J. Stewart: *Introduction to the Numerical Solution of Markov Chains*. Princeton University Press, Princeton, NJ, USA: 1994.
[30] J. R. Norris: *Markov Chains*. Cambridge University Press, Boston, MA, USA: 1998.
[31] R. A. Fisher: On the mathematical foundations of theoretical statistics. *Philosophical Transactions of the Royal Society A*, 1922, vol. 222, 309–368.
[32] S. Amari and H. Nagaoka: *Methods of Information Geometry*, Oxford University Press, UK: 2000.
[33] J. N. Kapur and H. K. Kesavan: *Entropy Optimization Principles with Applications*. Academic Press Inc., San Diego, CA, USA: 1992.
[34] I. Csiszár: Eine informationstheoretische ungleichung und ihre anwendung auf den beweis der ergodizitat von markoffschen ketten. *Magyar Tudományos Akadémia Matematikai Kutató Intézetének Közleményei*, 1963, vol. 8, 85–108.
[35] T. Morimoto: Markov processes and the H-theorem. *Journal of Physical Society of Japan*, 1963, vol. 18(3), 328–331.
[36] S. M. Ali and S. D. Silvey: A general class of coefficients of divergence of one distribution from another. *Journal of the Royal Statistical Society, Series B*, 1966, vol. 28(1), 131–142.

[37] S. Kullback and A. S. Leibler: On information and sufficiency. *Annals of Mathematical Statistics*, 1951, vol. 22(1), 79–86.

[38] W. Rudin: *Principles of Mathematical Analysis.* McGraw-Hill Publications, New York, NY, USA: 1976.

[39] K. Chatzikokolakis and K. Martin: A monotonicity principle for information theory. *Electronic Notes in Theoretical Computer Science*, 2008, vol. 218(22), 111–129.

[40] F. Nielsen and R. Nock: On the chisquare and higher-order chi distances for approximating *f-divergences. IEEE Signal Processing Letters*, 2013, vol. 21(1), 10–13.

[41] P. Kumar and S. Chhina: A symmetric information divergence measure of the Csiszar's *f*-divergences class and its bounds. *Computer and Mathematics with Applications*, 2005, vol. 49(4), 575–588.

[42] I. Csiszár: Information-type measures of difference of probability distributions and indirect observations. *Studia Scientiarum Mathematicarum Hungarica*, 1967, vol. 2, 299–318.

[43] I. Csiszár: Information measures: A critical survey. *Transactions of the Seventh Prague Conference on Information Theory, Statistical Decision Functions, Random Processes and of the 1974 European Meeting of Statisticians* (Prague, Czechoslovakia, August 18–23, 1974), Academia, Berkeley, CA, USA, 1974, pp. 73–86.

[44] R. A. Fisher: Theory of statistical estimation. *Proceedings of Cambridge Philosophical Society*, 1925, vol. 22, 700–725.

[45] A. Rényi: On measures of entropy and information. *Proceedings of the Fourth Berkeley Symposium on Mathematical Statistics and Probability*, University of California Press, Boston, MA, USA: 1961, vol. 1, pp. 547–561.

[46] F. Österreicher: Csiszár's *f*-divergences — Basic properties, *Research Report Collection*, 2002, Available online at: https://www.uni-salzburg.at/fileadmin/oracle_file_imports/246178.PDF

[47] K. Ferentimos and T. Papaiopannou: New parametric measures of information. *Information and Control*, 1981, vol. 51, 193–208.

[48] I. Vajda: On *f*-divergences and singularity of probability measures. *Periodica Mathematica Hungarica*, 1972, vol. 2, 223–234.

[49] P. K. Bhatia and S. Singh: On a new Csiszar's *f*-divergences measure. *Cybernetics and Information Technologies*, 2013, vol. 13(2), 43–57.

[50] J. Lin: Divergence measures based on the Shannon entropy. *IEEE Transactions on Information Theory*, 1991, vol. 37(1), 145–151.

[51] S. Kullback: *Information Theory and Statistics.* Dover Publications, New York, NY, USA: 1968.

[52] U. Kumar, V. Kumar and J. N. Kapur: Some normalized measures of directed divergence. *International Journal of General Systems*, 1986, vol. 13, 5–16.

[53] R. E. Blahut: *Principle and Practice of Information Theory.* Addison Wesley Publication Co., Reading, MA, USA: 1987.

[54] J. Aczél: Measuring information beyond communication theory. *Information Management Process,* 1984, vol. 20, 383–395.

[55] J. Havrda and F. Charvát: Quantification method of classification processes. Concept of structural α-entropy. *Kybernetika,* 1967, vol. 3(1), 30–35.

[56] A. Rényi: On measures of entropy and information. *Proceedings of the Fourth Berkley Symposium of Mathematical Statistics and Probability* (University of California June 20–July 30, 1960, Berkley, CA, USA: 1960), 1961, vol. 1, pp. 547–561.

[57] B. D. Sharma and D. P. Mittal: New and non-additive measures of entropy for discrete probability distributions. *Journal of Mathematical Sciences,* 1975, vol. 10, 28–40.

[58] J. N. Kapur: Generalized entropy of the order α and β. *The Mathematics Seminars, CMU,* 1967, vol. 4, 78–94.

[59] F. Liese and I. Vajda: On divergences and informations in statistics and information theory. *IEEE Transactions on Information Theory,* 2006, vol. 52(10), 4394–4412.

[60] A. R. F. Carvalho, J. Manuel, R. S. Tavares and J. C. Principe: A novel parametric distance estimator for densities and error bounds. *Entropy,* 2013, vol. 15, 1609–1623.

[61] A. Bhattacharyya: On a measure of divergence between two statistical populations defined by their probability distributions. *Bulletin of the Calcutta Mathematical Society,* 1943, vol. 35, 99–109.

[62] M. S. Nikulin: Hellinger distance. In: M. Hazewinkel: *Encyclopedia of Mathematics,* Springer-Verlag GmbH, Heidelberg, The Netherland: 2001.

[63] E. Hellinger: Neue Begründung der theorie quadratischer formen von unendlichvielen veränderlichen. *Journal für die reine und angewandte Mathematik,* 1909, vol. 136, 210–271.

[64] S. Chatterjee: Distances between probability measures. Available online at: *https://web.archive.org/web/20080708205758/http://www.stat.berkeley.edu/~sourav/Lecture2.pdf.*

[65] P. C. Mahalanobis: On the generalised distance in statistics. *Proceedings of the National Institute of Sciences of India,* 1936, vol. 2(1), 49–55.

[66] E. Levina and P. Bickel: The Earth Mover's distance is the Mallows distance: Some insights from statistics. *Proceedings Eighth IEEE International Conference on Computer Vision (ICCV 2001), (7–14 July 2001,* Vancouver, Canada), Vol. 2, pp. 251–256.

[67] V.M. Zolotarev: Lévy metric. In: M. Hazewinkel, *Encyclopedia of Mathematics,* Springer-Verlag GmbH, Heidelberg, Netherlands: 2001.

[68] G. J. Székely and M. L. Rizzo: A new test for multivariate normality. *Journal of Multivariate Analysis,* 2005, vol. 93(1), 58–80.

[69] G. J. Szekely and M. L. Rizzo: Energy statistics: statistics based on distances. *Journal of Statistical Planning and Inference,* 2013, vol. 143(8), 1249–1272.

[70] P. S. Neelakanta, S. Pandya T. V. Arredondo and D. De Groff: Heuristics of AI-based search engines for massive bioinformatic data-mining: An example of codon/non-codon delineation in a binary DNA sequence. *Presented in: 1st Indian International Conference on Artificial Intelligence* (IICAI-03). Hyderabad, India: December 18–20, 2003.

[71] P. S. Neelakanta, S. Pandya and T. V. Arredondo: Binary representation of DNA sequences towards developing useful algorithms in bioinformatics. *Proceedings of the 7th World Multi Conference on Systemics, Cybernetics and Informatics* (SCI 2003), Orlando, FL, USA: July 27–30, 2003, vol. VIII, pp. 195–197.

[72] S. B. Pandya: Binary Representation of DNA Sequences Towards Developing Useful Algorithms in Bioinformatics. *MSEE Thesis,* Florida Atlantic University, Boca Raton, FL, USA: 2003.

[73] L. R. Nemzer: A binary representation of the genetic code. *Biosystems,* 2017, vol. 155, 10–19.

[74] H. Cramér: On the composition of elementary errors. *Skandinavisk Aktuarietidskrift,* 1928, vol. 11, 141–180.

[75] R. A. Fisher: The use of multiple measurements in taxonomic problems. *Annals of Eugenics,* 1936, vol. 7, 179–188.

[76] R. A. Fisher: *Contributions to Mathematical Statistics.* John Wiley & Sons, New York, NY, USA: 1950.

[77] A. Cedilnik and K. Košmelj: Relations among Fisher, Shannon-Wiener and Kullback measures of information for continuous variables. In: A. Mrvar and A. Ferligoj: *Developments in Statistics, Metodološki zvezki,* vol. 17. FDV, Ljubljana, Slovenia: 2002, pp. 55–62.

[78] A. I. Khinchin: *Mathematical Foundations of Information Theory.* Dover Publications, Inc., Mineola, NY, USA: 1957.

[79] A. I. Khinchin: The entropy concept of probability theory. *Uspekhi Matematicheskikh Nauk,* 1963, vol. VIII(3), 3–20.

[80] C. Emmeche: The Sarkar challenge to biosemiotics: Is there any information in a cell? *Semiotica* (Special issue: *Biosemiotica,* edited by Thomas A. Sebeok), 1999, vol. 127(1/4), 273–293.

[81] S. Sarkar: Biological information: a skeptical look at some central dogmas of molecular biology. In: S. Sarkar: *The Philosophy and History of Molecular Biology: New Perspectives.* Kluwer Academic Publishers, Dordrecht-Boston, MA, USA: 1996, pp. 187–231.

[82] L. L. Gatlin: *Information Theory and Living Systems.* Columbia University Press, New York, NY, USA: 1972.

[83] J. M. Baldwin: A new factor in evolution. *The American Naturalist,* 1896, vol. 30(354), 441–451.

[84] J. B. Reece, L. A. Urry, M. L. Cain, S. A. Wasserman, P. V. Minorsky and R. B. Jackson: *Campbell Biology.* Benjamin Cummings, San Francisco, CA, USA: 2014.

[85] J. H. Postlethwait and J. L. Hopson: *Modern Biology.* Holt, Rinehart and Winston, Austin, TX, USA: 2000.

[86] G. B. Johnson: *The Living World.* McGraw-Hill, Higher Education, Columbus, OH, USA: 2015.

[87] A. Tozeren and S. W. Byers: *New Biology for Engineers and Computer Scientists.* Pearson Education, Inc., Upper Saddle River, NJ, USA: 2004.

[88] E. O. Voit: *A First Course in Systems Biology.* Galland Science, New York, NY, USA: 2013.

[89] G. J. Tortora, B. R. Funke and C. E. Case: *Microbiology: An Introduction.* Benjamin Cummings, San Francisco, CA, USA: 2012.

[90] P. A. Pevzner: *Computational Molecular Biology: An Algorithmic Approach.* The MIT Press, Cambridge, MA, USA: 2000.

[91] J. J. Ramsden: *Bioinformatics: An Introduction.* Springer (India) Private Limited, New Delhi, India: 2004.

[92] C. R. Legendy: On the scheme by which the human brain stores information. *Mathematical Biosciences,* 1967, vol. 1, 555–597.

[93] P. Kangueane: *Bioinformation Discovery — Data to Knowledge in Biology.* Springer International Publishing, New York, NY, USA: 2009.

[94] G. M. Tomkins, The metabolic code. *Science,* 1975, vol. 189, 760–763.

[95] E. Chargaff: Structure and function of nucleic acids as cell constituents. *Federation Proceedings,* 1951, vol. 10, 654–659.

[96] P. F. Baisnée and S. Hampson: Why complemantry DNA strands are symmetric? *Bioinformatics,* 2002, vol. 18, 1021–1033.

[97] S. J. Bell and D. R. Forsdyke: Deviations from Chargaff's second parity rule correlate with direction of transcription. *Journal of Theoretical Biology,* 1999, vol. 197, 63–76.

[98] GenBank Overview. Available online at: *https://www.ncbi.nlm.nih.gov/genbank/.*

[99] P. Bernaola-Galván, I. Grosse, P. Carpena, J. L. Oliver, R. Román-Roldán, and H. E. Stanley: Finding borders between coding and non-coding DNA regions by an entropic segmentation method. *Physical Review Letters,* 2000, vol. 85, 1342–1345.

[100] Codon Usage Database: *Homo sapiens* [gbpri]: 93487 CDS's (40662582 codons). Available online at: https://www.kazusa.or.jp/codon/cgi-bin/showcodon.cgi?species=9606.

[101] A. V. Oppenheim and R. W. Schafer: *Discrete-Time Signal Processing,* Pearson Education Ltd., 2013, Harlow, Essex, UK

[102] T. V. Arredondo, P. S. Neelakanta and D. De Groff: Fuzzy attributes of a DNA complex: development of a fuzzy inference engine for

codon- “junk” codon delineation. *Artificial Intelligence in Medicine,* 2005, vol. 35, 87–105.

[103] H. C. G. Leitão, L. S. Pessõa and J. Stolfi: Mutual information content of homologous DNA sequences. *Genetics and Molecular Research,* 2005, vol. 4(3), 553–562.

[104] D. C. Benson: Fourier methods for biosequence analysis. *Nucleic Acids Research,* 1990, vol. 18(21), 6305–6310.

[105] S. Tiwari, S. Ramachandran, S. Bhattacharya and R. Ramaswamy: Prediction of probable genes by fourier analysis of genomic sequences. *Computer Applications in Biosciences,* 1997, vol. 113, 263–270.

[106] B. P. Lathi: *Communication Systems.* John Wiley & Sons, New York, NY, USA: 1968.

[107] F. Balado: On the Shannon capacity of DNA data embedding. *Proceedings of 2010 IEEE International Conference on Acoustics, Speech and Signal Processing* (ICASSP 2010), Dallas Texas, USA: March 14–19, 2010, pp. 1766–1769.

[108] P. Moulin and R. Koetter: Data-hiding codes. *Proceedings of the IEEE,* 2005, vol. 93(12), 2083–2126.

[109] W. W. Gibbs: The unseen genome gems along the junk. *Scientific American,* November 2003, vol. 289(5), 53.

[110] P. C. Wong, K. Wong, and H. Foote: Organic data memory using the DNA approach. *Communications of the ACM,* 2003, vol. 46(1), 95–98.

[111] S. S. Pradhan, J. Chou and K. Ramachandran: Duality between source coding and channel coding and its extension to the side information case. *IEEE Transactions on Information Theory,* 2003, vol. 49(5), 1181–1203.

[112] R. J. Barron, B. Chen and G. W. Wornell: The duality between information embedding and source coding with side information and some applications. *IEEE Transactions on Information Theory,* 2003, vol. 49(5), 1159–1180.

[113] R. B. Ash: *Information Theory.* Dover, New York, NY, USA: 1965.

[114] R. G. Gallager: *Information Theory and Reliable Communication.* John Wiley and Sons, Hoboken, NJ, USA: 1968.

[115] S. I. Gel’fand and M. S. Pinsker: Coding for channel with random parameters. *Problems of Control and Information Theory,* 1980, vol. 9(1), 19–31.

[116] C. C. Kopezynski and M. A. T. Muskavitch: Introns excised from the delta primary transcript are localized near sites of delta transcription. *The Journal of Cell Biology,* 1992, vol. 119, 503–512.

[117] C. Wills: *Exons, Introns, and Talking Genes — The Science Behind the Human Genome Project.* Basic Books, New York, NY, USA: 1993.

[118] B. Patrusky: The intron story. *Mosaic,* 1992, vol. 23(3), 22–43.

[119] J. Bergman: The functions of introns: from junk DNA to designed DNA. *Perspectives on Science and Christian Faith,* September 2001,

vol. 53(3). Available online at: *http://www.asa3.org/ASA/PSCF/2001/PSCF9-01Bergman.html.*

[120] A. S. Wu and R. K. Lindsay: *A Survey of Intron Research in Genetics.* Proceedings of International Conference of Evolutionary Computation — The 4th International Conference on Parallel Problem Solving from Nature (PPSN IV 1996), (September 22–26, 1996, Berlin, Germany), pp. 101–110.

[121] S. B. Nagl: Protein evolution as a parallel-distributed process: A novel approach to evolutionary modeling and protein design. *Complex Systems,* 2000, vol. 12, 261–280.

[122] K. Lichtenecker and K. Rother: Die Herleitung des logarithmischen mischungsgesetzes aus allgemeinen Prinzipien der statioenaeren Stroemung. *Physikalische Zeitschrift,* 1938, vol. 32, 255–260.

[123] Codon Usage Database: *Drosophila melanogaster* [gbinv]: 42417 CDS's (21945319 codons). Available online at: https://www.kazusa.or.jp/codon/cgi-bin/showcodon.cgi?species=7227 .

[124] B. Ryabko, Z. Reznikova, A. Druzyaka and S. Panteleeva: Using ideas of Kolmogorov complexity for reading biological texts. *Theory of Computing Systems,* 2013, vol. 52(1), 133–147.

[125] R. A. Fisher: *Statistical Methods, Experimental Design, and Scientific Inference.* Oliver & Boyd Ltd., Edinburgh, Scotland, UK: 1956.

[126] W. J. Evens and G. R. Grant: *Statistical Methods in Bioinformatics — An Introduction.* Springer Science + Business Media, Inc., New Delhi, India: 2005.

[127] A. Isaev: *Introduction to Mathematical Methods in Bioinformatics.* Springer (India) Private Limited, New Delhi, India: 2006.

[128] Learn Science at Scitable: Single-nucleotide polymorphism/SNP. Available online at: www.nature.com.

[129] X. Zhu, S. Zhang, D. Kan and R. Cooper: Haplotype block definition an its application. *Proceedings of the Pacific Symposium on Biocomputing* (PSB-2004), Hawaii, USA: January 6–10, 2004, vol. 9, pp. 152–163.

[130] R. Schwartz, B. V. Halldorsson, V. Bafna, A. G. Clark and S. Istrail: Robustness of inference of haplotype block structure. *Journal of Computational Biology,* 2003, vol. 10, 13–19.

[131] T. C. Schulze, K. Zhang, Y. S. Chen, N. Akula, F. Sun and F. J. McMahon: Defining haptype blocks and tag single-nucleotide polymorphisms in the human genome. *Human Molecular Genetics,* 2004, vol. 13(3), 335–342.

[132] S. B. Gabriel, S. F. Schaffner, H. Nguyen, J. M. Moore, J. Roy, *et al.*: The structure of haplotype blocks in the human genome. *Science,* 2002, vol. 296, 2225–2229.

[133] K. Zhang, M. Deng, T. Chen, M. S. Waterman and F. Sun: A dynamic programming algorithm for haplotype block partitioning. *Proceedings of the National Academy of Science, USA,* 2002, vol. 99, 7335–7339.

[134] M. J. Daly, J. D. Rioux, S. F. Schaffner, T. J. Hudson and E. S. Lander: High-resolution haplotype structure in the human genome. *Nature Genetics,* 2001, vol. 29, 229–232.

[135] P. Taillon-Miller, I. Bauer-Sardina, N. L. Saccone, J. Putzel, T. Laitinen, *et al.*: Juxtaposed regions of extensive and minimal linkage disequilibrium in human Xq25 and Xq28. *Nature Genetics,* 2000, vol. 25(3), 324–328.

[136] H. Herzel and I. Groβe: Measuring correlations in symbols sequences. *Physica A,* 1995, vol. 216, 518–542.

[137] M. Berryman, A. Allison and D. Abbott: Mutual information for examining correlations in DNA. *Fluctuation and Noise Letters,* 2004, vol. 4(2), 237–246. Available online at: *https://www.researchgate.net/publication/263876564_MUTUAL_INFORMATION_FOR_EXAMINING_CORRELATIONS_IN_DNA.*

[138] S. V. Buldyrev, N. V. Dokholyan, A. L. Goldberger, S. Havlin, C.-K. Peng, H. E. Stanley and G. M. Viswanathan: Analysis of DNA sequences using methods of statistical physics, *Physica A,* 1998, vol. 249, 430–438.

[139] R. Mansilla, N. Del Castillo, T. Govezensky, P. Miramontes, M. Jose and G. Cocho: Long-Range correlations in the whole human genome. Cornell University Library, USA. pp. 1–8. Available online at: *http://arxiv.org/ftp/q-bio/papers/0402/0402043.pdf.*

[140] H. Herzel and I. Groose: Correlations in DNA sequences: The role of protein coding. *Physical Review E,* 1997, vol. 55(1), 800–810.

[141] U. Lagerkvist: "Two out of three": an alternative method for codon reading. *Proceedings of the National Academy of Sciences (USA),* 1978, vol. 75(4), 1759–1762.

[142] A. A. Komar: The Yin and Yang of codon usage. *Human Molecular Genetics,* 2016, vol. 25(R2), R77–R85.

[143] A. J. Hockenberry, M. I. Sirer, L. A. N. Amaral and M. C. Jewett: Quantifying position-dependent codon usage bias. *Molecular Biology and Evolution,* 2014, vol. 31(7), 1880–1893.

[144] P. Mazaheri, A. H. Shirazi, N. Saeedi, G. R. Jafari and M. Sahimi: Differentiating the protein coding and noncoding RNA segments of DNA using Shannon entropy. *International Journal of Modern Physics C,* 2010, vol. 21(1), 1–9.

[145] P. L. Kannappan: On Shannon entropy, directed divergence and inaccuracy. *Zeitschrift Wahrschilichkeitstheoie Verwandte Gebiete,* 1972, vol. 22, 95–100.

[146] P. L. Kannappan and P. N. Rathie: On generalized directed divergence function. *Czechoslovak Mathematical Journal,* 1974, vol. 24(1), 5–14.

[147] Z. Daróczy: Generalized information functions. *Information and Control,* 1970, vol. 16(1), 36–51.

[148] J. Aczel and Z. Daroczy: *On Measures of Information and their Characteristics.* Academic Press, New York, NY, USA: 1975.

[149] M. B. Bassat: *f*-entropies, probability of error, and feature selection, *Information and Control*, 1978, vol. 39, 227–242.

[150] C. H. Chen: *Statistical Pattern Recognition*. Hayden Book Co., Rochelle Park, NJ, USA: 1973.

[151] C. H. Chen: On information and distance measures, error bounds, and feature selection, *Information Sciences*,. 1976. vol. 10, 159–173.

[152] S. Guiasu: *Information Theory with Applications*. McGraw-Hill Book Co., New York, NY, USA: 1977.

[153] P. S. Neelakanta and D. Pappusetty: Bioinformatics-inspired algorithms for 2D-image analysis-application to synthetic and medical images. Part I: Images in rectangular format. *International Journal of Biomedical & Clinical Engineering,* 2012, vol. 1(1), 14–38.

[154] P. S. Neelakanta, E. M. Bertot, and D. Pappusetty: Bioinformatics-inspired algorithms for 2D-image analysis-application to synthetic and medical images. Part II: Images in rectangular format. *International Journal of Biomedical & Clinical Engineering,* 2012, vol. 1(1), 49–58.

[155] D. V. Gokhale and S. Kullback: *Information in Contingency Tables*. Marcel Dekker, New York, NY, USA: 1978.

[156] J. N. Kapur: A comparative assessment of various measures of directed divergence, *Advances in Management Studies*, 1984, vol. 3(1), 1–16.

[157] T. Kailath: The divergence and Bhattacharyya distance measures in signal selection. *IEEE Transactions on Communication Technology,* 1967, vol. 15(1), 52–60.

[158] R. W. Johnson: Axiomatic characterization of the directed divergences and their linear combinations. *IEEE Transactions on Information Theory*, 1979, vol. 25(6), 709–716.

[159] C. R. Rao and T. K. Nayak: Cross entropy, dissimilarity measures. *IEEE Transactions on Information Theory*, 1985, vol. 31(5), 589–593.

[160] C. R. Rao: Diversity and dissimilarity coefficients: A unified approach. *Theoretical Population Biology*, 1982, vol. 21, 24–43.

[161] C. R. Rao: Diversity: its measurement, decomposition, apportionment and analysis. *Sankhya: The Indian Journal of Statistics, Serial A*, 1982, vol. 44(pt. 1), 1–22.

[162] J. Lin and S. K. M. Wong: Approximation of discrete probability distributions based on a new divergence measure. *Congressus Numerantium*, 1988, vol. 61, 75–80.

[163] M. E. Hellman and J. Raviv: Probability of error, equivocation, and the Chernoff bound. *IEEE Transactions on Information Theory*, 1970, vol. 16(4), 368–372.

[164] S. Kullback: A lower bound for discrimination information in terms of variation. *IEEE Transactions on Information Theory*, 1967, vol. 13, 326–327.

[165] G. T. Toussaint: Sharper lower bounds for discrimination information in terms of variation, *IEEE Transactions on Information Theory*, 1975, vol. 21(1), 99–100.

[166] D. Kazakos and T. Cotsidas: A decision theory approach to the approximation of discrete probability densities. *IEEE Transactions on Pattern Analysis and Machine Intelligence*, 1980, vol. PAMI-2, issue 1, 61–67.

[167] T. T. Kadota and L. A. Shepp: On the best finite set of linear observables for discriminating two Gaussian signals. *IEEE Transactions on Information Theory*, 1967, vol. 13(2), 278–284.

[168] I. Vajda: Note on discrimination information and variation, *IEEE Transactions on Information Theory*, 1970, vol. 16, 771–773.

[169] I. Vajda: *Theory of Statistical Inference and Information*. Kluwer Academic Publishers, Dordrecht-Boston, MA, USA: 1989.

[170] H. Jeffreys: An invariant form for the prior probability in estimation problems. *Proceedings of the Royal Society of London, Serial A*, 1946, vol. 186, 453–461.

[171] G. T. Toussaint: On some measures of information and their application to pattern recognition. *Proceedings of Conference on Measures of Information and Their Applications*, Indian Institute of Technology, Bombay, India: August 1974.

[172] J. W. Van Ness: Dimensionality and the classification performance with independent coordinates. *IEEE Transactions on Systems, Man and Cybernetics*, 1977, vol. SMC-7, 560–564.

[173] A. K. C. Wong and M. You: Entropy and distance of random graphs with application to structural pattern recognition. *IEEE Transactions on Pattern Analysis and Machine Intelligence*, 1985, vol. PAMI-7(5), 599–609.

[174] K. Ramakrishnan and P. S. Neelakanta: DNA microarray data classification *via* Haralick parameters. *International Journal of Advance in Medical Science*, 2013, vol. 1(2), 19–28.

Section II
Informatics of Biomolecular Sequences

Framing informatic details of analytical and computational methods as warranted in bioinformatic studies refer to a synergistic interplay of information-theory, biological concepts, computer science, and medical/health studies. In essence, it is an art of applying information-processing techniques to comprehend the buried knowledge across diverse sets and distinct *strata* of biosystems. Hence, this section is designed to present (through a pair of chapters), the fundamentals of biomolecular informatics enclaving analytical and computational perspectives extended to biomolecular pairwise/multiple-sequences and next-generation sequencing.

Chapter 3

Biomolecular Informatics

As described in earlier chapters, bioinformatics refers to the pedagogy evolved to analyze and ascertain informatic details in biological sciences. Such details commence at the morsels of genes residing within biomolecular cells and stretch through the entire complexity of whole-life systems. Correspondingly, the gene-related composition of residues in a biosequence represents a statistical structure that poses informative (negentropy) as well as non-informative (posentropy) details. Elucidating such informatic details in biomolecular sequences forms the essential theme of this chapter.

> *"Lo! 'bioinformation' is plentiful for those who understand the simple rules of its acquisition"*
> *Quote from: "The Richest Man in Babylon" by George S. Clason (...with due apologies for morphing "money" into "bioinformation"!!)*

3.1 Molecular Biology: A Flash-Back Summary

As observed in prior chapters, biology is the science of living systems describing constituent details of the elements of life, and in its broad context, biological science depicts a feature-revealing subject on perpetually evolving and self-replicating biosystems of colossal complexity. The gamut of life-systems is made exorbitantly of a large extent of biomolecules, constituting the diverse features of complexity of "life", and these molecules sustain cellular functions, dictate reproductive ability, and formulate a genetic continuity. Also, the living entities bear the ability to capture energy from exogenous sources (such as sunlight and consumed nutritions) and use or dissipate it in different forms (such as heat and chemical energy) with a regimented consistency of thermodynamic considerations.[1,2]

Complementing the details on genomics and proteomics (or omics), the synergistic aspects of molecular, organal, and species-level subsets in biosystems constitute inner and extended framework of bioinformatics. Further as well known, the universe of living systems can be viewed with multitudes of diverse characteristics in excess of 30 million. Despite the associated palette of such diversity, living organisms possess common molecular patterns and exhibit similar *expressions of life*, and they use the same building-blocks in framing the essentials of life, namely, proteins, lipids, and carbohydrates. Additionally, in most of the living complex, identical type of information flows from nucleotide-based genes, transcribed and eventually translated into protein complex. The common mode of associated energy considerations refers to the chemistry of *adenosine triphosphate* (ATP). Prevailing in cells, the ATP represents a strongly bonded compound of high-energy storage, and its decomposition into lower levels of chemical bonding (such as *adenosine diphosphate*, or ADP) delivers required energy availed by biosystems and carry out the life functions as needed.

Concomitant to the suites of genomics and proteomics, the associated entities of biomolecular residues possess genetic information by their statistically implied biochemical features. The expressed genetic details of omics at microbiological level decide the crux of underlying bioinformatics, and relevant details can be ascertained *via* formulated analytical models, supplemented by uniquely tailored computational methods. The underlying analyses refer to genomics-based studies envisaged in acquiring knowledge on the structural and functional aspects of DNA/RNA sequences, and the associated fine-scale genetic-mapping enables ascertaining the functional dynamics of genes depicting various expressions of life. Thus, the analytics of genomics and related computational tasks provide *in silico* details on genetic pathways leading to understanding the functional aspects of the associated information, and this informatic framework bears a wide scope for use in the contexts of biology and in the applications of medical/health sciences. For example, finding answers to queries like, "why, in a given environment, some people get sick, while others do not" and finding relevant answers are medically pragmatic in finding newer and better ways of improving societal health and preventing of diseases.

Parallel to genomics details, analyzing the gamut of various proteomes in cellular contexts (as well as in certain sub-cellular domains, for example, in *viral proteome*) implies elucidating the vital chemistry of proteomics itself. Adjunct analytical efforts (supplemented by experimental/wet-lab studies) thereof govern the essence of bioinformatics in deciphering the complexity of primary and secondary structures (of proteins) as decided by the inlaid physico-chemistry considerations.

In essence, studies at omic levels, as above, help transcribing and translating the associated knowledge into a required information base on the heuristics of microbiology, which can be extended to healthcare strategies and in disease prevention methods (as will be detailed in Chapter 6 *via translational bioinformatic* considerations that explain the perspectives of *clinical bioinformatics*).

3.2 Some Extended Microbiological Details

As a logistic perusal of aforesaid studies on omics, thematically addressed in this chapter are analytical details and computational methods in the framework of information theory (implying the desired scope of biomolecular informatics).

Genomic-Level Analyses

The subject of sequence analysis in molecular biology includes a wide range of topics related to genomics as identified and listed below.

(a) Pairwise sequence comparison directed at finding the associated similarity so as to infer whether the compared sequences are related (meaning that they belong to homologous species).

(b) Identification of intrinsic features of a given biosequence pertinent to: (a) locations of certain *active* and *post translational modification sites/segments*; (b) presence of unique *gene-structures* having characteristic features; (c) making formatted structures of a given DNA structure into so-called *reading frames* and finding the associated *open-reading frame* (ORF), (which is most probably compatible for transcription through translation efforts); (d) inferring symmetry/asymmetry aspects of statistical distributions of the nucleotide residues in a pair of sequences (such as, forward (5′–3′) and reverse complementary (3′–5′) strands); (e) locating delineation boundaries between exons, introns, and regulatory elements in a biosequence; and (f) ascertaining crisp and fuzzy details at those delineation boundaries.

(c) Elucidating differences and variations in sequences arising as a result of *point mutations* and knowing the resulting *single nucleotide polymorphism* (SNP).

(d) Exploring evolution and genetic diversity *via* multiple set of homologous sequences of organisms and deciding on pertinent phylogenetic trees and phenetic variations.

Proteomic-Level Analyses

The proteomics-based analytics and necessary computational methods can be enumerated as follows.[3]

(a) Updating and modernizing traditional schemes of biochemical and pharmacological and/or genetically dissected reconstruction of protein-chemistry using the knowledge gained from analytical models and wet-lab studies.
(b) Establishing protein-related newer perspectives of functional connections between processes and pathways or functional modules existing actively in the same cell.
(c) Increasing the scope on individualized assays that indicate functions of a specific component in the proteins.
(d) Designing quantitative proteome analysis, "for high throughput quantitative profiling of proteins in complex mixtures".
(e) Developing computer algorithms and programming methods to assist sequence identification studies *via collision-induced dissociation* (CID)* *spectra*. Relevant technique can generate sequence information on several peptides from a protein, so as to enable "redundant and unambiguous identification of the protein from the database." (*CID: An activation method in the art of so-called tandem *mass spectrometry* (MS) of peptides and proteins involves energetic collisions using a neutral target gas.[4])
(f) Global analysis of protein expression — a method adopted "to study the steady-state and perturbation-induced changes in protein profiles".
(g) Framing proteome analyses (and related wet-lab and/or instrumented studies) designed as quantitative proteomics in finding the *ratio of abundance* (using chemical reagents like *isotope-coded affinity tags* (ICAT) and *tandem mass-spectrometry* (MS/MS).

In summary, proteomic studies in the context of biomolecular sequence analyses govern a devoted effort on sequence-related protein biochemistry exercised on a high-throughput scale, and the associated task "complements other functional genomics approaches, including microarray-based expression profiles, systematic phenotypic profiles at the cell and organism level, systematic genetics and small-molecule-based arrays." As indicated by the Centre for Proteomics of University of Antwerp,[5] the following is the summary on relevant list of commensurate areas.

(a) *Clinical and peptide-centric aspects of proteomics*: This refers to workflows on label-free or labeled analysis quantitative proteomics, characterization, and quantification details of *peptidomics.*
(b) *Protein-chemistry, proteomics, and epigenetic signaling*: Relevant studies include differential (chemo)proteomics for cancer research, post-translational modifications, and biomarker identification and related studies.
(c) *Systemic physiological and eco-toxicological research*: The conceived tasks in this context are mechanistic insights (mode-of-action) of (environmental) toxicants and toxicity estimation in environmentally persisting pharmaceutical pollutants.
(d) *Biomolecular and analytical MS*: These efforts include mechanistic insights on the development of novel separation, identification, and fragmentation techniques for proteomics, peptidomics, and metabolomics (*via* MS imaging of biomolecules and native MS and ion-mobility approaches for structural biology).
(e) *Translational neurobiology mechanistic considerations*: These efforts are concerned with development of novel separation methods, identification and fragmentation techniques for proteomics, quantitative spatiotemporal proteomics of neuro-degenerative diseases.

3.3 Analyzing Biosequences

As observed earlier, a biosequence represents a one-dimensional sequential stretch of biological residues (such as nucleotides, codons, amino acids (AAs)). In the frameworks of biosequence-related studies, a class of bioinformatic analytics and computational efforts exists as listed below.

(1) Genomics- and proteomics-related studies *via* biosequence informatics.
(2) Cellular-level cytogenetic studies *via* information-theoretics of biosystemics.

(3) Whole-body organismic-level studies governed by the associated information heuristics.

The focus of this chapter is mostly on the first topic of interest (pertinent to omics-specified informatics), and the other two topics are deliberated in other chapters.

Genomic and Proteomic Heuristics of Biosequence Analyses

As described in earlier chapters, a biomolecular sequence denotes a primary level of chains or strands of the DNA/RNA formed by a unique set of chemical compositions of nucleotides, namely, {A, C, T, G}. Furthermore, the characteristics of such sequences can be ascribed to their constituents, namely, the codons (exons and introns) and subsegments of residues (denoting the *untranslated regions* (UTRs) and *coding sequence* (CDS) parts, etc.).

In essence, the biological data sequence analysis refers to methods carried out on the constituents of a given, stand-alone sequence, or the analysis is done on a pair (and/or a multiple set) of DNA structures. The motivations for such DNA sequence analyses can be summarized as follows: (i) deducing the hierarchy of genomic information; (ii) rapid sequencing of complementary DNA (cDNA) libraries; and (iii) elucidating finer details of a query sequence, such as finding the associated *expressed sequence tags* (ESTs), which denote a subsequence of a *cloned DNA** (or cDNA) sequence. (*The *clones* consist of DNA that is complementary to mRNA, as such the ESTs represent portions of expressed genes.)

As indicated by Attwood,[6] "sequencing of entire genomes is a major achievement, but the meaning of the mass of accumulated data is only just beginning to be unraveled..." So, in the context of omic studies, "what is the state-of-the-art in sequence-structure-function bioinformatics?"[6]

A (bio)sequence as defined earlier denotes a one-dimensional (1D) arrangement/array of biomolecular residues like nucleotide bases or AAs, and such a sequence is an arrangement of identifiable entities in the form of a *string*. Relatedly, a *subsequence* of a string can be specified as an ordered sequence of characters (not necessarily consecutive from the main sequence).

Three major biosequence analysis strategies of bioinformatics are: (a) *pairwise sequence analysis*; (b) *subsequence analysis* within a given single global sequence; and (c) *multiple sequence analysis* (MSA).

Furthermore, biosequence analyses are pertinent to the following efforts: (i) comparing a given set of raw (unformatted) sequences with undefined/blindly specified start and/or stop sites of the residues and (ii) comparing a set of formatted sequences (such as, aligned set of sequences with the alignment done on the basis of certain optimal criteria). Alignment of two strings, say v (of n characters) and w (of m characters, with m not necessarily being the same as, n) refers to a two-row matrix, such that the first row contains characters of v in order; while the second row contains characters of w in order with spaces interspersed in the string at different locations. As a result, the characters in each string would appear in order, though not necessarily adjacent.

The objective of comparison (in both cases of unformatted and aligned sequences), refers to finding the underlying *sequence similarity*, which is defined as follows: it refers to the extent of distinguishable closeness between sequences being compared. In other words, sequence similarity is an assessment of statistical "distance" or "divergence" (Chapter 2) between sequences being compared; furthermore, while comparing a pair of sequences toward elucidating similarity features, the associated *matches* and *mismatches* of the residues at each site along the sequence length is specified as follows: The columns that contain the same character in both rows in a pair of sequences depict the "matches", whereas different characters contained in the column denote the "mismatches."

Considering pre-formatted sequences (depicting an aligned set of sequences *via* designated alignment procedures), the columns of the alignment may contain a *space* implying an insertion–deletion (*indel*) site with a space in the top row of a column representing an *insertion* and a space in the bottom row of the same column depicting a *deletion.*

Typically, the sequence comparison is done posterior to aligning the compared sequences. Relevant considerations of alignment procedures with reference to pairwise alignment and MSA procedures are outlined below, and more details on analytical pursuits and computational tools toward alignment and scoring schemes are described in Chapter 4.

Types of Sequence Alignment

As indicated above, pairwise sequence alignment implies a task enabling a pair of sequences adjusted against each other (with mutual lateral shifts as well as incorporating indels as needed), so that these sequences attain optimally the "best" score of similarity of residues at an observed site in the top and bottom sequences being compared. Ascertaining such similarity scores between the compared pair of sequences specifies implicitly their homologous properties.

The MSA refers to a procedure carried out across a multiple set of sequences, and the procedural logic conforms to those indicated for a pair of sequences. In essence, the primary step of MSA procedure includes first identifying a group of sequences that form gene families and next tracing the hereditary connections within such groups *via* conserved family characteristics across evolution. The identified and grouped sequences as above are then subjected to the alignment procedure involving the placements of indels as needed, so that the post-aligned sequences show (at least in some subsequence sites) high similarity scores confirming conserved domains (called *motif* regions) of homologous characteristics. The completed MSA effectively increases the so-called *signal-to-noise ratio* (SNR) within the sets of sequences being analyzed, which means, when the test sequences are (optimally) aligned to offer maximum similarity scores in the biologically significant regions (which may be diagnostic of either the structural or functional details of eventually ascertained protein complex), possibly conserved across the organisms of the test sequences being compared.

3.4 Similarity Between Biosequences

While establishing the similarity and/or dissimilarity features between biosequences, some pertinent details are furnished as follows.

Biosequence comparison: With reference to analyzing bioinformatic sequences, comparing biosequences involves varying scales of handling different types of sequences with various forms of information. The nucleotide and protein sequence comparison constitutes one of the basic cornerstones upon which the extraction of biologically interesting bioinformatic data are built. As defined earlier, a sequence consists of an ordered list of 1D symbols, which in bioinformatics refer to biologically relevant sources of genetic information. Such a sequence may represent information derived from an algorithmically-predicted sequencing process (or obtained *via* wet-lab experimentation). Regardless of the sources of biosequences, sequence comparison methods can be conceived with sequence patterns made of residues representing nucleotides and/or codon triplets. (Equally conceivable data sequences amenable for comparison efforts are binary data streams from a computer or alphanumeric text of a literature, etc.). Each version of such sequences signifies information-bearing entities compatible for intended analyses.

Homologous biosequences: With reference to a pair or a multiple set of biosequences, the associated genes between them are *homologous*, if they are related to a common ancestor,[7] and the homologous genes perform identical or similar functions in biosystems. Thus, a pair of biosequences can be tied to homologous genes of two species with a common ancestral root, for example, as in the case of humans *versus* mice.[7,8] More homological considerations in phylogeny are detailed in Chapters 9 and 10.

Notion of "statistical distance": A host of similarity assessment strategies have been evolved in the contexts of comparing aligned and/or unaligned biosequences. There are two possible avenues of sequence comparison in terms of: (i) *statistical distances* (in Euclidian sense) and (ii) *statistical divergence measures*. Relevant details and examples of distance measures and metrics of divergence are elaborated in Chapter 2, and outlined below are pertinent summary versions.

Distance measures: Given two character strings (e.g., representing biosequences), the measure of "distance" (u) between them can be assessed in terms of the underlying statistical features. Different metrics intended for measuring or computing the "distance," imply the associated norm of the distance, $||u||$ (as elaborated in Chapter 2).[9–20] Use of distance measures to ascertain the symmetry/asymmetry features of compared sequences can be demonstrated as follows.

For example, in order to ascertain symmetry/asymmetry profile between the complementary strings of a double-stranded DNA (dsDNA), Hamming distance (HD-measure) can be considered as described in Pandya,[21] Neelakanta *et al.*,[22] and Neelakanta *et al.*[23] and it can be applied on a pair of biological, test sequences. Furthermore, for relevant comparison, binary formatted sequences can be used.

As posed in Baisnee *et al.*,[24] a pertinent example of comparison follows the suite of the query, "Why are complementary DNA strands symmetric?" In the analysis pursued thereof, HD is used as a measure to establish the contrast between forward (5′–3′) and (complementary) reverse (3′–5′) strings of a DNA string. (It may be noted that binary formatted sequences are not *per se* popularly seen in the reported analyses of biosequence comparisons except in Pandya,[21] Neelakanta *et al.*,[22] and Neelakanta *et al.*[23]).

As elaborated in Pandya,[21] Neelakanta *et al.*,[22] Neelakanta *et al.*,[23] and Baisnee *et al.*,[24] ascertaining the symmetry or asymmetry details between forward and complementary strands of a test DNA (e.g., the *Saccharomyces cerevisiae*) with HD applied to binary formatted test strands is as follows: Considering forward and reverse test nucleotide strands of *S. cerevisiae*, they are first converted into their binary format using the conversions, say A-00, C-01, G-10, T-11, (which retains the WC pairing relation: A ↔ T and C ↔ G), then each sequence is compared against a common (binary-formatted) "junk" sequence to elucidate the associated symmetry/asymmetry details in the test strands. Relevant computational steps are outlined in Pandya,[21] Neelakanta *et al.*,[22] and Neelakanta *et al.*[23] with the procedure on generating binary formats of a test DNA sequence (e.g., *S. cerevisiae*[25]). Relevant statistics of DNA of the test species: *S. cerevisiae* adopted refers to occurrence frequency of the residues gathered from the GenBank. Hence, the test forward (5′–3′) and complementary (3′–5′) test strands are simulated with a specified length such that the occurrence of each nucleotide residue (of the set {A, T, C, G}) in the simulated strands appears with the statistics corresponding to the DNA of the test species: *S. cerevisiae*. The probability of occurrence of nucleotide bases conform to the following (approximate) values: $p(\text{A}) = 0.3$; $p(\text{T}) = 0.3$; $p(\text{C}) = 0.4$; and $p(\text{G}) = 0.4$. Likewise, the junk sequence of the same length simulated has all residues having equal probability of occurrence, that is, $p(\text{A}) = p(\text{T}) = p(\text{C}) = p(\text{G}) = 0.25$.

Next, determining the symmetry–asymmetry details between forward and complementary strands of the test DNA sequence is done *via* HD algorithm applied to the binary-formatted, simulated versions of forward and complementary test biosequence strands *versus* the junk sequence. The binary formatting involves converting the simulated sequences into binary strands by replacing, A-00, C-01, G-10, T-11. Relevant conversion also retains the WC matching, A ↔ T and C ↔ G. Furthermore, the simulated junk sequence is non-informative, inasmuch as all the four residues bear statistical invariancy of their occurrence. That is, the simulated "junk" sequence has nucleotide bases randomly dispersed along the length such that all the bases A, C, T, and G occur equally likely (Chapter 2), (each with a probability of ¼; as such, it is non-informative with totally of posentropy contents).

Evaluation of similarity/dissimilarity features of forward and reverse strands of the test sequence is done by comparing forward and reverse strands individually against the simulated junk sequence. Relevant computation refers to using HD algorithm to determine the statistical distance between (i) forward *versus* junk sequence and (ii) reverse *versus* junk sequence — all of same length. The assessed HD gives the contrast features of the forward strand against a reference (junk) sequence; likewise, HD assessed between complementary strand *versus* the same reference sequence yields the associated contrast features. Thus, the forward and reverse strands are independently compared against a reference entity. This comparison enables deducing relative symmetry/asymmetry profiles of the forward and complementary strands. The HD values can be estimated across the pair of strands being compared on window-by-window basis (where each window depicts a sub-segment of the whole length of the test strands). That is, by segmenting each sequence into a number of "windows," the HD estimation is done in each window. Such HD estimation implies *modulo-2 (XOR)* operation between the compared residues.

Hence, the results of HD values measured window-by-window across the simulated, forward and reverse DNA sequences (each *versus* the junk sequence) are presented in Fig. 3.1 *via* histogram representations on the population density (PD) spread of the HD computed, and polynomial curves are fitted to the histograms of both forward and complementary reverse strand data.

A statistical comparison of calculated measures of population density (PD) spread of HD measure yields details as follows (Fig. 3.1): Forward strand – Mean = 49.40; variance = 810.71, and standard deviation = 57.64, and

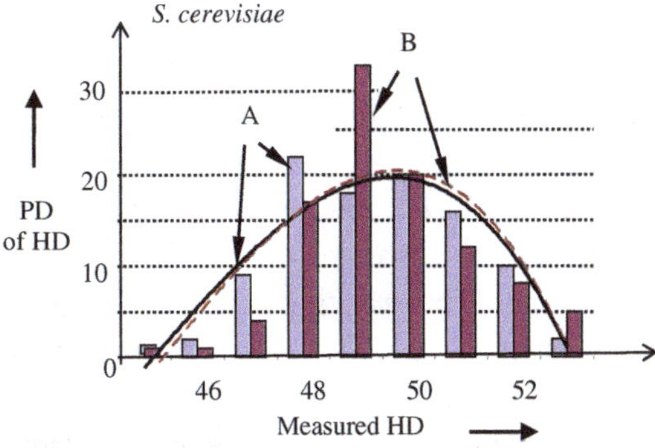

Fig. 3.1 Histogram — Population density (PD) spread of the estimated values of HD measure computed window-by-window. A: Forward (5′–3′) DNA strand and B: complementary (3′–5′) DNA strand of the test species: *Saccharomyces cerevisiae*.

for the complementary, reverse strand – Mean = 49.56; variance = 889.73, and standard deviation = 60.19. These results confirm the symmetry features of the strands compared as indicated in Pandya,[21] Neelakanta *et al.*,[22] and Neelakanta *et al.*[23]

The terms "contrast," "comparativeness," or dissimilarity between the strands can be synonymously viewed as the "distortion" or "skewness" that depicts the statistical difference between the sequences, therefore, the contrast can be specified in terms of the known attributes of the sequence characteristics. That is, should there be a perceivable difference in the statistics of sequences, a *cross-entropy* (or *mutual information* in Shannon's sense) should prevail and can be used as a metric to assess such differences. A relevant theorem and its proof are as follows: As regard to the context of a sequence $\{S^+\}$ denoting, for example, the forward (5′–3′) strand of a DNA and its reverse complement, (3′–5′) strand $\{S^-\}$ are statistically large (with their residue contents $N \to \infty$), then the element features of $\{S^+\}$ and $\{S^-\}$ should bear uncertainties by virtue of their probabilities assigned as $[q_1; (1 - q_1)] \in \{S^+\}$ and $[q_2; (1 - q_2)] \in \{S^-\}$]. (In binary representation of such sequences, q_1 and q_2 denote the probabilities of occurrence of 1's, and $(1 - q_1)$ and $(1 - q_2)$ are probabilities of occurrences of 0's.

In addition to HD measure and cross-entropy-based distances indicated above for contrast estimation across biosequences, other metrics in vogue (as outlined in Chapter 2) are (i) *Edit distance or Levenshtein distance (LD)* — It implies the distance (as defined and explained in Chapter 2) denoting the minimal number of operations or changes (insertions/deletions (*indels*) and substitutions) exercised to transform one sequence to another. For example, the edit distance may approximately specify the number of DNA replications taken place across two sequences. That is, the LD is a string metric for measuring the difference between two sequences. Simply, the LD between two words is the minimum number of single character edits (i.e., insertions, deletions, or substitutions) required to change one word into the other. (ii) *Mahalanobis distance (MD; and its variations)* — Given a statistical distribution of a variate x, the MD refers to a measure of distance between a specified value of x (say, x_s) and measuring how many standard deviations this specified value of x, namely, x_s is away from the expected mean of the distribution.[12,13] Mahalanobis measure expresses the *coefficient of likeness* and classically specified as the d^2 statistical distance norm. (iii) Another Euclidean measure of distance is based on the concept of dot product or inner product. That is, given a pair of vectors $\mathbf{x}_i$ and $\mathbf{x}_j$ of the same dimension, their inner product is $\mathbf{x}_i^{\mathrm{T}}\mathbf{x}_j$, (as described in Chapter 2). (iv) *Bhattacharyya distance measure*: Given a pair of entities (1 and 2) being compared in terms of their statistical distributions p and q, respectively, then the Bhattacharyya distance measure (B) is given by: $B_{i=1,2} = \Sigma_i \log_b(\rho_i)$, $(0 < B < \infty)$, where ρ_i depicts the Bhattacharyya coefficient specified explicitly as: $\rho_{i=1,2} = \Sigma_i[\log_b(p_i \times q_i)]^{1/2}$ with $(0 < \rho_{i=1,2} < \infty)$. A measure closely related to ρ_i is *Kolmogorov variational distance* (KV) defined as: $\mathrm{KV}_{i=1,2}$ equal to $[1/2\Sigma_i\,|\pi_1 \times p_i(x) - \pi_2 \times q_i(x)|]$, where π_1 and π_2 denote *a priori* probabilities of hypotheses H_1 and H_2 assumed as regard to the compared entities.

3.5 Sequence Contrasting

Notions of discriminant measures are alternative views of prescribing the statistical distance between biosequences being compared. They refer to sequence contrasting algorithms. Relevant measures identified in Chapter 2 are revisited here with the following subsections briefly describing their potentials in elucidating biosequence similarities. Relevantly, a *ratio discriminant measure* is often considered as a similarity or dissimilarity metric that depicts proportion of similar or dissimilar aspects of two entities in terms of a fraction. Classical versions of these measures are detailed in Chapter 2 and include: (i) Linear discriminant measures (LDMs) due to Fisher[26] and it denotes a *contrast function* for determining the distance between two statistical distributions (such as those of two biosequences being compared). Suppose, the pertinent discriminant function $h(X)$ has expected values (η_1, η_2) and variances (σ_1, σ_2), then a criterion set forth by a function of $(\eta_1, \eta_2; \sigma_1, \sigma_2)$ is maximized or minimized in order to determine the optimum coefficients of the linear discriminant function.[25] The Fisher LDM can be used as a metric to specify and identify the similarity and/or dissimilarity features between biosequences or the segments of a biosequence. (ii) Sum-ratio measure: the similarity or dissimilarity aspects of two sets of random entities can also be specified in terms a fraction involving a sum vector set on contrasting features as indicated in Chapter 2. Suppose $\{\mathbf{x}_i\}$ and $\{\mathbf{x}_j\}$ are entities being compared, the following ratio measure can be specified to indicate the relative fractions (of underlying similarity or dissimilarity): $r = [\Sigma_x\,|\mathbf{x}_j - \mathbf{x}_i|]/[\Sigma_x \mathbf{x}_{j \text{ or } i}]$ and $\bar{r} = [1 - r]$; and, lastly

(iii) the Socratic ratio measure, which implies a ratio measure denoting an *a priori* of the expected benefits (of admitting the uncertainty) relative to the expected benefits of perfect knowledge, and this associated relative details can provide similarity or dissimilarity aspects of two compared entities (Chapter 2).

In literature, mostly no aggressive pursuits seem to exist where the aforesaid discriminant measures are used as sequence contrast functions in the context of biological sequence analysis (except for some sparse indications furnished in Pandya,[21] Neelakanta *et al.*,[22] Neelakanta *et al.*[23]). As such, availing these discrimination measures forms an open-question towards biosequence comparison analyses.

3.6 Visual Comparison of Sequences

The dot-plot: Visual comparison of biosequences is done *via dot-matrix representation*, which enables assessing (sequence) similarities either *globally* or *locally* along the stretch of the strands being compared. Relevant dot-matrix representation[27,28] implies constructing a graph that depicts a two-dimensional plot, where sequences being compared form the two axes of the graph. That is, a dot marked at a coordinate (x, y) indicates that the subsequence at position x in the first sequence matches the subsequence at position y in the second according to some criterion as illustrated in Fig. 3.2.

The dot-plot in Fig. 3.3 depicts a statistical chart consisting of data points plotted on a fairly simple scale, typically using filled-in circles. (A code called *Dotlet* written by Junier and Pagni[29] of the Swiss Institute of

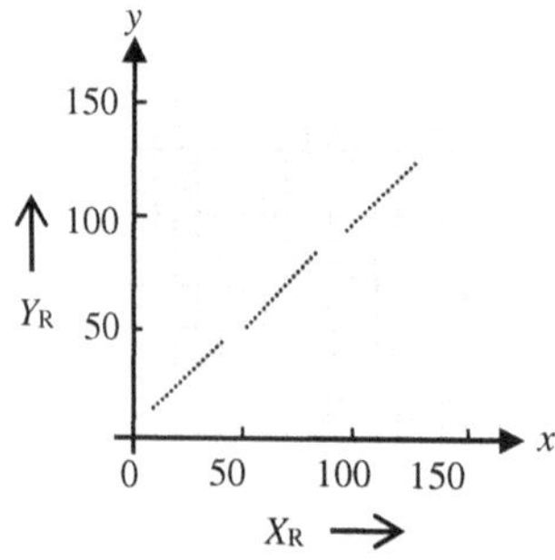

Fig. 3.2 A typical dot matrix representation constructed with X_R denoting the residues of α-chain and Y_R depicting the residues of β-chain residing in a window of human hemoglobin sequence.

Y_j ↓ / X_j →	G	T	G	A	C	C	G	C	T	A	A	C	C	T	C
G	o														
T		o							o						
T															
G	o						o								
C					o	o		o				o			
G	o		o				o								
A				o						o	o				
C					o			o				o			o
T		o							o						
G	o		o				o								
C						o		o				o			o
G			o				o								
G															
C					o										o
G															
T									o						

Fig. 3.3 Example of dot-matrix representation constructed for a given pair of nucleotide sequences *X* and *Y*.

Bioinformatics is a diagonal plot tool.) An example of dot-matrix is illustrated in Fig. 3.3 with reference to the following pair of nucleotide sequences:

X: G T G A C C G C T A A C C T C
Y: G T T G C G A C T G C G G C G T

For this problem, one can write an algorithm in MATLAB and corresponding *seqdotplot* (*Seq1*, *Seq2*) plots would enable visualizing the match between two sequences. Producing a dot-matrix is a two-step process. The first step involves using a compare program to perform the sequence comparison and save the results to a file. This data file is subsequently used as input to another program, which generates the graphics.

Thus, how similar are two sequences is a bioinformatics-driven query typically posed on judging the extent of similarity in an aligned pair of sequences. The simplest method thereof is to devise a dot matrix or the dot-plot indicated above, where one sequence is written out horizontally and the other sequence is written out vertically, along the top and side of an $[m \times n]$ grid with m and n denoting the lengths of the two sequences.

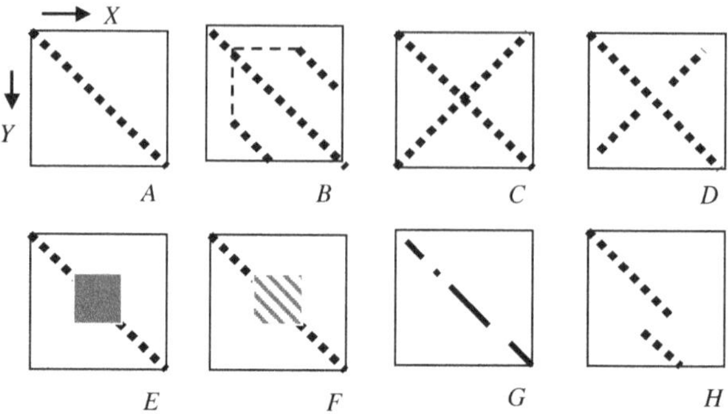

Fig. 3.4 Possible dot-plot patterns of two sequences X and Y, both of same lengths.

A dot is placed in a cell in the grid wherever the two sequences match. A diagonal line in the grid visually shows where the two sequences have sequence identity. In general, given two sequences *X* and *Y*, their dot-plot patterns can be of different types as illustrated in Fig. 3.4.

The patterns of Fig. 3.4 can be interpreted as follows.

(1) A continuous main diagonal implies a perfect similarity between the residues of the sequences compared.
(2) The minor diagonals duplicated as parallels to the main diagonal indicate repeated regions of residues at different parts of the sequences observed in the same reading direction.
(3) The perpendiculars to the main diagonal depict palindrome areas. In this example, the sequence is totally palindromic in the observed window area.
(4) Here, the pattern depicts a partially palindromic sequence. In the case of DNA sequences, such partial palindrome signifies a perfect match of the normal strand with its reverse complement, which is not uncommon for many transposable elements.
(5) The grey-lock on the main diagonal means finding repetitions (called *microsatellite repeats*) of the same residue in both sequences.
(6) Cluster of parallel lines shown imply tandem repeats of a larger motif in both sequences and are termed as *mini-satellite patterns* with the distance between the diagonals being equal between the distance of the motif.
(7) The interrupted diagonal denoting a discontinuous line-pattern signifies that the sequences *X* and *Y* being compared share a common source, that is, the sequences are homologous with a

common ancestor. The extent of interruptions seen increases with modifications imposed on the residues independent of the time of evolution and mutation rate.

(8) Partial deletions in X or insertions in Y amounting to "indel" being signified.[27,30]

In constructing the dot-plots, the entities of interest are as follows: (i) *Window size* depicting the number of nucleotides compare each time (usually odd number); (ii) *stringency* specifying the minimum number of nucleotides in the window must "match," so that a dot can be placed; (iii) *mismatch limit* implying the maximum number of nucleotides in the window that do not match so preventing the placement of a dot; and (iv) mismatch limit is equal to window size minus stringency. For example, in the following sequences AGAGACTC and AGAGTGTG, the window size is equal to 5 and the stringency is 3.

3.7 Sequence Comparison: Information-Theoretic Considerations

To compare and contrast any two biosequences or elucidate the similarity between two subsequences of a given chain of sequence, the underlying approach is to invoke and apply the concept of a contrast function *via* notions of distance and/or divergence measures. With reference to a sequence length of N residues, suppose this domain length is $n_1 < N$ for a sequence #1 and corresponding domain length is $n_2 < N$ for the other sequence #2. Furthermore, denoting M_1 and M_2 as the relative number of nucleotides of a given type of residue or symbol, respectively, in a query sequence (sequence #1) being compared against sequence #2 a total *vector space* V can be defined such that $\mathbf{V}_\ell = [(n_1/n_2) \times \mathbf{V}_{1\ell} + (n_2/n_1) \times \mathbf{V}_{2\ell}]$ with $\mathbf{V}_{1\ell} = \{M_{1\ell}\}$ and $\mathbf{V}_{2\ell} = \{M_{2\ell}\}$, where, for example, $\ell \ni$ {A, T, C, G}. and the probabilities of occurrence of these entries randomly prevailing across the sequence length are denoted by the set $\{q_{1\ell}\}$. Likewise, for the sequence #2, bearing a statistical distribution of its entries can be specified by the probabilities of occurrence of such entries, $\{q_{2\ell.}\}$.

Distance and statistical divergence measures (based on entropy-theoretic (*cross-entropy*) considerations or *relative* (*mutual*) *information* in Shannon's sense) are detailed in Chapter 2, and they can be specified as implicit measures of contrast function toward elucidating symmetry/asymmetry attributes of biological sequences. Furthermore, the family of *Csiszar's measures*[31,32] described in Chapter 2, can also be considered for the purpose of comparing the statistics of two strands of sequences. Hence, the host of divergence metrics (based on Csiszar's *f*-divergence concept) such as Kullback–Leibler (KL) measures, related metrics like Jeffery's or symmetrized KL measure (J) and Jensen–Shannon (JS) measure, etc., are compatible for use in sequence comparison efforts in bioinformatics.[31,32]

3.8 Complexity Metric Toward Sequence Comparison

A biological system well qualifies itself to be modeled as a complex system in view of the following heuristics: a complex system, in general, refers to a universe with a large number of interacting units. Such units when viewed in information-theoretic perspective could be seen to possess gross redundant features. Furthermore, a complex system is inherently stochastic in its extensive spatio-temporal universe and hence, some of its statistical features could manifest as patterns occurring more frequently than others, and likewise some other patterns could form rare elements of the same set. In information-theoretic perspectives (in Shannon sense) such more (or less) probabilistic occurrences of features specify the redundant (or non-redundant) aspects of the complex system. This attribute of information redundancy of a complex system can be adopted to devise a useful complexity metric, which can be adopted in bioinformatic sequence comparisons, for example, in ascertaining the border between the constituent subsets of codons and non-codons in the complex microbiological sequences by delineating the information profile across their borders.

Typically, a large-sized binary mixture can be regarded as a complex system where each constituent subset (of the mixture) depicts a conglomeration of elements occurring in a proportion as decided by some probabilistic distribution. Relevant to such a complex set, an algorithm can be developed to specify a metric that quantitatively evaluates the complexity (in terms of the associated information redundancy). An example of such a complex system is the binary mixture of codon and non-codon/junk codon constituents in a DNA structure.

As indicated by the author in Neelakanta *et al.*,[33] a measure of global complexity can be specified in terms of the maximum entropy associated with the disordered constituent entities (assuming each having a large population mixed in a specified proportion). In general, the codon and non-codon regions being compared can be regarded as parts of a total sequence length N as considered earlier. Designating the codon region by subscript 1 and non-codon region by the subscript 2, the fractional regions being compared are proportioned as n_1/N and n_2/N, respectively, (so that, $(n_1 + n_2) = N$ constitutes the whole universe of statistical mixture of codons and non-codons). Each of the triplets or phases of the nucleotide set $\ell \ni \{A, T, C, G\}$ occurs with a particular probability in a given genome type. For the regions 1 and 2, these probabilities can be specified as $P_{1,\ell}$ and $P_{2,\ell}$, respectively. For example, with reference to the genes of the species *Homo sapiens Escherichia coli, Methanococcus jannaschii, and Rickettsia prowazekii,* the associated set of probabilities $P_{1,\ell}$ specified in terms of relative frequencies of occurrence of 64 codon triplets are listed in Table 1A-7 (in Chapter 1) in the Appendix section of Chapter 1. Relevant sets of probabilities $P_{1,\ell}$ (denoting the relative frequencies of occurrence of 64 codon triplets) are exemplars of codon usage frequency (CUF) useful in the present context.

With reference to triplets or phases of nucleotides occurring with probabilities specified as $P_{1,\ell}$, and $P_{2,\ell}$, respectively, in the regions 1 and 2 as indicated above, the following concept of *statistical mixture theory* (due to Lichtenecker and Rother[34–37] based on effective medium theory) can be invoked to assess the probabilistic attribute of a mixture proportioned by a pair of attributes of the constituents. The underlying heuristics is outlined in Chapter 2 and summarized below: suppose a weighted probability r describes the effective statistical attribute of a mixture proportioned by a pair of attributes $\{\mu\}$ and $\{\nu\}$, and this weighted probability can be written as: $r(\theta) = P_{1\ell}^{\ \theta} P_{2\ell}^{\ 1-\theta}$ with $\theta = [n_1/(n_1 + n_2)]$, and a measure of the associated redundancy (R) of the mixture entropy can be deduced as follows: $R = 1 - H(r)/[H(r)]_M$, where $H(r) = -[r \times \log_e(r)]$, and $H(r)]_M$ refers to the maximum value of $H(r)$ over the fraction $(0 \leq \theta \leq 1)$ or $(1 \geq (1 - \theta) \geq 0)$ of the binary mixture constituents.

3.9 Implementation of Biosequence Comparison

Corresponding to the techniques of using a distance and/or a divergence measure in sequence comparisons, the discriminatory features between the items being compared can be either the parametric characteristics that describe the entities constituting the 1D patterns (such as the type and physico-chemical properties of the residues constituting a biosequence) or the associated entropy profiles (in information-theoretic sense) of the sequences.

For the purpose of comparing biosequences *via* underlying residue characteristics, first some molecular-biology concepts can be considered so as to assess the relative scoring, and comparison is done on the basis of such scores as indicated below.

(a) *Signature scoring*: A "signature" in the context of a DNA sequence refers to an amino acid distribution resulting from the expression of the DNA. Scoring on the signature within a sequence token involves assessing the probability of occurrence of the amino acid. The example of comparing the forward and reverse complement of a DNA sequence described earlier uses this strategy.
(b) *Length scoring*: In this effort, a target length is specified for a token sequence of a DNA, and any observed length in the scoring effort is assumed as a random variable around the target length within a specified set of upper and lower bounds.
(c) *Electronic charge and hydrophobicity scoring*: This refers to scoring based on the abstract notions on the extents (high or low) of charge content and/or hydrophobicity associated with a DNA sequence. Correspondingly, the physico-chemical measure of charge in an amino acid and hence, across in the relevant sequence is identified. Likewise, the profile of hydrophobicity index is specified for the sequence. The relative extent of such parameters indicate the measurable scoring metrics of comparison.
(d) *Amphipalic alpha-helix and/or beta-sheet scoring*: These are other possible sets of scoring functions and they correspond to hydrophobic and hydrophilic weightings that can be attributed to a given DNA sequence available at the translational step in the central dogma.

Biosequence Comparison: Scoring Measures and Metrics

Entropic details or information content can be adopted to score and discern the codon and non-codon sub-segments in a DNA sequence. Relevant method developed by Bernaola-Galván *et al.*[38] is known as *entropic segmentation method*. Pertinent segmentation technique for genomic sequences is based on assessing statistically distinct features associated with codon and non-codon segments. Evaluating and comparing such differing features of codon/non-codon segmentations is the theme outlined in Bernaola-Galván *et al.*[38]; and, comparing segmentations by applying randomization techniques is indicated in Haiminen *et al.*,[39] and the segmentation method (due to Bernaola-Galván *et al.*[38]) uses the Jensen–Shannon (JS) version of divergence metric to compute the cross-entropy between successions of adjoining (short) windows of a long DNA sequence, so as to get a profile of variation of the JS measure along the test strand; hence, the presence of codons and junk codons are discerned or delineated in the intervening region between 5′ and 3′ ends in the test DNA structure.

Apart from using the JS metric as above, the host of other divergence metrics can also be used to assess the discriminatory profiles between biosequence entities as considered in Arredondo *et al.*[40] It refers to finding feature distinctions across residual segments in a DNA. It corresponds, for example, to identifying and delineating codon (X) and non-codon (Y) segments in a given DNA sequence, where an identifiable boundary of separation prevails between an adjunct pair of a codon and a non-codon. Entropy segmentation method of Ref. 38 uses the JS metric for this purpose as indicated above. The delineation or identifiable separation between an adjunct pair of a codon and a noncodon, in general, could be a distinct (or crisp) boundary or when the codon and non-codon constituents exist within a subsequence as overlapping differential blocks (possibly caused by mutation), they may pose an imprecise or fuzzy demarcation of separation[40] X and Y.

The method adopted uses the associated codon statistics pertinent to a test DNA sequence of a given species. Relevant to scoring evaluated in the search processes indicated above, an information-theoretic measure can be invoked. For example, a cross-entropy measure such as Kullback–Leibler (KL) measure can facilitate the assessment of relative information between the differential blocks of subspaces, X, Y, and Z as illustrated in Fig. 3.5.

The search for the codon/non-codon border in practical circumstances also warrants ascertaining details across overlapping (codons and non-codons) regions as well. It can be done on the basis of using statistical discriminant (divergence metric based) scoring across the test DNA and fusing *artificial intelligence* (AI) techniques that offer a reduced search efforts and yet yield robust results in spite of the imprecise data structure of the search-space having domains containing fuzzy details as indicated by Arredondo *et al.* in Arredondo *et al.*[40] That is, a

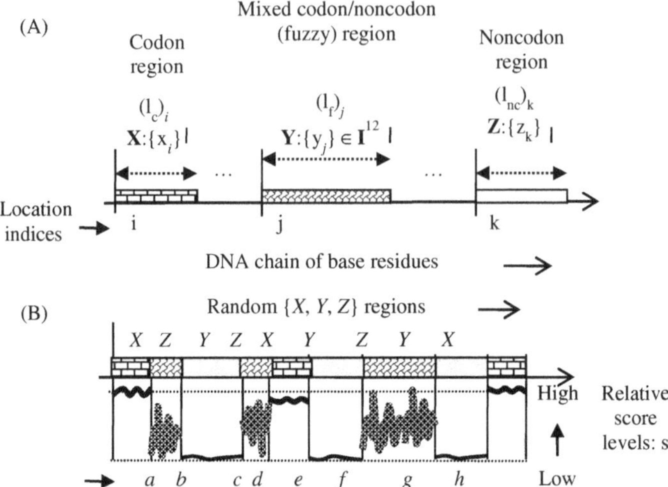

Fig. 3.5 (A) DNA chain of base residues is divided into regions of subsequences containing: Codon-only, non-codon-only, and mixed codon–non-codon occurrences. These subsequences are identified by location indices *i*, *k*, and *j*, respectively. (B) Randomly disposed subsequences along the sequence designated as: Codon subsequence set *X*: $\{\mathbf{x}_i\}$; non-codon subsequence set *Y*: $\{\mathbf{y}_k\}$; and, mixed/fuzzy subsequence set *Z*: $\{\mathbf{z}_j\}$ with corresponding lengths of event spaces $(\ell_c)_i$, $(\ell_{nc})_k$, and $(\ell_f)_j$, respectively. The transitions of *X*, *Y*, and *Z* are located at sites indexed as: $\{a, b, \ldots, h\}$ as shown, and the estimated divergence/distance measures correspond to ($s = C$, $s = NC$ and $s = FZ$) specified, respectively, for codon *X*, non-codon *Y*, and fuzzy *Z* regions, and $C > (s = FZ) > NC$.

fuzzy logic strategy can be pursued to elucidate a relational database portraying the fuzzy state of nucleotides, *via* a search process using a clustered database, and a description on fuzzy aspects of the codon/non-codon contents in a DNA sequence is presented. Hence, the method of applying the strategy of delineation in question pertinent to a large DNA dataset corresponding to human codon statistics[41] is furnished.

With reference to codon, non-codon, and fuzzy subsequences denoted by the sets: *X*: $\{\mathbf{x}_i\}$, *Y*: $\{\mathbf{y}_k\}$, and *Z*: $\{\mathbf{z}_j\}$, respectively, with event lengths $(\ell_c)_i$, $(\ell_{nc})_k$, and $(\ell_f)_j$ as in Figs. 3.5(A) and 3.5(B), the problem in hand is two-fold.

(1) To identify all those subsequences which are similar and label them by a single index. By aggregating such similar subsequences (of *X*, *Y*, and *Z* categories) would eventually reduce the vast search-space compatible for subsequent data-mining and related efforts.

(2) To determine the delineation border (as shown in Fig. 3.5(B)) between the crisp sets of codons and non-codons as well as, between overlapping codons and non-codons within a fuzzy subspace of the set, Z: $\{\mathbf{z}_j\}$. These transition sites are indexed in Fig. 3.5(B) as: $\{a, b, c, d, e, f, g, h\}$.

Hence, the method of constructing a fuzzy relational database of the subsequence set Z: $\{\mathbf{z}_j\}$ across a test DNA sequence refers to an aggregate of similar subsequences containing overlapping codon–non-codon regimes. The corresponding set of polynucleotide can be visualized as a fuzzy space, $\mathbf{I}^{12}$ with 64 codons residing at 2^{64} = 4096 corners of the 12-dimensional unit hypercube.

In contrast to codon regions, the so-called non-codons (or "junk" codons) that do not carry any known functional attributes may be regarded to exist with equally likely random occurrences, meaning that they are uniformly distributed with a probability of occurrence equal to 1/64. That is, within the sets Y: $\{\mathbf{y}_k\}$, the occurrence probabilities of the triplets have a uniform distribution, namely, $\{P_{2y}\}_k = 1/64$ and $\Sigma_{r=1}^{64} p_{2r} = 1$.

Now, considering the database design of aggregating similar fuzzy subsequences Z: $\{\mathbf{z}_j\}$, the relevant procedure involves clustering identical entities (such as the fuzzy subsequences) into a vector set, Z: $\{\mathbf{z}_j\}$ and prescribing each such set with a single index, j. It corresponds to an $\mathbf{I}^{12}$-dimensional vector space (where $j \in \mathbf{I}^{12}$ denotes the number of permissible index terms). (Each such set can be sorted out in the universe of DNA subsequences by ascertaining the weighting function attributed to it.) This relational database pertinent to $\mathbf{I}^{12}$-dimensional vector space, in general, can be represented by a table with columns containing "domains" that depict the "attributes" of the elements in the event space, and the rows correspond to labeled entries or "tuples" describing the elements of the relations.

The cluster process of generating $\mathbf{I}^{12}$-dimensional vector space Z: $\{\mathbf{z}_j\}$, in essence, corresponds to the endeavors of data acquisition, storage, and manipulation of knowledge therein in a manner compatible with human thinking. The motivation for the application of fuzzy set theory in the relevant design of databases and information storage/retrieval system *vis-à-vis* the $\mathbf{I}^{12}$-dimensional vector in question is to deal with imprecise information and exercise subjective expert opinion — the values of which can mostly be specified only in linguistic terms.

Furthermore, vague queries, such as "Which part of the DNA chain or segment is domineering in coding for the protein?" or "Which clusters of triplets along the DNA chain are significantly belong to the 'junk

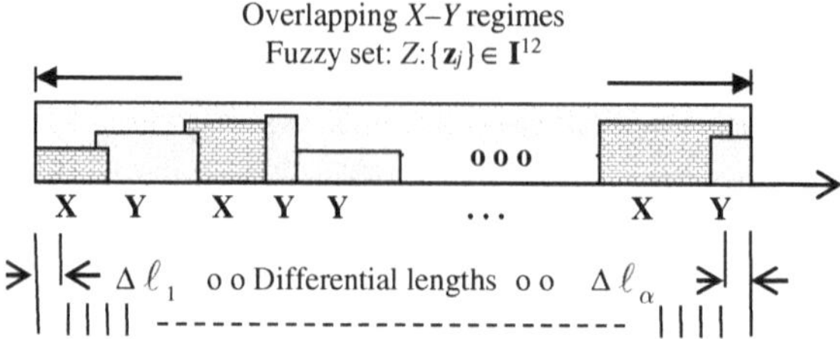

Fig. 3.6 Illustration of the overlaps of *X* and *Y* regions constituting the fuzzy set *Z*: $\{\mathbf{z}_j\}$. The fuzzy segment is divided into differential lengths across which a moving window-based scoring is performed.

set?'" capture the relevant concerns of DNA database users more often than precise queries. Therefore, in an attempt to reduce the vastness of DNA search space, the clustering process indicated refers to binding together such entities, which identically fall within the answers for a given imprecise or fuzzy query with a positional attribute index, *j*, (or *i* and *k* for crisp event spaces). Figure 3.6 illustrates the overlaps of *X* and *Y* regions constituting the fuzzy set *Z*: $\{\mathbf{z}_j\}$, and the fuzzy segment itself is divided into differential lengths ($\Delta\ell$) across which a moving window-based scoring can be performed.

The analysis concerning the codon/non-codon delineation in the universe of DNA chain under discussion has two steps as indicated earlier: (i) Aggregating similar subsequences (*X*, *Y*, and *Z* categories) and (ii) determining the border of separation of overlapping (fuzzy) codons and non-codons within each subsequence of $\{\mathbf{z}_j\}$ category. Once the aggregation or clustering of all $\{\mathbf{z}_j\}$-subsequences (as illustrated in Fig. 3.6, as fuzzy regions posing overlapping mix of codons–non-codons) is done, it will reduce the search space (where codon–non-codon border to be ascertained) *via* fuzzy queries into a single (conglomerated) set *Z*: $\{\mathbf{z}_j\}$. Such aggregation will lead to applying a single algorithm to any element sequence of that set size aggregating similar subsequences with fuzzy attributes and eventually searching for the border in each of this subsequence.

Referring to Figs. 3.5 and 3.6, the various boundaries across the subsequences of codon and non-codon regimes include crisp as well as fuzzy transition boundaries. The transitions as above can be identified in a search process using the considerations: described in Arredondo *et al.*,[40] and relevant moving-window simulations lead to coarse and fine searches of delineating boundaries. An exemplary computation on the coarse search refers to first simulating the DNA codon sequence of the

Human (X) of total length constituted by approximately say 6,000 base residues.[41] (It conforms to random occurrences of 64 triplets of {A, T, C, G} as per the probabilities (of occurrence) indicated earlier for the human DNA being available in Chapter 1, Table 1A-7.)

Corresponding to the same length of 6,000 base residues, a non-codon string (Y) is simulated, again made of the 64 triplets of {A, T, C, G}, but all the triplets occurring with the same probability of (1/64) as per uniform distribution. The lengths of codon and non-codon strands are then divided into 100 small window-segments (each containing 60 bases). (The size of the segment decides the resolution and accuracy of the final result).

With the two types of strings being emulated, the following computational steps are pursued: with reference to the pair of each window-segment in codon and non-codon strands, suppose {p(A), p(C), p(T), p(G)}, and {q(A), q(C), q(T), q(G)}, respectively, denote the probabilities of occurrence of the base residues {A, C, T, G}, a functional relation can be deduced window-by-window *via* by any one of the divergence metrics (such as KL measure) indicated earlier.

Alternatively, the confounding statistics across each pair of window-segment being compared can be utilized in deducing distance measure, ratio-discrimination measure, etc. by this moving-window search method, without any loss of generality. Simulations on coarse-search done (*via* moving-window method) as above, lead to results illustrated in Figs. 3.7 and 3.8 pertinent to the test human-codon sequence. Marked in

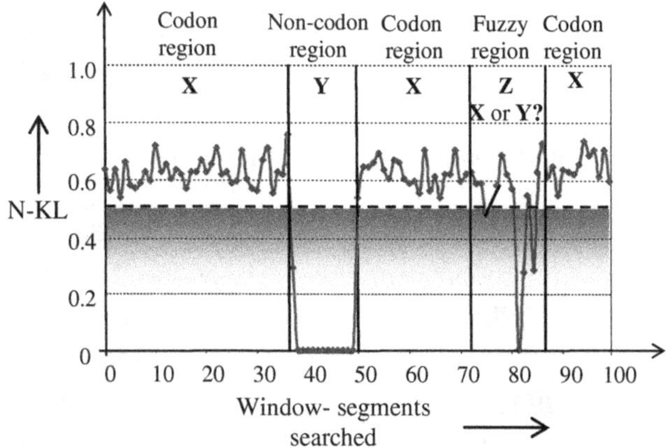

Fig. 3.7 Simulated results of normalized KL (N–KL) metric scores obtained by coarse search (*via* moving-window method) on 100 window segments across the (test) human DNA sequence to visualize the delineation boundaries of **X**, **Y**, and **Z** regions.

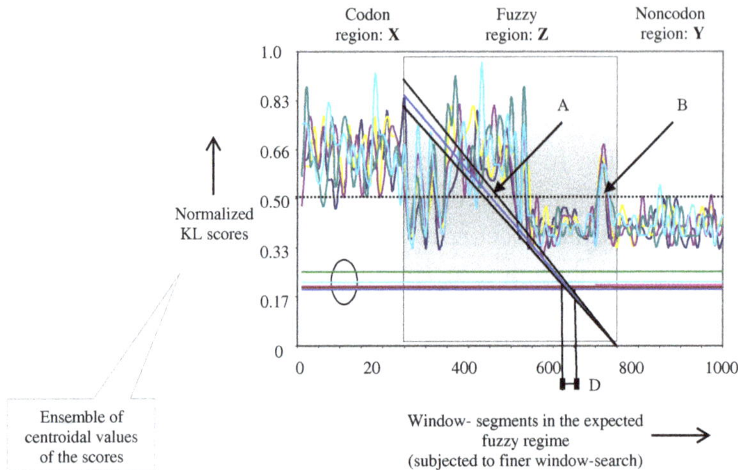

Fig. 3.8 Simulated results of finer search (done *via* moving-window method) across the regions around 350th through 700th locations with window size of 1,000, so as to resolve the finer structure of codon–non-codon statistics in this region of the human DNA sequence. The delineation boundary in this region is ascertained using the estimated KL measure scores and defuzzification procedure as described in the text. (A: Slant-lines constructed as described in the text; B: Ensemble of simulated scores; and, D: Defuzzified value of the required demarcation (shown as an error-bar) on ensemble results.)

Fig. 3.8 are crisp delineation boundaries between X and Y regions. However, when a fuzzy region is encountered and indicated as (X or $Y \rightarrow Z$), the delineation boundaries are unclear.

Furthermore, a closer look at the details in Fig. 3.8 around 35th and 50th window locations, though the delineating transitions appear to be crisp at these spots, a finer search may, however, reveal the fuzzy, overlapping aspects of codons and non-codons at those locations. Hence, in order to find the delineating location more precisely, the required finer search is as follows.

With reference to the stretch of fuzzy region Z in Fig. 3.8, where the states of codon–non-codon cannot be crisply specified with the estimated (normalized) KL score (in the scale of 0 to 1), the following defuzzification is exercised. The score level of 0 to 1 is first divided into three *membership levels* (of subjective reasoning and linguistic terms), such as (0–0.6): low; (0.6–0.8): medium; and (0.8–1.0): high. Here, the linguistic descriptions, namely (low, medium, high) refer to the relative score of the highest value being equal to 1 conforming to "all-codon"

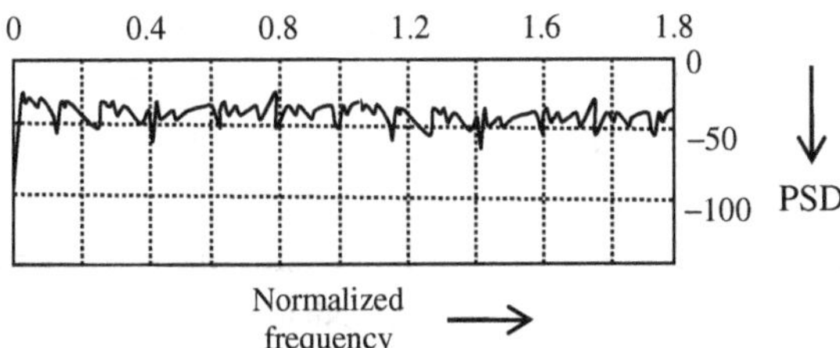

Fig. 3.9 Computed flat power spectral density (PSD: Expressed in dB/radian/sample) characteristics of the "noise" depicting the cross-correlation between the forward and reverse complementary strands of the human DNA *versus* normalized frequency (in π radian/sample).

conditions and the lowest value of the score equal to 0 refers to an "all-non-codon" condition.

Next, a segment-by-segment assignment of membership attribute is used toward a standard *defuzzification procedure*[42] using the *centroid method*, and the *centroidal level* is marked on the score scale of 0–1 as shown. Furthermore, a slant line is drawn from the highest score value obtained in each ensemble simulations indicated earlier, passing through the centroid value. A horizontal line drawn through the centroid value intersecting the slant line defines the defuzzified delineation location at a point, d_z along the *abscissa* of the windo-segments.[23,42]

3.10 Biosequence Comparison in the Spectral Domain

This section outlines the concepts of spectral domain visualization of a data sequence that can be applied in biosequence comparison studies. Essentially, the underlying considerations refer to the classical *statistical spectral analysis* applied to a data sequence, which can be viewed as occurrences of certain events/epochs, discretely or continuously along a spatial or temporal dimension. In biosequence contexts, such a data sequence represents epochs of occurrence of biological residues along the stretch of the sequence length. They can be subjected to spatial analyses, and the associated statistical features can be extracted in terms of the degree of randomness or variability in the spatial series. (It is similar to the time-series analyses of a temporal pattern of a dataset.)

Alternatively, such a spatial series of data can also be transformed into the spectral domain, which decomposes the spatial entity series into sine/cosine components. (This is akin to Fourier transform of a time-domain dataset), and the corresponding "spectral" aspects of these components are implicitly ascertained to elucidate the underlying features of the data sequence. The decomposition mentioned would involve attributing appropriate weighting functions on the sine/cosine components, and the spectral components, in essence depict the sum of such weighted functions. Outlined below are the salient details on the fundamentals of spectral analysis, and some of the commonly used methods of such analysis *vis-à-vis* biosequences are described.[43–61]

When a sequence is represented in its weighted form of sine or cosine components, its power spectrum estimates the distribution of the "power" profile of the data (signal) across the frequency for a given (a finite) record of the input sequence data. In real-world situations, there are two types of such signals corresponding to a given sequence, namely, (i) deterministic and (ii) random types. The deterministic entities are those whose values can be predicted at any input condition by some kind of a mathematical model. And, the random entities are characterized by the statistics attached with it. As such, they are implicated by the underlying probabilistic model. Hence, they are not amenable for any "deterministically" predictive models. Following the efforts due to Joseph Fourier, the concept of spectral expansion of a periodic function was proposed by Dirichlet. Relevant to such Fourier analysis, the heuristics that apply can be summarized as follows.[43–61]

- ❖ *Spectral analysis of functions*: Decomposition of a function into a set of weighted sine or cosine functions, called *spectral components.*
- ❖ *Spectral density or "spectrum"*: The weighting function on the decomposition represents the density of spectral components.
- ❖ *Signal extraction*: The technique of extracting information-bearing parts from measured data that may contain unwanted (noisy) artifacts.
- ❖ *Spectral filtering*: A process of filtering out noisy parts from the spectral components.
- ❖ *Statistical spectrum*: The spectral components viewed along with its noisy content.

Spectral analysis of biosequences is outlined in Chapter 2, and the method of constructing the periodogram is described. Relevant details refer to spectral domain biosequence comparison elucidating the associated similarity or dissimilarity relation between a pair of sequences, (for example, the forward and reverse complementary strands of a DNA), and the cross-correlation measure between their statistics is determined.

That is, considering two test sequences (namely, the forward and reverse complementary strands of a DNA), the constituent residues, namely, the 64 trinucleotides (codon triplets) of each sequence occur with the probabilities p_i and q_i, respectively, where $i = 64$ depicting. The set $\{p_i\}$ and $\{q_i\}$, for example, can be availed from the GenBank dataset for a given species, such as the human. Furthermore, $\sum_{64} p_i = \sum_{64} q_i = 1$.

In terms of p_i and q_i, the sequences can be cross-compared using correlation measure in spectral domain by a procedure narrated *via* a pseudocode outlined in Chapter 2. Relevant computed result on power spectral density (PSD) is reproduced below as Fig. 3.9. The results shown are consistent with the observed "noisy" PSD exhibiting "flat" spectral characteristics.

That is, the periodogram computed yielding the PSD corresponds to a (Gaussian) white noise, signifying that the two test sequences are highly correlated. In other words, the symmetry features of the forward and reverse complementary strands of a DNA are justified as indicated earlier. Furthermore, the similarity–dissimilarity relation between the test sequences being compared can also be addressed in terms of the cross-correlation measure. As stated before, the cross-correlation between two random processes decides its predictability or similarity between them.

Typically, the cross-correlation between two random sequences (that are almost similar), *versus* the window frame of observation of the random occurrences (as in a random event-space representing the residue statistics in a biosequence) follows somewhat a triangular shape, but with irregularities on the edges representing the noise or difference artifacts in the sequences.

Shown in Fig. 3.10 is the cross-correlation function (CCF) of the forward and reverse complementary strands of the human DNA with irregularities on the edges representing a "noise." (The spectrum assessed and shown in Fig. 3.9 corresponds to the PSD estimated on this noise.) Should the test sequences be exactly similar, the CCF would degenerate to an autocorrelation function (ACF), for which the computed details would correspond to an ideal triangular function of time-signal window length.

More details and additional examples on spectral domain analysis of biosequences are presented in Chapter 8 with reference to comparing

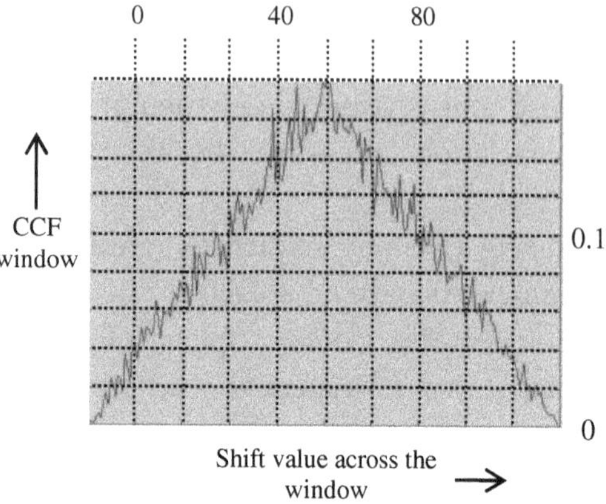

Fig. 3.10 Cross-correlation function (CCF) evaluated between the two test strands of the human DNA. It approximates to a triangular autocorrelation function (ACF) plot with noisy artifacts confirming highly correlated similarity existing between the test sequences.

the sequences of serovars of a virus. Thus, commensurate with the details furnished, the feasibility of spectral domain approach is promising in genomic signal processing effort in comparing biological sequences. The exhaustive studies related to the frequency domain analyses of genomic sequences can be found in Benson,[43] Benson,[44] Cheever *et al.*,[45] Zhou *et al.*,[46] Anastassiou,[47] Anastassiou *et al.*,[48] Rao and Swamy,[49] Jeng *et al.*,[50] De Sousa Vieira,[51] Chen *et al.*,[52] Isohata and Hayashi,[53] Herzel and Grosse,[54] Herzel *et al.*,[55] Lee and Luo,[56] Fukushima *et al.*,[57] Nyeo *et al.*,[58] Hanson,[59] Silverman and Linsker,[60] and Tiwari *et al.*[61]

3.11 Review Section: Examples and Problems

This section provides a tutorial that outlines the summary of this chapter followed by an exemplar set of solved exercises for self-study endeavors. Furthermore, do-it-yourself versions of rider problems (with solution hints/answers as necessary) are also assigned to enhance the scope to understand the subject matter deliberated across the chapter. In addition, project exercises compatible as term-papers, directed independent studies (DIS) and/or open-questions toward basic and/or advanced research pursuits are furnished.

Tutorial T-3.1

A Review Note on Genomics and the Central Dogma

The essence of genomic and proteomic considerations detailing genes, codons (exons and introns), AAs, and central dogma of molecular biology are explained earlier in Chapter 1. In short, the passage of central dogma conforms to transitions of DNA-to-RNA and eventual translation to protein. This involves three versions of RNA, namely, messenger RNA (mRNA), transfer RNA (tRNA), and ribosomal RNA (rRNA) as outlined in Chapter 1. As regard to the underlying details of central dogma, examples, problems, and project exercises are furnished in Chapter 1. Also, additional details on reading frames and ORFs are furnished therein. A review-reading on such genomic and proteomic considerations is useful in addressing the assignments of this section where more genomic and proteomic sequence-related examples and exercises are presented consistent with the contents of this chapter.

Tutorial T-3.2

A Review Note: Dot-Plot and Related Exercises

A basic visual approach method of comparing two sequences is known as the *dot-plot*.[62,63] It represents a graphical representation of comparing two sequences A and B (whose lengths can be different), and ideally, they are almost similar.

Suppose a rectangular matrix is constructed with residues of A mapped along x-axis and those of B along y-axis. Next, the following steps are carried out.

Step 1: The matrix is first filled with zeros.

Step 2: Each of its cells $\mathbf{x}_i\,\mathbf{y}_j$ (i = 1, 2, ... length of A; j = 1, 2, ... length of B) is assigned a value indicating level of similarity between the two residue (A_i and B_i) positions.

Step 3: *If* $A_i \equiv B_j$, cell value is equal to 1; *otherwise,* the cell value is equal to 0.

The following example in Fig. 3.11 is illustrative of using the above steps in constructing a dot-plot.

	T	T	G	C	A	A	A	C	G	C
T	**1**	1	0	0	0	0	0	0	0	0
T		**1**	0	0	0	0	0	0	0	0
G			**1**	0	0	0	0	0	1	0
C				**1**	0	0	0	1	0	1
A					**1**	1	1	0	0	0
A						**1**	1	0	0	0
A							**1**	0	0	0
C								**1**	0	1
G									**1**	0
C										**1**

Fig. 3.11 Dot-plot of a sequence constructed with itself: It is a symmetric matrix with the dots (1's) along its diagonal as shown.

Example Ex-3.1

Considering the given sequence (X) shown below, construct the dot-plot against itself.

X: T T G C A A A C G C

Solution

A rectangular matrix is constructed with residues of X mapped along x- and y-axes as in Fig. 3.12. Next, the steps indicated above are carried out.

Step 1: The matrix is first filled with zeros.

Step 2: Each of its cells $\mathbf{x}_i$ and $\mathbf{x}_j$ with $(i = j) = 1, 2$ (length of X) is assigned a value indicating level of similarity between the two residue (X_i and X_j) positions.

Step 3: Whenever $X_i \equiv X_j$, cell value is equal to 1; *otherwise,* the cell value is equal to 0 as illustrated in Fig. 3.11.

Example Ex-3.2

Considering two distinct sequences (x and y) shown below, construct a relevant dot-plot.

x: T T A C A C A C G C
y: C T A C A A A C G C

Solution/Result

	T	T	A	C	A	C	A	C	G	C
C				1		1		1		1
T	1	1								
A			1		1		1			
C				1		1		1		1
A			1		1		1			
A			1		1		1			
A			1		1		1			
C				1		1		1		1
G									1	
C				1		1		1		1

Fig. 3.12 Dot-plot of a sequence constructed with another sequence: It is a non-symmetric matrix as shown. (Unfilled cells contain, "0's").

Tutorial T-3.3

On Using "Dotlet" Program

Making of a dot-plot is simple as illustrated above when the sequence length is very short. In real-world biosequences, the length may stretch into hundreds of bps, and in such cases, use of computer programs is necessary For example, a popular program thereof is called "Dotlet" written by Marco Pagni and Thomas Junier of the Swiss Institute of Bioinformatics in Epalinges, Switzerland.[29] The Dotlet source code is available free of charge for academic users, and the distribution of Dotlet is facilitated (as indicated in Junier and Pagni[29]) *via* ftp://ftp.isrec.isb-sib.ch/pub/software/java/dotlet. The Dotlet output results conform to plots presented *via* a web browser.

Example Ex-3.3

Find the dot-plot of the sequence pairs: ID #s: P05049 and ID #s: P08246 using Dotlet program available on the web.

Solution

First sequence: ID #: P05049[64]

Description:	Serine protease snake EC = 3.4.21-
Source organism:	Drosophila melanogaster (Fruit fly) (NCBI taxonomy ID 7227) View Pfam proteome data
Length:	435 amino acids (AAs)
Reference Proteome:	✓

Second input sequence: ID #: P08246[63]

Description:	Neutrophil elastase EC = 3.4.21.37
Source organism:	Homo sapiens (Human) (NCBI taxonomy ID 9606) View Pfam proteome data
Length:	267 AAs
Reference Proteome:	✓

Dotlet program indicated earlier is intended to compare sequences by the diagonal plot method.[29] It is designed to be platform-independent and to run in a web browser, thus enabling majority of researchers to use it. This applet can be tested at: *http://www.isrec.isb-sib.ch/java/dotlet/Dotlet.html*, and its source code is available upon request[63] at *Thomas. Junier — Marco.Pagni@isrec.unil.ch.* Full documentation about *dotlet* is also available from the above URL.

Step 1: The query sequence data/details are available at EMBL-EBI,[64] EMBL-EBI.[65]

Step 2: Guest "login" of the dotlet opens the screen where the first and second query sequence data are specified as input data. Place ID #s: P05049 sequence data is typed or pasted as per interactive instructions. The screen display can then be captured. Similarly, ID #: P08246 sequence data are typed or pasted as per interactive instructions, and the screen display is captured.

Step 3: Performing the "click" on the screen display, dot-plots of the two sequences are displayed as — horizontal: seq_1 and vertical: seq_2. Matrix is availed with: Blosum 62; sliding window:15; zoom 1:1; score range 60–165 and grey scale: 0%–100%.

Tutorial T-3.4

Sequence Comparison via Similarity Concepts and Notions of "Distance": A Review Notes, Examples, and Exercises

Given two character strings, the measures and notions of "distance" described in Chapter 2 can be applied to assess similarity features between the strings of biosequences in terms of: (i) statistical distances in Euclidian-distance sense (such as with MD and its variations); (ii) Hamming distance concepts; and (iii) LD (edit distance) considerations. The following summary details review relevant topics of Chapter 2 for easy cross-reference.

Hamming distance: As indicated before, the Hamming distance between two strings of equal length refers to the number of positions at which the corresponding symbols are different, and it measures the minimum number of *substitutions* required to change one string into the other or the minimum number of *errors* that could have transformed one string into the other. It implicitly provides a similarity perspective between the entities being compared. The examples and problems of Chapter 2 are illustrative of the underlying considerations.

Example Ex-3.4

Given the following two binary sequences X and Y, plot the HD between the binary digits across the 0 to 72 binary residues divided into 18 windows (each with four residues). Hence, identify the location(s) of the substring pairs wherein similarity features are dominantly seen.

Each window is set to include four residues. The HD values for each window across X *versus* Y in Fig. 3.13 are calculated. Then, the window # *versus* HD value is plotted as shown in Fig. 3.14. Since the test strings are taken in binary format, the HD is decided by XOR operation across the residues one below the other in the strings, and the number of 1's are counted for each window in the resulting XOR output string, depicting the HD. The computed HD values show a minimum (zero value) at two segments with ID #8 and #14 in Fig. 3.13. These substring locations exhibit most common similarity features between the test binary stretches, X and Y.

X	0110	1110	0101	1010	0110	1100	0111	1111	0100
	1100	1101	0101	1100	0110	1110	0010	1001	1010
Y	1010	0010	1001	1011	1001	0110	0011	1111	1010
	1110	0110	1010	0001	0110	1000	1111	1100	0101

Solution

Window ID	1	2	3	4	5	6	7	8	9
X	0110	1110	0101	1010	0110	1100	0111	1111	0100
Y	1010	0010	1001	1011	1001	0110	0011	1111	1010
HD	2	2	2	1	4	2	1	**0**	3

Window ID	10	11	12	13	14	15	16	17	18
X	1100	1101	0101	1100	0110	1110	0010	1001	1010
Y	1110	0110	1010	0001	0110	1000	1111	1100	0101
HD	1	3	4	3	**0**	2	3	2	4

Fig. 3.13 Estimated HD values in each window.

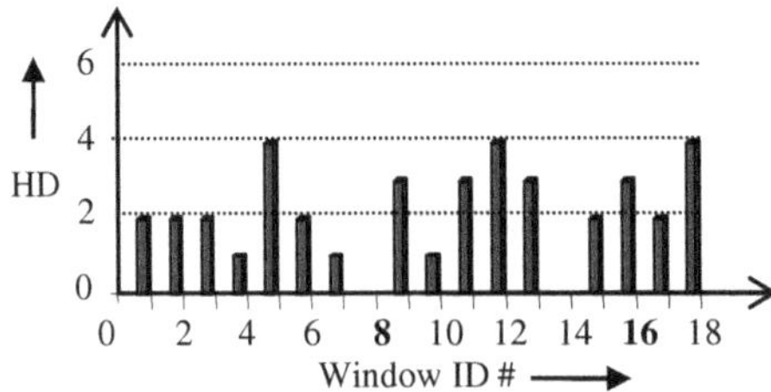

Fig. 3.14 Window ID number *versus* estimated HD values.

Example Ex-3.5

For the two test binary sequences *X* and *Y*, indicated in Example Ex-3.4, plot Kullback–Leibler (KL) measure between the test strings. Hence, confirm the most similar substring locations between them as decided *via* HD measure in the previous problem.

Solution

Window ID #	1	2	3	4	5	6	7	8	9
X	0110	1110	0101	1010	0110	1100	0111	1111	0100
$p(0)$	0.50	0.25	0.50	0.50	0.50	0.50	0.25	10^{-3}	0.75
$p(1)$	0.50	0.75	0.50	0.50	0.50	0.50	0.75	1.00	0.25
Y	1010	0010	1001	1011	1001	0110	0011	1111	1010
$q(0)$	0.50	0.75	0.50	0.25	0.50	0.50	0.50	10^{-3}	0.50
$q(1)$	0.50	0.25	0.50	0.75	0.50	0.50	0.50	1.00	0.50
KL	0.00	0.27	0.00	0.07	0.00	0.00	0.07	0.00	0.07

Window ID #	10	11	12	13	14	15	16	17	18
X	1100	1101	0101	1100	0110	1110	0010	1001	1010
$p(0)$	0.50	0.25	0.50	0.50	0.50	0.25	0.75	0.50	0.50
$p(1)$	0.50	0.75	0.50	0.50	0.50	0.75	0.25	0.50	0.50
Y	1110	0110	1010	0001	0110	1000	1111	1100	010 1
$q(0)$	0.25	0.50	0.50	0.75	0.50	0.75	10^{-3}	0.50	0.50
$q(1)$	0.75	0.50	0.50	0.25	0.50	0.25	1.00	0.50	0.50
KL	0.07	0.07	0.00	0.07	**0.00**	0.27	2.80	0.00	0.00

Fig. 3.15 Estimated (normalized) window-by-window KL measure values between the test strings, *X* and *Y*.

Again, by selecting a window size of four residues, the KL measure is computed for each window and the estimated measure, namely, KL = (KL1 + KL2) *versus* window # across the test strings is estimated (Fig. 3.15).

By depicting $p(0)$ as the probability of 0 in a chosen window and $q(1)$ as the probability of 1 in that window, KL1 and KL2 are determined as follows: KL1 = $(p(0)\ \log_e[(p(0)/q(1)])_{\text{window \#1}} + \cdots$ and KL2 = $(q(1)\ \log_e[(q(1)/p(0)])_{\text{window \#1}} + \cdots$. Hence, KL = $(1/4) \times [p(0) \times \log_e[(p(0)/q(0)] + P(1) \times \log_e[P(1)/q(1)] + q(0) \times \log_e [q(0)/p(0)] + p(1) \times \log_e[q(1)/p(1)]$. (For the cases in which $p(\cdot) = 0$, a very small value, say 0.000001, can be used to avoid overflow error in the computations.)

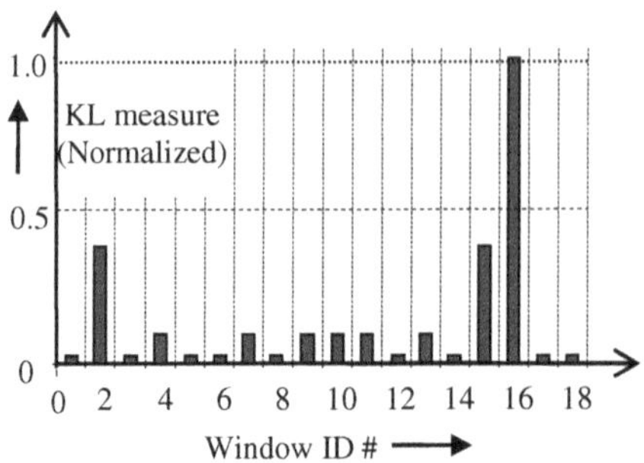

Fig. 3.16 Window ID number *versus* estimated (normalized) KL measure.

For the given two test binary sequences X and Y, the plot of the estimated (normalized) Kullback–Leibler (KL) measures *versus* the window numbers are shown in Fig. 3.16 where it can be observed that the KL measure discerns the test sequences with better resolution than the HD measure advocated (on the same pair of sequences), that is, the most similar substring locations between X and Y as decided by KL metric are subregions (with KL measure $\rightarrow 0$), when considered in the entropic plane.

Tutorial T-3.5

Project Exercises in Biosequence Comparison with Distance Measures

This note on project exercises refers to applying analytical concepts and computing methods to solve a problem comprehensively on biomolecular sequences.

The exercise also involves writing a report on the project and elaborates results obtained with necessary discussions conclusions and extension feasibilities. The following exemplars can serve as guidelines to develop more pertinent pedagogy toward classroom efforts and research endeavors. Such pursuits can culminate in eventual directed independent and thesis/dissertation studies.

Project-Exercise PE-3.1

This exercise refers to developing a project on applying and computing the distance or divergence metrics in comparing a given pair of sequences.

Suppose a human codon string (designated as X) of 6400 residues is emulated as per codon statistics with codon probabilities of occurrence $p_{i = 1,2,\ldots, 64}$ (as in Chapter 1: Table 1A-7, *Homo sapiens*: Relative *CUF* of the 64 triplets). In parallel, suppose a junk codon sequence (designated as Y) with triplets attributed with equiprobable probabilities, $q_{i = 1,2,\ldots, 64}$ (each equal to 1/64) is also emulated to depict a purely non-codon sequence.

The problem involved is to develop a computer code to deduce the distance or divergence measure values between X and Y across the stretch of sequence length and plot the deduced values as a function of a set of segmented windows along the sequence. The sequence (X) of a test (human) DNA codon composition, (say of length 6400 residues) is first emulated to fill randomly $(1/64) \times 6400 = 100$ locations with 64 triplets as per the proportions of occurrence dictated by the probability values listed (as in Chapter 1: Table 1A-7 — *Homo sapiens*: Relative *CUF* of the 64 triplets), and corresponding to the same length of 6400 residues, a junk sequence (Y) of triplets equally proportioned and constituted by residues randomly placed along the sequence length (as per uniform distribution) is also emulated.

Next, the stretches of sequences X and Y are divided into, say, 100 windows each containing 64 resides. In the gaps of the segmented regimes of windows of X and Y, the frequencies of occurrence of the triplets encountered are noted, and in terms of the relevant values obtained, window-by-window of X and Y, the underlying statistics are contrasted pursuing a moving-window procedure, using any one of the distance and/or divergence metrics. The values computed *versus* the designated window are tabulated and plotted. Exemplary computed results obtained with relevant computations are illustrated in Fig. 3.17.

They correspond to computed data (pertinent to an ensemble average of 100 runs on the test DNA string) obtained with Jensen–Shannon (JS) metric plotted (Fig. 3.17) as a functions of the pointer positions across the test DNA string: The test codon sequences adopted have codon statistics (i.e., relative *CUF* of the 64 triplets) of the following species: *Homo sapiens*; (b) *E. coli*; (c) *Methanocaldococcus jannaschii*; and (d) *Rickettsia prowazekii*. Relevant codon usage probabilities are accessed from Codon Usage Database[41] and are listed in Chapter 1, Table 1A-7.

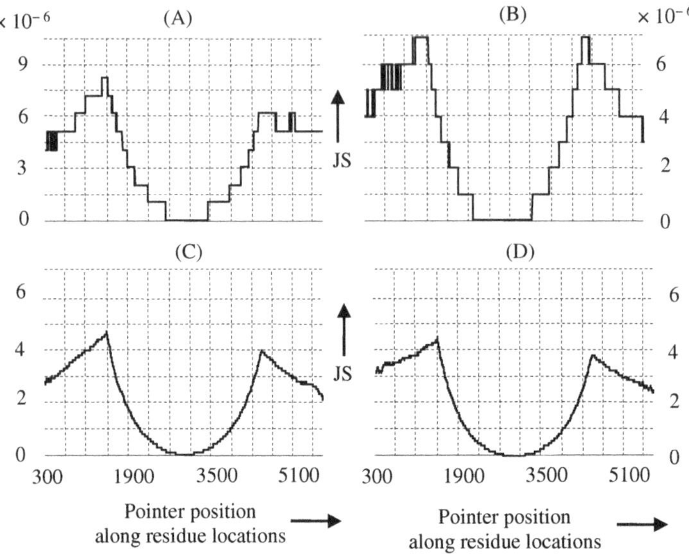

Fig. 3.17 Computed (ensemble average of 100 runs on the test DNA strings) of Jensen–Shannon (JS) measures *versus* the pointer positions along the residues across test DNA strings of the species: (A) *Homo sapiens*; (B) *Escherichia coli*; (C) *Methanocaldococcus jannaschii*; and (D) *Rickettsia prowazekii*.

Project-Exercise PE-3.2

Repeat Project-Exercise PE-3.1 by applying relevant concept and computations on other distance or divergence metrics with codon sequence data of actual biological sequences. That is, *in lieu* of using an emulated biological sequence, a real-world biosequence should be accessed from a database and contrasted against an emulated junk sequence.

The actual biological sequence data can be accessed from relevant genome data bank, for example: The NCBI database (www.ncbi.nlm.nih.gov/), the European Molecular Biology Laboratory (EMBL) database,[64–66] and the DNA Database of Japan (DDBJ) database.[67] These databases collect all publicly available DNA, RNA, and protein sequence data and make it available for free.

Tutorial T-3.6

Physico-Chemical Properties of Biological Sequence Residues: A Tutorial Note, Examples, and Exercises

Comparing biosequences *via* the underlying physico-chemical characteristics of sequence residues indicated in an earlier section amounts to the usage of electronic charge and hydrophobicity scoring based on the abstract notions on the extents (high or low) of charge-content and/or hydrophobicity associated with a DNA sequence. The relative extent of other parameters, like amphipalic alpha-helix and/or beta-sheet, also would enable scoring on sequence comparison.

Example Ex-3.6

This example enunciates a study performed to ascertain the energetics profile of a test ssDNA pertinent to human parvovirus B-19 virus[68,69] using the concept of conformational free-energy that prevails between nearest-neighbor (NN) residues on the sequence, and hence, the advocated procedure implies comparing sub-segments of the test sequence so as to ascertain the segment-by-segment contrasts of energetics profile across the whole sequence.

Structurally, the test ssDNA indicated contains a small, linear genome of 5.6 kbp length, which harbors two identical *inverted terminal repeats* (ITRs) (that serve as an origin of DNA replication in the host cell). The nucleotide sequence of B-19 *parvoviridae* (NC_000883) is available in NCBI Genbank database.

This viral ssDNA assumes invariably a hairpin format (for structural stability) and this hairpin form consists of a base-paired stem structure plus a loop sequence having unpaired or mismatched nucleotides. Knowing the profile of hairpin structures in such viral DNAs is pertinent to: (i) Understand virus replication process (ii) in drug synthesis applications, where a relevant compound being sought, may act as a binding agent in a specific DNA inhibiting the replication of certain viruses.

In general, as regard to the framework of overall folding of DNA structure into bulge and hairpin formats, exclusive research in bioinformatic perspectives is often needed (adjunct to wet studies) so as to ascertain their energetics profile responsible for the stability of the genome backbone. Such stability dynamics relies on free-energy minimization specified by NN parametric attributes of base-pairs in the test sequence. That is, the stability in question conforms to the rules stipulated by each base-pair depending only on the most adjacent pairs with the associated total free-energy being the sum of each contribution of the neighbors.

The underlying considerations are as follows: Watson–Crick base-pairs are significant motifs, whose thermodynamic aspects can be well represented by the NN model that indicates the stability of the base-pairs being dependent on the identity of the adjacent pairs. Known generally as *individual NN* (or INN) model, it implies a preferential stacking of energetically conducive pairs with loop-initiation leading to an eventual bulged and/or hairpin structure. Totally, the associated conformational free-energy is constituted by paired and unpaired nucleotide stacking in the bend as well as by the nucleotides at the loop. The free-energy increments of the base-pairs in the sequence are counted as stacks of adjacent pairs. For example, the consecutive CG base-pairs are worth about (–3.3 kcal/mol).[70,71] The loop region formed normally has unfavorable increments called *loop initiation energy* that largely reflect an entropic cost expended in constraining the nucleotides within the loop.

For example, the hairpin loop made of four nucleotides may have an initiation of energy as high as +5.6 kcal/mol. Mostly, the unpaired nucleotides in the loop contribute favorable energy increments. From the literature indicated above, a set of approximate conformational free-energy can be gathered on the basis of NN considerations.

Suppose the A, C, T, G bases are translated with 0, 1 scoring (with 1 indicating WC pair and 0 denoting non-WC (NWC-pairing). Hence, by considering, –3.3 kcal/mol as zero reference (depicting the lowest energy level), Table 3.1 can be constructed, which depicts the data on relative energy values compiled from Mathews *et al.*,[71] and Xia *et al.*,[72,73] for NN composition of four adjacent WC and or NWC as shown. That is, indicated in Table 3.1 are four adjacent matched or unmatched WC pairs along the sequence structure with the assigned scores of 1 (denoting WC and 0 for pairs NWC).

Apart from genetic information (in Shannon sense) inherent to the test ssDNA, the adjacent nucleotide neighbors that prevail in the sequence dictate the associated Gibbs free-energy profile (at a given temperature) enabling the stability of the single strand. In the example indicated, an ensemble set of four nucleotide neighbors occurring sequentially along the test genome is considered, and the associated energy values are specified by a scoring scheme. Hence, the variation of (scored) energy levels along the sequence is ascertained as illustrated in Fig. 3.18. The existence and extent of minimum energy levels depicts the stability feature.

Furthermore, with reference to the stretch of 5′-end of 383 nucleotides shown in Fig. 3.19, it can be observed that the sequence tends to

Table 3.1 Arrangement of possible four adjacent WC and NWC constituents *versus* the associated free-energy values: Depiction of four WC- and/or non-WC neighbors and the corresponding (approximate) relative levels of free-energy (Scoring scheme: WC-1 and NWC-0).

Arrangement of four WC or NWC constituents with two in the center and two as left- and right-side neighbors scored by the scheme: WC-1 and NWC-0).

Rightside residues NN-R	Middle residues CE		Leftside residues NN-L	EV
1	1	1	1	0
0	1	1	1	0
1	1	1	0	0
0	1	1	0	0
1	1	0	1	1.2
1	1	0	0	2.2
0	1	0	1	2.7
0	1	0	0	6.6
1	0	1	1	1.2
1	0	1	0	2.2
0	0	1	1	2.7
0	0	1	0	6.6
1	0	0	1	4.3
1	0	0	0	8.9
0	0	0	1	8.9
0	0	0	0	10.0

NN-R: Neighbor on the rightside of centre element

CE: Center element

NN-L: Neighbor on the leftside of centre element

EV: Energy values (in kcal/mol specified relative to −3.3 kcal/mol depicting the lowest energy level and taken as zero reference level)

fold into a bulge around 365th base. The sets, {TCTG**a**} and {**t**GTCT}, on either side of the bulge, constitute what is known as the ITRs of palindromes. The bases (**a** and **t**) shown in lower case bold fonts depict the *closing pairs* in the loop. (The bases **a**TTTGG**t** in the bulge can flip-flop to a complement set of bases, namely **t**AAACC**a**; that is, the bulge can format itself in two alternative "flip" or "flop" orientations.)

Indicated in Table 3.2 is a pseudocode relevant to the exercise indicated above. It describes the underlying computations of the energetic profile using INN-model applied to a query DNA sequence (of Parvovirus B-19).

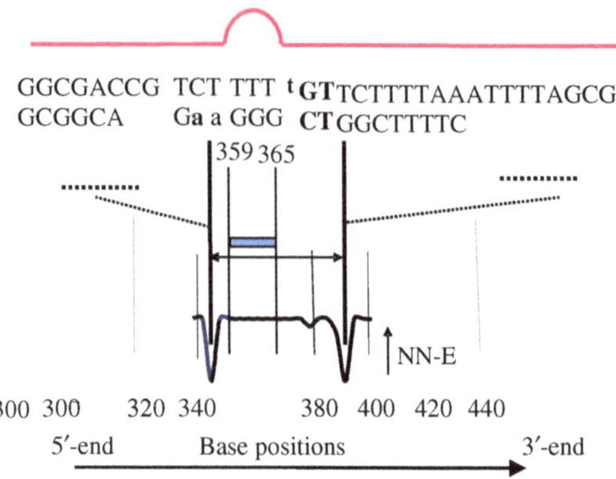

Fig. 3.18 Illustration of NN-based, computed EV profile of the bulged, test viral ssDNA sequence of B-19 virus.

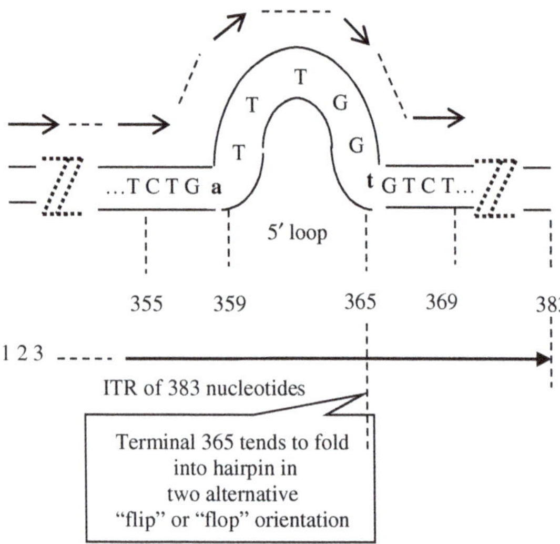

Fig. 3.19 Illustration of the start of 5′-end of the genomic sequence of B19V and the associated bulge formation.

Table 3.2 A pseudocode describing computation of the energetic profile *via* INN-model applied to a query DNA sequence (of the virus: Parvovirus B-19).

```
Initialize
%% This exercise involves a computation that refers to ascer-
   taining the bulged structural features of the ssDNA strand
   of B-19 virus at 5′-end
  Input
    → In this computation, only nucleotides from base number
      300-500 (at the 5′-end) are considered:
  5′-end  ...(1 ccaaatcaga tgccgccggt cgccgccggt aggcgggact
  tccggtacaa gatggcggac..... gaaatgacgt aattgtccgc catcttgtac
  cggaagtccc gcctaccggc ggcgaccggcggcatctgat ttgg 5581) ...
  3′-end
  [Human parvovirus B-19, complete genome NCBI Reference
  Sequence: NC _ 000883.1]

    → Test sequence from positions 300-500 (corresponding to
      5′-end) is posted as a string in terms of {A, T, G, C}
      residues
    → The test sequence is then converted into a
      [1 × n] matrix with each letter of {A, T, G, C} repre-
      senting matrix element
       ← This is named as: window-1 (W1)
    → Likewise, the reverse (3′-5′) of the test sequence is
      considered (omitting the dangling end) and converted
      into a [1 × n] matrix with each letter of {A, T, G, C}
      representing matrix element.
       ← This is named as: window-2 (W2)
Compare
  → Elements residing in W1 and W2

     ← Each element in W1 matrix is compared against the W2
       matrix in the same column
     → Comparison implies looking for (C ↔ G or G ↔ C) or (A
       ↔ T or T ↔ A) CW match
Perform
  → Scoring in binary format

  If
  (C ↔ G or G ↔ C) or (A ↔ T or T ↔ A) are seen across W1 and
  W2,
    Then
    The score is indicated as 1
          or else
          The score is set as 0
```

(Continued)

Table 3.2 (*Continued*)

```
      ← This scoring is continued across the entire window
        segments being compared, and 1s and 0s generated are
        stored in a [1 × n] matrix
Construct
   → NN-concept based energetic profile for the nucleotide
     stretch 300 to 500.
       ← These energetic values are assigned by considering two
         center elements (of 1 - 0 as per scored W1 versus W2
         mapped in terms of WC NWC ...); and, the resulting
         segments of four constituents made of {WC, NWC} each
         with one neighbor on each side of the central pair are
         identified. Pertinent EV value is assigned for each
         arrangement of [NN-R CE NN-L] as indicated in Table 3.1
           → The energy value so assigned is stored in a matrix
Print
      ← Positions in the window segment 300-500 versus the EV
        values obtained
Plot
      → Positions in the window (indexed as 300 to 500) versus
        EV value estimated in each segment is plotted as an x-y
        graph
Result/Output
   → Example: As in Neelakanta et al.[74]
   End
```

Project-Exercise PE-3.3

Repeat Example Ex-3.6 to ascertain the energetics profile of a test ssDNA pertinent to human parvovirus B-19 virus[68,69] at the 3′-end stretching across the residue positions 5401–5594 so as to demonstrate the existence of a hair-pin bend in this range of residues. Develop an appropriate computer code (MATLAB or otherwise).[74] (*Solution hint*: Figure 3.20 shown is an exemplar result of relevant exercises of the assigned project.)

Example Ex-3.7

Considering another physico-chemical parameter, it refers to a set of numerical representations of values that conform to the so-called *electron–ion interaction-pseudopotentials* (EIIP) assigned to the nucleotides in a DNA sequence as listed in Nair and Sreenadhan.[75] For a quick

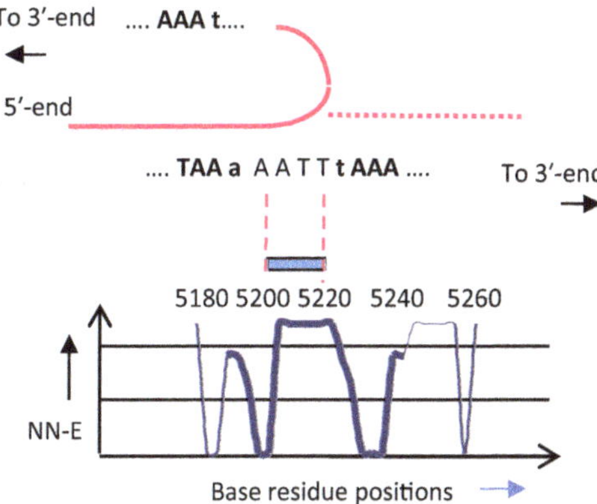

Fig. 3.20 Illustration of the start of 3′-end NN-E profile of the genomic sequence of B19V and the associated hair-pin bend formation.

reference, the EIIP values of nucleotide bases[49,75] are as follows: Base A — EIIP value = 0.1260; Base T — EIIP value = 0.1335; Base G — EIIP value = 0.0806; and Base C — EIIP value = 0.1340.

For example, considering the following character sequence: (... A A A G T A G C ...), it can be correspondingly denoted *via* EIIP values as: (... 0.1260, 0.1260, 0.1260, 0.0806, 0.1335, 0.1260, 0.0806, 0.1340 ...). In general, the EIIP values signify the physico-chemistry of effective interaction between the ions *via* the electrons leading to specify relative interionic forces of unpaired entities. The physical quantities, such as the energy associated with the bases, for example can be presented in the concise form of effective electron–ion interaction (pseudopotential) or hydropathy index (HI) as in Table 3.3. Similar to numerical representation of nucleotides as indicated above, the amino acid residues can also be specified by a physico-chemical numerical entity, such as hydrophobicity (or hydropathy) index (HI). Shown in Table 3.3 are lists of such indices.

With reference to numerically represented biosequences, the following example can be considered. It refers to comparing two arbitrary amino acid sequences *X* and *Y*:

X: 5′... A R Q C T K W H G P R T N ... 3′
Y: 5′... E D Q Y T N W G C L F V S ... 3′

The numerical values of hydrophobicity can be assigned to each amino acid residue in the sequences in a normalized format with 0 to 1 values by considering the lowest value (–4.5) of HI taken as 0 reference. The

Table 3.3 AAs sorted by increasing hydropathy index (HI) (based on previous table due to Kyte and Doolittle.[76]

AA	R	K	N	D	Q
HI	−4.5	−3.9	−3.5	−3.5	−3.5

AA	S	T	G	A	M
HI	−0.8	−0.7	−0.4	1.8	1.9

AA	E	H	P	Y	W
HI	−3.5	−3.2	−1.6	−1.3	−0.9

AA	C	F	L	V	M
HI	2.5	2.8	3.8	4.2	1.9

significance of such plots will be understood when the underlying data are transformed into spectral domain as explained in the following subsection.

Biosequence Comparison in the Spectral Domain

The concepts of spectral domain visualization of a data sequence can be applied in biosequence comparison studies. The underlying considerations refer to the classical *statistical spectral analysis* applied to a data sequence, which can be viewed as occurrences of certain events/epochs, discretely or continuously along a spatial or temporal dimension.

In biosequence contexts, a data sequence (such as EIIP values or HI) represents epochs pertinent to biological residues along the stretch of the sequence length. They can be subjected to spatial analyses, and the associated statistical features can be extracted in terms of the degree of randomness or variability in the spatial series. (It is similar to the time-series analyses of a temporal pattern of a dataset.)

The use of Fourier methods for biosequence analysis is described in Kyte and Doolittle,[76] Mabrouk,[77] and Chatterjee *et al.*[78] Furthermore, detection of similarities between DNA sequences is illustrated in Tiwari *et al.*[61] by enforcing *Fast-Fourier transform* (FFT) to ascertain the correlation between DNA sequences using complex-plane encoding. The Fourier transform method is adopted to distinguish coding and non-coding subsequences in a complete genome. A comprehensive review on genomic signal-processing method is addressed in Rao and Swamy,[49] Jeng *et al.*,[50] and De Sousa Vieira[51] where the local texture information in genome structures is extracted by Fourier spectral mapping of test sequences.

In extended contexts of using Fourier transform methods in the analysis of genomics and proteomics, *digital signal-processing* (DSP) techniques, where spectrograms are proven to be powerful tools for DNA sequence analyses providing local frequency information. Specifically, the so-called *short-time Fourier transform* (STFT) appears to provide a useful localized measure of frequency content in the spatial sequence pattern. This STFT is consistent with traditional *discrete Fourier transform* (DFT). It is applied to genome sequence analysis using sliding-window technique.

Thus, homologous DNA sequences can be implicitly specified by their power spectrum attributes.[54] Furthermore, to reflect the differences in coding structure of nucleotides, the power spectral analysis of DNA sequences is useful (e.g., with reference to the classification of bacteria), and the power spectra of DNA sequences can be presented as self-organizing maps. Such power-spectrum approach is regarded as intuitive as well as effective in reducing the dimension of the complete DNA sequences of a clustered set of species. An exclusive thesis on FFT analysis is applied to DNA sequences due to Hanson.[59] As an illustrative example, the following pseudocode in Table 3.4 is presented to outline the computation of spatial frequency characteristics of a test RNA sequence using STFT method.

A practical application of spectral domain approach in biosequence comparisons will be presented in Chapter 8 with reference to comparing viral sequences.

Project-Exercise PE-3.4

With reference to the exercise indicated earlier toward depicting two arbitrary amino acid sequences, *X* and *Y*, in terms of numerical HI values, deduce their corresponding spectral domain representations *via* STFT method and plot the spectra obtained.

X: 5′... A R Q C T K W H G P R T N ... 3′
Y: 5′... E D Q Y T N W G C L F V S ... 3′

Table 3.4 A pseudocode to compute spatial frequency characteristics of a test RNA sequence using short-time Fourier transform (STFT) method.

```
%% To compute spatial frequency characteristics of a test RNA
   sequence using STFT method
Initialize
    Identify electron-ion interaction-pseudopotentials (EIIPs)
    values, which refer to chemistry-specified electron-ion
    interaction potential for each base as listed in Table 3.1
Input
  → Test nucleotide sequences from (5′-end) to (3′-end) is
    posted as a string (numbered from I = 1 to N)
Construct
  Step I
        → Spatial frequency spectrum for the test sequence of
          the test DNA sequence based on EIIP values using STFT
Call
        ← EIIP values of nucleotide bases from Table 3.1 and
          assign them appropriately to each base encountered
          along the entire test sequence and store it in a matrix
Construct
        ← A string of EIIP values replacing {A, T, G, C} on the
          sequence
%% The STFT of the test sequence of EIIP string is then
   obtained as follows:
Suppose an analog numerical signal f(x) with x of varying
amplitude is specified (e.g., as in a DNA sequence with EIIP
values assigned to each base). Then the STFT is defined as
follows:
```

$$f(x) \leftrightarrow F(f) = \sum_{m=1}^{R=120} [f(x-m) \times w(m) \times \exp(-m\omega)] \quad f(x) \leftrightarrow F(f) = \sum_{m=1}^{R=120} [f(x-m) \times w(m) \times \exp(-m\omega)] \qquad \text{Eq. (3.1)}$$

```
Perform
        ← STFT on the test EIIP String by evaluating Eq. (3.1)
          → The whole test sequence is divided into N equally
            spaced intervals or frequency (N can be arbitrarily
            taken to match the window size improvised).
            Furthermore, ω = 2πk/N, with the substitution of
            value of k as dictated by the resolution sought.
            The subsequence spanning the window is denoted by
            w[m] with the window length (size) being R ≤ N
Result
        ← Plot the computed spatial frequency characteristics
          of the test RNA sequence
      End
```

Project-Exercise PE-3.5

A physico-chemical parameter refers to a set of approximate conformational free-energy that can be gathered on the basis of NN considerations. In Example Ex-3.7 relevant application of NN considerations was advocated to compare and contrast the sub-segments of a test sequence so as to ascertain the energetics profile across the whole sequence. The underlying aspects of energetic considerations stems from the thermodynamics of base-pairs in the biosequence as considered in Xia *et al.*,[73] and Neelakanta *et al.*[74] Expand the heuristics of the associated details.

3.12 Concluding Remarks

As a part of bioinformatic narrations on biosequences, the pertinent art of information-processing involves comparing biosequences of the diverse sets of biosystems as necessary. This chapter provides an overview on relevant bioinformatic considerations *vis-a-vis* genomic and proteomic characterizations of the central dogma of microbiology. Following introductory notes thereof, methods of comparing raw genomic sequences (without aligning) are indicated with examples and relevant exercises. Such efforts lead to deeper understanding of the subject presented across other chapters of this book. More relevant details can be seen in Arredondo[79] and additional information on biosequence analyses/comparison methods (specific to viral sequences) are furnished in Chapter 8 and in Chatterjee.[80]

References

[1] G. B. Johnson: *The Living World*. McGraw-Hill Education, New York, NY, USA: 2008.

[2] N. A. Campbell: *Biology*. The Benjamin/Cummings Publishing Company, Inc., Menlo Park, CA, USA: 1987.

[3] R. Aebersold: Quantitative proteome analysis: methods and applications. *The Journal of Infectious Diseases*, 2003, vol. 187(Supplement 2), 5315–5320.

[4] J. M. Wells and S. A. McLuckey: Collision-induced disassociation (CID) of peptides and proteins. *Methods of Enzymology*, 2005, vol. 402, 148–185.

[5] Centre for Proteomics, University of Antwerp (VITO): Available online at: *https://www.uantwerpen.be/en/research/publications-and-expertise/core-facilities/core-facilities/centre-for-proteomics/about-us/*.

[6] T. K. Attwood: Genomics: the babel of bioinformatics. *Science,* 2000, vol. 290(5491), 471–473.

[7] W. Makalowski, J. Zhang and M. S. Boguski: Comparative analysis of 1196 orthologous mouse and human full-length mRNA and protein sequences. *Genome Research,* 1996, vol. 6(9), 846–857.

[8] T. F. Smith: Comparison of biosequences. *Advances in Applied Mathematics,* 1961, vol. 2(4), 482–489.

[9] M. M. Deza and E. Deza: *Encyclopedia of Distances.* Springer Science-Business Media, Heidelberg, Germany: 2010.

[10] M. D. Malkauthekar: Analysis of Euclidean distance and Manhattan distance measure in face recognition. *Proceedings of the Third International Conference on Computational Intelligence and Information Technology* (CIIT 2013). Mumbai, India: 18–19 October 2013, pp. 503–507.

[11] M. A. Khamsi and W. A. Kirk: "§1.4 The triangle inequality in $\mathbb{R}^n$". In: *An Introduction to Metric Spaces and Fixed Point Theory.* Wiley-IEEE Press, Hoboken, NJ, USA, 2001.

[12] P. C. Mahalanobis: On the generalised distance in statistics. *Proceedings of the National Institute of Sciences of India,* 1936, vol. 2(1), 49–55.

[13] P. C. Mahalanobis: Analysis of race mixture in Bengal. *Journal of Asiatic Society of Bengal,* 1925, vol. 23, 301–333.

[14] S. Dasgupta: The evolution of the D^2-statistic of Mahalanobis. *Indian Journal of Pure and Applied Mathematics,* 1995, vol. 26(6), 485–501.

[15] A. Bhattacharyya: On a measure of divergence between two statistical populations defined by their probability distributions. *Bulletin of the Calcutta Mathematical Society,* 1943, vol. 35, 99–109.

[16] T. Kailath: The divergence and *Bhattacharyya* distance measures in signal selection. *IEEE Transactions on Communication Technology,* 1967, vol. 15(1), 52–60.

[17] J. A. Adell and P. Jodra: Exact Kolmogorov and total variation distances between some familiar discrete distributions, *Journal of Inequalities and Applications,* 2006, vol. 2006, 1–8.

[18] A. Kolmogorov: Sulla determinazione empirica di una legge di distribuzione. *Giornale dell Istituto Italiano degli Attuari,* 1933, vol. 4, 83–91.

[19] N. Smirnov: Table for estimating the goodness of fit of empirical distributions. *Annals of Mathematical Statistics,* 1948, vol. 19, 279–281.

[20] R. W. Hamming: Error detecting and error correcting codes. *The Bell System Technical Journal,* 1950, vol. 29(2), 147–160.

[21] S. B. Pandya: Binary representation of DNA sequences towards developing useful algorithms in bioinformatics. *MSEE Thesis,* Florida Atlantic University, Boca Raton, FL, USA: 2003.

[22] P. S. Neelakanta, S. Pandya and T. V. Arredondo, Binary representation of DNA sequences towards developing useful algorithms in bioinformat-

ics. *Proceedings of the 7th World Multi Conference on Systemics, Cybernetics and Informatics* (SCI 2003), Orlando, FL, USA: July 27–30, 2003, vol. VIII, pp. 195–197.

[23] P. S. Neelakanta, S. Pandya, T. V. Arredondo and D. De Groff: Heuristics of AI-based search engines for massive bioinformatic data-mining: an example of codon/non-codon delineation in a binary DNA sequence. *Presented in: 1st Indian International Conference on Artificial Intelligence* (IICAI-03), Hyderabad, India: December 18–20, 2003.

[24] P. F. Baisnee, S. Hampson and P. Baldi: Why are complementary DNA strands symmetric? *Bioinformatics,* 2002, vol. 18(8), 1021–1033.

[25] G.-C. Yuan, Y.-J. Liu and M. F. Dion: Genome-scale identification of nucleosome positions in *S. cerevisiae. Science,* 2005, vol. 309, 626–630.

[26] R. A. Fisher: The use of multiple measurements in taxonomic problems. *Annals of Human Genetics,* 1936, vol. 7(2), 1469–1809.

[27] A. J. Gibbs and G. A. McIntyre: The diagram, a method for comparing sequences. Its use with amino acid and nucleotide sequences. *European Journal of Biochemistry,* 1970, vol. 16, 1–11.

[28] E. L. Sonnhammer and R. Durbin: A dot-matrix program with dynamic threshold control suited for genomic DNA and protein sequence analysis. *Gene,* 1995, vol. 167, 1–2.

[29] T. Junier and M. Pagni: Dotlet: diagonal plots in a web browser. *Bioinformatics Applications Note,* 2000, vol. 16(2), 178–179. (Distributed *via*: *ftp://ftp.isrec.isb-sib.ch/pub/software/java/dotlet.*)

[30] J. Schultz: Introduction to dot-plots. Available online at: *http://www.code10.info/index.php%3Foption%3Dcom_content%26view%3Darticle%26id%3D64:inroduction-to-dotplots%26catid%3D52:cat_coding_algorithms_dotplots%26Itemid%3D76.*

[31] P. S. Neelakanta: *Information Theoretic Aspects of Neural Networks.* CRC Press, Inc., Boca Raton, FL, USA: 1999.

[32] J. N. Kapur and H. K. Kesavan: *Entropy Optimization Principles with Applications.* Academic Press Inc., San Diego, CA, USA: 1992.

[33] P. S. Neelakanta, T. V. Arredondo and D. De Groff: Redundancy attributes of a complex system: application to bioinformatics. *Complex Systems,* 2003, vol. 14, 215–233.

[34] K. Lichtenecker: Die Dielektrizitätskonstante natürlicher und künstlicher Mischkörper. *Physikalische Zeitschrift,* 1926, vol. 27, 115–158.

[35] P. S. Neelakanta: *Handbook of Electromagnetic Materials: Monolithic and Composite Versions and Their Applications.* CRC Press, Boca Raton, FL, USA: 1995.

[36] K. Lichtenecker and K. Rother: Die herleitung des logarithmischen mischungsgesetzes aus allegemeinen prinzipien der station-aren stromung. *Physikalische Zeitschrift,* 1931, vol. 32, 255–260.

[37] T. Zakari, J.-P. Laurent and M. Vauclin: Theoretical evidence for "Lichtenecker's mixture formulae." *Journal of Physics D: Applied Physics,* 1998, vol. 31, 1589–1594.

[38] P. Bernaola-Galván, I. Grosse, P. Carpena, J.-L. Oliver, R. Román-Roldán and H. E. Stanley: Finding borders between coding and noncoding DNA regions by an entropic segmentation method. *Physical Review Letters,* 2000, vol. 85, 1342–1345.

[39] N. Haiminen, H. Mannila and E. Terzi: Comparing segmentations by applying randomization techniques. *BMC Bioinformatics,* 2007, vol. 8, 171–178.

[40] T. V. Arredondo, P. S. Neelakanta and D. De Groff: Fuzzy attributes of a DNA complex: development of a fuzzy inference engine for codon-"junk" codon delineation. *Artificial Intelligence in Medicine,* 2005, vol. 35(1–2), 87–105.

[41] Codon usage database: Data Source - NCBI-GenBank Flat File Release 160.0 [June 15 2007]. Available online at: *https://www.kazusa.or.jp/codon/*

[42] W. van Leekwijck and E. E. Kerre: Defuzzification: criteria and classification. *Fuzzy Sets and Systems,* 1999, vol. 108(2), 159–178.

[43] D. C. Benson: Digital signal processing methods for biosequence comparison. *Nucleic Acids Research,* 1990, vol. 18(10), 3001–3006.

[44] D. C. Benson: Fourier methods for biosequence analysis. *Nucleic Acids Research,* 1990, vol. 18(21), 6305–6310.

[45] E. A. Cheever, G. C. Overton and D. B. Searls: Fast Fourier transform-based correlation of DNA sequences using complex plane encoding. *Computer Applications in the Biosciences,* 1991, vol. 7(2), 143–154.

[46] Y. Zhou, L. Zhou, Z. Yu and V. Anh: Distinguish coding and noncoding sequences in a complete genome using Fourier transform. *Proceedings of the Third International Conference on Natural Computation* (ICNC 2007), Haikou, Hainan, China: 24–27 August 2007, pp. 295–299.

[47] D. Anastassiou: Genetic data processing. *IEEE Signal Processing,* 2001, vol. 18(4), 8–20.

[48] D. Anastassiou, H. Liu and V. Varadan: Variable window binding for mutually exclusive alternative splicing. *Genome Biology,* 2006, vol. 7, Article R2.

[49] K. D. Rao and M. N. S. Swamy: Analysis of genomics and proteomics using DSP techniques. *IEEE Transactions on Circuits and Systems I (Regular Papers),* 2008, vol. 55(1), 370–378.

[50] C.-C. Jeng, I.-C. Yang, K.-L. Hsieh and C.-N. Lin: Bacteria classification on power spectrums of complete DNA sequences by self-organizing map. *Neural Information Processing– Letters and Reviews.* 2005, vol. 9(3), 53–58.

[51] M. De Sousa Vieira: Statistics of DNA sequences: a low-frequency analysis. *Physical Review E,* 1999, vol. 60(5), 5932–5937.

[52] Y. H. Chen, S. L. Nyeo and J. P. Yu: Power-laws in the complete sequences of human genome. *Journal of Biological Systems,* 2005, vol. 13(2), 105–115.

[53] Y. Isohata and M. Hayashi: Analyses of DNA base sequences for eukaryotes in terms of power spectrum method. *Japanese Journal of Applied Physics,* 2005, vol. 44(2), 1143–1146.

[54] H. Herzel and I. Grosse: Measuring correlations in symbolic sequences: *Physica A,* 1995, vol. 216, 518–542.

[55] H. Herzel, O. Weiss and E. N. Trifonov: 10–11 bp periodicities in complete genome reflect protein structure and DNA folding. *Bioinformatics,* 1999, vol. 15(3), 187–193.

[56] W. J. Lee and L. F. Luo: Periodicity of base correlation in nucleotide sequence. *Physical Review E,* 1997, vol. 56(1), 848–851.

[57] A. Fukushima, T. Ikemura, M. Kinouchi, T. Oshima, Y. Kudo, H. Mori and S. Kanaya: Periodicity in prokaryotic and eukaryotic genomes identified by power spectrum analysis. *Gene,* 2002, vol. 300(1–2), 203–211.

[58] S. L. Nyeo, I. C. Yang and C. H. Wu: Spectral classification of archaeal and bacterial genomes. *Journal of Biological Systems,* 2002, vol. 10(3), 233–241.

[59] R. W. Hanson: Fast Fourier transform analysis of DNA sequences, *BA Thesis,* The Division of Mathematics and Natural Sciences, Reed College, USA: May 2003.

[60] B. D. Silverman and R. Linsker: A measure of DNA periodicity. *Journal of Theoretical Biology,* 1986, vol. 118, 295–300.

[61] S. Tiwari, S. Ramachandran, S. Bhattacharya and R. Ramaswamy: Prediction of probable genes by Fourier analysis of genomic sequences. *Bioinformatics,* 1997, vol. 13(3), 263–270.

[62] A. Coghlan: A Little Book of R for Bioinformatics Release 0.1. Wellcome Trust Sanger Institute, Cambridge, U.K.: 2013. Available online at: http://www.cs.ukzn.ac.za/~hughm/bio/docs/a-little-book-of-r-for-bioinformatics.pdf.

[63] M. Pagni and T. Junier: Dotlet – SIB Swiss Institute of Bioinformatics, Epalinges, Switzerland. The Dotlet source code is available free of charge for academic users. The distribution is in ftp:*//ftp.isrec.isb-sib.ch/pub/software/java/dotlet.*

[64] EMBL-EBI site on protein/P05049. Available online at: *http://pfam.xfam.org/protein/P05049.*

[65] EMBL-EBI site on protein/*P08246*. Available online at: *http://pfam.xfam.org/protein/P08246.*

[66] EMBL database. Available online at: *www.ebi.ac.uk/embl/.*

[67] DDJ database. Available online at: www.ddbj.nig.ac.jp/.

[68] E. D. Heegaard and K. E. Brown: Human parvovirus B19. *Clinical Microbiology* Review, 2002, vol. 15, 485–505.

[69] Human parvovirus B19, complete genome. Available online at: *http://www.ncbi.nlm.nih.gov/nuccore/356457872* (and NCBI Reference Sequence: NC_000883.1).

[70] A. D. Baxevanis and B. F. Ouellette: *Bioinformatics: A Practical Guide to the Analyses of Genes and Proteins*. John Wiley and Sons, Hoboken, NJ, USA: 2005, pp. 146–147.

[71] D. H. Mathews, J. Sabina, M. Zuker and D. H. Turner: Expanded sequence dependence of thermodynamic parameters improves prediction of RNA secondary structure. *Journal of Molecular Biology*, 1999, vol. 288, 911–940.

[72] T. Xia, J. A. McDowell and D. H. Turner: Thermodynamics of nonsymmetric tandem mismatches adjacent to G-C base pairs in RNA. *Biochemistry*, 1997, vol. 36, 12486–12497.

[73] T. Xia, J. SantaLucia, Jr., M. E. Burkard, R. Kierzek, S. J. Schroeder, J. Xiaoqi, C. Cox and D. H. Turner: Thermodynamic parameters for an expanded nearest-neighbor model for formation of RNA duplexes with Watson-Crick base pairs. *Biochemistry*, 1998, vol. 37, 14719–14735.

[74] P. S. Neelakanta, S. Chatterjee and G. A. Thengum-Pallil: Computation of entropy and energetics profiles of a single-stranded viral DNA. *International Journal Bioinformatics and Applications*, 2011, vol. 7(3), 239–261.

[75] A. S. Nair and S. P. Sreenadhan: A coding measure scheme employing electron-ion interaction pseudopotential (EIIP). *Bioinformation*, 2006, vol. 1(6), 197–202.

[76] J. Kyte and R. F. Doolittle: A simple method for displaying the hydropathic character of a protein. *Journal of Molecular Biology*, 1982, vol. 157(1), 105–132.

[77] M. S. Mabrouk: A study of the potential of EIIP mapping method in exon prediction using the frequency domain techniques. *American Journal of Biomedical Engineering*, 2012, vol. 2(2), 17–22.

[78] S. Chatterjee, P. S. Neelakanta and M. Pavlovic: A cohesive analysis of DNA/RNA sequences via entropy, energetics and spectral-domain methods to assess genomic features across single viral diversity. *International Journal of Bioinformatics Research and Applications*, 2015, vol. 11(4), 281–307.

[79] T. V. Arredondo: Studies on Information-Theoretics Based Data-Sequence Pattern-Discriminant Algorithms: Applications in Bioinformatic Data Mining. *Ph.D. Dissertation*, Florida Atlantic University, Boca Raton, FL, USA: December 2003.

[80] S. P. Chatterjee: Bioinformatic Analysis of Viral Genomic Sequences and Concepts of Genome-specific Rational Vaccine Design. *Ph.D. Dissertation*, Florida Atlantic University, Boca Raton, FL, USA: May 2013.

Chapter 4

Pairwise and Multiple Biosequence Analyses and Next-Generation Sequencing

This chapter deals with heuristics of biomolecular sequences extended to address the following bioinformatic tasks: (i) developing dynamic-programming for meaningful alignment of pairwise and multiple biosequences; (ii) prescribing thematic schemes on scoring the integrity of alignments performed; and, (iii) investigating the details of biomolecular sequence via high-throughput sequencing (HTS) schemes (also known as next-generation sequencing, or NGS).

> *"Everything in the world is going to get sequenced" ... and such sequenced entities may show disparities upon comparison, far "any meaningful differences (even) in identical twins, behaviors or personalities are likely to have been acquired, not innate..."*
>
> *(Greg Lucier and Jeffrey Kluger)*

4.1 Introduction

Addressed in this chapter are specific details deliberated in two parts on biosequence alignment techniques: related comparison methods and high-throughput sequencing (HTS) schemes.

Part I describes alignment and subsequent comparison of pairwise- and/or multiple-sequence entities encountered in microbiological contexts. Part II is an exclusive study on sequencing details in specific instances of biomolecular contexts. Such studies termed as HTS refer to *next-generation sequencing* (NGS) useful in generating hypotheses for functional studies in bioinformatics.

4.2 Part I: Biomolecular Sequence Alignment and Comparison

As observed in earlier chapters, the science of biology covers the microcosm details of living organisms at the omic level, which expands into macrocosmic elaborations of the whole living society. The omic information manifests as genetic aspects of the biosequences, and the underlying considerations of aligning and comparing a pair- and/or multiple-biosequences are described in Part I of this chapter. Relevant tasks imply fundamental efforts in biosequence analyses facilitating *global* as well as *local* alignments, and the aligned pair of query sequences enable the assessment of underlying similarity and/or dissimilarity between them (with adequate integrity).

An extended perspective of pairwise alignment and comparison of biosequences refers to *multiple sequence alignment* (MSA) carried out to elucidate comparative details between multiple sequences of a set of organisms. Such comparisons enable deciding similarity/dissimilarity features between multiple sequences[1] as well as determining biologically significant *motifs* (across multiple query sequences) that identify certain conserved features of the species *vis-à-vis* specific biological functions.

Tutorial T-4.1

Revisiting the omics ... Inasmuch as different organisms conform to possessing distinct sets of biological information tied to their omic details, the associated genomic structures of nucleotide residues are comprised of repetitive sequence entities of three types, namely: (i) *Unique, or single-copy, sequences* (that make the major component constituting about 30–75% of the chromosomal DNA in most organisms), (ii) *highly repetitive sequences* constituting approximately 5–45% of the genome depending on the species. (Some of these sequences may also depict *satellite DNA* consisting of very large arrays of tandem repeating, non-coding DNA), and (iii) *middle-repetitive sequences* depicting components that constitute about 1–30% of a eukaryotic genome and includes sequences that are repeated from a few times to 10^5 times per genome. Biosequences and their expressed details across the genetic constituents represent a domain of statistical features, and elucidating the associated stochastic profiling of biosequences is an implied effort in biosequence analyses. Adopting statistical features in sequence comparisons and scoring warrants strategies such as *hidden Markov method* (HMM) and relevant models.[2]

On Aligning Biomolecular Sequences

Sequence alignment? It is a way of arranging the sequences of DNA, RNA, or protein so as to identify regions of similarity that may be a consequence of functional, structural, or evolutionary relationships between the sequences being aligned, and it enables analyzing the evolutionary and functional relationships between molecular sequences. The alignment method involves arranging a given pair or set of biosequences (placing one below the other) such that in the post-arrangement disposition, the sequences in question will exhibit optimally maximum extent of similarity across the residues one below the other. In all, sequence alignment enables finding the following: (i) the "best" overall (global) alignment between entire stretches of any given pair or multiple query sequences and (ii) alignment of sub-segment (local) regions of compared entities (falls into two categories, namely, *global* and *local alignment* procedures).

Assessing similar features between sequences implies two avenues, namely (i) Similarity assessment by considering unformatted raw sequences (without doing any alignments between them) in terms of their constituent residues and (ii) comparing aligned sequences, first by performing an arrangement procedure (as indicated above) and then compare them subsequently as regard to their similar and/or dissimilar characteristics.

Global and Local Sequence Alignments

Computational algorithms advocated in sequence alignment problems denote formally, a "correct" method, and once such an alignment is done, the quality of alignment procedure done is evaluated by a *scoring function*. Whenever an alignment yields the maximum possible score, it is said to be globally optimal or true (correct) alignment procedure.[3]

The global alignment is based on *dynamic-programming*, for example, the so-called *Needleman–Wunsch* (NW) *algorithm*,[4] developed in 1970. It is intended to align protein or nucleotide sequences. It is widely used in optimal global alignment tasks, as will be explained in detail later with examples.

Whenever two sequences have only a few regions of similarity separated by divergent regions of variable length, the similar regions may not be aligned using global optimization strategy (because the gap penalties incurred would be prohibitive). Hence, it is desirable to have a way to identify aligned sub-segments whose score is *locally optimal*. A local sequence alignment procedure intended exclusively for this purpose refers to another version of dynamic programming based on, for example, the so-called *Smith–Waterman* (SW) *algorithm*.[5,6] It determines similar regions between two strings of nucleotide or protein sequences. *In lieu* of viewing the entire (total) sequence length, the SW algorithm compares sub-segments of all possible lengths and optimizes the alignment of such subsequences *via* a similarity measure.

Using aligned sequences, the integrity of alignment is decided by typical scoring methods like *substitution matrix* and *gap-scoring schemes*, and improved alternatives (with reference to SW algorithm) exist possessing better scaling[7] and more accuracy.[8]

Pairwise Alignment

As observed earlier, sequence alignment in bioinformatic contexts refers to arranging the sequences of DNA, RNA, or protein so as to identify regions of similarity implied by consequential aspects of functional, structural, or evolutionary relationships between the sequences leading to viable biological inferences.[9] The underlying aspects of aligning biosequences are as follows: two given sequences are brought into a state of alignment by writing them (one below the other) across in two rows, and the alignment ideally renders the placement of most of identical or similar characters (called *matches*) to be one below the other in the rows. The rest of non-identical pairs of characters in the rows seen (at a site or column) denote the *mismatches*. Thus, aligned sequences of nucleotide or amino acid residues are typically represented as rows within a matrix, and gaps could be inserted in either or both sequences between the residues as necessary so that the resulting matches are optimally maximized. Such matched regions in the sequences being compared could be seen as subsequences depicting conserved and biologically significant entities of bioinformatic interest.

Considering the two possible biosequence alignment schemes, namely, (i) *global* and (ii) *local* alignments,[10] the global alignment refers to aligning every residue against every other sequence. Such alignments are most useful when the sequences in the query set are similar and of almost equal size. A classical global alignment technique, as stated earlier is the NW algorithm,[4] is widely used for optimal global alignment of high quality. It essentially divides a large problem (say, given the full sequence) into a series of smaller problems and uses the solutions to the smaller problems to reconstruct a solution to the larger problem. Such a method refers to *dynamic programming* (also known as *dynamic optimization*).[11]

Dynamic programming methods also enable local alignments using techniques, such as the so-called SW algorithm. The local alignments are "more useful for dissimilar sequences that are suspected to contain regions of similarity or similar sequence motifs within their larger sequence context", and relevantly, the SW algorithm is adopted as a general local alignment method. Based on dynamic programming, the SW method performs local sequence alignment leading to finding similar (local) segments between two strings, for example, nucleotide or protein sequences. Local alignment implies the following: *In lieu* of considering the entire or the total sequence length, the SW algorithm compares segments of all possible lengths and optimizes the similarity measure thereof. It was originally proposed by Smith and Waterman in 1981.[5] Also based on dynamic programming algorithm, the SW method enables a guaranteed finding of optimal local alignment with respect to the scoring system (such as the substitution matrix and the gap-scoring scheme).[7,12]

There are also hybrid methods, known as *semiglobal* or *"glocal"* (meaning, *global-local*) methods used as alignment procedures that explicitly include the start and end of one or the other sequence. This is important, for example, whenever the downstream part of one sequence overlaps with the upstream part of the other sequence. In such cases, neither NW algorithm nor SW algorithm could be compatible, inasmuch as, "a global alignment would attempt to force the alignment to extend beyond the region of overlap, while a local alignment might not fully cover the region of overlap." Another situation where the semiglobal alignment is useful is when one sequence is short (for example, a gene sequence) and the other is very long (for example, a chromosome sequence). In that case, the short-sequence should be globally aligned; whereas only a local alignment is desired for the long sequence.[13] The semiglobal/local (glocal) alignment is thus a method useful in detecting overlap regions in a pair of sequences. The glocal alignment on a pair of two nucleotide sequences u and v is illustrated below with a simple example.

Example Ex-4.1

Statement of the Exercise

Considering the following two sequences (*u* and *v*), perform the glocal alignment between them *via* visual comparison:

u: TGTCTGTGGGTGG

v: TGCTTG

Solution

Required glocal (semiglobal) alignment can be done as follows:

```
u: T G T C T G - T G G G T G G
   |  |       |  |
v: T G C T T G
```

Global, Local, and Glocal Alignments: A Comparison

Commensurate with the earlier definitions,[14–17] a comparative illustration of global, local, and glocal alignments is shown in Figs. 4.1A and 4.1B. Given two query sequences *u* and *v*, the two categories of alignments, namely, (A) global and (B) local types, are shown in Fig. 4.1A, and shown in Fig. 4.1B is the glocal version (C).

Referring to Fig. 4.1A, the global alignment procedure performs an end-to-end alignment of *u* and *v*. In contrast, the local alignment finds

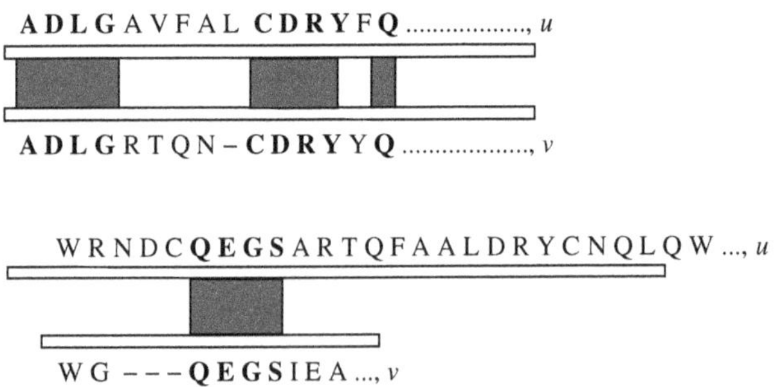

Fig. 4.1A Methods of biosequence alignment: global and local.

(A) Global — This is an end-to-end alignment of *u* and *v*.

(B) Local — This refers to finding the highest scoring alignment between *u′* (a substring of *u*) and *v′* (a substring of *v*). It is an alignment method useful in finding similar regions in strings that may not be globally similar.

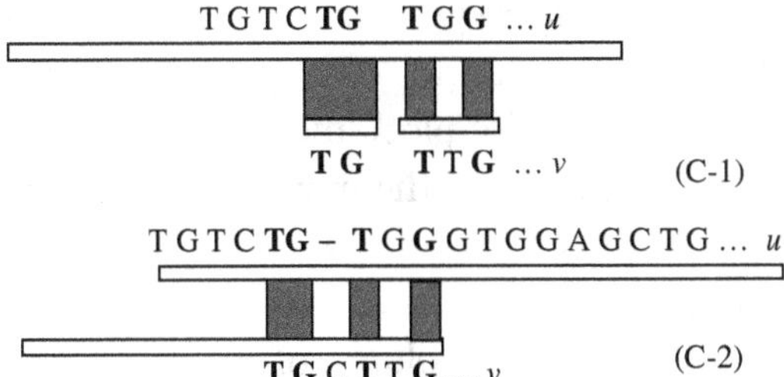

Fig. 4.1B Methods of biosequence alignment — C-1 and C-2: semiglobal/glocal, assuming that gaps at the beginning or end of *u* or *v* are free.

(C-1) This is called cost-free ends of fitting glocal alignment. It is useful whenever one string is significantly shorter than the other, and the same basic idea applies for the "overlap" variant as well.

(C-2) This is useful in finding overlaps between strings and overlap glocal alignments.

the highest scoring alignment between u' (a substring of u) and v' (a substring of v). It is an alignment method useful in finding similar regions in strings that may not be globally similar.

The glocal or semiglobal algorithm is based on modified NW algorithm. An illustrative example of semiglobal/glocal alignment is presented in Fig. 4.1B, where methods of biosequence alignment are indicated as C-1 and C-2. They refer to semiglobal/glocal exercises assuming that gaps at the beginning or end of u or v are free, and (C-1) is called a method with *cost-free ends* of fitting glocal alignment, which is pursued whenever one string is significantly shorter than the other, and the same basic idea applies for the "overlap" variant as well. The illustration of (C-2) is useful in finding overlaps between strings and overlap glocal alignments. Such global alignments imply an end-to-end alignment of sequences u and v. The inputs required for this are two potentially similar sequences u and v presumably having biologically significant relatedness.

Dynamic Programming of Pairwise Alignment

Global alignment via NW algorithm exercised via dynamic programming: The NW algorithm is a matrix-based scoring system that reveals the best-matching pathway for globally aligning two sequences. The underlying steps are as follows: the first step is to use both test sequences to frame an $[n \times m]$ matrix by placing each of the sequences in the x- and y-directions. Then the letters in each row/column are analyzed by indicating matches and mismatches with a numerical value. Matches are awarded, while mismatches are penalized. Typically, "1" is placed in the cell of the matrix for a match and "0" is placed for a mismatch. (This method is similar to the "dot-plot" method except instead of dots, numerical values are used.) After the matrix of 1's and 0's in the cells is created, a row and column of "dummy" zeros are added at the ends of the x- and y-axes, and the *leading cell* is defined as the last nth × mth cell in the matrix (prior to the row and column of "dummy" zeros is added).

The second step is to implement scoring algorithm by starting at the *leading cell* and looking at three of the abutting cells. Suppose the cell at the ith row/jth column designated as (i, j), has to be updated with a fresh score, that is, it is going to be replaced with a new score as $S(i, j)$ and the numerical value that currently resides in that cell is denoted as Old $S(i, j)$. The letter i is the row index and the letter j is the column index. The maximum value is noted from this cell by checking either in the diagonal direction $S(i - 1, j - 1)$ or along the horizontal direction $S(i - 1, j - k)$ or along the vertical direction $S(i - r, j - 1)$. It is then added to the Old $S(i, j)$ value, and the cell is then updated with this new $S(i, j)$ value. The updating procedure as above can be specified by:

$$S(i,j) = \text{Old}\, S(i,j) + \max \begin{cases} S(i-1, j-1) \\ S(i-1, j-k) \\ S(i-r, j-1) \end{cases} \qquad \text{Eq. (4.1)}$$

where $k, r = 1, 2, 3, \ldots, n$ or m.

After updating all the cells as per Eq. (4.1) in the matrix, a path is traced based on the maximum values obtained in the matrix. "A path is traced back from the highest scoring position (which, by definition, must occur at the N-*termini* because of the cumulative nature of the summation process) to the origin." Relevantly, the path is traced back to the *leading cell* by following the next least maximum value compared to the current maximum value in a diagonal manner. The diagonal direction is preferred; however, occasionally the next least maximum value might not be in the diagonal direction but in the vertical or horizontal direction. When this occurs, a gap is inserted into one of the sequences depending on the direction of the path. For example, horizontal direction means a gap in sequence in the v-direction. If the next least maximum value is in the diagonal cell and is also equal to the value in the horizontal and/or vertical cells, the diagonal path is always chosen because it indicates a match or indel instead of a gap. Some sequences may produce "islands." This happens when the next least maximum value is in the vertical $S(i + 1, j)$ and horizontal direction $S(i, j + 1)$ is equal and greater than the value in the $S(i + 1, j + 1)$ cell. Island paths will eventually converge and usually produce two paths of equal score.

The algorithm outlined above can be implemented *via* computer program. Performing the trace-back procedure (even with possibility of islands), involves ascertaining the best scoring path. Relevant program should determine the path for two different scenarios: if the island paths are equal in length, only one path is chosen since the scores of the two paths will be equal. If the island paths are unequal in length, then the path with the highest score is chose. This is determined by comparing the values at: $S(i, j + 2)$, $S(i + 1, j + 2)$, $S(i + 2, j + 1)$, and $S(i + 2, j)$. (The NW procedure is exemplified later in the review section of this chapter with examples and exercises.)

Biosequence similarity search is an important application in modern molecular biology.[18] Relevant search algorithms are designed to identify sets of sequences whose extensional similarity implies a common evolutionary origin or function. BLAST is popularly used as a similarity search tool for biosequences and the program is designed to compare query sequences to a database, and *blastn* is a version of BLAST that searches DNA sequences.[18]

Tutorial T-4.2

A summary of biosequence comparison and alignment tools: As stated earlier, the *pairwise sequence alignment* (PSA) is used to identify regions of similarity that may indicate functional, structural, and/or evolutionary relationships between two biological sequences (protein or nucleic acid). Relevantly, in modern bioinformatics immersed in the big-data territory, user-friendly platforms are exploited that facilitate pairwise (as well as, multiple) genome comparison involving the management of huge datasets and high computational demands. For example, Artemis comparison tool (ACT) is a Java application for displaying pairwise comparisons between two or more DNA sequences. It can be used to identify and analyze regions of similarity and difference between genomes and to explore conservation of synteny in the context of the entire sequences and their annotation. It can read complete EMBL, GENBANK, and GFF entries or sequences in FASTA or raw format. There are also tools that are useful for small-scale genomic comparisons, typically in the order of 2–20 genomes. Most of such tools are intended for assembled data. An exemplar of pairwise comparison tool is the LAST, which can find similar regions between sequences.[19] It can find similar regions between sequences and it can align them.

Extended efforts of sequence comparison involve multiple sequence comparisons. Some salient versions of multiple alignment and consensus sequence tools are ClustalW, Clustal Omega,[20] MAFFT,[21] webPRANK,[22] GUIDANCE,[23] SALIGN,[24] AlignMe,[25] and PRALINE.[26] Implicitly, the MSA results assess the shared evolutionary origins of the *taxa* contributing those sequences.[27,28] Further, MSA based on conservation of protein domains of motifs lead to specifying the forms of the proteins, such as tertiary and secondary structures.[29,30] Moreover, Altschul outlined a practical method[31] that uses pairwise alignments to constrain the *n*-dimensional search space. The outcome of MSA, in essence, is pertinent to ascertaining the following in the compared set of multiple sequences: (i) the presence of conserved segments, that is, the motifs; (ii) the details on fingerprints, blocks, and profiles; and (iii) formulating the associated regular expressions and/or fuzzy regular expressions.

Briefly stating, the conserved domains (motif sections) in a set of multiple sequences can be regarded as distinct functional and/or structural units of a protein. The following is an example of multiple sequences containing a motif (shown in bold upper case alphabets):

... cc**TAATC**gt ...
... gg**TAATC**ga ...
... at**TAATC**gt ...
... gc**TAATC**gg ...

The above set of multiple sequences share an identical section in the middle, namely, TAATC.

The motifs are formatted in different forms, and a sequence pattern can be represented by a method that uses *regular expressions* (or *regex*, for short). Regular expressions are essential to identify, replace, or modify text, words, patterns, or characters. In general, applications of regex include allowing site URLs "to look pretty," remove all punctuation from a sentence, filtering *RSS feeds**, or other data. (*RSS feeds is a technology enabling streamlined, algorithm-free formats that could make a tool for reading online.)

Regex in short is one of the formats of the motif patterns and is advocated as in the PROSITE.[32] That is, the PROSITE patterns allow representing simple motifs by a user-friendly and easily readable format. Relevant syntax is standardized by the PROSITE databank, which stores known motifs in different formats available to the scientific community. For example, the aforesaid example of the motif TAATC can be represented like this, T-A-A-T-C. In regex format, each position in the sequence is represented by the nucleotide A, T, G, or C, and one can use "*x*," which is a special character denoting *"at this position, there can be any base,* A, T, G, or C" with each position in the sequence is separated by "-". Now, considering a set of sequences related to the four indicated above, relevant regex can be specified as follows: for the set of motif section (shown above in bold upper case alphabets), the following regex pattern can be specified: T-A-[AC]-T-C, meaning that the third alphabet can be either A or C. Similarly, the PROSITE syntax enables writing the regex for more complex motifs having repeats of letters (with variable length). For example, *A*(2, 4) means that the letter *A* can repeatedly exist from 2 to 4 times. The pattern syntax rules are described in the PROSITE database and summarized below: (The original version of this document can be found at the PROSITE website.[32]) The standard IUPAC one-letter codes are used to represent the amino acids.

1. For a position where *any* amino acid is accepted, the symbol "x" is used.
2. By listing the acceptable amino acids at a given position, the ambiguities are indicated between square brackets "[ ]." For example: [ALT] stands for: Ala or Leu or Thr.
3. When the amino acids that are not accepted at a given position, relevant ambiguities are indicated by listing the amino acids between a pair of curly brackets "{ }": For example: {AM} can stand for any amino acid except Ala and Met.
4. A hyphenation ("-") is used to separate each element in a pattern from its neighbor.
5. Repetition of an element of the pattern can be indicated by following that element with a numerical value, or if it is a gap ("x"), by a numerical range between parentheses. Examples:
   - (i) x(3) ↔ x-x-x.
   - (ii) x(2, 4) ↔ x-x or x-x-x or x-x-x-x.

     (*Note*: A range can be specified only with "x"; that is, by writing, T(2,4) is not a valid pattern element).
   - (iii) C(3) ↔ C-C-C.

6. A pattern may either start with a "<" symbol or, respectively, end with a ">" symbol, whenever the pattern is restricted to either the N- or C-terminal of a sequence. In some rare cases, ">" can also occur inside square brackets for the C-terminal element. For example:
   (i) "F-[GSTV]-P-R-L-[G>]" means considering either "F-[GSTV]-P-R-L-G" or "F-[GSTV]-P-R-L>".
   (ii) TACHYKININ, PS00267
   Tachykinin family signature (PATTERN) with the consensus pattern: F-[IVFY]-G-[LM]-M-[G>].
   (iii) PYROKININ, PS00539
   Pyrokinins signature (PATTERN) with the consensus pattern: F-[GSTV]-P-R-L-[G>].

There are also extended syntax versions as in cases where a pattern may consist of one-letter amino acid codes only without any ambiguous residues nor specifying the "-". Hence, one can directly copy and paste peptide sequences into the text field. (For example, V-A-T-K-E can be written as VATKE.)

In searching all sequences that may not contain a certain amino acid, for example Cys, <{C}*> can be used. A tutorial on regular expression can be found at REG EX,[33] and some relevant examples are presented below.

Example Ex-4.2

Translate the following cases of regex into their expanded forms.

(i) [GC]-x-V-x(3)-{HD}

Solution: Pattern translation: [Gly or Cys]-any-Val-any-any-any-{any but His or Asp}.

(ii) <L-x-[ST](3)-x(0,1)-Q

Solution: In the N-terminal of the sequence ("<"), the pattern is translated as: Leu-any-[Ser or Thr]-[Ser or Thr]-[Ser or Thr]-(any or none)-Gln.

(iii) <{Y}*>

Solution: Pattern description: It depicts all sequences, which do not contain any Tyrosine.

(iv) WWRIFHLRNI

Solution: Pattern description: It depicts all sequences which contain the subsequence "WWRIFHLRNI".

The details on single- and three-letter amino acid code representation can be availed from of Chapter 1.

4.3 Mutational Considerations in Biosequence Comparisons

As mentioned earlier, multiple sequence comparisons reveal conserved sites as well as, point out evolutionary modifications that might have occurred across the compared entities, and presumably, such observed changes might have largely taken place as a result of *mutation.*

Mutations: Small- and Large-Scale Versions

Mutations (caused by an agent or substance called *mutagen*) imply significant interest in biosequence analyses, and they are classified in several different ways as well as, sorted by their effect on the structure of DNA or in a chromosome. In general, mutations are separated into two major groups, *small-* and *large-scale mutations* each with multiple specific categorizations as detailed below.

Small-scale mutations: Here the underlying mutational effects are at the molecular level of the DNA, changing the nucleotide base pairs in a normal sequence. These types of mutations could occur during the process of DNA replication during either meiosis or mitosis. In small-scale mutations, one or few nucleotides of a gene are affected.

Large-scale mutations: This version of mutation affects a higher level of the genetic material, for example, a change in the chromosome. Large-scale mutations may be further classified as follows: (1) amplifications (or gene duplications), (2) deletions of large chromosomal regions, and (3) chromosomal inversions.

The small-scale mutations could imply *point mutation,* which involves substituting an individual base with another. It is a type of mutation that

Table 4.1 Illustrative examples of three types of point mutations to a codon.

Levels as per central dogma	No mutation	Point mutation: Silent mutation	Nonsense mutation	Missense mutation: Conservative type	Non-conservative type
DNA level	TTC	TTT	ATC	TCC	TGC
mRNA level	AAG	AAA	UAG	AGG	ACG
Protein level	Lys	Lys	STOP	Arg	Thr

causes a single nucleotide base substitution, insertion, or deletion of the genetic material, DNA or RNA. Some common substitutions are as follows A for C; A for G; C for T; G for T, A for T; and G for C. A *point mutant* is an individual that is affected by a point mutation. There are three versions of point mutation as identified in Table 4.1 *via* illustrative examples of epochs at the levels specific to the central dogma. Outlined below are explanatory details on such point mutation depicting a single-base modification (such as substitution, insertion, or deletion of the genetic material, DNA or RNA).

Substitution mutation: This change renders exchanging one base for another (i.e., a change in a single "chemical alphabet set {A, C, T, G}," e.g., switching an A to a G). Whenever a substitution takes place, it can (i) change a codon to another that encodes a different amino acid, and this may cause a noticeable change in the protein produced. (For example, sickle-cell anemia is caused by a substitution in the beta-hemoglobin gene, which alters a single amino acid in the protein produced); (ii) or, alternatively, the substitution-specific change may imply a codon changing to another that would encode the same amino acid causing no change seen in the protein produced. Such substitutions are known as *silent mutations*; and (iii) the substitution mutation may also change an amino-acid-coding codon into a single "stop" codon and cause an incomplete protein. This can have serious effects since the incomplete protein probably may not be functionally useful. An example is as follows: GTGGAG changes into: GTGGCG.

Insertion: This version of mutation involves *extra base pairs* being inserted into a new site in the DNA. For example, ATGGAG → ATGGTGGAG.

Deletion: Suppose a section of DNA is lost or deleted. It refers to deleting a segment of a sequence, for example, as in: TTGGAG → TTAG.

Frameshift: A *frameshift mutation* depicts the mutation due to addition or deletion of a base pair. Inasmuch as protein-coding DNA is divided into trinucleotides, insertions and deletions can alter a gene so that its encoded message no longer remains correctly parsed (in central dogma sense). That is, relevant frameshift is an error that occurs at the DNA level causing the codons to be parsed incorrectly, usually generating truncated proteins, such as "... aef ate ats at ...," which are meaningless and noninformative toward intended protein translations.

Transposition: This implies moving a segment of the sequence from one location to the other in the overall ordered set of residues.

Duplication: This refers to observing a repeat segment in a section of the sequence one or more times.

Repeat induced point (RIP) mutations: RIP is a recurring point mutation depicting the genome, for example, the defense mechanism seen in fungi that hypermutates repetitive DNA. It is suggested in Hood *et al.*[34] that RIP limits the accumulation of transposable elements.

Transition and transversion mutations are two versions of DNA substitution mutations. Whenever interchanges of two-ring purines (A ↔ G) or of one-ring pyrimidines (C ↔ T) take place, it refers to transitions involving bases of similar shape, and transversions denote interchanges of purine for pyrimidine bases. This version of mutation, thus, involves exchange of one-ring and two-ring structures.

As a result of the underlying molecular mechanisms involved, the transition mutations are generated at higher rates of occurrence than transversions. Further such transitions are less likely to result in amino acid substitutions (due to "*wobble*" position), and as such, they are more likely to exist as "*silent substitutions*" in populations as *single nucleotide polymorphisms* (SNPs). Approximately, two out of three SNPs could be transitions. Transitions could be caused as a result of *oxidative deamination* and *tautomerization.*[35,36] More complex models[37,38] on mutational considerations take into account a variety of biological phenomena and lead to accurate estimates of evolution-theoretics of phylogeny, as will be explained later in Chapters 9 and 10. The transition

Table 4.2 Illustration of a simple transition matrix of nucleotides in terms of score values 0 and 1.

	A	C	T	G
A	1	0	0	0
C	0	1	0	0
T	0	0	1	0
G	0	0	0	1

probability matrices that define evolutionary models contain relative rates of all possible replacements (due to mutations) established to calculate the probabilities of a change from any nucleotide to any other nucleotide,[39] and a transition matrix in its simple form can be depicted in terms of score value 1 and 0 as indicated in Table 4.2.

A substitution models adopted in the estimation of evolutionary distances is due to Jukes and Cantor.[40] This model starts from the assumptions that all substitutions are independent, all sequence positions are equally subject to change, substitutions occur randomly among the four types of nucleotides, and that no insertions or deletions have occurred. Based on these assumptions, the authors derived an equation for estimating evolutionary distances from observed dissimilarity: $d_{AB} = -(3/4) \times \log_e[1 - (4/3) \times f_{AB}]$. Here, f_{AB} denotes the dissimilarity (expressed in terms of the fraction of observed differences) between sequences A and B, and i is the estimated evolutionary distance (denoting the fraction of expected substitutions between sequences A and B).

Several other equations for the estimation of evolutionary distances also have been proposed. For example, Kimura[41] has provided a method for inferring evolutionary distances based on a model of evolution in which transitions and transversions may occur at different rates, and *Kimura 2-parameter* (K2P) *model* has equal base frequencies, one transition rate and one transversion rate.[41] The other Kimura model is known as *Kimura 3-parameter (K3P) model* with variable base frequencies, equal transition rates, and two transversion rates.[42]

Substitution models in vogue in which the four different are not in equal proportions or where a bias in the direction of change is accounted for refers to *Tajima-Nei (TrN) model*,[43] and *Tamura-Nei (TrN) model*[44] is specified with variable base frequencies, equal transversion rates, and variable transition rates. It is indicated for the estimation of the number of substitutions in the control region of mitochondrial DNA in human and chimpanzee. Further, addressed in Tamura[45] is the model where the estimation of the number of nucleotide substitutions when there are strong transition-transversion and (G + C) content biases exist is explained, and the estimation of evolutionary distances between nucleotide sequences is detailed in Zharkikh[46] *via* a symmetrical model (*SYM model*).

In the contexts of substitution matrix adopted, while developing phylogenetic hypotheses, modeling of evolution is done with *maximum likelihood* (ML)* algorithms. (ML* refers to a statistical method for estimating random variable parameters of a population, (such as the expected mean and variance, from a set of sample data, which enables selecting as estimates those parameter values maximizing the probability of obtaining the observed data). For example, the following models thereof can be identified: (i) *Felsenstein model*: This is conceived with variable base frequencies and all substitutions are equally likely events[47] and (ii) *Hasegawa–Kishino–Yano* (HKY) *model,* which is constructed with variable base frequencies, one transition rate and one transversion rate.[48] The GTR model indicated above is a variable base frequency, symmetrical substitution matrix model outlined in Lanave *et al.*,[49] Tavaré,[50] Rodkigljfk *et al.*,[51] and Yang.[52] In addition to models having rates changing from one nucleotide to another, there are other versions that describe rate variation among sites in a sequence. Hence, a commonly used model thereof is known as *Gamma distribution (G) model,* where a gamma-distributed rate variation among sites is presumed.

The Jukes and Cantor, one-parameter substitution model, indicated before is the simplest type available for estimating the number of nucleotide substitutions per site and is probably still the most used one. The Kimura model provides a method to infer the evolutionary distance, in which the transitions and transversions are treated separately. This evolutionary distance (d_{AB}) is given by: $d_{AB} = -(1/2)\times\log_e\left((1-2P-Q)\times\sqrt{(1-2Q)}\right)$, where P is the fraction of sequence positions differing by a transition and Q is the fraction of sequence positions differing by a transversion as implied by the variables α and β in the matrix shown in Table 4.3.

In the substitution models as above, the base-frequency parameters describe frequencies of nucleotide bases averaged over all sequence sites and over the phylogenetic tree constructed. These parameters can be considered to represent constraints on the occurrence frequency of nucleotides due to the observed effects, such as overall GC content, and

Table 4.3 Illustrative matrix of Kimura model.

	A	T	C	G
A	—	β	β	α
T	β	—	α	β
C	β	α	—	β
G	α	β	β	—

they can be regarded as weighting factors in a model depicting how certain bases are more likely to arise when substitutions take place. The relative tendencies of bases being substituted for one another can be denoted by base exchangeability parameters. Relevant parameters "represent a measure of the biochemical similarity of bases"; hence, as stated earlier, transitions (i.e., C $\leftrightarrow$ T or A $\leftrightarrow$ G) typically occur more often than transversions (namely, C $\leftrightarrow$ G).[53,54] Modeling rate heterogeneity among sequence sites so as to describe the rate at each site can be done *via* random draw statistics of gamma probability distribution as indicated in Yang.[55]

Large-Scale Mutations

As stated earlier, whenever the stretch of chromosome is affected by mutational changes, it amounts to a *large-scale mutation.* Some large-scale mutations may effect only single chromosomes, while there are others that may occur across nonhomologous pairs. The large-scale mutations in the chromosome are analogous to small-scale mutations in DNA, but the difference is that for large-scale mutations, the entire genes or sets of genes are altered (rather occurring only at a single nucleotide of the DNA). Mutation in a single chromosome is very likely to occur due to some error that had taken place in the DNA replication stage of cell growth; as such, it could occur during meiosis or mitosis.

Mutations in multiple chromosomes may occur more likely in meiosis during the crossing-over that occurs during the prophase I. Typically, large-scale mutations involve deletion, duplication, inversion, insertion, translocation, and non-disjunction types. Pertinent versions of mutation and its derivatives are as follows: As stated earlier, mutations result in a change(s) in DNA, usually in the sequence, the number of copies of a sequence that are present, how the DNA is arranged, or the DNA disposition (namely, at which chromosome). Observable changes in the case of large-scale mutations refer to the following methods being considered in the designs toward evolving single-strand and/or "duplex" versions relevant to a query sequence.

Deletion: This type of mutation involves the loss of one or more genes from the parent chromosome depicting a large-scale, single-chromosome mutation.

Duplication: This mutation refers to the addition of one or more genes that are already present in the chromosome, and this is also a single-chromosome mutation.

Inversion: When a segment of the sequence is inverted (conforming to a complete reversal of one or more genes within a chromosome), the pertinent genes are retained as such in post-inversion.

Insertion: This type of large-scale mutation incorporates insertions in multiple chromosomes. That is, one or more genes are removed from one chromosome and inserted into another nonhomologous chromosome.

Translocation: If the chromosomes swap one or more genes with another chromosome, it implies translocation involving multiple nonhomologous chromosomes.

Non-disjunction: The class of nondisjunction mutations does not cause any errors in DNA replication or crossing-over. On the other hand, such mutations occur during the *anaphase* and *telophase* when the chromosomes are not separated properly into the new cells.

Effects of Mutations

Having seen the types of mutations, their effects can be identified as follows: such effects may range from insignificant influence to total unviability of a cell. The mutations, in general, could affect the proteins made during protein synthesis, but not all mutations will significantly impact the final protein complex. Also, the small-scale and large-scale mutations distinctly pose the end results. The associated effects can be identified as follows: (i) *conservative substitution,* which refers to a nucleotide mutation, altering an amino acid sequence encoded for a protein; relevantly, this mutation implies substitution of one amino acid with another (that has a side-chain with similar charge/polarity

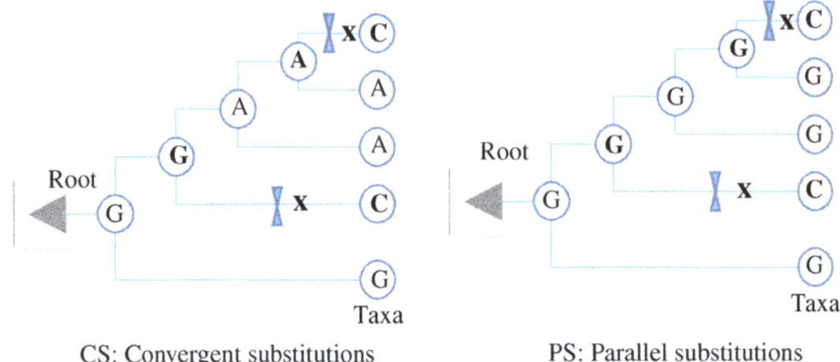

Fig. 4.2 Examples of CS — convergent substitutions and PS — parallel substitutions.

characteristics). The size of the side-chain may also be an important consideration. Mostly, conservative mutations, however, are unlikely to alter profoundly the structure or function of a protein (with exceptions). (ii) *Non-conservative substitution*: This version of mutation results in the substitution of one amino acid within a polypeptide chain with an amino acid with a different physico-chemical property, such as polarity/charge group. (iii) *Convergent and parallel substitutions*: Comparing orthologous proteins pertinent to a given set of species, convergent substitutions (CS; at a particular site) refer to independent changes seen different with reference to ancestral amino acids *versus* the same derived amino acid.

In summary, considering orthologous proteins from a given set of species, the CS at a particular site refer to independent changes from different ancestral amino acids to the same derived amino acid. In Fig. 4.2, (a) refers to the CS, where a change occurs in one species from **A** in the ancestral state to **C** in the derived state, and in another species, the change seen refers to from **G** to **C**. (Relevant CS changes are marked with bold letters and the routes are marked with: **x**). Next considering (Fig. 4.2), (b) the parallel substitutions at a site refer to independent changes from the same ancestral residue **G** to the same derived **C**, and these two changes (from **G** to **C**) occurred in two different species.

The parallel substitutions are denoted by bold letters and the routes are marked with: **x**. In sets of closely related species, parallelism is generally more common than convergence inasmuch as, at any given site, close relatives are likely to share the same ancestral state prior to the occurrence of independent substitutions.[55] Lastly, *coincidental substitutions* may also prevail with the occurrence of two substitutions at the same nucleotide site in two homologous sequences.

Example Ex-4.3

Statement of the Problem

Suppose a hypothetical initial strand is ... AAAAGGGGTTTTGACC ... Perform an insertion version of mutation with a subsequence inserted at an arbitrary location in the initial strand.

Solution
For the assumed strand, the insertion version of mutation, for example, with a subsequence "CCCC" at an arbitrary location will result in the following mutated strand:

... AAAAGGCCCCGGTTTTGACC ...

Example Ex-4.4

Statement of the Problem

Assuming different types of mutational changes as indicated, evaluate the resulting outcomes on ancestral query sequences specified.

Presumed versions of mutational changes	Result on the query sequence segment: ... AAAAGGGGTTTTGACC ...
No mutational change: Query sequence is retained as it is	... AAAAGGGGTTTTGACC ... → ... AAAAGGGGTTTTGACC ...
Single substitution: C → A	... AAAAGGGGTTTTGACC... → ... AAAAGGGGTTTTGAAA ...
Multiple sequential Substitutions: G → A → T	... AAAAGGGGTTTTGACC... → ... AAAAAAAATTTTAACC ... → ... TTTTTTTTTTTTTTCC ...
Back-substitution: C → T → C	... AAAAGGGGTTTTGACC ... → ... AAAAGGGGTTTTGATT ... → ... AAAAGGGGTTTTGACC ...

Coincidental substitutions

Presumed mutation	Result
With reference to two homologous sequences, two substitutions at the same nucleotide site: T → G	Homolog sequence : Y1 ... A A A T A A A ... Homolog sequence : Y2 ... C A A T A A A ... Coincidental substitutions are shown in bold: Y1* → ... A A A **G** A A A ... Y2* → ... C A A **G** A A A ...

Example Ex-4.5

Statement of the Problem

With reference to the following hypothetical cases of biosequence segments, outline briefly the possible parallel and CS.

Case (i): Given a DNA sequence segment: *Z*: ... G A A A C A A T... of a homolog species indicate examples of parallel substitutions at a specific site.

Solution

Parallel substitution refers to, for example, depicting an independent change at a site of the same ancestral amino acid to the same derived amino acid: For example, T → C or G. Hence, resulting parallel mutations are shown as follows:

Z: ... G A A A C A A **T**...
*Z*1*: → ... G A A A C A A **C** ...
*Z*2*: → ... G A A A C A A **G** ...

Case (ii): Given two different ancestral amino acids, *Z*1: ... A A T G A T... and *Z*2 : ... A A T..., perform hypothetical CS.

Solution

This refers to independent changes from different ancestral amino acids to the same derived amino acid. For example, with reference to two different ancestral AAs, *Z*1 and *Z*2 considered, the derived residue remains the same, T:

*Z*1: ... A A T G A **T** ...
*Z*2: ... A A **T** ...

Problem P-4.1

Statement of the Problem

A segment is presumably mutated at a location (shown underlined as GGGG) in a hypothetical strand:

... TTAAGGGGGGCCTTTTGAAA ...

(a) Write down the eventual resulting strand with the following additional mutations happening in succession at the site marked as

GGGG: (*Note*: The answer may depend on subjective selection as required): (1) Inversion of some (arbitrary) subsequence part; (2) deletion of some (arbitrary) subsequence part; (3) transpose of some (arbitrary) subsequence part; (4) duplication of a base-pair in the query sequence; and (5) point mutation of one base into another in the query sequence.

(b) Write down the corresponding final resulting duplex strands.

Problem P-4.2

Statement of the Problem

Given a query codon sequence:

X: 5′-TAC GGA TCG AAT GCT CCC GTA ATC-3′

Suppose the following mutations have occurred sequentially.

A single-point mutation, deletion of a triplet, and duplication of a triplet twice in succession.

Suppose the resulting complementary strand is found to be the following:

Y: 3′-ATG CCT AAC TTA CGG CAT CAT CAT TAG-5′

Trace the associated mutational changes occurred from *X* to *Y*.

Problem P-4.3

Statement of the Problem

Construct a matrix of the set {A, C, T, G} to illustrate the characteristic of the transition and transversion mutations.

(*Hint*: You may assume a score of 100% to depict the element of the matrix pertinent to no mutation, and use prorated percentages to represent other elements illustrating the transition and transversion characteristics. The spontaneous base substitutions ratio of transitions to transversions is approximately 2:1. Therefore, each transition should have a probability of two-third while each transversion has one-third probability.)

4.4 Review Section: Examples and Exercises

Example Ex-4.6

Statement of the Problem

This example is indicated to illustrate the differences in the postalignment results pertinent to global *versus* local alignments performed on a pairs of sequences. Suppose a pair of hypothetical amino acid sequences *X* and *Y* subjected to alignment procedure is as follows:

X: AGPSSKQNGKPSSRIWDN
Y: ANITKSAGKPAIMRLGDD

The results of global and local alignments of *X* and *Y* are shown below to understand the underlying differences. (The results shown are obtained using NW and SW algorithms as explained in later examples, and performing those algorithms on *X* and *Y* given above will be indicated as problem exercises.)

1. Global alignment: Result of aligning *X* and *Y* over their entire lengths *via* NW algorithm. When the alignment is completed, both sequences are of same length.

```
X: A G P S S K Q N G K P S - S R I W D N
   |         |     | | |       |     |
Y: A N - I T K S   A G K P A I M R LG D D
```

2. Local alignment: Result of aligning *X* and *Y* *via* SW algorithm showing the longest or best subsequence pair that has maximum similarity.

```
X: - - - - - - - -N G K P - - - - - - - - -
                    |   | |
Y: - - - - - - - -A G  K P - - - - - - - - -
```

Example Ex-4.7

Statement of the Problem

The results of *global* and *glocal* alignments of two nucleotide sequences *u* and *v* are presented below to understand the underlying differences.

u: TGTCTGTGGGTGG
v: TGCTTG

The results of global and glocal alignments of *u* and *v* are shown below to understand the underlying differences.

1. Global alignment

```
u: T G T C T G T G G G T G G
   | |         |       |   |
v: T G - C - - T - - - T - G
```

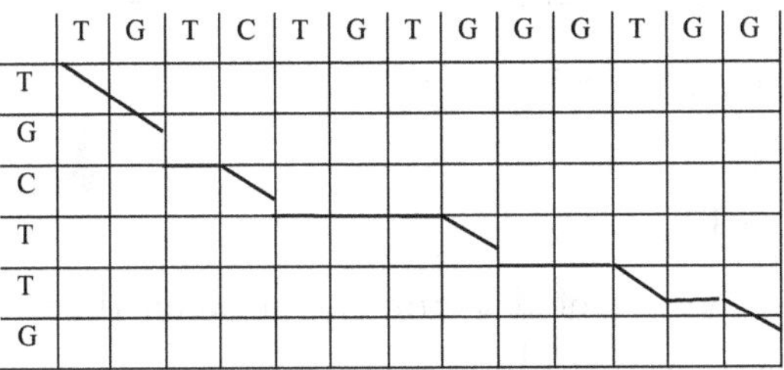

2. Glocal (semiglobal) alignment

```
u: T G T C T G - T G G G T G G
           | |   |     |
v:         T G C T T G
```

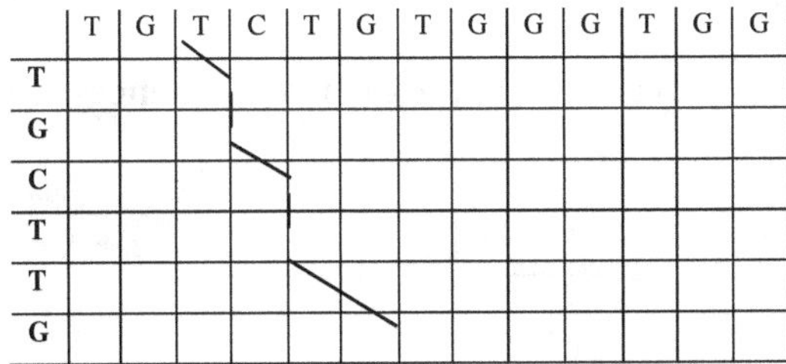

The result of (1: Global alignment) shown below is obtained using NW algorithm, and the semiglobal algorithm is also based on NW algorithm modified as follows: once the NW algorithm-based updating the values of the underlying matrix is completed, the trace-back is started at the greatest element of the last row of the alignment matrix (scores matrix) or last column, if there are more rows than columns. This is in contrast with NW algorithm where the starting of the trace-back commences from the absolute last cell (leading cell) of the matrix.

Implementation of NW Algorithm

As explained before, global alignment of sequences is based on dynamic programming. Relevant NW algorithm can be adopted to align protein or nucleotide sequences. Pertinent optimal global alignment can be understood with the examples and exercises furnished below.

Example Ex-4.8

Statement of the Problem

Given a set of sequence pairs, *x* and *y*:

x: C T C G T
y: C T A A G T

Determine the "best" global alignment between them *via* trace-back procedure using NW algorithm.

Solution
Global alignment refers to aligning sequences over their entire length resulting in sequences of the same length. The global alignment of the sequences can be determined using the NW algorithm. The underlying steps are furnished below with respect to the sequence pair *x* and *y* indicated above.

Step 1: Construct a matrix [*x*, *y*] for the two sequences as shown below.

x \ *y*	C	T	C	G	T
C					
T					
A					
A					
G					
T					

Step 2: Initialization: Matrix cells representing residue identities are scored 1 and cells representing mismatches are scored 0.

y \ x	C	T	C	G	T
C	1	0	1	0	0
T	0	1	0	0	1
A	0	0	0	0	0
A	0	0	0	0	0
G	0	0	0	1	0
T	0	1	0	0	1

Step 3: Add "dummy" columns (I_2, I_1, ...) and rows (J_2, J_1, ...) at the end of the matrix as illustrated below and fill these columns and rows with zeros (as shown). Mark the *leading cell,* LC. (This the last corner cell on the right-side of the matrix designated as (i = I_3, j = J_3)th cell, as indicated.)

		I_7	I_6	I_5	I_4	I_3		
		C	T	C	G	T	Dummy	
J_8	C	1	0	1	0	0	I_2	I_1
J_7	T	0	1	0	0	1	0	0
J_6	A	0	0	0	0	0	0	0
J_5	A	0	0	0	0	0	0	0
J_4	G	0	0	0	1	0	0	0
J_3	T	0	1	0	0	1 LC	0	0
		J_2	0	0	0	0	0	0
Dummy		J_1	0	0	0	0	0	0

Step 4: Starting from the leading cell, LC (designated as (i = I_3, j = J_3)th cell), perform the following procedures.

- Move to (I_2, J_2)th cell (diagonally going downward) and apply the following algorithm to update the score value in LC.
  Algorithm: Update the (leading) cell entry by adding the maximum value encountered at A: (i – 1, j – 1)th cell or along the three tracks, B or C as shown in Fig. 4.3.
- Observed maximum value at A or along B and C: 0.
- Existing value in the LC: 1.
- Hence, the updated value for the leading cell is (1 + 0) = 1.

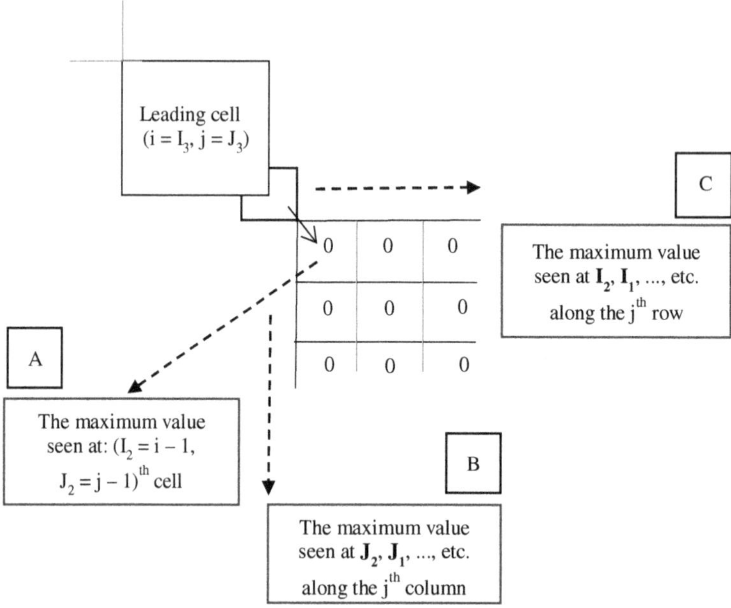

Fig. 4.3 Illustration of implementing NW algorithm. Updating cell score entry *via* diagonal, horizontal, and vertical track pursuits.

Step 5: Use the same procedure and algorithm as above to update the value for each cell in the entire matrix pursuing the track route indicated below.

(a) After updating the value in the LC, consider the other cells one-by-one along the row J_3 following the track moving leftward and encountering the cells: I_4, I_5, . . . , and I_7.

		I_7	I_6	I_5	I_4	I_3		
		C	T	C	G	T	Dummy	
J_8	C	1	0	1	0	0	I_2	I_1
J_7	T	0	1	0	0	1	0	0
J_6	A	0	0	0	0	0	0	0
J_5	A	0	0	0	0	0	0	0
J_4	G	0	0	0	1	0	0	0
J_3	T	0	1	0	0	1	0	0
						LC		
		J_2	0	0	0	0	0	0
Dummy		J_1	0	0	0	0	0	0

Corresponding updated values are shown in the matrix indicated above.

(b) Next considering the cells one-by-one, $I_3, I_4, \ldots$, and I_7, encountered along the upper row J_4 (by following again the track moving leftward), use the same procedure and algorithm as above to update the existing value in each cell.

		I_7	I_6	I_5	I_4	I_3		
		C	T	C	G	T	Dummy	
J_8	C	1	0	1	0	0	I_2	I_1
J_7	T	0	1	0	0	1	0	0
J_6	A	0	0	0	0	0	0	0
J_5	A	0	0	0	0	0	0	0
J_4	G	1	1	1	2	0	0	0
J_3	T	0	1	0	0	1	0	0
						LC		
		J_2	0	0	0	0	0	0
Dummy		J_1	0	0	0	0	0	0

Corresponding updated values are shown in the matrix indicated above.

		I_7	I_6	I_5	I_4	I_3		
		C	T	C	G	T	Dummy	
J_8	C	1	0	1	0	0	I_2	I_1
J_7	T	0	1	0	0	1	0	0
J_6	A	0	0	0	0	0	0	0
J_5	A	2	2	2	1	0	0	0
J_4	G	1	1	1	2	0	0	0
J_3	T	0	1	0	0	1	0	0
						LC		
		J_2	0	0	0	0	0	0
Dummy		J_1	0	0	0	0	0	0

(c) Likewise, considering the cells one-by-one, $I_3, I_4, \ldots$, and I_7, encountered along the next upper row J_5 (by following the track, moving leftward) and using the same procedure and algorithm indicated above the existing value in each cell is updated. The updated values are shown in the matrix indicated above.

(d) The aforesaid procedure is repeated for the cells one-by-one, $I_3, I_4, \ldots$, and I_7, encountered along the next upper rows J_6 through J_7 (by following the track, moving leftward) and by using the procedure and algorithm indicated above, the existing value in each cell is updated. Corresponding updated values are shown in the following matrices:

		I_7	I_6	I_5	I_4	I_3		
		C	T	C	G	T	Dummy	
J_8	C	1	0	1	0	0	I_2	I_1
J_7	T	0	1	0	0	1	0	0
J_6	A	0	2	2	1	0	0	0
J_5	A	2	2	2	1	0	0	0
J_4	G	1	1	1	2	0	0	0
J_3	T	0	1	0	0	1	0	0
						LC		
		J_2	0	0	0	0	0	0
Dummy		J_1	0	0	0	0	0	0

		I_7	I_6	I_5	I_4	I_3		
		C	T	C	G	T	Dummy	
J_8	C	1	0	1	0	0	I_2	I_1
J_7	T	0	3	2	1	1	0	0
J_6	A	0	2	2	1	0	0	0
J_5	A	2	2	2	1	0	0	0
J_4	G	1	1	1	2	0	0	0
J_3	T	0	1	0	0	1	0	0
						LC		
		J_2	0	0	0	0	0	0
Dummy		J_1	0	0	0	0	0	0

(e) Lastly, the procedure repeated for the cells one-by-one, $I_3, I_4, \ldots$, and I_7 encountered along the next upper row J_8 (by following the track, moving leftward) completes the updating with the values shown in the following matrix.

		I_7	I_6	I_5	I_4	I_3		
		C	T	C	G	T	Dummy	
J_8	C	4	2	3	1	0	I_2	I_1
J_7	T	0	3	2	1	1	0	0
J_6	A	0	2	2	1	0	0	0
J_5	A	2	2	2	1	0	0	0
J_4	G	1	1	1	2	0	0	0
J_3	T	0	1	0	0	1	0	0
						LC		
		J_2	0	0	0	0	0	0
Dummy		J_1	0	0	0	0	0	0

(f) The best global alignment is then determined using the back tracing method as follows.

Starting from the lead cell (LC), trace an upward diagonal path. This is done regardless of the cell corresponding to a match or not.

	C	T	C	G	T
C	4	2	3	1	0
T	0	3	2	1	1
A	0	**2**	**2**	1	0
A	2	**2**	**2**	1	0
G	1	1	1	2	0
T	0	1	0	0	1
					LC

(g) This trace path is continued until an *island* is met. An island refers to a set of four cells with three or more entries that are identical, as shown in bold entries below.

	C	T	C	G	T
C	4	2	3	1	0
T	2	3	2	1	1
A	2	**2**	**2**	1	0
A	2	**2**	**2**	1	0
G	1	1	1	2	0
T	0	1	0	0	1

(h) From the island, there are two possible paths that can be pursued.
(A) First vertical, then diagonal

	C	T	C	G	T
C	4	2	3	1	0
T	2	3	2	1	1
A	2	**2**	**2**	1	0
A	2	**2**	**2**	1	0
G	1	1	1	2	0
T	0	1	0	0	1

(B) First horizontal, then diagonal:

	C	T	C	G	T
C	4	2	3	1	0
T	2	3	2	1	1
A	2	**2**	**2**	1	0
A	2	**2**	**2**	1	0
G	1	1	1	2	0
T	0	1	0	0	1

Back tracing: A Summary.

1. Start from lead cell.
2. Go diagonal from the cell to the next regardless of the entry if the new cell corresponds to a match or not.
3. Continue on diagonal trace until an "island" (shown with bold score entries) is reached. (The island can be identified as the set of four cells with three or more entries are identical in the four cells.)
4. Across the island, go "vertical" and continue on diagonal track or go "horizontal" and then continue on the diagonal trace, as feasible.

   The choice of picking one of the two paths depends on the number of cells ahead of the island, whichever path that has more cells traced ahead should be the selected path. (In the present example, taking the vertical path leads to four cells, whereas the horizontal path has only one ahead. Therefore, the vertical path is selected.)

(i) Now, the given sequences can be aligned as per the following rules.

I. A vertical track implies introducing a gap in the *X* (upper) sequence at its site.

II. A horizontal track implies introducing a gap in the *Y* (bottom) sequence at its site.

Relevant to present example of sequences *X*: CTCGT and *Y*: CTAAGT, the trace-back is illustrated below.

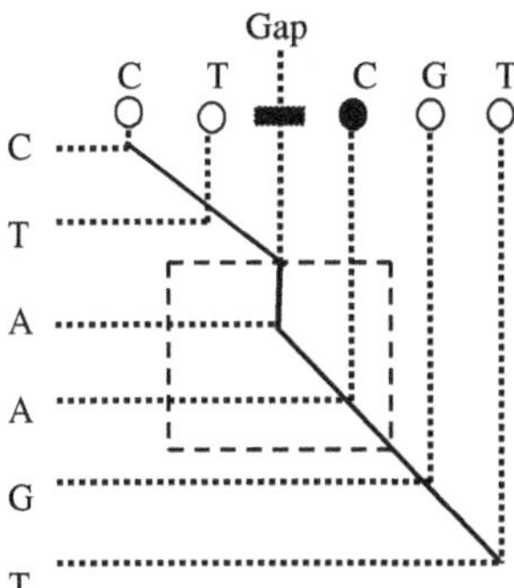

Hence, the aligned sequences are written as follows, with the gap introduced on sequence *x:*

x: C T – C G
| | |
y: C T A A G T

(j) In summary, the NW algorithm involves: (i) Setting up the matrix, (ii) updating the scores in the matrix cells, and (iii) identifying the optimal alignments *via* a trace-back suite.

The four aspects of trace-back procedure are as follows: (a) Encountering identity of residues at a site between the sequences *x* and *y*: Stay tracking along the diagonal; (b) mismatch of residues encountered at a site between *x* and *y*: Stay tracking along the diagonal; (c) gap in top (*x*) sequence corresponds to tracking vertically in the island; and (d) gap in the bottom (*y*) sequence corresponds to track horizontally in the island. Furnished below is an illustrative tutorial on trace-back procedure relevant to NW algorithm.

Tutorial T-4.3

A note on trace-back ... Based on vertical or horizontal trace pursuit at the island indicated above, the following can be specified as alignment rules.

- Outside the island where the diagonal pursuit is done, it implies a region where "identity" or "mismatch" of residues across the compared sequences exist without any gaps to exist at those sites.
- Within the island, if a vertical pursuit is done, it implies a "gap" to be introduced at the site of the upper sequence, for example, in *U* of a hypothetical pair of sequences, *U* and *V*.

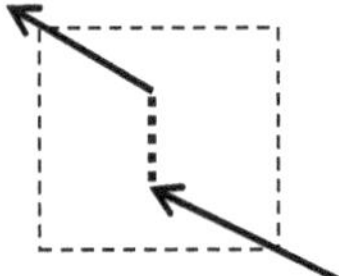

```
U: A A G – C T G
   | |     | |
V: A A T C G T G
```

- Within the island, if a horizontal pursuit is done, it implies a "gap" to be introduced at the site of the bottom sequence, for example, in *V* of a hypothetical pair of sequences, *U* and *V*.

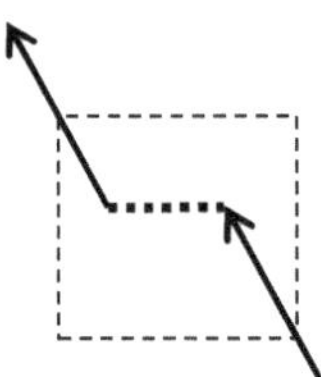

```
U: G A G C C T A
   | |     | |
V: G A T – G T A
```

Example Ex-4.9

Statement of the Problem

Given a set of sequence pairs, *X* and *Y* as indicated below. Determine in each case the best global alignment *via* trace-back using NW algorithm

X: G A G C A *Y*: G A T T C A

Solution

Initialize scoring matrix:

	G	A	G	C	A
G	**1**	0	**1**	0	0
A	0	**1**	0	0	**1**
T	0	0	0	0	0
T	0	0	0	0	0
C	0	0	0	**1**	0
A	0	**1**	0	0	**1**

Successive application of NW algorithm as indicated in the previous example in updating the cell entries leads to the following result. (The original initiated entries are shown in bold.)

	G	A	G	C	A	Dummy	
G	4 +	2 θ	3 +	1 θ	0 0	0	0
A	2 θ	3 +	2 θ	1 θ	1 1	0	0
T	2 θ	2 θ	2 θ	1 θ	0 0	0	0
T	2 θ	2 θ	2 θ	1 θ	0 0	0	0
C	1 θ	1 θ	1 θ	2 +	0 0	0	0
A	0 θ	1 +	0 θ	0 +	1 +	0	0
	0	0	0	0	0	0	0
Dummy	0	0	0	0	0	0	0

Lead cell

The updated final score matrix is as follows:

	G	A	G	C	A
G	4	2	3	1	0
A	2	3	2	1	1
T	2	2	2	1	0
T	2	2	2	1	0
C	1	1	1	2	0
A	0	1	0	0	1

Next back tracing procedure is applied as described in the previous example. It leads to an island where the pursuit of vertical or horizontal track is adjudicated as explained before and illustrated below.

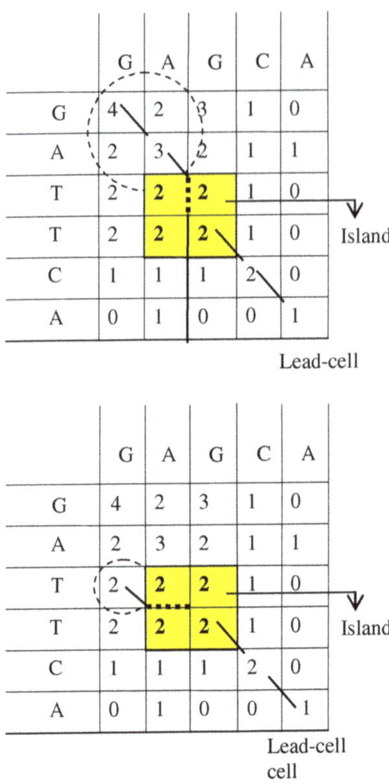

Among the two track pursuits (vertical and horizontal) shown above, if the horizontal track is chosen, it permits only one cell ahead to be traced (unlike, the vertical track leading to four cells as shown). Therefore, the vertical track pursuit is chosen giving the following solution on alignment:

GA – G C A
| | | |
GA T T C A

Example Ex-4.10

Statement of the Problem

Given the sequence pairs:

x: W F G Q E T S A I S
y: S F T Q F S E D A I

Perform NW algorithm based comparison between the two sequences and elucidate the optimal pathway in aligning them globally.

Solution
Step 1: Develop a score matrix for the two sequences and do initialization of cells representing the identities with the score value 1 and the cells representing mismatches with score value 0. The constructed initial matrix is shown below.

		x									
		W	F	G	Q	E	T	S	A	I	S
y	S	0	0	0	0	0	0	1	0	0	1
	F	0	1	0	0	0	0	0	0	0	0
	T	0	0	0	0	0	1	0	0	0	0
	Q	0	0	0	1	0	0	0	0	0	0
	F	0	1	0	0	0	0	0	0	0	0
	S	0	0	0	0	0	0	1	0	0	1
	E	0	0	0	0	1	0	0	0	0	0
	D	0	0	0	0	0	0	0	0	0	0
	A	0	0	0	0	0	0	0	1	0	0
	I	0	0	0	0	0	0	0	0	1	0

Step 2: Add "dummy" rows and columns at the left end of the matrix and fill these columns and rows with zeros. Name the columns (*i*: I_{12}, $I_{13}, \ldots, I_3$) and rows (*j*: $J_{12}, J_{11}, \ldots, J_3$) as shown. Identify the corner most cell (last cell of the matrix; not including dummy rows/columns) and designate it as the leading cell, LC specified with the coordinate (*i*, *j*).

		I_{12}	I_{11}	I_{10}	I_9	I_8	I_7	I_6	I_5	I_4	I_3	I_2	I_1
		W	F	G	Q	E	T	S	A	*I*	S	Dummy	
J_{12}	S	0	0	0	0	0	0	1	0	0	1	0	0
J_{11}	F	0	1	0	0	0	0	0	0	0	0	0	0
J_{10}	T	0	0	0	0	0	1	0	0	0	0	0	0
J_9	Q	0	0	0	1	0	0	0	0	0	0	0	0
J_8	F	0	1	0	0	0	0	0	0	0	0	0	0
J_7	S	0	0	0	0	0	0	1	0	0	1	0	0
J_6	E	0	0	0	0	1	0	0	0	0	0	0	0
J_5	D	0	0	0	0	0	0	0	0	0	0	0	0
J_4	A	0	0	0	0	0	0	0	1	0	0	0	0
J_3	I	0	0	0	0	0	0	0	0	1	0	0	0
J_2	Dummy	0	0	0	0	0	0	0	0	0	0	0	0
J_1		0	0	0	0	0	0	0	0	0	0	0	0

Step 3: As described in the previous example, starting from the leading cell (LC: *i*, *j*), move to the $S(i-1, j-1)$ cell (downward and diagonal). Then update the LC score value, $S(i, j)$ by adding the maximum of one of the three following observed value as per NW algorithm.

- $S(i-1, j-1)$.
- Maximum of $S(i-1, j-2)$, $S(i-1, j-3)$, . . . (along I_2, I_1, . . .), $S(i-1, j-n)$.
- Maximum of $S(i-2, j-1)$, $S(i-3, j-1)$, ... (along J_2, J_1, . . .), $S(i-n, j-1)$.

This procedure is repeated for the entire cells one-by-one, encountered along the columns I_3, I_4, . . . , and I_{12} and rows J_3, J_4, . . . , and J_{12} and the updating of the scores completed leading to the values shown bold in the following matrix.

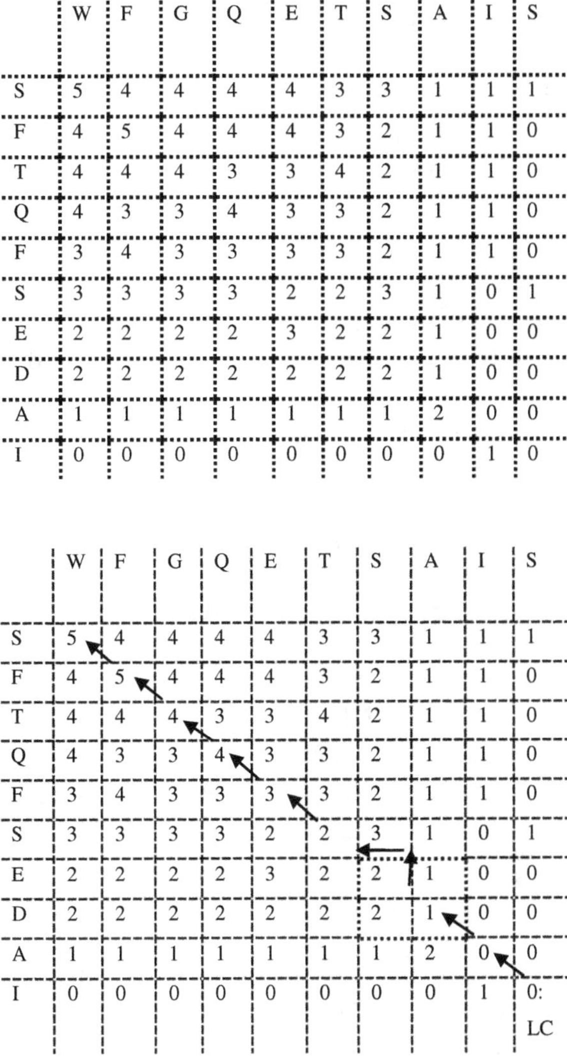

	W	F	G	Q	E	T	S	A	I	S
S	5	4	4	4	4	3	3	1	1	1
F	4	5	4	4	4	3	2	1	1	0
T	4	4	4	3	3	4	2	1	1	0
Q	4	3	3	4	3	3	2	1	1	0
F	3	4	3	3	3	3	2	1	1	0
S	3	3	3	3	2	2	3	1	0	1
E	2	2	2	2	3	2	2	1	0	0
D	2	2	2	2	2	2	2	1	0	0
A	1	1	1	1	1	1	1	2	0	0
I	0	0	0	0	0	0	0	0	1	0

	W	F	G	Q	E	T	S	A	I	S
S	5	4	4	4	4	3	3	1	1	1
F	4	5	4	4	4	3	2	1	1	0
T	4	4	4	3	3	4	2	1	1	0
Q	4	3	3	4	3	3	2	1	1	0
F	3	4	3	3	3	3	2	1	1	0
S	3	3	3	3	2	2	3	1	0	1
E	2	2	2	2	3	2	2	1	0	0
D	2	2	2	2	2	2	2	1	0	0
A	1	1	1	1	1	1	1	2	0	0
I	0	0	0	0	0	0	0	0	1	0: LC

Step 4: Using the final score matrix, a trace-back from the LC is done via diagonal pursuit and resorting to vertical or horizontal path whenever an island is encountered. Shown below is the track prescribed thereof to the present problem:

The aligned sequence is therefore:

```
W F G Q E T S - - A I S
  |   |   |     | |
S F T Q F - S E D A I
```

Problem P-4.4

The following pair of sequences is indicated in Attwood and Parry-Smith[56] and g relevant global alignment *via* NW algorithm is described. The present exercise is to obtain the final score matrix given in Attwood and Parry-Smith[56] for the test pair and verify the result on alignment as posted:

u: A D L G A V F A L C D R Y F Q
v: A D L G R T Q N C D R Y Y Q

Final gapped alignment shown in Attwood and Parry-Smith[56] is:

```
u: A D L G A V F A L C D R Y F Q
   | | | |           | | | |   |
v: A D L G R T Q N - C D R Y Y Q
```

Problem P-4.5

Statement of the Problem

Illustrate the application of NW algorithm in comparing two global sequences with steps indicated *via* a pseudo-code as regard the following two query sequences (of alphabets):

v: (... PDJGPVFPJCDRYFQ...)
w: (... PDJGRTQNCDRYYQ...).

The problem is to compare and align them globally and compute the similarity score *via* NW algorithm.

Implementation of SW Algorithm

As mentioned earlier, local alignment of sequences is based on dynamic programming using SW algorithm, which can be adopted to align protein and/or nucleotide sequences. The underlying optimal local alignment can be understood with the examples and exercises furnished below.

Example Ex-4.11

Statement of the Problem

Given a set of sequence pairs, *u* and *v* shown below, establish their best local alignment using SW algorithm.

u: ...W R N D C Q E G S A...
v: ...W G Q E G S I E A...

Solution
Application of SW algorithm toward aligning *u* and *v* and elucidating the local alignment conforms to the procedure with following steps.

(1) Construct an initial matrix framed with *u* and *v* residues as shown below.
(2) Add a set of edge elements *x*: "0" along the right-most column and the top-most row of the matrix as shown. Inasmuch as the first row and first column cannot be an end-point of any alignment (*x*: 0's) are introduced as indicated so as to serve as a dummy placeholder.
(3) Next, populate the cells corresponding to identical matches (of residues of *u* and *v* at the cell site) scored with entries of "1"s likewise, and cells corresponding to mismatches of residues scored with entries of "0"s

	u	W	R	N	D	C	Q	E	G	S	A
v	x	0	0	0	0	0	0	0	0	0	0
W	0	**1**	0	0	0	0	0	0	0	0	0
G	0	0	0	0	0	0	0	0	**1**	0	0
Q	0	0	0	0	0	0	**1**	0	0	0	0
E	0	0	0	0	0	0	0	1	0	0	0
G	0	0	0	0	0	0	0	0	**1**	0	0
S	0	0	0	0	0	0	0	0	0	**1**	0
I	0	0	0	0	0	0	0	0	0	0	0
E	0	0	0	0	0	0	0	**1**	0	0	0
A	0	0	0	0	0	0	0	0	0	0	**1**

(4) Highlight all the match-score entries (1's) bold. Relevant to each cell having match score entry (1) and marked bold, pursue the following three possible tracks to populate the rest of the cells with updated entries.

(a) Diagonal track: Suppose the cell having a match score entry (1) and marked bold. It is designated with a coordinate $(i - 1, j - 1)$ and the diagonal tracking is done downward toward the cell: (i, j) as illustrated in Figs. 4.4A and 4.4B. Suppose the score value at $(i - 1, j - 1)$ is $S(i - 1, j - 1)$, then the score on cell $S(i, j)$ is decided as follows:

$$S(i, j) = S(i - 1, j - 1) + 1.0,$$

if a similarity (match) of u and v exists at the site $(i - 1, j - 1)$; otherwise,

$$S(i, j) = S(i - 1, j - 1) - 0.3,$$

if a dissimilarity (mismatch) of u and v exists at the site $(i - 1, j - 1)$.

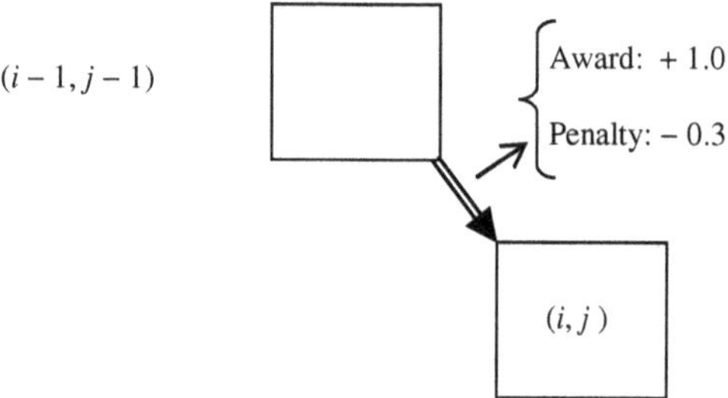

Fig. 4.4A Illustration of implementing NW algorithm. Updating cell score entry *via* diagonal track pursuit.

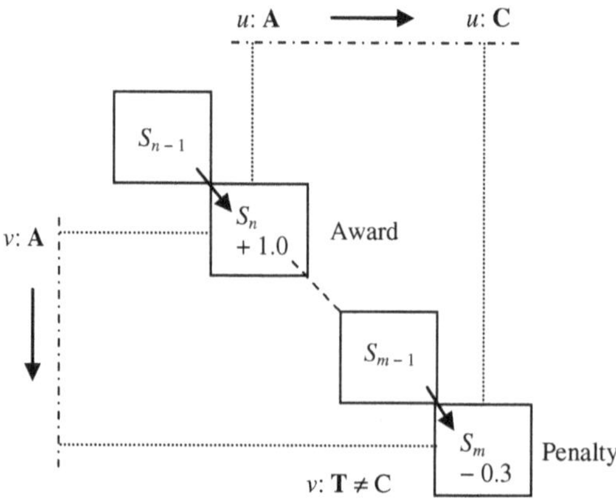

Fig. 4.4B Illustration of implementing NW algorithm. Updating cell score entry *via* diagonal track pursuit.

Updating of scores in Fig. 4.4A with addition of 1.0 implies an "award" given to similarity match observed, and subtracting 0.3 refers to "penalty" given to dissimilarity-(mismatch) observed. (The value 0.3 is an approximation of 1/3 depicting the degree-of-freedom).

in Fig. 4.4B, said diagonal track score update procedure is illustrated with an example: An *n*th cell with an existing score S_n gets an award of (+ 1) due to the identity of residues (A $\leftrightarrow$ A) of *u* and *v* and its updated score, therefore, becomes (S_n + 1). On the other hand, considering the *m*th cell as shown, with a score S_m takes a penalty of (−0.3) due to the mismatch of residues (C $\leftrightarrow$ T) of *u* and *v*, and, as such, its updated score becomes (S_m – 0.3).

(5) The diagonal pursuit as per the above step is exercised at all those cells that shoe the score entry of "1's" as confirmed in the initialization, and all the relevant cells in the diagonal pursuits are updated with the new scores. This diagonal path of updating the score is terminated when the computed score value becomes negative. At that cell and subsequently, the score entries along the diagonal path are rendered as "0's." Further, this diagonal cell score-filling is discontinued when a cell having a positive score value (possibly, 1 as registered in the initialization) is encountered *en passé.*

(6) The next step involves performing the following two algorithms with a horizontal pursuit along a row toward right and a vertical pursuit along a column downwards as illustrated in Fig. 4.4C so as to populate the cells with updated entries.

Suppose (i, j) is any cell considered. Then, the horizontal track along the row from this cell as shown leads to a set of sequential set of cells whose entries are updated with scores using the following algorithm: $S(i, j + k) = [S(i, j) - (1.0 + 0.3 \times k)]$, $k = 1, 2, \ldots$

The *k*-value denoting the *k*th cell as shown is ended, when a negative value of the score results in, and thereupon, the subsequent cells are filled with 0 scores. This horizontal-track based filling is continued and eventually stopped when a match (identity)-value of 1 of the initial matrix is encountered on the row.

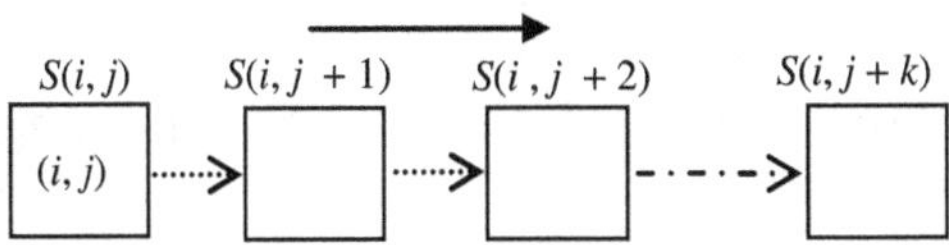

Fig. 4.4C Illustration of implementing NW algorithm. Updating cell score entry *via* horizontal track pursuit.

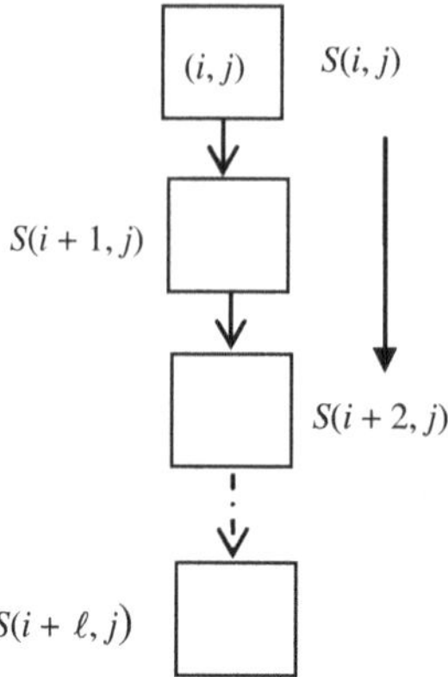

Fig. 4.4D Illustration of implementing NW algorithm. Updating cell score entry *via* vertical track pursuit.

Next, considering again a cell (i, j), the vertical track along the column from this cell as shown in Fig. 4.4D leads to a set of sequential cells downward whose entries are updated with scores using the following algorithm:

$$S(i + l, j) = [S(i, j) - (1.0 + 0.3 \times l)],\ l = 1, 2, \ldots.$$

The ℓ-value denotes the ℓth cell as shown is ended, when a negative value of the score results, and thereupon fill the subsequent cells with 0 scores. This vertical-track-based filling is continued and eventually stopped when a match-value of 1 of the initial matrix is encountered on the column. In the above procedures, the values of k and ℓ denote penalty lengths that specify the extent of penalties being imposed on the scores of the cells consistent with the possible deletions.

While proceeding along k or ℓ, if a negative value of the computed score results, then the corresponding cell and the rest seen subsequently are filled with 0 scores. (It implies that there is no alignment similarity up to the current cell position.)

Furthermore, score-filling horizontally (along the row) or vertically (along the column) is terminated when a cell with identity (similarity) score of 1 (registered in the initialization) is seen ahead *en passé.*

Thus, commencing from each of the cell $(i - 1, j - 1)$ with initialized score entry $S(i - 1, j - 1) = 1$, the updating of score values

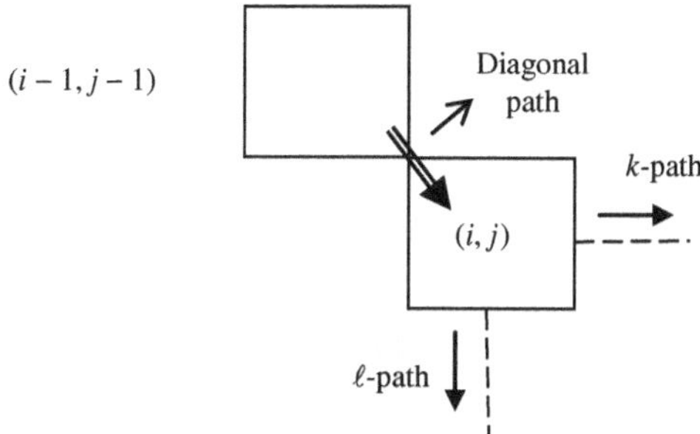

Fig. 4.4E Summary of NW algorithm: updating cell score entry *via* diagonal track pursuit followed by horizontal and vertical paths.

is done *via*: (a) Diagonal passage from cell $(i - 1, j - 1)$ to (i, j), (b) horizontal path-lengths, k (along the row), or (c) vertical path lengths, ℓ (along the column) as illustrated in Fig. 4.4E.

(7) Now, considering the alignment exercise in hand, the diagonal, horizontal, and vertical scoring procedures indicated above are performed in order to update the cell scores using the aforesaid algorithms pertinent to SW scheme of local alignment. The updated scores as computed are shown bold in the following matrix. The scored out values are the existing scores, and all the pertinent diagonal pursuits are marked with arrows.

	u	W	R	N	D	C	Q	E	G	S	A
v	x	0	0	0	0	0	0	0	0	0	0
W	0	1	0	0	0	0	0	0	0	0	0
G	0	0	**0.66** ~~0~~	0	0	0	0	0	1	0	0
Q	0	0	0	**0.33** ~~0~~	0	0	1	0	0	**0.66** ~~0~~	0
E	0	0	0	0	0	0	0	**2** ~~1~~	**0.66** ~~0~~	**0.33** ~~0~~	**0.33** ~~0~~
G	0	0	0	0	0	0	0	**0.66** ~~0~~	**3** ~~1~~	**1.66** ~~0~~	**1.33** ~~0~~
S	0	0	0	0	0	0	0	**0.33** ~~0~~	**1.66** ~~0~~	**4** ~~1~~	**2.66** ~~0~~
I	0	0	0	0	0	0	0	0	**1.33** ~~0~~	**2.67** ~~0~~	**3.66** ~~0~~
E	0	0	0	0	0	0	0	1	**1** ~~0~~	**2.33** ~~0~~	**2.33** ~~0~~
A	0	0	0	0	0	0	0	0	**0.66** ~~0~~	**2.0** ~~0~~	**2.0** ~~1~~

	u	W	R	N	D	C	Q	E	G	S	A
v	x	0	0	0	0	0	0	0	0	0	0
W	0	1	0	0	0	0	0	0	0	0	0
G	0	0	0.66	0	0	0	0	0	1	0	0
Q	0	0	0	0.33	0	0	1	0	0	0.66	0
E	0	0	0	0	0	0	0	**2**	0.66	0.33	0.33
G	0	0	0	0	0	0	0	0.66	**3**	1.66	1.33
S	0	0	0	0	0	0	0	0.33	1.66	**4**	2.66
I	0	0	0	0	0	0	0	0	1.33	2.66	3.66
E	0	0	0	0	0	0	0	1	1.00	2.33	2.33
A	0	0	0	0	0	0	0	0	0.66	2	2

(8) Alignment *via* trace-back: The trace-back refers to commencement of a trace bottom-up from the largest score value observed on the final score matrix and proceeding diagonally upward as illustrated below relevant to scores highlighted bold.

Hence, the final result on local alignment is as follows:

```
QEGS
||||
QEGS
```

Example Ex-4.12

Statement of the Problem

Given a set of sequence pairs, *u* and *v* shown below, determine the best local alignment *via* trace-back method using SW algorithm.

Given query pair of sequences:

u: AASTHECWCTWH
v: AASRNPSCWTTWHT

Solution

Step 1: As indicated in the previous example, the starting point for the SW algorithm is to construct the matrix with given residue sequences and initialize it with edge elements to *x*: 0 denoting the placeholder that accommodates the condition specified as follows: The first row and the first column of the matrix cannot form the endpoint of any specified alignment.

Step 2: Next, the cells in the matrix representing with identities of residues (between u and v) are scored 1; and, rest of the cells representing mismatches are scored 0. Shown below is the resulting matrix after Steps 1 and 2. (The mismatch values of "0"s are omitted in the matrix illustration for clarity.)

Step 3: Reference to the initialized matrix as above, the cells are filled with updated scores following the algorithm indicated in the last example.

Step 4: Using the final score matrix, the trace-back is performed starting from the highest value (4.3), as shown. Relevantly, the diagonal pursuit is continued along the path having the highest values until an "island" is met. (The description of an island is given earlier in the exercise pertinent to NW algorithm.)

	u	A	A	S	T	H	E	C	W	C	T	W	H
v	x	0	0	0	0	0	0	0	0	0	0	0	0
A	0	1	1										
A	0	1	1										
S	0			1									
R	0												
N	0												
P	0												
S	0			1									
C	0							1		1			
W	0								1			1	
T	0				1						1		
T	0				1						1		
W	0								1			1	
H	0					1							1
T	0				1						1		

	u	A	A	S	T	H	E	C	W	C	T	W	H
v	*x*	*0*	*0*	*0*	*0*	*0*	*0*	*0*	*0*	*0*	*0*	*0*	*0*
A	*0*	**1**	1	0	0	0	0	0	0	0	0	0	0
A	*0*	1	**2**	0.7	0.3	0	0	0	0	0	0	0	0
S	*0*	0	0.7	**3**	1.7	1.3	1	0.7	0.3	0	0	0	0
R	*0*	0	0.3	1.7	**2.7**	1.3	1	0.7	0.3	0	0	0	0
N	*0*	0	0	1.3	1.3	**2.3**	1	0.7	0.3	0	0	0	0
P	*0*	0	0	1	1	1	2	0.7	0.3	0	0	0	0
S	*0*	0	0	1	0.7	0.7	0.7	1.7	0.3	0	0	0	0
C	*0*	0	0	0	0.7	0.3	0.3	**1.7**	1.3	1.3	0	0	0
W	*0*	0	0	0	0	0.3	0	0	**2.7**	1.3	1	1	0
T	*0*	0	0	0	1	0	0	0	0	**1.3**	2.3	1	0.7
T	*0*	0	0	0	1	0.7	0	0	0	0	**2.3**	2	0.7
W	*0*	0	0	0	0	0.7	0.3	0	1	0	0	**3.3**	2
H	*0*	0	0	0	0	1	0.3	0	0	0.7	0	0	**4.3**
T	*0*	0	0	0	1	0	0.7	0	0	0	1.7	0	0

Here, the trace is directed horizontal and then vertical direction as shown and then pursued diagonal to 2.3 and further. It implies introducing a gap between H and E residues of *u* and *v*, respectively. Hence, the aligned sequences are written as follows:

```
A A S T H - E C W C T W H
| | |   :     | |   | | |
A A S R N P S C W T T W H
```

The final updated matrix is indicated above.

Example Ex-4.13

Statement of the problem

With reference to the following pair of sequences *u* and *v*, perform SW-algorithm-based comparison and elucidate locally significant, common regions of similarity.

u: W Y G Q E Q S Y I Q
v: W Y T Q E T S D I Q

Solution

Step 1: Pertinent to implementing SW algorithm, construct the initial score matrix with edge elements x: 0 being a placeholder as was done in the prior examples.

	u	W	Y	G	Q	E	Q	S	Y	I	Q
v	x	0	0	0	0	0	0	0	0	0	0
W	0	1									
Y	0		1						1		
T	0										
Q	0				1		1				1
E	0					1					
T	0										
S	0							1			
D	0										
I	0									1	
Q	0				1		1				1

		I_0	I_1	I_2	I_3	I_4	I_5	I_6	I_7	I_8	I_9	I_{10}
			W	Y	G	Q	E	Q	S	Y	I	Q
J_0		x	0	0	0	0	0	0	0	0	0	0
J_1	W	0	1	0	0	0	0	0	0	0	0	0
J_2	Y	0	0	2	0.7	0.3	0	0	0	1	0	0
J_3	T	0	0	0.7	1.7	0.3	0	0	0	0	0.7	0
J_4	Q	0	0	0.3	0.3	2.7	1.3	1	0.7	0.3	0	1.7
J_5	E	0	0	0	0	1.3	3.7	2.3	2	1.7	1.3	1
J_6	T	0	0	0	0	1	2.3	3.3	2	1.7	1.3	1
J_7	S	0	0	0	0	0.7	1	2	4.3	3	2.7	2.3
J_8	D	0	0	0	0	0.3	0.7	1.7	3	4	2.7	2.3
J_9	I	0	0	0	0	0	0.3	1.3	2.7	2.7	5	3.7
J_{10}	Q	0	0	0	0	1	0	1.3	2.3	2.3	3.7	6.0

Next, the cells representing identities are scored 1 and those representing mismatches are scored 0. The resulting matrix is shown below with the omission of 0 scores on mismatches for clarity.

Step 2: The cells are populated with updated scores *via* SW algorithm applied first to diagonal direction, then followed by row- and column-wise pursuits exercised at each cell showing the entry 1 (match scores). Hence, the resulting final score matrix is as indicated above.

Step 3: The trace-back pathway conforms to the passage that accumulates most matches. The common regions of similarity can be determined by referencing these matches as illustrated below.

		I_0	I_1	I_2	I_3	I_4	I_5	I_6	I_7	I_8	I_9	I_{10}
			W	Y	G	Q	E	Q	S	Y	I	Q
J_0		x	0	0	0	0	0	0	0	0	0	0
J_1	W	0	**1**	0	0	0	0	0	0	0	0	0
J_2	Y	0	0	**2**	0.7	0.3	0	0	0	1	0	0
J_3	T	0	0	0.7	**1.7**	0.3	0	0	0	0	0.7	0
J_4	Q	0	0	0.3	0.3	**2.7**	1.3	1	0.7	0.3	0	1.7
J_5	E	0	0	0	0	1.3	**3.7**	2.3	2	1.7	1.3	1
J_6	T	0	0	0	0	1	2.3	**3.3**	2	1.7	1.3	1
J_7	S	0	0	0	0	0.7	1	2	**4.3**	3	2.7	2.3
J_8	D	0	0	0	0	0.3	0.7	1.7	3	**4**	2.7	2.3
J_9	I	0	0	0	0	0	0.3	1.3	2.7	2.7	**5**	3.7
J_{10}	Q	0	0	0	0	1	0	1.3	2.3	2.3	3.7	**6**

The aligned sequence is as follows with the residues shown bold could be the locally significant aligned pairs of interest.

```
W Y G Q E Q S Y I Q
| |   | |   |   | |
W Y T Q E Q S D I Q
```

Problem P-4.6

Statement of the Problem

Illustrate application of SW algorithm using a pseudo-code with the following example of two query sequences, *u* and *v*:

u: (... VSTVVLENPGLGRALS...)
v: (... MSTVVTPNPGLGKAS...)

The problem is to compare and align them and compute the local similarity score *via* SW algorithm.

Problem P-4.7

Statement of the Problem

The following pair of sequences is considered in Attwood and Parry-Smith[56] by Attwood and Parry-Smith and their local alignment is deduced *via* SW algorithm. The present exercise is to obtain the final score matrix given in Attwood and Parry-Smith[56] for the test query pair and verify the result on alignment as posted:

u: A D L G A V F A L C D R Y F Q
v: A D L G R T Q N C D R Y Y Q

The final score matrix, after the calculation of all the scoring SW parameters is presented in Table 6.9 of Attwood and Parry-Smith.[56] The resulting aligned sequences are shown below with the highest score-yielding, locally significant substring:

```
u: A D L G A V F A L C D R Y F Q
   | | | |           | | | |   |
v: A D L G R T Q N - C D R Y Y Q
```

With reference to the above sequences (*u*, *v*), as they are studied in the original paper of Smith and Waterman,[5] the final score matrix indicated differs from that evaluated in Attwood and Parry-Smith.[56] However, the score values in Smith and Waterman[5] lead to the same alignment as in Attwood and Parry-Smith[56] furnished above.

In Attwood and Parry-Smith[56] an additional iteration on scoring is done as per SW algorithm, and it appears to change simply the "level" of numerical scores across the matrix; however, their relative magnitude remains almost unchanged. As such, the modified scores indicated in Attwood and Parry-Smith[56] leads to the same alignment as in Smith and Waterman[5] without modification.

Now, the question arises on the discrepancy observed in Smith and Waterman,[5] Sellers,[6] Altschul and Erickson,[7] Hogeweg and Hesper,[8] Mount,[9] Polyanovsky *et al.*,[10] Sniedovich,[11] Gotoh,[12] Brudno *et al.*,[13] Batzoglou *et al.*,[14] Rose,[15] Shuffle-LAGAN Algorithm,[16] Brudno *et al.*,[17] Krishnamurthy *et al.*,[18] Last Genome-scale Sequence Comparison,[19] Clustal Omega,[20] Mafft,[21] WebPRANK,[22] Penn *et al.*,[23] Braberg *et al.*,[24] Stamm *et al.*,[25] Simossis and Heringa,[26] Karun *et al.*,[27] Masomian *et al.*,[28] Carrillo and Lipman,[29] Wang and Jiang,[30] Altschul *et al.*,[31]

PROSITE website,[32] REG EX,[33] Hood *et al.*,[34] Collins and Jukes,[35] Ebersberger *et al.*,[36] Hillis *et al.*,[37] Huelsenbeck,[38] Lio and Goldman,[39] Jukes and Cantor,[40] Kimura,[41] Kimura,[42] Tajima and Nei,[43] Tamura and Nei,[44] Tamura,[45] Zharkikh,[46] Felsenstein,[47] Hasegawa *et al.*,[48] Lanave *et al.*,[49] Tavaré,[50] Rodkigljfk *et al.*,[51] Yang,[52] Brown *et al.*,[53] Keller *et al.*,[54] Yang,[55] and Attwood and Parry-Smith[56] as outlined above on score matrices. The author of this book has the opinion that the result indicated in Attwood and Parry-Smith[56] is possibly due to the consequence of additional modification (recursion) exercised by Smith and Waterman on the SW algorithm described before. Pertinent modification attempted is as follows.

Referring to the sequences u and v in the Example Ex-4.10, the SW algorithm is summarized as follows: commencing from each of the cell $(i - 1, j - 1)$ with initialized score entry $S(i - 1, j - 1) = 1$, the updating of score values is done *via*: (i) diagonal passage from cell $(i - 1, j - 1)$ to (i, j), (ii) horizontal path-lengths, k (along the row), or (iii) vertical path lengths, ℓ (along the column), and the score-filling horizontally (along the row) or vertically (along the column) is terminated when a cell with identity (similarity) score of 1 (registered in the initialization) is seen ahead *en passé*. Exercising the three steps as above, Table 6.9 of Attwood and Parry-Smith[56] is obtained and based on the scores a trace-back is suggested in Attwood and Parry-Smith[56].

In Attwood and Parry-Smith[56], an additional iteration seems to have been exercised. That is, after the three pursuits (diagonal, horizontal, and vertical) updating of score entries are done, all the cells that correspond to identity of residues in u and v are noted (depicting the updated similarity scores at those cells). With these current values of scores at those locations, the aforesaid steps (i), (ii), and (iii) are iterated, and the resulting updated values of scores would then correspond to those indicated in the publication.[56] If a trace-back is done on these scores, however, the same alignment results as in Smith and Waterman,[5] Sellers,[6] Altschul and Erickson,[7] Hogeweg and Hesper,[8] Mount,[9] Polyanovsky *et al.*,[10] Sniedovich,[11] Gotoh,[12] Brudno *et al.*,[13] Batzoglou *et al.*,[14] Rose,[15] Shuffle-LAGAN Algorithm,[16] Brudno *et al.*,[17] Krishnamurthy *et al.*,[18] Last Genome-scale Sequence Comparison,[19] Clustal Omega,[20] Mafft,[21] WebPRANK,[22] Penn *et al.*,[23] Braberg *et al.*,[24] Stamm *et al.*,[25] Simossis and Heringa,[26] Karun *et al.*,[27] Masomian *et al.*,[28] Carrillo and Lipman,[29] Wang and Jiang,[30] Altschul *et al.*,[31] PROSITE website,[32] REG EX,[33] Hood *et al.*,[34] Collins and Jukes,[35]

Ebersberger *et al.*,[36] Hillis *et al.*,[37] Huelsenbeck,[38] Lio and Goldman,[39] Jukes and Cantor,[40] Kimura,[41] Kimura,[42] Tajima and Nei,[43] Tamura and Nei,[44] Tamura,[45] Zharkikh,[46] Felsenstein,[47] Hasegawa *et al.*,[48] Lanave *et al.*,[49] Tavaré,[50] Rodkigljfk *et al.*,[51] Yang,[52] Brown *et al.*,[53] Keller *et al.*,[54] Yang,[55] and Attwood and Parry-Smith.[56] So, it is opined here that the second iterated efforts as above could be redundant, inasmuch as this additional procedure does not alter the final result on alignment sought.

Problem P-4.8

Statement of the Problem

(a) Perform SW algorithmic procedure on the test sequences *u* and *v* given above using the procedure indicated in Examples Ex-4.10 through Ex-4.13 consistent with the description in Attwood and Parry-Smith[56] and obtain the final score matrix.
(b) Repeat (a) using the modified procedure explained above consistent with the description in Smith and Waterman[5] and obtain the final score matrix.

Compare the results and show that the trace-backs on both of them lead to the same result on local alignment sought.

Problem P-4.9

Statement of the Problem

For the following pair of sequences *u* and *v* obtain the relevant local alignment *via* SW algorithm by obtaining the final score matrix as per the method: (1) as in Attwood and Parry-Smith[56] illustrated *via* Examples Ex-4.10 through Ex-4.13 by the modified procedure in Smith and Waterman[5] outlined earlier.

u: A C A G C C U C G C U U A G
v: A A U G C C A U U G A C G G

Solution hint: The local alignment is:

... G C C – U C G ...
... G C C A U U G ...

Problem P-4.10

Statement of the Problem

Loss-of-function mutations are significantly more frequent than gain-of-function mutations. Why is this so? (Mutations are more likely to reduce or eliminate gene function than to enhance it. As stated in the text, "By randomly changing or removing one of the components of a machine, it is much easier to break it (i.e., loss of function) than alter the way it works, that is, gain of function.)

Tutorial T-4.4

Transition and transversion mutations: a review summary: definitions of "transition mutations" and "transversion mutations" are as follows.

Transitions: Change from a purine to a purine or a pyrimidine to a pyrimidine. Examples: A to G; G to A; C to T; T to C. Transversions: Change from a purine to a pyrimidine or vice versa. Examples: A to C or T; G to C or T; C to A or G; T to A or G. The frequency of spontaneous transitions is significantly higher than the rate of spontaneous transversions.

Proposed explanation for why transitions are more common than transversions: in a normal double-stranded piece of DNA, purines are always paired with pyrimidines and *vice versa*. Because purines and pyrimidines are different-sized molecules (made of two or one ring, respectively), the consistent pairing results in a consistent distance between the two sugar-phosphate backbones. If two purines or two pyrimidines pair, the double-helix will be distorted, and thus be easily recognized by the various repair systems. For a transversion to occur, it is necessary for two purines or two pyrimidines to pair (for at least one round of replication). This will result in a distortion and repair. Thus, transversions are not likely to occur because they are easy to detect by repair systems. Whereas transitions do require mispairing, they do not cause the severe distortions of the double-helix and are thus more likely to "escape" the repair systems.

Problem P-4.11

Statement of the Problem

Shown below is a list of statements (1-to-11) and types of mutations. For each type of mutation, write the letter(s) of all that apply to that type of mutation *vis-à-vis* the statements marked as: (a-to-h). Justify the identified results. (*Note*: Each statement may be used more than once and each type of mutation may have more than one correct statement.)

Types of Mutations

(1) A mutation that changes UUA to UUG 2; (2) mutation that gives methionine instead of leucine; (3) created by the addition of a nucleotide to a coding region; (4) stop codon is read as an amino acid; (5) chemically similar amino acid is replaced by the mutation; (6) mutation that changes CCU to ACU; (7) deleting a nucleotide in a coding region gives this type of mutation; (8) mutation does not alter the peptide; (9) mutation changing UAU to UAG; (10) premature termination codon is responsible for this mutation, and (11) a chemically different amino acid is replaced by the mutation.

Mutation types: (a) Missense mutation: 2, 5, 6, 11. (b) Silent mutation: 1, 8. (c) Frameshift mutation: 3, 7. (d) Nonsense mutation: 9, 10. (e) Synonymous mutation: 1, 8. (f) Suppressor mutation: 4.

Example Ex-4.14

Statement of the Problem

Shown below is a partial peptide sequence for the wild type and three mutant alleles of a gene, PET1. Each mutant is presumed as a result of a single point mutation.

Wild Type: M I R M D K W ...
Mutant 1: M I Q N G ...
Mutant 2: M I R M G K W ...
Mutant 3: M I S M D K W ...

Deduce the exact DNA sequence of the coding strand of the wild-type allele using the amino acid sequence of the wild-type and the three mutants.

Solution

Each mutation enables finding the correct codon possible in depicting the amino acids. (Therefore, specifying alternate codons for an amino acid (except for the K (lys) codon, which could be either AAA or AAG) is not recommended.)

Wild Type: M I R M D K W ...
ATG ATC AGA ATG GAT AA(G/A) TGG
Mutant 1: M I Q N G ...
ATG ATC CAG AAT GGA TAA
Mutant 2: M I R M G K W ...
TG ATC AGA ATG GGT AA(G/A) TGG
Mutant 3: M I S M D K W ...
ATG ATC AG(C/T) ATG GAT AA(G/A) TGG

In the above example exercise, what type of mutation (transition *versus* transversion *versus* indel; missense *versus* nonsense *versus* frameshift) has occurred in each mutant?

Solution hint

Mutant 1: Insertion of either a C in the first base of the third codon or a T, C, or A in the third base of the second codon; Frameshift mutation.

Mutant 2: A to G transition in the second base of the fifth codon; Missense mutation

Mutant 3: A to C or T transversion in the third base of the third codon; Missense mutation.

4.5 Multiple Sequence Alignment

Multiple sequence alignment (MSA) refers to aligning three or more biological sequences (protein or nucleic acid) of similar length. From the output, the associated homology of the species of the query sequences can be inferred; also the evolutionary relationships between the sequences can be studied. Alignment of pairwise sequences studied in earlier sections implies a process of determining the most-likely source within the genome sequence for the observed DNA sequencing read, whenever the knowledge of the species from which the test sequence has come from. Relevantly, sequencing reads may also be aligned to other (multiple) genomes assuming that the evolutionary distance between such genomes is appropriate. (Unlike MSA, the pairwise alignments discussed above are used to identify regions of similarity that may indicate functional, structural, and/or evolutionary relationships between two biological sequences.)

The pairwise alignments included in the multiple alignment form a new matrix that is used to produce a hierarchical clustering. If it is different from the first one, iteration of the process can be performed. Relevant method is illustrated by an example concerning a global alignment of 39 sequences of cytochrome *c* by Corpet.[57] Macromolecules, representing residues of both nucleic acids and proteins, can be sequenced by using techniques that are automatized in modern context for fast outcomes. The increasing bulk of sequences being analyzed and aligned, it is imperative that high-level data processors and computers are needed supported by adequately compatible algorithms and computer programs.

However, MSA-specific approaches have emerged with: (i) limiting the problem to three short sequences; (ii) or specifying problem only to closely related sequences; (iii) resorting to using a predetermined evolutionary tree; and (iv) finding common subsequences (like 5–8) and selecting the best pairwise alignments from the scores of all pairwise comparisons.

Since the sequences in the already aligned set preserve their relative structure, finding the proper order is crucial. Largely, the algorithms proceed in a sequential way, for example, Feng and Doolittle[58] utilize a clustering technique and the closely related subsets of sequences are prealigned prior to the final alignment. The algorithm due to Corpet,[57] the clustering method indicated is much simpler without prealignment of clusters.

Example Ex-4.15

Statement of the Problem

This example summarizes the algorithm due to Ref. 57 indicated for two sequences A and B that have lengths, m and n. Suppose they are denoted by $A(i)$ and $B(j)$ with ith and jth elements in the respective sequences. To every pair of elements $A(i)$, $B(j)$, a weight $w(i, j)$ can be assigned from a suitable matrix $[D]$ (e.g., the mutation data matrix of Dayhoff[59]) for amino acids (with a suitable constant added to make all matrix entries non-negative): $w(i, j) - D[A(i), B(j)]$. The value B of w are not stored but computed as needed from stored values of the matrix. Next, method due to Needleman and Wunsch[4] and of Murata *et al.*[60] is to work backward from the cell (m, n), calculating the maximum total value S for paths from each cell. Suppose $S(i, j)$ is the maximum, overall paths from cell (i, j) to the bottom or the right-side of the sums of values w over the cells in the path minus "g" times the number of gaps in the path. This gap penalty, "*g*" is independent of the gap length, as suggested by the results of Barton and Sternberg.[61] Subsequently, follow the descriptive procedure in Corpet[57] and learn about the MSA *via* clustering proposed.

Tutorial T-4.5

Methods of MSA Implementation

The most widely used approach to MSAs uses a heuristic search known as *progressive technique* (also known as the *hierarchical* or *tree method*) developed by Feng and Doolittle[58] cited earlier. Such a progressive alignment builds up a final MSA by combining pairwise alignments beginning with the most similar pair and progressing to the most distantly related. All progressive alignment methods require two stages: (1) the first stage in which the relationships between the sequences are represented as a tree, called a *guide tree* and (2) the second step in which the MSA is built by adding the sequences sequentially to the growing MSA according to the guide tree. The initial *guide tree* is determined by an efficient clustering method such as *neighbor-joining* (NN) or UPGMA and may use distances based on the number of identical two letter subsequences (as in FASTA rather than a dynamic programming alignment). Details on NN and UPGMA are presented in Chapter 10.

Progressive alignments are not, in general, guaranteed as being globally optimal. A primary hindrance is that whenever errors incur at any stage in growing the MSA, such errors are carried forward through to the final step. Progressive alignments also perform not so well when all of the sequences in the set are distantly related. Modern progressive methods often modify their scoring function with a secondary weighting function that assigns scaling factors to individual members of the query-set in a nonlinear fashion. This is done on the basis of their phylogenetic distance from their nearest neighbors. This correction accounts for nonrandom selection of sequences offered to the alignment program.

Progressive alignment methods are preferably efficient on a large scale of many sequences in the order (100s to 1000s). Relevant alignment services can be availed from publicly accessible web servers; as such, the users need not locally install pertinent applications.

Most popularly used progressive alignment refers to the Clustal family, (e.g., the weighted variant *ClustalW*). Access is provided to ClustalW by a large number of web portals including GenomeNet, EBI, and EMBNet. However, variations in user interface may prevail across different portals or implementations that can make different parameters accessible to the user. Another local alignment program is LALIGN, which finds multiple regions of local. Details on this tool can be found in LALIGN.[62] The resulting alignment and phylogenetic tree are used as a guide to produce new and more accurate weighting factors.

Iterative Methods

These are methods developed to enable MSAs while reducing the errors inherent in progressive methods. These are classified as "iterative" since they work similar to progressive methods but iteratively realign the initial sequences as well as adding new sequences to the growing MSA. A reason why progressive methods are so strongly dependent on high-quality initial alignment is that such alignments are invariably included into the final result; that is, once a sequence is aligned into the MSA, its alignment is not considered any more. Relevant approximation improves the efficiency at the cost of accuracy. Iterative methods on the contrary can return to previously calculated pairwise alignments (and/or sub-MSAs) incorporating subsets of the query sequence so as to optimize a general objective function such as finding a high-quality alignment score.

Subtly distinct iteration methods have also emerged, which can be availed in software packages, and relevant reviews and comparisons are useful but, in general, choosing a "best" technique is not advisable. Examples of such tools are as follows: (i) a hill-climbing algorithm adopted to optimize the MSA alignment score is used in the software package PRRN/PRRP. It iteratively corrects both alignment weights as well as locally divergent (or "gappy") regions of the growing MSA. Furthermore, the PRRP performs best in the contexts of refining an alignment previously constructed by a faster method.

DIALIGN is another iterative program, which takes a unique approach concentrating on local alignments between subsegments or sequence motifs without introducing a gap penalty. The alignment of individual motifs is then achieved *via* a matrix representation as in dot-matrix plot used in a pairwise alignment. An alternative method that uses fast local alignments as anchor points or "seeds" for a slower global-alignment procedure is adopted in the suite of the tool called CHAOS/DIALIGN.

A third popular iteration-based method is known as *MSA by log-expectation* (MUSCLE), which is a version improved on progressive methods. It has a more accurate distance measure to assess the relatedness of two sequences, and the distance measure is updated between iteration stages. The original version of MUSCLE contained only 2–3 iterations depending on whether refinement was enabled.

Consensus Methods

In these methods, it is attempted to elucidate the optimal MSA when multiple different alignments of the same set of sequences are specified. The two consensus methods based tools available are M-COFFEE and MergeAlign. They use MSAs generated by seven different methods so as to implement consensus alignments. MergeAlign is capable of generating consensus alignments from any number of input alignments generated using different models of sequence evolution or different methods of MSA. There is also a default option for MergeAlign to infer a consensus using alignments generated using 91 different models of protein sequence evolution.

The multiple alignment programs designed (or modified) can search though large databases to find homologous sequences and short-read alignment programs are generally used for the alignment of DNA sequence from the species of interest to the reference genome assembly of that species. The underlying differences could be subtle, but consequently, the final algorithm design and implementations (along with specified assumptions on the number of expected mismatches and species polymorphism rate plus the technology error rate, etc.) lead to considerations of evolutionary substitutions. With the advent of rowing number of such implementations for short-read sequence alignments, the methods most commonly used refer to the following: (i) *Hash table-based implementations*, in which the hash may be created using either the reference genome or the set of sequencing reads and (ii) *Burrows–Wheeler transform* (BWT)*-based methods*, which initially create an efficient index of the reference genome assembly in a way that facilitates rapid searching in a low-memory footprint. Both methods as above can be applied to color-space (*SOLiD*) reads or base-space (Illumina, 454) reads with pertinent features toward such color-space capacity are included in the design of the alignment program.

MSA procedure and programs involve a multistep procedure in accurately mapping the sequence. Using the underlying heuristics relevant efforts swiftly identify a small set of places in the reference sequence where the location of the best mapping is most likely to be found. Once a smaller subset of possible mapping locations is located/identified, then a slower and precise alignment algorithm (such as SW technique) is exercised on the limited subset in question.[63] This short-segment search is a more computationally feasible method than running an accurate alignment algorithm toward a full search of all possible residue places where the sequence may map is computationally infeasible. All alignment programs include a "mapping policy" and the following algorithmic details can be indicated as exemplars.

Hash-Based Alignment Technique

The generation of conceived alignment programs designed for short-read alignment in NGS machines is based on a *hash-table data structure* for indexing and scanning the sequence data. Here, the "hash table" refers to a common data structure that can index the complex and non-sequential data so that a rapid searching is facilitated.[63] This is compatible and appropriate in DNA sequencing reads that may very unlikely possess every possible combination of nucleotides as well as probabilistically contain duplicates. Tools like MAQ8, SOAP9 SHRiMP10, ZOOM11, BFAST are examples that use hash-based algorithms as listed in Hash-based algorithms for MSA[64]; also, MOSAIK is another option that belongs to this category of tools.[65–67]

Salient MSA Tools

Concerning a variety of MSA tools outlined earlier, summarized below (in alphabetical order) are some salient tools that can be availed *via* programmatic access to bioinformatics from EMBL-EBI (updated in 2017 or later). If any issues are encountered in availing them, the EMBL-EBI support can be sought.

Clustal Omega: This MSA tool uses seeded-guide trees and HMM profile-profile techniques to generate alignments. It is suitable for medium to large alignments.

Kalign: This tool is relatively very fast, and its focused effort is on local regions, and it is used for large alignments.

MAFFT: This tool is based on Fast Fourier Transforms (FFTs) and it is compatible for use in medium to large MSAs.

MUSCLE: This tool is regarded as an accurate MSA tool, mostly with proteins and it is suitable for medium alignments.

MView: This tool is adopted to transform the results due to sequence similarity search into a MSA or reformat a MSA using its program.

T-Coffee: It is a consistency-based MSA tool. It enables mitigating pitfalls of progressive alignment methods. It is suitable for small alignments.

WebPRANK: It is a new version tool of the EBI, which is a phylogeny aware MSA program. It uses evolutionary details so as to place insertions and deletions as necessary.

MSA Applications

Identification of motifs (which represent conserved patterns of a set of nucleotides and/or amino acids possessing biologically significant details in the sequences being compared), etc., evolving corresponding gene information-processing in eventual making of a protein complex, stochastic profiling, and scoring of biosequences, elucidating statistical and entropic features of the microbiological complex and application of hidden Markov modeling (HMM) in sequence comparison and alignment tasks as needed are still details that offer scope for open-questions and extended research on MSA-related issues.

Tutorial T-4.6

A note on HMM and MSA: The HMM refer to probabilistic models that assign likelihood of occurrences of entirely feasible combinations of gaps, matches, and mismatches across a set of multiple sequences. Relevant assignment of likelihood details enables determining the most likely alignment of multiple sequences in question or a set of such possible MSAs. Hence, a single highest-scoring output can be realized *via* HMM, and a family of possible alignments so produced can be evaluated subsequently for biological significance. Both global and local alignments are feasible *via* HMM pursuits. HMM-based methods are relatively recent and improvements seen in computational methods and speed, allow realization of MSAs for sequences that may contain overlapping regions.

HMM-based methods involve representing an MSA in the form of a *directed acyclic graph*. It is also known as a *partial-order graph*, which consists of a series of nodes representing possible entries in the columns of an MSA. With the partial-order graphical representation of a MSA, a column that is absolutely conserved (with all the sequences in the MSA sharing a particular character at a particular position) is coded as a single node, and this node is designated "to have as many outgoing connections, as there are possible characters in the next column of the alignment."

In HMM-specific heuristics of MSA, the "observed states" are individual alignment columns, and the "hidden states" represent the presumed ancestral sequence from which the sequences in the query set are hypothesized to have descended. Software adopted for HMM-based methods are noted for their scalability and efficiency thought; proper usage of HMM approach needs cautious steps and it is more complex than using common progressive methods. Examples of software on HMM-based methods are as follows: *partial-order alignment* (POA) tool; a generalized method adopted in *Sequence Alignment and Modeling System* (SAM) package and HMMER. HHsearch is another software used in the detection of remotely related protein sequences based on the pairwise comparison *via* HMM.[68,69]

Problem P-4.12

Statement of the Problem

Apply the heuristics of HMM in the contexts of biosequence analyses using the parallel concept of HMMs prescribed in "time-series" applications indicated in Zucchini and MacDonald.[70] Another citable reference on HMM is Bhar and Hamori.[71]

4.6 Remarks on Part I

In all, the details deliberated in Part I include topics of biomolecular entities in relevance to elucidating the analytical aspects of comparing pairwise and/or multiple sequences. Such comparisons require first performing a meaningful alignment and then exercising an appraisal of the compared entities *via* scoring. Furthermore, an outline description on MSA and related tools are also furnished.

4.7 Part II: Next-Generation Sequencing

As indicated earlier, the so-called NGS refers to the global theme of HTS, and this part (Part II) of the chapter is conceived to explain details on complex NGS scenarios and their state-of-the-art considerations.

In evolving bioinformatic platforms intended to understand the genome, the transcriptome and the epigenome of any organism, related sequencing applications are largely dictated by the way sequencing libraries are prepared. There are a number of standard library preparation kits that offer pertinent protocols for sequencing the whole genomes and mRNA posing targeted regions, such as the whole *exomes*, custom-selected regions, protein-binding regions, and more. Relevantly, in addressing specific objectives, several novel protocols that isolate such specific regions of the genome each associated with a given biological function have been developed. Thus, came into being in 2007 is the art of the NGS in modern microbiology.

NGS (also synonymously known as *high-throughput sequencing* or HTS) refers to a desirable scheme devised to enable a variety of modern sequencing technologies in bioinformatics. Relevant technologies supply quick and cost-effective sequence details abundantly, and they are better than prior technique of traditional Sanger sequencing; hence, NGS has revolutionized the thematic pursuits of existing genomics leading to an advanced framework of molecular biology.

The concept behind NGS technology is similar to *capillary electrophoresis* (CE)-*based Sanger sequencing* wherein the bases of a small fragment of DNA are sequentially identified from signals emitted (when each fragment is re-synthesized from a DNA template strand), and NGS extends this process across millions of reactions in a massively parallel fashion, rather than being limited to a single or a few DNA fragments. The implied scientific advances enable rapid sequencing of large stretches of DNA base pairs that span the entire genomes with the latest instruments capable of producing hundreds of giga bases of data in a single sequencing run.

The state-of-the-art NGS is sprouting as an expanding universe with *raison d'être* that will place multitude sequencers across "tens of thousands of labs worldwide." This would concurrently mean feasible emergence of a lot of expensive instruments and a plentiful of consumables. Furthermore, significant percent of life scientists currently using traditional genomics technologies may switch to adopt NGS over the near future. This will set a trend on much greater demand for bioinformatics specialists, a need for specialized workforce and an expanded scope for neoteric perspective in bioinformatics.

4.8 Sanger and Next-Generation Sequencing

The CE-based Sanger sequencing mentioned earlier and the consequential NGS are compared in Table 4.4.

The associated tasks in NGS processes can be explained with the following example: Suppose a *genomic deoxyribonucleic acid* (gDNA), (which depicts a chromosomal DNA, in contrast to extra-chromosomal DNAs like plasmids) is considered. Most organisms have the same gDNA in every cell; however, only certain genes are active in each cell so as to allow for designated cell functions and differentiation within the body.

Table 4.4 CE-Based Sanger *versus* NGS methods.

CE-based Sanger sequencing	Next-generation sequencing
Library preparation is more involved Each sample should contain a single template, requiring purification from single bacterial, yeast colonies, or phage plaques	Library preparation is more streamlined Each sample can consist of a population of DNA molecules not requiring clonal purification
Sequencing is complete within days to weeks, depending upon the size of the genome being sequenced	Sequencing can be completed within hours, regardless of genome size

In the NGS process, a single gDNA sample is first fragmented into a library of small segments that can be uniformly and accurately sequenced in millions of parallel reactions. The newly identified strings of bases (called *reads*) are then reassembled using a known reference genome as a *scaffold* (implying a resequencing task) or, in the absence of a reference genome (as in the so-called *de novo* sequencing being described later), the full set of aligned reads would reveal the entire sequence of each chromosome in the gDNA sample.

NGS (or HTS) has been revolutionizing the scope of molecular biology in modern times.[72] With the advent of this new technology, it is now practical to carry out a "panoply of experiments at an unprecedented low-cost and high-speed" efforts in microbiology. Relevant high-throughput DNA sequencing technologies are based on a set of sequential procedures involving: (a) immobilization of the DNA samples onto a solid support; (b) cyclic sequencing reactions using automated fluidic devices; and (c) detection of molecular events by imaging.

NGS involves sequencing whole genomes, transcriptomes (depicting the set of all messenger RNA molecules in one cell or a population of cells. It differs from the so-called *exome,* in that it includes only those RNA molecules found in a specified cell population and usually includes the amount or concentration of each RNA molecule in addition to the molecular identities) and small non-coding RNAs. The outcome of NGS enables the description of methylated regions and identification of the protein complex *vis-à-vis* DNA interaction sites and detection of genomic structural variations. Generating within days of giga bases of sequence information for each of such huge bandwidth systems implies rapid development of bioinformatic applications for NGS data analysis, challenging.[73,74]

Computational-Infrastructure of HTS

NGS is highly influential in the molecular biology. Commensurate with such NGS/HTS technology, schemes of experiments at a reasonably low-cost and high-speed have emerged. As stated before, the underlying efforts include sequencing whole genomes, transcriptomes, and small non-coding RNAs as well as enunciating descriptive annotations on methylated regions and identification of protein (including DNA interaction sites and detection of structural variation). Making of giga bases on sequence information for each of bioinformatic applications within a short time makes the NGS data analysis posing inevitable challenges. The domains of analyses *vis-ả-vis* HTS include "distinguishing between "counting" and "reading" applications. Consistent with widely used such applications and associated basic sequencing concepts, various software programs exist as outlined in Normand and Yanai,[75] Metzker,[76] Normand and Yanai.[77]

NGS/HTS Technologies: Stages of Template Preparation

There are two methods used in preparing templates for NGS reactions namely, (i) clonally amplified templates originating from single DNA molecules and (ii) single DNA molecule templates. These methods robustly produce a representative, non-biased source of nucleic acid material from the genome under investigation. In general, relevant methods involve randomly breaking genomic DNA into smaller sizes, and using these fragments either fragment templates or mate-pair templates are formed. Then the template is attached or immobilized to a solid surface or support. Immobilization of spatially separated template sites allows thousands to billions of sequencing reactions to be performed simultaneously.

Sequencing and imaging: The basic differences in sequencing clonally amplified and single-molecule templates are as follows: the clonal amplification leads to a population of identical templates, (each of which has undergone the sequencing reaction). Relevantly, the observed signal upon imaging refers to a consensus of the nucleotides or probes added to the identical templates for a given cycle.

Cyclic Reversible Termination

DNA sequencing using reversible terminators, as one sequencing by synthesis strategy, has garnered a great deal of interest due to its popular application in the second-generation high-throughput DNA sequencing technology.[78] The *reversible termination sequencing* technology is a sequencing-by-synthesis approach that infers the sequence of a template by stepwise primer elongation. It is popularized as a second-generation sequencing technology on the Illumina platform*. (*Illumina** refers to a comprehensive solution for the NGS workflow, from library preparation to data analysis. Relevant library preparation kits are available for all NGS methods, including WGS, exome sequencing, targeted sequencing, RNA-Seq, and more.[79]) Sequencing using *cyclic reversible termination* (CRT) (also known as "sequencing by synthesis"), refers to a NGS architecture that infers the sequence of a template by stepwise primer elongation using fluorescently tagged triphosphates that have their 3-OH group blocked.

Sanger's method: In the initially perceived conceptual stages of genomic sequencing, the so-called *Sangers method* was ushered in, and the relevant method involves the following steps: First, the DNA test sample is divided into four separate sequencing reactions, containing the set of all the four standard deoxynucleotides (dATP, dGTP, dCTP, dTTP) plus a DNA polymerase. Each of the individual reaction is added with only one of the four dideoxynucleotides (ddATP, ddGTP, ddCTP, or ddTTP). In the next step, the template DNA is extended into the process (from the bound primer) and the resulting DNA fragments are heat denatured, and the fragments are separated by size using gel electrophoresis technique. The process as above is repeatedly performed using a denaturing polyacrylamide-urea gel having each of the four reactions run in one of four individual lanes (designated as lanes A, T, G, C).

Faster sequencing method (*via* NGS): This version of faster sequencing method employs micro- and nano-technologies with reduced size of sample components, reducing reagent costs and massively enabled parallel-sequencing reactions. The NGS procedure involves highly multiplexed, simultaneous sequencing, and the underlying analysis accommodates millions of samples. Hence, the NGS became popular and became commercially available since 2005 (e.g., *Solexa* sequencing technologies). Subsequently, a variety of sequencing methods came into being and progressed continually at astounding rates and numbers. NGS has the ability to process millions of sequence reads in parallel rather than 96 at a time (1/6 of the cost). In contrast, Sangers method bears demerits as regard to its fidelity, read length, infrastructure cost, and in handling large volume of data.

Considering the wide range of NGS applications, the associated key technical characteristics across the existing sequencing methods can be summarized as follows.

A. *de novo sequencing method*: This refers to sequencing and establishing a new genome or transcriptome. "(Such) new sequences can provide novel insight into the biology of an organism."
B. *Method of DNA/RNA-protein interactions* (ChIP-seq): In genomics, measuring the levels of DNA and RNA in a cell is important, and yet, there are applications wherein apart from the level of particular sequences, knowing how these sequences are interacting with the proteome is also critically emphasized. (Such interactions can grossly influence the regulation of gene expressions.)
C. *Metagenomics*: As indicated before, metagenomics is the study of environmental samples where multiple microbial genomes are analyzed simultaneously. It differs from the technique of isolating and cultivating individual species prior to sequencing their genomes. The merit of metagenomics is its ability in allowing the discovery and study of microbial genomes.
D. *Method of methylation sequencing*: In molecular biology, the methylation state of DNA (specifically that of the nucleotide base, cytosine) could influence the gene expression. Citable example is the mammalian cells, where higher levels of methyl CpG islands around the *transcription start-site* (TSS) have been observed to be associated with transcriptional silencing (in spite of more complex patterns in other regions of the genome are feasible).
E. *Resequencing method*: This application of NGS as regard to human samples is used in determining the genomic variations of a sample in relation to a common reference sequence. It involves the generated sequence being aligned to the reference sequence and mined for SNPs and CNVs in addition to genomic rearrangements and indels.
F. *Method of targeted resequencing*: The targeted resequencing method is a modified resequencing NGS application where only a small subset of the genome is sequenced, such as the exome, a particular chromosome, a set of genes or a specific region of interest. Focused sequencing on such a subset of the genome can be of reduced costs, in general.
G. *Transcriptome sequencing method*: This sequencing approach includes a broad spectrum of applications from a simple mRNA

profiling to the discovery and analysis of the entire transcriptome, (including both coding mRNA and non-coding RNA, for example, miRNA, small RNAs, and lincRNAs). Pertinent applications are collectively known as *RNA-Seq*, and they are quite popular in NGS platforms.

4.9 Genomic and Post-Genomic NGS Applications

As described in the previous sections, technical advances achieved in the development of molecular cloning, Sanger sequencing, polymerase chain reaction (PCR), and oligonucleotide microarrays augmented the speed and capacity of sequencing with the gross inclusions of complete organismal genomes. Relevant NGS platforms have surmounted unimaginably the economy of NGS methods and improvements in accuracy and read length in genomic and post-genomic biology. Analysis of microarray data plus the assembly and annotation of complete genome sequences achieved with conventional sequencing data, the scope of NGS data bears an enlarged scale *via* informatics tools that assemble, map, and interpret enormous quantities of relatively or extremely short nucleotide sequence data. Bioinformatics approaches thereof today envisage several genomics and functional genomics applications of NGS.[79,80]

In all, major bioinformatics applications for dealing with NGS including genome mapping, *de novo* assembly, detection of SNPs and editing sites, transcriptome analysis, ChIP-Seq, small RNA characterization, and epigenomics studies.[81–86]

NGS Applications: A Review

A review on NGS applications can be found in Ref. 87 The high-throughput, next-generation DNA sequencing technologies address a diverse range of problems in biological and medical sciences. The required scale and efficiency of sequencing in vogue provide extreme support in the analysis of genomes and in the knowledge base of how proteins interact with nucleic acids. Relevant highlights on NGS applications, their importance and related considerations are summarized below.

- Detecting rare and subclonal mutations has been enabled via accurate NGS yielding high-throughputs, which are otherwise limited in standard techniques posing difficulty in distinguishing sequencing errors as explained in Applications of NGS.[88]

- Profiling cancer transcriptome at the juncture of clinical translation implies the art of cancer genome sequencing required in cancer research.[89]
- NGS can support public health surveillance systems so as to improve the early detection of emerging infectious diseases.[90]
- Ancient genomes are capable of informing the history of human adaptation *via* direct tracking of changes observed in genetic variant frequency across different geographical locations and temporal era.[91]
- Errors of technical nature may prevent the interpretation of NGS data posing challenges in relevant clinical applications. Hence, reference standards for NGS are developed.[92]
- Knowing genetic variations of the human Y-chromosome is crucial in the pedagogy of human evolution, population history, genealogy, forensics, and male medical genetics. The NGS has enabled understanding the details on human Y-chromosome variation in the genome-sequencing era as described in Hardwick *et al.*[93]

4.10 Review Section: Tutorials and Exercises

Tutorial T-4.7

NGS in forensic contexts: An example of NGS application refers to exercises in functional forensics. That is, exclusive to forensic studies, newer genomic sequence analyses are constantly sought in order to deduce (forensically relevant) markers of pertinent biological test sequences. Specifically, *short-term repeats* (STR), *single nucleotide polymorphism* (SNPS) across a variety of genetic loci in the cytogenetic framework are viably considered in such efforts.

Relevant algorithmic suites in forensic contexts of genomic sequence analysis are expected to identify independently the discriminatory features on length of allele, sequence content, base composition, and visible markers. In essence, as compared to (ideal) the reference sequence, a forensically acquired test sequence will be dissected so as to know its base composition in the highly polymorphic-repeat stretch. Even in the context of scant or degraded DNA, it is presumed that one of the three proposed approaches can identify and/or offer kinship relation even with a small fragment of short nucleotides.

Another objective is to frame independent algorithmic suites, so as to arrive at cohesively common and differential genomic details toward annotation and indication for robust forensic evidential conclusions. Such cohesive information can be sought with reference to ancestry informative SNPs useful toward investigative leads. Concurrently, a haplotyping kinship, phylogenetics, and mixed state of test sequences. A related effort is to decide the common (motif) features between the test sequence or a set of test sequences collected at the forensic site *versus* the reference sequence using necessary alignment procedure and prescribing thereof a regular expression for the data obtained in the individual method as well as in the combined method. Development of NGS software for applications in forensic contexts has been a grossly exploited trait in bioinformatics with large scope for exploration. Newer schemes are often looked for so as to achieve fast, robust, and accurate outcomes of test sequence analyses.

Problem P-4.13

Statement of the Problem

This problem refers to defining the variability in many human genomes in terms of exemplar disease genetics, rare alleles and structural variants, and elaborating such details as a term-paper in the perspectives of NGS.

An example thereof refers to cytogenetic aspects of such pathological states. In recent multi-megabase sequences of Y-chromosome, DNA is ascertained from large population samples, with the consequent unbiased ascertainment of variation. Relevant investigations reveal the complexity of changes in Y-chromosome lineage distributions and frequencies over time, and genealogical studies show the male-line inheritance of the Y-chromosome which makes it an appropriate tool for studies of male family history, and pertinent NGS technologies can be applied on Y-Chromosome variation studies implicated, for example, in a single simple heritable disorder. The NGS has enabled understanding the details on human Y-chromosome variation in the genome-sequencing era as described in Hardwick *et al.*,[93] Jobling and Tyler-Smith,[94] and Bonasio *et al.*[95]

Problem P-4.14

Statement of the Problem

This exercise is about developing a term paper that reviews details on NGS applications in epigenomic research covering various types of epigenomic data afforded by NGS and including some of the outcomes of the epigenomics projects. Hence, open-questions as regard to the following are addressed: the "slew of epigenetic data has revolutionized our perception of the human genome." In general, not all "epigenetically marked" regions of the genome are likely to be functional, and analysis of transcriptional and epigenetic data has indicated that almost half of the genome is involved in carrying out specific biochemical functions. However, where such biochemical functions fit in the NGS framework and how they are regulated remain still an open question.

Another fundamental problem is that the causal relationships among various epigenomic events, such as DNA methylation, histone modifications, transcription factor (TF)-binding, and gene expression, etc., are currently not well understood. Although correlations between these events have been well documented in different contexts, it is not known whether one is necessary and/or sufficient to observe another.

Before the potential of NGS in mapping the epigenome is attained, several technical challenges need to be overcome. For instance, techniques, like ChIP-seq, used for profiling transcription factor binding and histone modifications, require a large amount of starting material, ranging up to 5 million cells.

Epigenomics: Following the protocols of the microbiological central dogma, epigenetics imply how genes are read by cells, and subsequently how they produce proteins.

Metagenomics is a study of genetic material recovered directly from environmental samples).

Problem P-4.15

Statement of the Problem

Ebola and Zika epidemics have driven the society toward continuous surveillance, rapid diagnosis, and real-time tracking of emerging infectious diseases. Relevantly, a fast, affordable sequencing of pathogen genomes is a crucial project in public health microbiology laboratory with highly well-resourced settings. Pertinent genomic diagnostics and epidemiology warrant digital disease detection platforms that can serve as open and global digital pathogen surveillance systems.

Considering human, animal, and other environmental health cohesively, such a genomics-based system is profoundly potential to improve public health especially in settings that lack robust laboratory capacity. That is, the role of genomics in rapid outbreak response and the challenges that need to be tackled for genomics-informed pathogen surveillance has become a global reality. Hence, a conceivable project is to develop a term paper on the pros and cons on the underlying issues in futuristic needs, developments, and visions of NGS-based genomics-informed pathogen surveillance.

Problem P-4.16

Statement of the Problem

A class of genome-sequencing projects refers to efforts of ancient human genome-sequencing projects targeted with genomic-scale ancient DNA datasets covering ancient human and archaic hominin (like Neanderthal valley individuals). Relevant datasets identify and track the spatiotemporal trajectories of genetic variants associated with human adaptations *vis-à-vis* morphing environments and rural/agricultural lifestyles. Co-evolving pathogens and the evidence of adaptive introgression of genetic variants (from archaic hominins to humans) as well as the emerging ancient genome datasets of domesticated animals and plants indicate evolutionary truths on human and the consequences of human behavior. Such datasets seem to be superior to modern genomic data and/or the fossil and archaeological records.

Problem P-4.17

Statement of the Problem

A possible scheme of NGS and application in the contexts of robust genomic sequencing (such as in forensic analysis platform) can be conceived in the *in silico* information-theoretic framework. The associated algorithmic strategy could rely on forging together three independent sequence analyses pursuits based on entropy, energetics, and Fourier spectral details comparing test sequence residues *versus* reference genomic sequence data. Pertinent three prong deployment of entropy, energetics, and spectral analyses in sequence analyses are elaborated in other chapters (Chapters 3 and 8) with illustrative examples. The underlying algorithmic suite will be compatible for complex NGS scenarios (encountered as in forensic cases).

4.11 Remarks on Part II

Thus, Part II of this chapter is a conceived effort to explain the heuristics of NGS scenarios. The details and implications of NGS show new directions on doable information-theoretics-based *in silico* framework pertinent robust genomic sequencing platforms.

References

[1] J. Shao, X. Yan and S. Shao: SNR of DNA sequences mapped by general affine transformations of the indicator sequences. *Journal of Mathematical Biology*, 2013, vol. 67(2), 433–451.

[2] A. Siepel and D. Haussler: Combining phylogenetic and hidden Markov models in biosequence analysis. *Journal of Computational Biology*, 2004, vol. 11(2–3), 413–428.

[3] S. F. Altschul and M. Pop: Sequence Alignment. Chapter 20.1, in Handbook of Discrete and Combinatorial Mathematics (Eds. K. H. Rosen, D. R. Shier and W. Goddard), CRC Press/Taylor & Francis Group, 2017, Boca Raton, FL: USA. Available online at: *https://www.ncbi.nlm.nih.gov/books/NBK464187/*

[4] S. B. Needleman and C. D. Wunsch: A general method applicable to the search for similarities in the amino acid sequence of two proteins. *Journal of Molecular Biology*, 1970, vol. 48(3), 443–453.

[5] T. F. Smith and M. S. Waterman: Identification of common molecular subsequences. *Journal of Molecular Biology*, 1981, vol. 147, 195–197.

[6] P. H. Sellers: On the theory and computation of evolutionary distances. *SIAM Journal on Applied Mathematics*, 1974, vol. 26(4), 787–793.

[7] S. F. Altschul and B. W. Erickson: Optimal sequence alignment using affine gap costs. *Bulletin of Mathematical Biology*, 1986, vol. 48, 603–616.

[8] P. Hogeweg and B. Hesper: The alignment of sets of sequences and the construction of phyletic trees: an integrated method. *Journal of Molecular Evolution*, 1984, vol. 20, 175–186.

[9] D. M. Mount: *Bioinformatics: Sequence and Genome Analysis.* Cold Spring Harbor Laboratory Press, Cold Spring Harbor, NY, USA: 2004.

[10] V. O. Polyanovsky, M. A. Roytberg and V. G. Tumanyan: Comparative analysis of the quality of a global algorithm and a local algorithm for alignment of two sequences. *Algorithms for Molecular Biology*, 2011, vol. 6(1), 25.

[11] M. Sniedovich: *Dynamic Programming: Foundations and Principles*. CRC Press/Taylor & Francis, Boca Raton, FL, USA: 2010.

[12] O. Gotoh: An improved algorithm for matching biological sequences. *Journal of Molecular Biology*, 1982, vol. 162, 705.

[13] M. Brudno, S. Malde, A. Poliakov, C. B. Do, O. Couronne, I. Dubchak and S. Batzoglou: Glocal alignment: finding rearrangements during alignment. *Bioinformatics*, 2003, vol. 19(Supplement 1), i54–i62.

[14] S. Batzoglou, L. Pachter, J. P. Mesirov, B. Berger and E. S. Lander: Human and mouse gene structure: comparative analysis and application to exon prediction. *Genome Research*, 2000, vol. 10, 950–958.

[15] Freiburg RNA Tools Teaching-Needleman-Wunsch Available online at: *http://rna.informatik.uni-freiburg.de/Teaching/index.jsp?toolName=Needleman-Wunsch* and Teaching-Smith-Waterman Available online at: *http://rna.informatik.uni-freiburg.de/Teaching/index.jsp?toolName=Smith-Waterman.*

[16] M. Brudno, S. Malde, A. Poliakov, C. Do, O. Courone, I. Dubchak and S. Batzoglu: Glocal alignment: finding rearrangements during alignment. Special Issue on the Proceedings of the ISMB 2003, *Bioinformatics*, vol. 19, 54i–62i, 2003 Available online at: *http://lagan.stanford.edu/lagan_web/citing.shtml.*

[17] M. Brudno, C. Do, G. Cooper, M. F. Kim, E. Davydov, A. Sidow and S. Batzoglou: LAGAN and Multi-LAGAN: Efficient tools for large-scale multiple alignment of genomic DNA, *Genome Research*, 2003, vol. 13(4), 721–731.

[18] P. Krishnamurthy, J. Buhler, R. Chamberlain, M. Franklin, K. Gyang, A. Jacob and J. Lancaster: Biosequence similarity search on the Mercury system. *The Journal of VLSI Signal Processing Systems for Signal, Image, and Video Technology*, 2007, vol. 49(1), 101–121.

[19] Last Genome-scale Sequence Comparison: Tutorial-CRBC. Available online at: *http://last.cbrc.jp/doc/last-tutorial.html.*

[20] EMBL-EBI (Wellcome Genome Campus, Cambridgeshire, UK): Clustal Omega – Available online at: *https://www.ebi.ac.uk/Tools/msa/clustalo/*

[21] Mafft: A sequence alignment program. Available online at: *https://mafft.cbrc.jp/alignment/software/.*

[22] WebPRANK — Available at: *https://www.ebi.ac.uk/goldman-srv/webprank/.*

[23] O. Penn, E. Privman, H. Ashkenazy, G. Landan and D. Graur: GUIDANCE: a web server for assessing alignment confidence scores. *Nucleic Acids Research*, 2010, vol. 38(Web Server Suppl_2 Issue), W23–28.

[24] H. Braberg, B. M. Webb, E. Tjioe, U. Pieper, A. Sali and M. S. Madhusudhan: SALIGN: a web server for alignment of multiple protein sequences and structures. *Bioinformatics*, 2012, vol. 28(15), 2072–2073.
[25] M. Stamm, R. Staritzbichler, K. Khafizov and L. R. Forrest: AlignMe — a membrane protein sequence alignment web server. *Nucleic Acids Research*, 2014, vol. 42(W1), W246–W251.
[26] V. A. Simossis and J. Heringa: PRALINE: a multiple sequence alignment toolbox that integrates homology-extended and secondary structure information. *Nucleic Acids Research*, 2005, vol. 33(Web Server issue), W289–W294.
[27] K. Karun, D. Kurian and Y. S. Sheeja: Biological sequence alignment — A review. *International Journal of Engineering Research and Application*, 2017, vol. 7(12 Part 1), 18–23.
[28] M. Masomian, R. Abd Rahman, A.-B. Salleh and M. Basri: Analysis of comparative sequence and genomic data to verify phylogenetic relationship and explore a new subfamily of bacterial lipases. *PLoS One*, vol. 11(3), e0149851. doi:10.1371/journal.pone.0149851.
[29] H. Carrillo and D. J. Lipman: The multiple sequence alignment problem in biology. *SIAM Journal of Applied Mathematics*, 1988, vol. 48(5), 1073–1082.
[30] L. Wang and T. Jiang: On the complexity of multiple sequence alignment. *Journal of Computational Biology*, 1994, vol. 1(4), 337–348.
[31] S. F. Altschul, W. Gish, W. Miller, E. W. Myers and D. J. Lipman: Basic local alignment search tool. *Journal of Molecular Biology*, 1990, vol. 215, 403–410.
[32] Prosite: Database of protein domains, families and functional sites: Available online at: *https://prosite.expasy.org/index.html*.
[33] Regular-Expressions.info: RegexBuddy. Available online at: *https://www.regular-expressions.info/*.
[34] M. E. Hood, M. Katawczik and T. Giraud: Repeat-induced point mutation and the population structure of transposable elements in Microbotryum violaceum. *Genetics*, 2005, vol. 170(3), 1081–1089.
[35] D. W. Collins and T. H. Jukes: Rates of transition and transversion in coding sequences since the human-rodent divergence. *Genomics*, 1994, vol. 20(3), 386–396.
[36] I. Ebersberger, D. Metzler, C. Schwarz and S. Pääbo: Genome wide comparison of DNA sequences between humans and chimpanzees. *American Journal of Human Genetics*, 2002, vol. 70(6), 1490–1497.
[37] D. M. Hillis, J. P. Huelsenbeck and C. W. Cunningham: Application and accuracy of molecular phylogenies. *Science*, 1994, vol. 264(5159), 671–677.
[38] J. P. Huelsenbeck: Performance of phylogenetic methods in simulation. *Systematic Biology*, 1995, vol. 44, 17–48.
[39] P. Lio and N. Goldman: Models of molecular evolution and phylogeny. *Genome Research*, 1998, vol. 8(12), 1233–1244.

[40] T. H. Jukes and C. R. Cantor: Evolution of protein molecules. In H. N. Munro: *Mammalian Protein Metabolism*. Academic Press, New York, NY, USA: 1969, pp. 21–123.
[41] M. Kimura: A simple method for estimating evolutionary rates of base substitutions through comparative studies of nucleotide sequences. *Journal of Molecular Evolution*, 1980, vol. 16(2), 111–120.
[42] M. Kimura: Estimation of evolutionary distances between homologous nucleotide sequences. *Proceedings of National Academy of Sciences* (USA), 1981, vol. 78, 454–458.
[43] F. Tajima and M. Nei: Estimation of evolutionary distance between nucleotide sequences. *Molecular Biology and Evolution*, 1984, vol. 1, 269–285.
[44] K. Tamura and M. Nei: Estimation of the number of nucleotide substitutions in the control region of mitochondrial DNA in humans and chimpanzees. *Molecular Biology and Evolution*, 1993, vol. 10(3), 512–526.
[45] K. Tamura: Estimation of the number of nucleotide substitutions when there are strong transition-transversion and G+C-content biases. *Molecular Biology and Evolution*, 1992, vol. 9(4), 678–687.
[46] A. Zharkikh: Estimation of evolutionary distances between nucleotide sequences. *Journal of Molecular Evolution*, 1994, vol. 39(3), 315–329.
[47] J. Felsenstein: Evolutionary trees from DNA sequences: a maximum likelihood approach. *Journal of Molecular Evolution*, 1981, vol. 17(6), 368–376.
[48] M. Hasegawa, H. Kishino and T. Yano: Dating of the human-ape splitting by a molecular clock of mitochondrial DNA. *Journal of Molecular Evolution*, 1985, vol. 22(2), 160–174.
[49] C. Lanave, G. Preparata, C. Saccone and G. Serio: A new method for calculating evolutionary substitution rates. *Journal of Molecular Evolution*, 1984, vol. 20, 86–93.
[50] S. Tavaré: Some probabilistic and statistical problems in the analysis of DNA Sequences. *Lectures on Mathematics in the Life Sciences*, 1986, vol. 17, 57–86.
[51] F. Rodkigljfk, J. L. Oiner, A. Martin and J. R. Medina: The general stochastic model of nucleotide substitution. *Journal of Theoretical Biology*, 1990, vol. 142, 485–501.
[52] Z. Yang: PAML 4: Phylogenetic analysis by maximum likelihood. *Molecular Biology and Evolution*, 2007, vol. 24(8), 1586–1591.
[53] W. M. Brown, E. M. Prager, A. Wang and A. C. Wilson: Mitochondrial DNA sequences of primates: tempo and mode of evolution. *Journal of Molecular Evolution*, 1982, vol. 18, 225–239.
[54] I. Keller, D. Bensasson and R. A. Nichols: Transition-transversion bias is not universal: A counter example from grasshopper pseudogenes. *PLoS Genetics*, 2007, vol. 3(2), e22. doi:10.1371/journal.pgen.0030022.

[55] Z. Yang: Maximum likelihood phylogenetic estimation from DNA sequences with variable rates over sites: approximate methods. *Journal of Molecular Evolution,* 1994, vol. 39, 306–314.

[56] T. K. Attwood and D. J. Parry-Smith: *Introduction to Bioinformatics.* Pearson Education Ltd., Essex, UK: 1999.

[57] F. Corpet: Multiple sequence alignment with hierarchical clustering. *Nucleic Acids Research,* 1988, vol. 16(22), 10881–10890.

[58] D.-F. Feng and R. F. Doolittle: Progressive sequence alignment as a prerequisite to correct phylogenetic trees. *Journal of Molecular Evolution,* 1987, vol. 25, 351–360.

[59] M. O. Dayoff (Ed.): *Atlas of Protein Sequence and Structure,* vol. 5. National Biomedical Research Foundation, Washington, D.C., USA: 1965, pp. 345–358.

[60] M. Murata, J. S. Richardson and J. L. Sussman: Simultaneous comparison of three protein sequences. *Proceedings of the National Academy of Sciences, USA.,* 1985, vol. 82, 3073–3077.

[61] G. J. Barton and M. J. Sternberg: A strategy for rapid multiple alignment of protein sequences. *Journal of Molecular Biology,* 1987, vol. 198(2), 327–337.

[62] EMBL-EBI - LALIGN: Pairwise Sequence Alignment. Available online at: *https://www.ebi.ac.uk/Tools/psa/lalign/.*

[63] S. Batzoglou: The many faces of sequence alignment. *Briefing in Bioinformatics,* 2005, vol. 6(1), 6–22.

[64] M. Helal, F. Kong, S. CA. Chen, F. Zhou, D.E. Dwyer, J. Potter and V. Sintchenko: Linear normalised hash function for clustering gene sequences and identifying reference sequences from multiple sequence alignments. *Microbial Informatics and Experimentation,* 2012, 2:2. Available online at: *https://www.ncbi.nlm.nih.gov/pmc/articles/PMC3351711/.*

[65] W-P. Lee, M. P. Stromberg, A. Ward, C. Stewart, E. P. Garrison and G. T. Marth: MOSAIK: A hash-based algorithm for accurate next-generation sequencing short-read mapping. *PLoS One,* vol. 9(3): e90581. https://doi.org/10.1371/journal.pone.0090581. Available online at: *https://journals.plos.org/plosone/article?id=10.1371/journal.pone.0090581.*

[66] M. Burrows and D. J. Wheeler: A block-sorting lossless data compression algorithm. Technical SRC Report 124, May, 10, 1994, Digital Equipment Corporation, Palo Alto, CA, USA. Available on-line at: *https://www.hpl.hp.com/techreports/Compaq-DEC/SRC-RR-124.pdf.*

[67] P. Ferragina and G. Manzini: Opportunistic data structures with applications. *Proceedings of the 41st Symposium on Foundation of Computer Science* (IEEE Computer Society, FOCS 2000), Redondo Beach, CA, USA: November, 12–14, 2000, pp. 390–398.

[68] R. Hughey and A. Krogh: Hidden Markov models for sequence analysis: Extension and analysis of the basic method. *CABIOS*, 1996, vol. 12(2), 95–107.

[69] C. Grasso and C. Lee: Combining partial order alignment and progressive multiple sequence alignment increases alignment speed and scalability to very large alignment problems. *Bioinformatics*, 1994, vol. 20(10), 1546–1556.

[70] W. Zucchini and I. L. MacDonald: *Hidden Markov Models for Time Series — An Introduction Using R.* CRC Press, Boca Raton, FL, USA: 2009.

[71] R. Bhar and S. Hamori: *Hidden Markov models — Applications to Financial Economics.* Kluwer Academic Publishers, Boston, MA, USA: 2004.

[72] N. Rodriguez-Ezpeleta, M. Hackenberg and A. M. Aransay (Eds.): *Bioinformatics for High Throughput Sequencing.* Springer, New York, NY: 2012.

[73] E. A. Trachtenberg and C. L. Holcomb: Next-Generation HLA Sequencing Using the 454 GS FLX System. *Methods in Molecular Biology*, 2013, vol. 1034, 197–219.

[74] M. Morey, A. Fernández-Marmiesse, D. Castiñeiras, J. M. Fraga, M. L. Couce and J. A. Cocho: A glimpse into past, present, and future DNA sequencing. *Molecular Genetics and Metabolism*, 2013, vol. 110(1–2), 3–24.

[75] R. Normand and I. Yanai: An introduction to high-throughput sequencing experiments: design and bioinformatics analysis. *Methods in Molecular Biology*, 2013, vol. 1038, 1–28.

[76] M. L. Metzker: Sequencing technologies — the next generation. *Nature Reviews Genetics*, 2010, vol. 11(1), 31–46.

[77] R. Normand and I. Yanai: An introduction to high-throughput sequencing experiments: design and bioinformatics analysis. *Methods in Molecular Biology*, 2013, vol. 1038, 1–26.

[78] F. Chen, M. Dong, M. Ge and R. Mu: The history and advances of reversible terminators used in new generations of sequencing technology. *Genomics, Proteomics & Bioinformatics*, 2013, vol. 11(1), 34–40.

[79] Illumina: An Introduction to Next Generation Sequencing Technology. Available online at: *https://www.illumina.com/documents/products/illumina_sequencing_introduction.pdf*

[80] D. S. Horner, G. Pavesi, T. Castrignano, P. D. De Meo, S. Liuni, *et al.*: Bioinformatics approaches for genomics and post genomic applications of next-generation sequencing. *Briefing in Bioinformatics*, 2009, vol. 11(2), 181–197.

[81] W. J. Ansorge: Next-generation DNA sequencing techniques, *New Biotechnology*, 2009, vol. 25, 195–203.

[82] S. C. Schuster: Next-generation sequencing transforms today's biology. *Nature Methods*, 2008, vol. 5(1), 16–18.

[83] V. R. Mardis: Next-generation DNA sequencing methods. *Annual Review of Genomics and Human Genetics*, 2008, vol. 9, 387–402.
[84] E. R. Mardis: The impact of next-generation sequencing technology on genetics. *Trends in Genetics*, 2008, vol. 24(3), 133–141.
[85] J. Shendure and H. Ji: Next-generation DNA sequencing. *Nature Biotechnology*, 2008, vol. 26(10), 1135–1145.
[86] D. A. Wheeler, M. Srinivasan, M. Egholm, Y. Shen, L. Chen, *et al.*: The complete genome of an individual by massively parallel DBA sequencing. *Nature*, 2008, vol. 452(7189), 872–876.
[87] R. Bonasio, S. Tu and D. Reinberg: Molecular signals of epigenetic states. *Science*, 2010, vol. 330(6004), 612–616.
[88] Applications of Next-generation Sequencing: Series: 01 January 2018 — *Nature Reviews Genetics* — Review Articles. Available online at: *https://www.nature.com/collections/jmgqdxpvsk.*
[89] J. J. Salk, M. W. Schmitt and L. A. Loeb: Enhancing the accuracy of next-generation sequencing for detecting rare and subclonal mutations. *Nature Reviews Genetics*, 2018, vol. 19, 269–285.
[90] M. Cieślik and A. M. Chinnaiyan: Cancer transcriptome profiling at the juncture of clinical translation. *Nature Reviews Genetics*, 2018, vol. 19, 93–109.
[91] J. L. Gardy and N. J. Loman: Towards a genomics-informed, real-time, global pathogen surveillance system. *Nature Reviews Genetics*, 2018, vol. 19, 9–20.
[92] S. Marciniak and G. H. Perry: Harnessing ancient genomes to study the history of human adaptation. *Nature Reviews Genetics*, 2018, vol. 18, 659–674.
[93] S. A. Hardwick, I. W. Deveson and T. R. Mercer: Reference standards for next-generation sequencing. *Nature Reviews Genetics*, 2017, vol. 18, 473–484.
[94] M. A. Jobling and C. Tyler-Smith: Human Y-chromosome variation in the genome-sequencing era. *Nature Reviews Genetics*, 2017, vol. 18, 485–497.
[95] R. Bonasio, S. Tu and D. Reinberg: Molecular signals of epigenetic states. *Science*, 2010, vol. 330(6004), 612–616.

Section III
Bioinformatic Resources and Translational/Clinical Informatics

A primary topic of interest in deliberating the notions of bioinformatics refers to understanding the associated resources comprised of databases, information networks, and computational tools. The global set of databases are improvised to store and disseminate the exhaustive bioinformatic data generated and disseminating associated details via modern information networking enable global sharing of such data as needed, in the contexts of bioinformatic studies, and computational tools are prescribed toward thereof to analyze the data and related data mining efforts.

Adjunct to assessing the details of omics across biosequences and studying the concepts of cytogenetics plus the notions of whole-life bioinformatics, the associated schemes of analytical and computational heuristics can be expanded to address the so-called translational research informatics (TRI), and they can be farther stretched into knowing the pedagogy of clinical bioinformatics and health science paradigms.

Hence, this section is designed to elaborate the aforesaid considerations via two pertinent chapters entitled as follows.

Chapter 5: Bioinformatic Resources: Databases, Information Networks, and Tools

Chapter 6: Translational and Clinical Bioinformatics

Chapter 5

Bioinformatic Resources: Databases, Information Networks, and Tools

This chapter is devoted to identify and describe various resources, associated information networks, and computational tools adopted in bioinformatics.

"Resources are hired to give results..."
(ad perpetuam rei memoriam)
Amit Kalantri, Wealth of Words

5.1 Bioinformatic Resources

The world of bioinformatics is conceived today as a resourceful framework comprised of extensive databases, elaborate information networking, and a plethora of computational tools. Relevantly, a large-scale of biological data is comprehensively hosted across a vast number of databases, and they are viably accessed through a global network of connectivity facilitated for data exchange and retrieval purposes. Also, exclusive versions of computational tools have been developed as resources commensurate with underlying analytical bioinformatic exercises.

5.2 Bioinformatic Databases

Bioinformatic databases represent a host of "electronic filing cabinets for storing vast amount of information"[1] and they correspond to elaborate servers located worldwide and interconnected *via* computer-specific networks. In simple terms, bioinformatic databases are repositories — search-enabled *via* compatible (search) engines. For example, the so-called *Entrez* is a *Global Query Cross-Database Search System* conforming to a federated search engine or web portal. Relevant Entrez global query facilitates knowing detailed views on gene and protein sequences, as well as on chromosome maps,[2] etc. It permits users to search several discrete health-science-related databases at the *National Center for Biotechnology Information* (NCBI) website including the *National Library of Medicine* (NLM) of the *National Institutes of Health* (NIH) operated by US Department of Health and Human Services. Data can be submitted to these databases and accessed *via* the *world wide web* (WWW). The primary archival of genomic and proteomic databases that hold extensive omic details are as follows.

Genomic Primary Databases

Biological raw data are stored in public genomic databanks on primary DNA sequences. Such genome information resources identified below refer to DNA sequence data repositories, including primary produces as well as a range of specialist genome information resources.

(a) *GENBANK*: This sequence-specific database is a bioinformatic open-access repository containing an annotated collection of all publicly available nucleotide sequences and their protein translations. This database is produced and maintained by NCBI as a unit of *International Nucleotide Sequence Database Collaboration* (INSDC). The NCBI is a part of the (NIH) in the United States.[3,4] The GenBank entry includes a concise description of the query sequence with scientific name and taxonomy of source organisms, bibliographic references, and a table of feature details.[5,6] It comprehensively lists areas of biological significance, such as coding regions (like coding sequence [CDS]) and corresponding protein translations, transcription units, repeat regions, and sites of mutations or residue modifications.

Though GenBank has the public-domain feature of "free" availability, it can be searched using NCBI's *Basic Local Alignment Search Tool* (BLAST). However, it lacks peer-reviewed sequences of *type strains* and sequences of *non-type strains*. In contrast, the commercial databases potentially contain high-quality filtered sequence data with limited number of reference sequences. The type strains refer to descendants of the *original isolates* that formed the basis for species descriptions, as defined by the Bacteriological Code, and they exhibit all of the relevant phenotypic and genotypic properties cited in the original published taxonomic circumscriptions. That is, type strain is defined in the "International Code of Nomenclature of Bacteria" as "nomenclatural type of the species." It denotes the "reference point" to which all other strains are compared to know whether they belong to that species. When conducting a catalogue search, in the list of strains found, a type strain has a green tick mark in the column "type strain."[7]

On daily basis, the GenBank data is exchanged with the data library of *European Molecular Biology Laboratory* (EMBL) and the *DNA Data Bank of Japan* (DDBJ) ensuring a worldwide coverage and enabling access through NCBI's retrieval system, Entrez. Complete bimonthly releases and daily updates of the GenBank database are available through file-transfer protocol, or FTP.[8] The GenBank and its related retrieval and/or analysis services can be started by logging into the site at NCBI Homepage.[9] Database sequences can be classified and queried using a comprehensive sequence-based taxonomy available at The NCBI taxonomy homepage,[10] developed by NCBI, EMBL, and DDBJ with necessary assistance from external advisers and curators.

As per *Release 209* (dated August 2015), the following specific divisions of GenBank support details on sequencing corresponding to various taxonomic groups: INV: *invertebrates*; MAM: *other mammals*; VRT: *other vertebrates*; WGS: *whole genome shotgun data*; PHG: *phages*; BCT: *bacteria;* PLN: *plants*; TSA: *transcriptome shotgun data*; VRL: *viruses*; ENV: *environmental samples*; HTG: *high-throughput genomic*; PAT: *patented sequences*; GSS: *genome survey sequences*; SYN: *gynthetic*; PRI: *primates*; EST: *expressed sequence tags*; ROD: *rodents*; HTC: *high-throughput cDNA*; UNA: *unannotated;* and STS: *sequence tagged sites.*[11]

(b) *EMBL:* As indicated above, EMBL (or *European Molecular Biology Laboratory*) is a research institution for molecular biology supported by several member states, created in 1974. Research at EMBL covers the broad spectrum of molecular biology ranging from basic research in molecular biology to molecular medicine. Today, EMBL is considered as the flagship laboratory for life sciences, and it is home for big data in biology directing so that ample applications of bioinformatics are comprehended *via* the art of storing, sharing, and analyzing biological data.

Established in 1993, a bioinformatics outstation of the EMBL is the *Wellcome Genome Campus* (owned by Welcome Trust), and the *Wellcome Trust Sanger Institute* (WTSI) in the United Kingdom is also a site of the European Bioinformatics Institute (EBI).[12,13] It supports field of research in genomic bioinformatics hosted by the *Medical Research Council's Rosalind Franklin Center.*[14]

(c) *DDBJ:* This refers to the *DNA Data Bank of Japan.*[15,16] It is a repository of bioinformatic data, collected, annotated, and released to the public in Japan. It hosts comprehensive number of entries and serves a number of bases. Consequent to human genome sequencing with remarkable advances in the research across life sciences, the paradigm of DDBJ shifted from the gene-scale to the genome-scale, and the Japanese effort on *Genome Information Broker* (GIB)[17] has further included more complete microbial genome and *Arabidopsis* genome data. A database on human genome of the DDBJ, the *Human Genomics Studio,*[18] can provide a continuous set of sequences in any one of the 24 human chromosomes. Both GIB and human genome studio (HGS) are updated as needed, incorporating newly available data and retrieval tools.

Genomic Secondary Databases

A sequel to primary databases containing nucleotide and amino acid sequence details, the results from the entries posted in such repositories are expanded in the so-called *secondary databases.* These contain details either manually curated or automatically generated entities and contain information on the following: *conserved sequence, signature sequence* (referring to short *oligonucleotides* of unique sequence pertinent to a certain ribosomal RNA of a specific class of prokaryotes), and *active-site residues* of a set of proteins subjected to multiple sequence analyses. The details available from curated databases are selective and are of high quality (e.g., as in SwissProt).

Information-Specific Databases

Depending on the type of information stored, the bioinformatic databases can be uniquely identified. Essentially, the following are pertinent databases.[19]

(i) *DNA sequence database*

It is a repository of all available nucleotide and/or amino acid sequences relevant to DNA and protein sequencing tasks. This sequence database is useful in homology searches toward retrieving related sequences in the phylogeny. The DNA sequence database is relevant to informatic details at genomic (DNA) level, transcription level involving RNA, and at proteomic level. The associated data can be relevant to complete (full stretch of the) sequence and/or concerned with partial segments, such as *expressed sequence tag* (EST), *sequence tagged site* (STS), or "all of the rest sequences", namely, the *single nucleotide polymorphism* (SNP) sites where variation in a nucleotide is present to some appreciable degree (>1%) within a population.

(ii) *Genomic single sequence database (GSS)*
GSS is a genome survey database containing single pass genomic sequences (i.e., sequences of bases in one strand of a *double-strand DNA* (dsDNA), *exon trapped genomic sequences* (of adapter splice sites adjacent to exons), and the family of repeated sequences (known as *Alul-PCR sequences*). Such repeated sequences are interspaced (on an average of 4–5 kbp) throughout in about 5% of human genome and in rodent DNA. GSS enables mapping the genome sequences independent of mRNA. It provides a preliminary global view of a genome containing both coding and non-coding segments of the DNA plus the repetitive details of the genome unlike ESTs.

GSS lacks long-range continuity inasmuch as its fragmentary nature makes it "harder to forecast gene and marker order." In detecting repetitive sequences in GSS data, for example, all the repeats may not be detected, since it is difficult to recognize the repetitive parts that are longer than the reads. Exemplary types of data contained in the GSS division are: (a) Random "single pass read" set of GSSs that are generated along single pass read by random selection. (b) The so-called end sequences of Cosmid/bacterial artificial chromosome (BAC)/yeast artificial chromosome (YAC) that enable sequencing the genome from the end side. (c) *Exon-trapped sequence* used in identifying the genes in cloned DNA by recognizing and "trapping" the carrier containing exon sequence of DNA. (d) Alu PCR sequences relevant to the Alu repetitive element*. (An *Alu repetitive element** is a short stretch of DNA and it a member of so-called *short interspersed elements, or* SINE, in mammalian genome). (e) *Transposon-tagged sequences* that involve analyzing the function of a particular gene sequence by replacing it or by causing a mutation and analyze the consequent results and effects. The associated transposable element could be a simple, *insertion sequence* (IS) or a *transposon* that has one or several characterized genes.

(iii) *High-throughput genomic (HTG) sequence database*
HTG is a repository of unfinished DNA sequences existing in the *contigs* greater than 2 kbp. The HTG sequences division was established to accommodate the growing need to enable unfinished genomic sequence data rapidly available to the bioinformatic community. It conforms to a coordinated effort among the International Nucleotide Sequence databases, namely DDBJ, EMBL, and GenBank.

The unfinished DNA sequences of HTG are generated by the high-throughput sequencing centers using traditional clone-based Sanger sequencing. Relevant sequence data are accessed *via* BLAST homology searches "against either the 'htgs' database or the 'month' database, which includes all new submissions for the prior month."[20]

(iv) *Draft genome sequence*
A single "pass"/"run" of sequencing is likely to produce a sequence containing errors as a result of the protocol in the steps involved, namely, breaking the DNA into readable smaller fragments, availing the sequence (e.g., *via* second-generation shotgun sequencing), adopting the method to read the signal produced, and reassembling the sequence data from the fragments into a continuous whole sequence.[21]

(v) *Sequence-tagged sites database (dbSTS)*
The *sequence-tagged site* (STS) refers to a relatively short and amicably *Polymerase Chain Reaction* (PCR)-amplified sequence (200–500 bp). Specifically amplified by PCR, the STS is detected in the presence of all other genomic sequences, and their locations in the genome are mapped. Olson *et al.*[22] introduced the concept of STS. Relevantly, ascertaining the locations of the known map of single-copy DNA sequences and observing, such locations serving as markers for genetic with physical mapping of genes along the chromosome, have visible impact in PCR performed in *human genome research* (HGR). The STS has the merit (over other mapping landmarks) as regard to the following: testing for the presence of a particular STS can be completely described as information in a database, and whenever copies of the marker are needed, the STS in the database can be looked up, specified primers are synthesized, and the PCR is run (under specified conditions) leading to amplifying the STS from genomic DNA.

(vi) *SNP database (dbSNP)*
The *single nucleotide polymorphism database* (dbSNP) is an open source of archive for data on genetic variations within and across different species.[23] This repository is developed and hosted by the National Center for Biotechnology Information (NCBI) in collaboration with the National Human Genome Research Institute (NHGRI). Notwithstanding the description designating "single nucleotide" polymorphisms, this database actually hosts a gamut of molecular variations, namely, (1) SNPs, (2) *short deletion and insertion polymorphisms* (indels/DIPs), (3) microsatellite markers or *short tandem repeats* (STRs), (4) *multinucleotide polymorphisms* (MNPs), (5) heterozygous sequences, and (6) named variants. The dbSNP was created in September 1998 to supplement the archival collections of publicly available nucleic acid and protein sequences in GenBank, NCBI. The dbSNP is a very rich collection of data, and relevant list of organisms plus the number of submissions for each is available at SNP,[24] Koboldt.[25]

(vii) *Expressed sequence tag database (dbEST)*
EST is a short subsequence of a cDNA sequence used to identify gene transcripts, ESTs available in public databases (such as GenBank) covering all species. *Database EST* (dbEST) is a division of Genbank established in 1992. Such GenBank data in dbEST are not curated and directly submitted by laboratories worldwide. The dbEST[26] is a division of GenBank that contains sequence data and other information on "single-pass" cDNA sequences or EST from a number of organisms.

(viii) *Restriction enzyme database*
The *restriction enzyme database* (REBASE) refers to a "comprehensive database of information about restriction enzymes and related proteins." This database on restriction enzymes and DNA methyltransferases has elaborate set of references, sites of recognition and cleavage, sequences and structures. It also holds information on the commercial availability details on each enzyme. Almost since 1980, REBASE has been in vogue as a biological database with regular descriptions of the resource deliberated in Roberts,[27] and Roberts and Macelis.[28] In short, REBASE is comprehensive and has archival data containing published and unpublished references, recognition and cleavage sites, isoschizomers, commercial availability, methylation sensitivity, crystal, and sequence data.

(ix) *Other sequence databases*

(a) *On raw sequence data*: The set of data of a clone that has yet to be assembled and linked to other clones via tilting path conforms to raw sequence data, for example, relevant to BAC or YAC clone. The *sequence read archive* (SRA) stores such raw sequencing data and alignment information from high-throughput sequencing platforms. For example, NCBI's short read archive is located in SRA.[29] It is a central depository for data from short read experiments.

(b) *On paired-end sequence data*: Paired-end sequencing is a method that allows sequencing both ends of a fragment and generate high-quality, alignable sequence data. It also facilitates detection of genomic rearrangements and repetitive sequence elements, as well as gene fusions and novel transcripts. The paired-end reads are more amenable to align to a reference, and as such, the quality of the entire dataset improves. *next-generation sequencing* (NGS) systems aim at paired-end sequencing.[30]

(c) *On seed sequence data*: *MicroRNA* (miRNA) refers to a small non-coding RNA molecule (containing about 22 nucleotides) found in plants, animals, and some viruses. Relevant microRNA (miRNA) is a small non-coding RNA molecule (typically about 22 nts found in plants, animals, and some viruses) that functions in RNA silencing and post-transcriptional regulation of gene expression. With reference to miRNAs that regulate the gene expression by binding to the mRNA, a *seed sequence* is essential for the binding of the miRNA to the mRNA. Such a *seed sequence* or *seed region* is a conserved *heptametrical sequence*, which is mostly situated at positions 2–7 from the 5′-end of the miRNA. Regardless of whether the base-pairing of miRNA and its target mRNA matches perfectly or not, the seed sequence has to be perfectly complementary. There exists a *miRBase Registry* that provides miRNA gene searchers with unique names for novel miRNA genes prior to publication of results. Relevant references are miRBase,[31] Griffiths,[32] and Griffiths *et al.*[33]

Proteomic Primary and Secondary Databases

These refer to protein-related primary databases identified as *protein information resources (PIR), Martinsried Institute for Protein Sequence*

(MIPS), SWISS-PROT, TrEMBL, and NRL-3D. The primary protein database details are described below and later the secondary and tertiary level databases are outlined.

Proteins as indicated in earlier chapters are generally comprised of one or more functional regions, commonly termed as domains. The presence of different domains in varying combinations in various proteins gives rise to a diverse repertoire of proteins found in nature. In view of the associated varying characteristics of functions and structures in such distinct sets of proteins, there exist three stratified types of protein sequence data representations, namely, *primary, secondary,* and *tertiary.*

The primary protein database is concerned with protein structure information and available as the *protein data bank* (PDB), created in the 1970s. Biologists and biochemists around the world provide submissions that are placed in PDB for free are accessible via the Internet through the websites of PDB's member organizations, namely, PDB in Europe (PDBe), PDB Japan (PDBj), and Research Collaboratory for *Structural Bioinformatics Protein Data Bank* (RCSB). The PDB is overseen by an organization called the *Worldwide Protein Data Bank* (wwPDB).[34]

The PDB is the key resource on structural biology, such as structural genomics. Other databases also use protein structures deposited in the PDB. For example, SCOP and CATH are databases that classify protein structures, while PDBsum provides a graphic overview of PDB entries using information from other sources, such as gene ontology. Corresponding *primary protein sequence databases* refer to PIR of *National Biomedical Research Foundation* (NBRF at Georgetown University, USA).[35,36]

The *protein identification resource*[37] provides efficient online computer system designed for the identification and analysis of protein sequences and their corresponding CDSs are established. In 2002, PIR along with its international partners, EBI, and Swiss Institute of Bioinformatics (SIB), expanded to form UniProt, a single worldwide database of protein sequence and function, and the underlying units are as follows: PIR-PSD, Swiss-Prot, and TrEMBL databases. PIR provides the scientific community with a single, centralized, authoritative resource for protein sequences and functional information.[38] The consortium of PIR is made of NBRF, *International Protein Information Database of Japan* (JIPID), and MIPS.

The MIPS group (at the Max-Planck-Institute for Biochemistry, Martinsried near Munich, Germany) collects, processes, and distributes protein sequence data within the framework of the tripartite association of the PIR-International Protein Sequence Database (PIR-IPSD). MIPS cater about 50% of the data input to the PIR-IPSD. Through its WWW server, the MIPS[39] permits Internet access to sequence databases, homology data, and to yeast genome information.[40]

NREF, or *nonredundant reference database,* is maintained by PIR. The PIR-NREF provides "a timely and comprehensive collection of protein sequences, currently consisting of more than 1,000,000 entries from PIR-PSD, SWISS-PROT, TrEMBL, RefSeq, GenPept, and PDB. The PIR website indicated earlier[35] connects "data analysis tools to underlying databases for information retrieval and knowledge discovery, with functionalities for interactive queries, combinations of sequence and text searches, and sorting and visual exploration of search results. The FTP site provides free download for PSD and NREF biweekly releases and auxiliary databases and files."[38]

The PIR database is divided into four sections: PIR1, PIR2, PIR3, and PIR4 with PIR1 and PIR2 accounting for 99% of all entries. PIR1 entries are fully "classified by superfamily assignment, fully merged with respect to other entries in PIR1, and extensively annotated." In PIR2, many entries are merged, classified, and annotated as fully as typical PIR1 entries. "Entries in PIR3 are not classified, merged or annotated. PIR3 comprises less than 1% of the total database and serves as a temporary holding tank for new entries." To include sequences identified as not naturally occurring or expressed (e.g., "known pseudo-genes, unexpressed ORFs, synthetic sequences, and non-naturally occurring fusion, crossover or frameshift mutations"), the PIR4 came into being.[41] The following are notable protein-specific databases.

SWISS-PROT database is a protein amino acid sequence database (of University of Geneva). It is a curated protein sequence database intended to provide a high level of annotation (such as the description of the function of a protein, its domain structure, posttranslational modifications, variants). (The curated database implies that changes and additions are done on the data by authorized curators.) The SWISS-PROT has a minimal extent of redundancy and carries extensive integration with other databases. The database includes comprehensively the number and scope of model organisms; cross-references to additional databases; a variety of new documentation files.[42]

TrEMBL database is an unannotated supplement to SWISS-PROT. It consists of entries in SWISS-PROT-like format derived from the translation of all CDSs in the EMBL nucleotide sequence database, except when the CDS is already included in SWISS-PROT. In short, TrEMBL stores the protein sequences that are yet to be added to the Swiss-Prot database. It is updated faster than the Swiss-Prot. It can be divided into two major parts: (a) SP-TrEMBL: In this part, those sequences are incorporated, which will eventually be a part of Swiss-Prot. Mainly FUN, HUM, INV, MAM, MHC, ORG, PHG, PRO, ROD, UNC, VRL, and VRT families sequences are included and (b) REM-TrEMBL: This part of TrEMBL is never added to the Swiss-Prot database. And it involves sequences of immunoglobins and T-cell receptors, synthetic sequences, patent application sequences, small fragments, CDS not coding for real proteins. The TrEMBL is divided into two major parts: (i) SP-TrEMBL: In this part, those sequences are incorporated, which will eventually be a part of Swiss-Prot.

NRL-3D database: The PIR and the PDB (of Brookhaven National Laboratory protein data bank) databases intended, respectively, for primary sequences and three-dimensional structures of proteins. However, the sequences in them are not redundant. Furthermore, the PIR programs are not amenable for use directly on PDB files in order to access primary sequences. This is because the formats of these two databases are different. The NRL-3D reported in Pattabiraman *et al.*[43] refers to a construction sequence structure database evolved from the sequences, chain identification, and the residue numbers of proteins in the PDB.

Protein Structural-Level Databases

The primary level of structural details of proteins includes formulated data on structural formats of the protein complex. They are repositories depicting the next level of protein databases populated with experimentally derived data (such as nucleotide sequence, protein sequence, or macromolecular structure). Relevant details are framed as per regions of local regularity features seen in the protein-making sequences (conforming to well-conserved motifs), and they lead to specifying structural formats, such as α-helices and β-strands of the protein complex. Examples of primary protein databases are European Nucleotide Archive (ENA), GenBank, DDBJ, ArrayExpress Archive and GEO, and PDB as will be described later.

Secondary databases contain derived data from the results of analyzing the primary data. (They are also referred to [wrongly] as "curated databases" inasmuch as primary databases are also curated so as to ensure that their data are consistent and accurate.) They obtain information from numerous sources, including other databases (primary and secondary), controlled vocabularies, and the scientific literature. They are highly curated, often using a complex combination of computational algorithms and manual analysis and interpretation to derive new knowledge from the public record of science. The details of such secondary level of protein data enclave *regular expressions, fingerprints, blocks, profiles,* etc. They are stored as *patterns.*

Secondary-level protein databases are *PROSITE, Profiles, PRINTS, Pfam, BLOCKS,* and *IDENTIFY* described below. There are also other repertoires of tertiary-level protein structural details of the domain or the module. These bases are *SCOP, CATH,* and *PDBsum* as described later.

PROSITE: Created by Amos Bairoch in 1988, PROSITE's protein database has entries describing protein families, domains, and the functional sites, as well as amino acid patterns and the associated profiles in them Hulo *et al.,*[44] Hulo *et al.,*[45] and Sigrist *et al.*[46] Furthermore, PROSITE offers tools for protein sequence analysis and motif detection, and it is part of the so-called ExPASy proteomics analysis servers.

PROSITE intend to contain patterns and profiles specific for more than thousand protein families or domains. Each of relevant signatures has annotated documentation providing background information on the structure and function of the proteins involved.

PROFILE: This is a discriminator adopted in protein family characterization *via* MSA and comparison. That is, profiles depict model protein families and domains evolved by converting MSAs into *position-specific scoring systems* (PSSMs). Given a set query multiple sequences, the amino acids at each position in the alignment are scored according to the probability of their occurrence and *substitution matrices* (such as BLOSUM matrices) are adopted to include the evolutionary distance weighting on the scored positions. Each profile in the compendium of profiles available at *Swiss Institute for Experimental Cancer Research* (ISERC, Lausanne) has separate files on data and family annotations formatted toward PROSITE compatibility and relevant documentation.

PRINTS: In the context of protein database, the PRINTS refers to a conglomeration of so-called *fingerprints.* The PRINTS database offers both a detailed annotation resource for protein families, as well as it is a diagnostic tool for newly determined sequences.[47,48] A fingerprint represents a group of conserved motifs observed in a MSA.[49] The motifs collectively form a characteristic signature for the aligned protein family. The motifs need not be contiguous in sequence, but they may collectively define and depict molecular binding sites or interaction surfaces in 3D space. Thus, PRINTS represents a compendium of protein motif fingerprints, which is a group of motifs excised from conserved regions of a sequence alignment.

Relevant details on PRINTS concerning resource description, database format, content of the release with details on updates and growth, database distribution, and applications can be seen in Attwood *et al.*[50] The diagnostic strength of fingerprints is their ability to distinguish sequence differences at the clan, superfamily, family, and subfamily levels.

Pfam: This is yet another database of protein families that includes their annotations and MSAs that are generated using the *hidden Markov models* (HMMs).[51,52] That is, in the relevant large collection of protein families, each family is represented by MSAs subjected to HMM.

The entries of Pfam are classified as follows: (i) Family representing a collection of related protein regions. (ii) Domain depicting a structural unit. (iii) Repeat representing a short unit, which is unstable in isolation, but it forms a stable structure when multiple copies are present. (iv) Motifs denoting short units found outside the globular domains. (v) Coiled-coil illustrating regions that predominantly contain "coiled-coil motifs" (with α-helices that are coiled together in bundles of 2–7). (vi) Disordered: regions that are conserved, yet are either shown or predicted to contain bias sequence composition and/or are intrinsically disordered (non-globular). Related Pfam entries are grouped together into clans; the relationship may be defined by similarity of sequence, structure, or profile HMM.

BLOCKS: This database contains multiple alignments of conserved regions in protein families. It is searchable *via* e-mail/WWW servers at the site BLOCKS WWW Server.[53] This database is said to classify protein and nucleotide sequences. In bioinformatic contexts, several known proteins are grouped into families according to functional and sequence similarities. In general, similarity of the proteins across the sequences in each family could be far from uniform.

The description of a protein family *via* its conserved regions portraying the characteristics of the family and distinctive sequence features could reduce the noisy details. Relevant databases of conserved features of protein families are used to classify sequences from proteins, cDNAs, and genomic DNA. An example of database constructed *via* a fully automated method with sequences of protein families consists of ungapped multiple alignments of short regions, called *blocks*. Searching the relevant *Blocks database* with a sequence query facilitates detection of one or more blocks representing a family.[54]

Searching of the Blocks database is done using protein or DNA sequence queries, and the results are returned with measures of significance for both single and multiple block hits. Blocks database is also useful for derivation of amino acid substitution matrices (the so-called Blosum series to be described later) and other sets of parameters. Relevant WWW and e-mail servers enable access to the database and associated functions, including a block maker for sequences provided by the user.[55]

EMOTIF and *IDENTIFY:* In Nevill-Manning *et al.*,[56] a method is described to discover conserved sequence motifs from families of aligned protein sequences, and the relevant implementation is indicated *via* a computer program called EMOTIF available in Web site on motif.[57]

Such a computer program generates a set of "motifs with a wide range of specificities and sensitivities for a given aligned set of protein sequences." Furthermore, it can exhaustively generate motifs that describe possible subfamilies of a protein superfamily. Relevant database contains several motifs having a probability of matching a false positive that range from 10^{-10} to 10^{-5}.

In assigning function to genes in newly sequenced genomes *via* highly specific search and comparison methods, relevant procedure involves "first identifying all *open-reading frames* (ORFs) or coding regions in the genome and translating them into putative protein sequences." Corresponding protein sequences can then be compared with "(i) databases of individual protein sequences, (ii) databases of protein consensus sequences, or (iii) families of aligned proteins", and the remaining unassigned proteins can also be compared with known protein folds or structures by using sequence–structure alignment and/or threading methods. As regard to highly specific and sensitive protein sequence motifs representing conserved biochemical properties and biological functions, the EMOTIF database is available with a collection of more than 170,000.[58]

The *emotif-maker algorithm* was developed by Nevill-Manning *et al.*[59] Relevant protein motifs are derivable across thousands of sequence alignments *via* BLOCKS+ and/or PRINTS databases. The EMOTIF protein pattern database is available in Web site on motif.[57] EMOTIF enables a systematic and objective method to determine sequence motifs using aligned sets of protein sequences. The EMOTIF generates many possible motifs over a wide range of sensitivity and specificity (unlike other methods prescribed to find a single "best" motif optimized at one level of sensitivity and specificity). Typically, the EMOTIF can generate extremely specific motifs with fewer than false prediction per 10^{10} tests, and it includes more sensitive motifs across all members of a family.

SCOP: An exclusive database known as *Structural Classification of Proteins* (SCOP) has been developed to provide "a detailed and comprehensive description of the relationships of known protein structures."[60]

Links between SCOP and PDB-ISL are feasible. The PDB-ISL is a library containing sequences homologous to proteins of known structure, and using pairwise sequence comparison methods, sequences of proteins of unknown structure can be matched to distantly related proteins of known structure in PDB-ISL so as to find the homologs. SCOP database and its associated files are freely accessible from a number of WWW sites mirrored from Web-site on SCOP.[61]

CATH: The *CATH Protein Structure Classification* is a medium developed in 1997 towards semiautomatic, hierarchical classification of protein domains.[62] CATH shares several features with its principal counterpart SCOP, with discernible differences as well.

The four levels of CATH hierarchical classification are: (i) Class pertinent to the overall secondary-structure content of the domain. (ii) Architecture providing details on high structural similarity but no evidence of homology. (iii) Topology implying a large-scale grouping that share particular structural features. (iv) Homologous superfamily details demonstrating evolutionary relationship. (Hence, the acronym CATH.) The website at: Web-site on CATH[63] allows user-friendly entry to the classification allowing for both browsing and downloading of data.[64]

PDBsum: An overview of the contents of each 3D macromolecular structure deposited in the PDB can be seen in pictorial format in the PDBsum database. It offers an illustrative suite of the molecule(s) that make up the microbiological structures, namely, the protein chains, DNA, ligand, and metal ions, as well as the schematic diagrams of interactions between them. Developed around 1995 by Laskowski and others at University College London, the PDBsum is maintained in the laboratory at the EBI.[65]

The 3D geometrical views (including the main, bottom, and right views) of each structure in the PDBsum database include the image of the structure (depicting the complex framework of molecular components).[66] Furthermore, 3D view of molecules and their interactions within PDBsum is feasible *via* molecular graphics software, such as *RasMol* and *Jmol*.[67,68]

Thus, the biological databases depict vast information resources containing details on all available biosequences. The stored and disseminated details conform to application-specific search efforts pertinent to: (i) *Annotation search*; (ii) *homology search*; (iii) *pattern search*; (iv) *prediction search*; and (v) *comparison search*.

Composite Databases

A conglomerated set of resources of primary databases refer to composite databases. They are conceived to enable easier querying and searching of multiple resources. Though several databases are combined in framing a composite database, the associated data redundancy (in composite databases) is kept to a minimum.

An example of composite database is known as the OWL. It is a composite of SwissProt, PIR, GenBank, and NRL-3D databases. OWL conforms to a comprehensive, nonredundant composite protein sequence database, which is "an amalgam of data from six publicly-available primary sources, and is generated using strict redundancy criteria."[69]

The so-called *non-redundant database* (NRDB) is a composite of the following sources: PDB sequences, SWISS-PROT, SWISS-PROTupdate, PIR, GenPept, and GenPeptupdate. This database is thus similar in content to OWL, but contains more up-to-date information. Strictly speaking, NRDB is not nonredundant, but nonidentical, that is, only identical sequence copies are removed from the database. As such, NRDB is larger and less efficient to search than OWL.

Another example composite database is the BioSilico, which is an integrated database of Ligand, Enzyme, EcoVyc, and MetaCyc resources. It is a web-based database system that facilitates the search and analysis of metabolic pathways, and LIGAND, ENZYME, EcoCyc, and MetaCyc

Table 5.1 Division codes for EMBL, GenBank,[3,4] and DDJB[15,16] databases.

Division	Code	Division	Code
Bacteriophages	PHG	Organelles	ORG
Construct/contig	CON	Other vertebrates	VRT
Expressed sequence tags	EST	*Mus musculus*	MUS
Fungi	FUN	Plants	PLN
Genome survey	GSS	Prokaryotes	PRO
High-throughput cDNAs	HTC	Rodents	ROD
High-throughput genome	HTG	Synthetic	SYN
Human	HUM	Sequence tagged sites	STS
Invertebrates	INV	Unclassified	UNC
		Viruses	VRL

are heterogeneous metabolic databases integrated into BioSilico in a systematic way.[70]

Biosequence Databases

Exclusively, the databases of biosequences can be grouped into two categories, namely, *nucleotide sequence database* and *protein sequence database.*

That is, considering the nucleotide sequence version of databases, the entries thereof are grouped into divisions based on taxonomy, with an exception of a few, such as *sequence-tagged site* (STS), *high-throughput genome* (HTG) sequence, *high-throughput cDNAs* (HTC), GSS *genome survey sequences* (GSS), *construct* (CON), and *patent* (PAT) divisions. Division codes for EMBL, GenBank,[3,4] and DDJB[15,16] databases are listed in Table 5.1

5.3 Biosequence Data Analyses

With reference to computer applications in bioinformatics, analyses of *data sequence patterns* refer to a primary effort. A data sequence pattern, in general, represents a set of one-dimensional random occurrences of data entities in space (or events in time) containing certain useful information (inherently built within the stochastic patterns of occurrence of data entities). Relevant algorithms of sequence analysis refer to deciphering these patterns and extracting the built-in information.

Referring to the science of biology, stochastically decided evolutionary forces act on genomes causing a (random) pattern of features that manifest in the form of biosequences (such as the DNA/RNA) described in prior chapters, and in relevant biosequence analysis, the following tasks can be indicated as regard to omic contexts.

(i) *Assessing similarity features between a pair of biosequences*: For example, (a) elucidating and verifying the inherent similarity features between a forward sequence of a *dual-stranded DNA* (dsDNA) and its reverse complimentary strand, and (b) comparing a given sequence pattern to assess its similarity with genomic sequences of other species (such as *model species* or otherwise) in order to achieve gene predictions on *coding DNA sequence* (CDS), conserved regions, regulatory regions, etc.

(ii) *Formulating algorithms that enable locating specific genomic regions*: For example, delineating or locating transitions across a heterogeneous mix of *coding* and *non-coding* regions in a query DNA sequence.

(iii) *Identifying the ESTs*: This task is done on a RNA sequence (of a known protein) so as to determine the sequence part that exclusively expresses the genomic information processed in realizing the protein. EST decisions are also needed in the context of knowing the wet lab-based, *complementary DNA* (cDNA) synthesized.

(iv) *Deducing the reading frames and identifying the ORFs*: This effort is to identify the sequence with the correct "start" to "stop" codon implied by translation of genetic code toward protein-making. The process of elucidating such gene-specific predictive decisions on sequence patterns *ab initio* is a vital genomic analytical pursuit with *ad hoc* computational needs.

(v) *Analyzing unique sequence features across the so-called single-stranded DNA (ssDNA)*: Visualizing the locales of Watson–Crick paired stem regions in the ssDNA or RNA as well as other secondary structural formats (such as, bulges, hair-pin bends, etc., for example, in viral DNA/RNA), has certain biological significance emphasizing the need for relevant computer-based distinct analyses.

(vi) *Aligning a pair of biosequences*: This is done to score the extent of meaningful comparisons with high degree of integrity.

(vii) *Aligning multiple sequences*: Families of similar sequences contain information on sequence evolution in the form of specific conserved patterns, almost at all sequence positions. Comparing such multiple sequences, alignments is done by (a) Building sequence profiles or HMM to perform more sensitive homology searches consistent with a sequence profile containing information about the variability of every sequence position. (b) Improving structure prediction methods (*secondary structure prediction*). (c) Determining active site residues and residues specific for subfamilies. (d) Predicting (protein–protein) interactions. (e) Analyzing SNPs in search of genetic sources of diseases.

(viii) *Formulating differential algorithms for local and global alignment of sequences*: Exclusive to "local" and "global" that represent (local) sub-segments and total stretches of biosequences, respectively, specific alignment procedures are formulated as bioinformatic exercises.

(ix) *Devising sensitive scoring schemes*: Posterior to the alignment efforts envisaged on biological sequences, sensitive *scoring schemes* are devised in order to deduce the extents of alignment and/or non-alignment states of the compared sequences and assess the efficacy of alignment procedure pursued.

(x) *Identifying motif sections*: *Motifs* denote a set of multiple sequences depicting regions of shared similarity across a multiple set of sequences. Such motifs possess conserved features of biological significance in conformance with homologous aspects of species (sequences) being compared.

(xi) *Enabling classification and assessment of biological macro-complexity*: This is done at molecular, organelle, and species levels beyond the microscopic details.

(xii) *Making decisions on classifying "informative" and "non-informative" parts*: Deducing redundancy aspects of biological sequences.

(xiii) *Extracting informative details of bioinformatic interest*: Literature survey on annotated details.

(xiv) *Analyzing the internal (segmental) structure*: *Gaining "glocal" details* of two subsequences being compared.

(xv) *Visualizing the inner details* of 1-, 2-, 3D features across complex protein structures.

5.4 Searching for a Database

With the advent of computer proliferation, biological, biomedical, and bioinformatic details have seen hyperbolic surges in their growth, and pertinent *in silico* data have faced systematic as well as organized passages through stupendous efforts of storage and sharing in the recent saga of timely established databases around the stretching edges of the globe.[71] Commensurately, the number of Internet- and WWW-accessible databases has been rapidly growing on annual basis and such biomedically relevant databases are widely identified in reputed literature. For example, *Nucleic Acids Research* has been publishing relevant journal as annual database issues since 1996.

Biological database search efforts could refer to almost all biosequences available in bioinformatic resources comprised of data compilation, organization, analysis and dissemination. Search efforts could involve several application-specific functions as summarized in Table 5.2. Relevantly, Attwood and Parry-Smith have presented in Attwood and Parry-Smith[14] excellent synopses toward building a sequence search protocol *via* an interactive web tutorial.

Pertinent to general considerations in searching for a database, the following pseudocodes (presented as Tables 5.3 and 5.4) are presented so as to facilitate a search systematically consistent with a bioinformatic query in hand.

The bioinformatic databases thus can be searched by querying on related details pertinent to them, as well as the system tools built in the databases or relevantly prescribed to analyze the input, and related entries can be invoked to deduce sequence-specific details in terms their characteristic features.

In summary, considering the efforts of bioinformatics of organizing bioinformational data availed *via* databases of different categories and

Table 5.2 Function-specific search efforts.

Search type	Searching function
Annotation search	For keywords, specific features and authors
Homology search	For homologous similar sequences
Pattern search	For queries on occurrences of pattern
Prediction search	For knowledge database search toward secondary structure predictions, fold recognition, etc.
Comparison search	For comparing gene families and polygenetic trees

Table 5.3 A roadmap on searching databases for a specified bioinformatic query: A pseudocode on searching databases *versus* an identified bioinformatic query.

```
%% Given a bioinformatic query, searching database(s) to find
   the objective(s) for which the search is made
Initialize
  → Specify the objective vis-á-vis bioinformatic query posed
    and the result sought pertinent to omic contexts: Genomic
    and/or proteomic
     → List all relevant bioinformatic databases: Genomic
       and/or proteomic
  Start
       → Search query? … Specify
       → Needs genome databases? Yes or no
  If no, Go To Search B
          → Otherwise, continue with Search A

Perform
Search: A
       → Access genome databases
       ← Repository of whole genome nucleotide sequences of
         various organisms: Prokaryotes, eukaryotes, and
         virus
    → Data type: On a variety of genomes, sequence maps with
      contigs*, integrated genetics, and physical maps plus
      annotated genes information
       ← Examples: (a) Entrez genome of NCBI provides genome
         sequence data on: Archaea, bacteria, eukaryotes,
         virus, viroids, and plasmids whole genome nucleo-
         tide sequences of various organisms: Prokaryotes,
         eukaryotes and virus, and (b) Genome Information
         Broker (GIB) – a database on complete genome
         sequence
%% Contig* denotes a set of overlapping DNA segments jointly
   represent a consensus region of DNA. Contigs can also
   denote overlapping physical segments (fragments) contained
   in clones
Go To  List the Results

Search B
  → Search query: Needs sequence databases? Yes or no
     → If no, Go To Search C
       → Otherwise, continue with Search B
```

(Continued)

Table 5.3 (*Continued*)

Perform Search B

→ Access sequence databases

← Nucleotide sequence databases or protein sequence database

→ Nucleotide sequence databases

← GenBank, EMBL-bank, DDBJ

→ Protein sequence databases

← Swissprot, Entrez protein, Integr8, Proteome, FASTA

Go To List the Results

Search C

→ Search query: Needs bibliographic databases? Yes or no?

→ **If** no, **Go To Search D**

→ Otherwise, Continue with Search C

Perform Search C

→ Access bibliographic databases

← Databases containing archival literature on bioinformatic topics

→ (a) PubMed (of NCBI): National Library of Medicine (NLM)

→ (b) Medline (of NLM)

Go To List the Results

Search D

→ Search query: Needs Microarray databases? Yes or no?

→ **If** no, **Go To Search E**

→ Otherwise, Continue with **Search D**

Perform Search D

→ Access microarray databases

← Data accessed from wet-lab experiments *vis-à-vis* genomic DNA, mRNA, protein molecules

← Data accessed, for example, from non-microarray technology (like SAGE), mass spectrometry, peptide profiling

→ Transcriptome data

← Example databases: Gene Expression Omnibus (GEO), Gensat, ArrayExpress, Cancer Gene Expression Database (CGED), Human gene Expression Index (HuGE)

Table 5.3 (*Continued*)

Go To List the Results

Search: E

→ Search query: Needs metabolic databases? Yes or no?
→ **If** no, **Go To Search F**
→ Otherwise, continue with **Search E**

Perform Search E

→ Access metabolic databases
← Data on biochemical pathways and enzymes of various organisms
← Examples: KEGG and MetaCyc
← Organism-specific example databases: EcoCyc, Flybase, and CCDB

Go To List the **Results**

Search: F

→ Search query: Needs chemical databases? Yes or no
→ **If** no, **Go To Search G**
→ Otherwise, Continue with **Search F**

Perform Search F

→ Access chemical databases
← Data on chemical details on various molecules
← Examples: PubChem (NCBI)

Go To List the Results

Search: G

→ Search query: Needs structure databases? Yes or no?
→ **If** No, **Go To Search H**
→ Otherwise, continue with **Search G**

Perform Search G

→ Access structure databases
← Data on 3D forms of proteins and relevant amino acids
← Data conforms to: Crystallographic, NMR, and X-ray explorations
← Examples: Nucleic Acid Database (NDB), Structural Classification of RNA (SCOR), Protein Data Bank (PDB)

Go To List the Results

(*Continued*)

Table 5.3 (*Continued*)

```
Search H
  → Search query: Needs disease databases? Yes or no?
  → If no, Go To Search I
      → Otherwise, continue with Search I
Perform Search I
  → Access: Disease databases
  ← Data on disease-specific pathology of complex diseases
    and disorders
      ← Examples: Online Mendelian Inheritance in Man (OMIM),
        Genetic Association Database (GAD)
Go To List the Result
Search I
  → Search query: Needs enzyme databases? Yes or no?
      → If no, Go To Result
          → Otherwise, continue with Search I
Perform Search I
  → Access enzyme databases
  ← Data on structure, function and chemistry of assorted
    enzymes, and related pathways
      ← Examples: EC Enzyme Database, Enzyme Nomenclature
        Database (ExPASy), REBASE
Go To List the Result
        Result
  ← List, compile, and tabulate the results of each search
    and format it as needed in the query
      End
```

Table 5.4 A pseudocode on searching databases *vis-à-vis* an example of bioinformatic query.

```
%% Pseudocode: A search protocol on the database vis-à-vis a
   specified bioinformatic query example
Initialize
  ← Protocol of searching the database consistent with a
    bioinformatic query
```

Table 5.4 (*Continued*)

Input

→ Query DNA sequence
← Example sequence:
5′....ATCCGG ...TGC... 3′

Determine/Obtain

← Six reading-frame translations

Next

Perform

→ Primary protein sequence similarity

Search

← Use (for example): OWL database search sequence option
← Compile the results

%% OWL is a Java library with a set of command-line tools intended for the analysis of biological macromolecules. It enables analyzing protein sequences and structures *via* built-in algorithms and interfaces to external tools.

← *Via* OWL search, the primary protein sequence identity is established through identifying the ORF and pin-pointing identical matches to the query/probe sequence *vis-à-vis* the database entries
← Such similarity search can also be established, for example, across: nrdb, (SP+SPTrEMBL) databases entries
→ Set of homologues of the probe sequence is identified

Continue

Next

→ Secondary protein structure of the queried probe?

Perform

→ Database search for secondary protein pattern
← Examples of databases: PROSITE, Profiles, PRINTS, Blocks, Pfam, etc.
→ Family relationships, locales of structural, and/or functional sites are obtained

Next

→ **If**

- … the results obtained indicate a known protein structure

(*Continued*)

Table 5.4 (*Continued*)

```
        then:

            - Compile the results and Continue

Otherwise ---Go to Next (alternative search Pursuit: Search A)

Continue
    Perform
    → Protein structure classification via ...
       → Database query with databases, like SCOP and CATH
          → Information on structural class, secondary
            structure, ligand-binding, etc. are availed
             → Compile the results
Next
  Search A

    → Alternative search pursuit when the results obtained
      indicate an unknown protein structure

  Perform
    → Protein fold pattern library search
       → Database query with examples of databases: SCOP
         and CATH
          → Compatible folds for the structure of the probe
            sequence are identified
             → Compile the results
Next
    Results → Collect, compile, and format the results

                End
```

retrieving of such data for analytical and computational pursuits, the databases are evolved as general purpose types as well as specialized databases intended to target definite analyses. Relevantly, a summary on the managed versions of the databases and their build up are as follows.

Understanding the inclusions in the basic biological databases, incorporation of new data types, integration of data and tools, and exercising relational database management involves the following: (i) defining the labyrinth of data stored data accessed *via* data flood; (ii) routing the repository data toward need-based acquisition and collection of nucleotide sequence data; (iii) enabling routes for acquisition and collection of amino-acid sequence data; (iv) directing specific efforts to global analysis of gene expressions; and (v) facilitating target discovery and validations in applications, such as toxicogenomic and pharmacogenomic suites. The next stratum of bioinformatic database refers to the set of knowledge bases relevant to protein families and motifs databases (PROSITE, BLOCKS, PFAM, etc.). Supplementing these are classification databases such as (i) clusters of orthologous groups of proteins (COGs) intended for genome-scale analysis of protein functions and evolutions[72]; (ii) structure-specific databases (like CATH and SCOP); (iii) pathway databases (KEGG, EcoCyc, etc.); and (iv) genome and genome-diversity databases (...OMIM...). Furthermore, grouped exclusively are computational databases intended for a automated genome annotation (Pedant) and calculated structures databases (3D Crunch), and the databases can also be categorized on the basis of data retrieval and user interface considerations involving browsing, text-based query (Entrez, SRS, etc.), sequence-based query (BLAST, FASTA, etc.), and data mining.

5.5 Bioinformatic Information Networks

The core resource of bioinformatics is the set of databases outlined above. Without the implementations of, and efficient access to, such databases (containing genomic and proteomic information), the implied scope of bioinformatics is not objectively conceived. In essence, bioinformatics is largely relying on the concept of data mining, which can be identified as an extraction of a small subset of relevant information from a huge repository of structured data, and, such data mining tasks cannot be accomplished without the dissemination and sharing of the stored information in the databases. In this section, outlined are basic details on the functionality of information network that provide the web of telecommunication linkage across the entire plethora of bioinformatic databases worldwide.

The bioinformatic centers and data centers represent a network of global communication and resource centralization dynamically managed across the world. In Attwood and Parry-Smith,[14] details concerning the significance of information networks in bioinformatic contexts in the perspectives of associated, limitations, capabilities, benefits of various network hardware structures are elaborated. In the relevant context of next-generation telecommunication, the authors in Neelakanta,[73] and Neelakanta and Baeza[74] have explained the underlying infrastructure and related communication protocol details of the Internet and connectivity details.[73,74]

Basic Definitions in Information Network Operations

Whether the passage of information is intended for general telecommunication or exclusively concerned with bioinformatic data, the underlying information network facilitated toward worldwide interconnectivity is a shared entity, simply dubbed as the Internet. It works on a communication protocol, known as *transmission control protocol/ Internet protocol,* abbreviated as TCP/IP. Relevant information flow is comprehended *via* packet transmission of data taken in binary format of binary digits or bits as described in Neelakanta,[73] Neelakanta and Baeza,[74] and MIT.[75]

IP-address, Internet services support *via file transfer protocols* (FTP), *worldwide web* (WWW). It is a hypermedia-based information system, related a *uniform resource identifier* (URI/URL) representing a web page, image, video, or other piece of content, hyperlinks enabling, the universal resource locator, or URL, salient network-related details, which can be found in the references such as Neelakanta,[73] Neelakanta and Baeza,[74] MIT,[75] and Ian and Norman.[76]

5.6 Bioinformatic Tools

As described in the earlier chapters, DNA sequences constitute the primary data of sequencing projects and are made available with necessary annotations. Given a raw DNA sequence, several layers of bioinformatic analyses are necessary in order to arrive at an eventual set of annotated protein sequence data. Such analyses conform to: (i) establishing the correct order of sequence contiguously so as to obtain a single continuous sequence; (ii) defining the so-called ORFs by ascertaining the translation and transcription initiation sites, *promoter sites,* and *splice sites* (of introns and exons) across the set of six possible reading frames that can be specified by the raw version of a trinucleotide sequence; (iii) translating the DNA sequence into a protein sequence using the ORF; and (iv) comparing a query DNA sequence with known protein sequences in order to verify the associated homologous details.

Sequence Analysis Tools

These correspond to performing a detailed analysis on any given query sequence toward elucidating physico-chemistry-based exclusive segments (e.g., *amphipathic, hydropathic,* or *hydrophilic*) regions, *CpG-islands* (CGIs), identification of mutational changes and related evolutionary analysis, etc. The CGIs denote short, interspersed DNA sequences that deviate significantly from the average genomic pattern by being GC-rich, CpG-rich, and they correspond to predominantly *non-methylated* and regions of compositional biases. Essentially, results of such sequence analyses enable knowing the specific function of the query sequence. The *DNA methylation* refers to adding methyl groups to DNA so as to modify the function of the DNA. It is an epigenetic mechanism used by cells to control gene expression. When observed in a gene promoter, DNA methylation typically acts to repress gene transcription.

A comparison of the genomes of genetically close organisms reveals genes responsible for specific properties of the organisms (e.g., infectivity). Protein interactions can be predicted from conservation of *gene order* or *operon organization*[(i)] in different genomes. Also the detection of *gene fusion*[(ii)] and *gene fission*[(iii)] (i.e., one protein is split into two in another genome) events helps to deduce protein interactions. The items indexed (i), (ii), and (iii) as above, are outlined as follows.

(i) *Gene order* (or *operon organization*): An *operon* is a functioning unit of genomic DNA containing a cluster of genes under the control of a single promoter. Operons occur primarily in prokaryotes and also in some eukaryotes, including nematodes such as *Caenorhabditis elegans* and the fly, *Drosophila melanogaster.* An operon is a regulatory gene made up of three basic DNA components, namely, the *promoter,* the *operator,* and the *structural genes.* The operon gene order or organization is optimized for ordered protein complex assembly. (ii) *Gene fusion*: It refers to a process by which, the complete or partial sequences of two or more distinct genes are fused into a single chimeric gene or transcript, as a result of DNA or RNA derived rearrangements. This phenomenon is widespread and has been observed across all domains of life. Here, the chimeric genes (or those made of parts from different sources) stem from the combination of portions of one or more CDSs to produce new genes. (iii) *Gene fission*: Gene fusion and fission are two principal processes in molecular evolution. One consequence of genome remodeling in evolution is the modification of genes, either by fusion with other genes (implying gene fusion) or by segregating into several parts meaning gene fission. Homologous sequences pertinent to related species (diverged set of organisms from a common ancestor root) exhibit a degree of similarity between them that can be located in regions (called motifs). Correspondingly, the extent of similarity can be measured and the status of homology can be verified. Thus, relevant tools can identify similarities between query sequences (of unknown structure and function), as well as between database sequences whose structure and function are already determined.

Sequence Comparison Tools

In modern practice of DNA sequence analysis, there are a number of tools that are available for sequence comparison. The simplest type produces a *dot-matrix representation* of sequence similarities.[77]

Homology and similarity search tools

A degree of similarity between multiple homologous sequences pertinent to related species (diverged set of organisms from a common ancestor root) can be observed in regions (designated motifs) along the stretch of aligned sequences. Relevant extent of similarity measures the status of homology. Tools have been developed to identify the similarities between query sequences (of unknown structure and function), as well as between database sequences whose structure and function are already determined.

Protein function analysis tools

Exclusively designed, this set of programs allows comparisons of a given protein sequence against secondary (or derived) protein databases. The secondary database entries contain information such as motifs, signatures, and protein domains. Relevantly, such a comparative assessment of a query sequence will show highly significant hits against different sequence patterns of the databases, when the sequence details tally to a large extent. Based on this comparison, the unknown biochemical function of the query protein can be inferred. The protein sequences are also analyzed to infer the function, which can be predicted whenever a sequence of a homologous protein with known function is found. Such homology searches are significant in bioinformatics applications, and efficient search methods have been developed thereof. Furthermore, in studies related to homologous proteins, distinguishing the underlying subclasses, namely, *orthologous versus paralogous* sequences facilitates functional annotation in the comparisons of whole genomes. Also, valuable information about the subcellular location of a protein is viably derived from protein sequence analyses by detecting the features like *glycosylation* and *myristylation*, as well as other sites and predicting the so-called *signal peptides* present in the amino acid sequence.

Structural analysis tools

In order to assess the unknown details on the structures (of a protein), a comparison against known structure databases can be done using this class of structural analyses tools. Typically, the function of a protein is related to its structure, and structural homologs tend to share similar functions. The analysis tools of this category are designed to cope with protein's 2D/3D structures.

Sequence alignment tools

Apart from sequence comparison efforts described earlier, another important task of primary interest in bioinformatics refers to sequence alignment designed to find the following: (i) the "best" overall (global) alignment between two given sequences; (ii) alignment of the smaller (local) regions of similarity; and (iii) aligning a multiple set of sequences.

Assessing the extent of residue-based similarity or dissimilarity details between sequences is one of the primary tasks in bioinformatics. The effort of elucidating and verifying the inherent similarity features between the biosequences can be done with two possibilities as follows: (i) by subjecting the raw query sequences as they are and comparing them for similarity assessment in terms of their constituent residues and (ii) by "aligning" the sequences first (prior to comparison) by a procedure of shifting one sequence with respect to other and incorporating insertions and deletions (*indels*) of residual constituents and/or gaps as necessary in the sequences being compared, so that distinct segments are identifiably seen across the sequences posing the "best" possible similarity between them. Then the sequences are placed one below the other with the identified sub-segments with similar features framing a column of a matrix (i.e., they are brought into a state of alignment with visual depictions of the alignment portraying a matrix of columns having all residues being identical or a few with differing characters (identified as indels and/or gaps) in one or more of the sequences. The indels and gaps (denoted by hyphens) imply the mutational states across the multiple sequences subjected to alignment.

There are two versions of sequence alignment: (i) alignment between a pair of sequences and (ii) alignment across a set of multiple sequences.

Relevant alignment procedure pursued thereof implies in short that the query sequences representing pair or multiple sequences (of DNA, RNA, or protein residues) are eventually brought into a state of alignment, then certain identifiable segments of these aligned sequences pose optimal similarity between them, seen in terms of their functional, structural, or evolutionary relationships.

Relevant to sequence alignment indicated above, there are two versions namely, *global alignments* and *local alignments*. The global alignment conforms to a global optimization procedure that enforces the alignment across the entire span of all the query sequences. In contrast, the local alignments identify alignments only in the locally significant segments of similarity (within the long sequences that are typically products of widely divergent species). Computational algorithms advocated in sequence alignment problems aim at getting formally "correct" results. Furthermore, inasmuch as the number of constituent residues across the query sequence is stupendously large, the programming is divided into progressive steps. The associated programming thereof (known as *dynamic programming*) implies dividing first the large problem (e.g., a full sequence) into a succession of smaller problems, and then the procedure uses the solutions acquired from those smaller problems to reconstruct a formal solution to the larger problem. (Exist also are other efficient methods such as *heuristic algorithms* or *probabilistic methods* designed for large-scale database search with limited guarantee in finding best matches.)

Furthermore, in a simplistic approach, an overall comparison of the relationship between two sequences toward visual alignment can be done by the dot-matrix representation of the residues in the sequences; however, it does not seem compatible to visualize the state of alignment between extremely large sequences.

In exercising alignment between sequences involves lateral shifting of residues of one sequence against the other by introducing indels/gaps as needed in order to make similar subregions of the sequences that are placed one below the other and line up in columns, vertically juxtapositions. Once such an alignment is done, it is also necessary to have some way to evaluate the quality of such alignment procedure envisaged. Hence, standard scoring function is needed to specify the efficacy of alignment performed. Considering the aligned subsegments placed one below the other and treating the arrangement as a matrix of residues incurring column-wise, "*credit*" or "reward" scores are assigned for each aligned character in a column and added up. Likewise, "penalty" scores are assigned and subtracted for each observed indel or gap (that was deliberately introduced for alignment). The resulting net value of the score, viewed across the *scoring matrix* conforms to score entry for every possible compared symbols/residues. As such, the net score represents the quality of the result due to the alignment procedure.

Global Alignment Tools

The global alignment tool intended to align protein or nucleotide sequences is based on dynamic programming. Relevantly, the so-called *Needleman–Wunsch* (NW) algorithm[14,78] was developed in 1970. The NW algorithm and the tool imply optimal matching algorithm toward global alignment technique illustrated in detail in Chapter 4 using examples. Unlike NW algorithm that maximizes the similarity in the alignment suite, the other possibility is to minimize the dissimilarity viewed in terms of the *edit-distance* (also known as *Levenshtein distance*) described in Chapter 2 and as Sellers[79] in 1974 has pointed out that the maximum and minimum approaches toward sequence alignment are equivalent to each other.

Local Alignment Tools

Whenever the sequences being compared have a few regions of similarity separated by divergent regions of variable length, such similar regions will never be aligned using the global optimization strategy, because the gap penalties incurred would be prohibitive. Therefore, it is desirable to identify sub-segments that can be aligned to projecting a score, which is locally optimal. Such aligned pairs of segments (with local-optimality) imply that, by changing the boundaries of local block(s) under comparison, it would render the score to decrease. In local alignment tools, the same scoring function scheme adopted in global alignment strategy can be used and aligning the local segments also needs imposing gaps as needed.

A local sequence alignment procedural tool refers to the so-called *Smith–Waterman* (SW) algorithm,[14,80] which determines similar regions between two strings or nucleotide or protein sequences. As stated above, *in lieu* of viewing the entire (total) sequence length, the SW algorithm compares sub-segments of all possible lengths and optimizes the alignment of such subsequences *via* a similarity measure. The SW algorithm is a variation of NW algorithm and it also pursues dynamic programming.

It offers a guaranteed optimal local alignment with respect to the scoring system being used. Typical scoring methods used are substitution matrix and gap-scoring scheme explained in Chapter 4. Improved alternatives of SW algorithm imply better scaling and being more accurate.

Tools for MSA

Often a need arises in bioinformatics to test a set of many sequences that may have similar features. Hence, MSA is advocated in aligning three or more biological sequences, generally protein, DNA, or RNA. Typically, the input test set of query sequences bear presumably a common evolutionary relationship sharing a genetic lineage and are assumed to have descended from a common ancestor (root). The resulting MSA allows inferring the associated sequence homology through phylogenetic analysis (Chapters 9 and 10) performed to assess the shared evolutionary origins and progress of the query sequences. MSA exercise can be done *via* a number of databases from where the alignment results for entries containing similar sequences against a reference (query) sequence can be availed (Examples of such programs are the BLAST and the FASTA. The underlying alignment procedure in such database programs are not *per se* a straightforward MSA tooling in the strict sense. The procedure involves performing in essence, sequence alignment pair-by-pair, across the entire set of multiple sequences. In such progressive alignment methods, the first step establishes the relationships between the sequences represented as a tree (called a guide tree) and in the second step, the MSA is built by adding the sequences sequentially to the growing MSA as per the guide tree. The initial guide tree can be decided by a clustering strategy, such as neighbor-joining (NN) or *Unweighted Pair Group Method with Arithmetic Mean* (UPGMA) methods described in Chapter 10.

Progressive alignments are not, however, guaranteed to be globally optimal due to accumulating errors at stages *en passant* in the growing alignments of multiple sequences. Furthermore, poor performance can be expected when the sequences in the set are distantly related. So, modern progressive methods modify the scoring function with secondary weighting by assigning scaling factors to individual members of the query set in a nonlinear fashion based on their phylogenetic distance from their nearest neighbors. There are other MSA techniques based on HMM, genetic algorithms/simulated annealing, phylogeny-aware method, etc. Once a MSA is established (whether made of DNA or amino acids), it can yield information pertinent not only in a single sequence but also can be used to compare a number of very similar sequences to see where they differ. As mentioned above, MSA can be used as input to phylogenetic analysis programs in order to study the evolutionary relationships between sequences (and even between organisms). They can also pinpoint segments (*motifs*) that are particularly conserved or particularly divergent between related sequences. This offers implicit information on the evolutionary processes undergone by those sequences. Furthermore, such alignments at the protein level (adopted as input to suitable protein-modeling software) can help to understand and predict the structure of the protein in terms of how the individual sequences of the multiple set characterize that protein structure, better than such details availed from a single sequence. The multiple sequence programs used in practice are described in Chapter 4. In short, MSA is helpful in assessing sequence conservation of protein domains, tertiary and secondary structures, and even individual amino acids or nucleotides.

Sequence Alignment: Database Tools

Consistent with pairwise and/or MSAs described above, there are a number of tools that exist in practice. Sequence alignment analysis packages are compatible with biological databases, and the associated algorithms enable database search and perform alignment as needed *via* a simple web interface. Stand-alone analysis suites thereof are also available as public and commercial packages.

BLAST: This BLAST comes under the category of homology and similarity tools. It is a set of search programs designed for the Windows platform and is used to perform fast similarity searches regardless of whether the query is for protein or DNA. Comparison of nucleotide sequences in a database can be performed. Also a protein database can be searched to find a match against the queried protein sequence. NCBI has also introduced the new queuing system to BLAST (Q BLAST) that allows users to retrieve results at their convenience and format their results multiple times with different formatting options. Depending on the type of sequences to compare, there are different programs.

- blastp compares an amino acid query sequence against a protein sequence database.
- blastn compares a nucleotide query sequence against a nucleotide sequence database.

- blastx compares a nucleotide query sequence translated in all reading frames against a protein sequence database.
- tblastn compares a protein query sequence against a nucleotide sequence database dynamically translated in all reading frames.
- tblastx compares the six-frame translations of a nucleotide query sequence against the six-frame translations of a nucleotide sequence database.

FASTA: This renders homology search for all sequences. It is an alignment program for protein sequences created by Pearson and Lipman in 1988.[81] The program is one of the many heuristic algorithms proposed to speed up sequence comparison. The basic idea is to add a fast prescreen step to locate the highly matching segments between two sequences, and then extend these matching segments to local alignments using more rigorous algorithms such as Smith–Waterman.

EMBOSS: The European Molecular Biology Open Software Suite, or EMBOSS, is a software-analysis package. It can work with data in a range of formats and also retrieve sequence data transparently from the Web. Extensive libraries are also provided with this package, allowing other scientists to release their software as open source. It provides a set of sequence-analysis programs, and also supports all UNIX platforms.

Clustalw: It is a fully automated sequence alignment tool for DNA and protein sequences. It returns the best match over a total length of input sequences, be it a protein or a nucleic acid.

RasMol: It is a powerful research tool to display the structure of DNA, proteins, and smaller molecules. Protein Explorer, a derivative of RasMol, is an easier to use as program.

PROSPECT: This refers to *PROtein Structure Prediction and Evaluation Computer ToolKit,* which is a protein-structure prediction system that employs a computational technique called protein threading to construct a protein's 3D model.

PatternHunter: This is based on Java, can identify all approximate repeats in a complete genome in a short time using little memory on a desktop computer. Its features are its advanced patented algorithm and data structures, and the java language used to create it. The Java language version of PatternHunter is just 40 KB, only 1% the size of Blast, while offering a large portion of its functionality.

COPIA: The Consensus Pattern Identification and Analysis (COPIA) is a protein structure analysis tool for discovering motifs (conserved regions) in a family of protein sequences. Such motifs can be then used to determine membership to the family for new protein sequences, predict secondary and tertiary structure and function of proteins, and study evolution of the sequences.

Standard Scripts in Bioinformatics

JAVA: Bioinformatic research centers exist all over the world, ranging from private units to academic settings as well as a wide range of hardware, software tools, and operating systems are used in those places. Specifically, among standard scripts, Java is a key version playing dominant role in bioinformatic research and studies. Relevantly, in simulation technologies, Java faces increasing adoption, for example, as in computer-based applications of Physiome Sciences and in Pattern Hunter of Bioinformatics Solutions. Furthermore, BioJava and BioCorba[82] refer to open-source projects in bioinformatics and such open-source tools are freely available for download on the Internet. *BioJava,* in short, is a large collection of classes for solving bioinformatics problems.

Perl: This is another open-source software, which has been successfully used in a variety of diverse bioinformatic tasks involving text-processing, system administration, web programming, web automation, graphical user interface (GUI) programming, games programming, and code generation. Also, its use is found in genetics and etymological research, as well as in testing and quality assurance exercises of biotechnology. Furthermore, string manipulation, regular expression matching, file parsing, data format inter-conversion, etc. are common text-processing tasks performed in bioinformatics, and Perl excels in such tasks and adopted by code or tool developers. Thus, developers have designed some individual modules for this purpose, *via* coordinated BioPerl projects.[83] BioPerl is a Perl interface to bioinformatics and other biological modeling and data analysis pursuits. It is a collection of object-oriented modules that enable effective lifescience data analysis.

Bioinformatics Projects: An Overview

As pointed out across the chapters of this book, research projects big and small are conceived as interesting and inquisitive efforts in the existing as well as in the next-generation of bioinformatics. Both in academic domain and in bioengineering/biotechnology industrial sectors, there are an extensive number of such internationally conceived and aggressively pursued projects and more such efforts are identified *via* focused studies. Identified in the next section are examples and exercises outlining a review on pertinent topics.

BioJava project: As mentioned above, this is a dedicated effort developed to provide Java tools for processing biological data, which includes objects for manipulating sequences, dynamic programming, file-parsers, simple statistical routines, etc. Relevantly, *Java3D* refers to a protein viewer in the context of a CSI 4900 project.

BioPerl project: It is a task of international association of developers focused on making Perl tools for exclusive applications in bioinformatics. The project is intended to be an online resource for modules, scripts, and web links for developers of Perl-based software.

BioXML project: This came into being as a part of the BioPerl project and depicts a resource to gather XML documentation, DTDs, and XML aware tools for biology in one location.

Biocorba project: Emergence of interface objects facilitated interoperability between bioperl and other perl packages (like Ensembl and the Annotation Workbench). Using biocorba, objects written within bioperl can communicate with objects written in biopython as well as in biojava. Details on biocorba project can be availed in Website on biocorba project.[82] Furthermore, the Bioperl BioCORBA server and client bindings are available in the bioperl-corba-server and bioperl-corba-client bioperl CVS repositories, respectively.[83]

Ensembl project: This is an ambitious automated-genome-annotation project at EBI. Much of Ensembl's code is based on bioperl, and Ensembl developers, in turn, have contributed significant pieces of code to bioperl. In particular, the bioperl code for automated sequence annotation has been largely contributed by Ensembl developers. Ensembl and its capabilities are described in its website cited Ensembl website.[84]

Bioperl-db project: This effort refers to a newly conceived project[85] with the objective to transfer some of the capabilities of Ensembl for integration with bioperl syntax and enabling a stand-alone MySQL database to the bioperl code-base. Most of the bioperl objects can map directly to tables in the bioperl-db schema. As such, object data such as sequences, their features, and annotations can be easily loaded into the databases. Likewise, one can query the database in a variety of ways and retrieve arrays of Seq objects.

Biopython and biojava projects: Objectively, these are open-source projects parallel to bioperl. But, their code is implemented in python and java, respectively. With the advent of interface objects and biocorba, java or python objects can be written and accessed by a bioperl script; alternatively, it is possible to call bioperl objects from java or python code.[83]

5.7 Review Section

Consistent with the topics and details addressed in this chapter on repositories, information networking, and tools of bioinformatics, this section is written to present and review the subject-matter *via* relevant examples, tutorials, and self-study problems/exercises.

Example Ex-5.1

Problem Statement and Approach

In case one is interested in knowing the basics of "human insulin" and related details in the contexts of bioinformatics-based medical and health sciences. The search can be anchored, for example, on Google search-engine by typing the key-word "human insulin." This will supply a humongous number of "hits" on the topic. To mention, a search using a keyword "human insulin" could yield over a million hits on results that provide information such as the following: (a) details narrating what insulin refers to Raptis and Dimitriadis,[86] Nicole and Smith,[87] Mandal,[88] Joshi *et al.*,[89] and Bell *et al.*[90]; (b) human insulin as defined/explained in medical dictionary; (c) types of insulin, in vogue; (d) information on recombinant human insulin; and (e) making of insulin using bacteria and related details.

Relevant archival of details such as in Bell *et al.*[90] with cross-references in PubMed and MEDLINE can also be availed.[91]

Example Ex-5.2

Problem Statement and Approach

This is an extended web-based search exercise on details of a specific, (and/or partially known) topic, for example, knowing information on *amino acid (AA) sequence of human insulin.* A Google search thereof could show as many as couple of hundred hits with citations on annotated details, for instance, as in Raptis and Dimitriadis,[86] Nicole and Smith,[87] Mandal,[88] Joshi *et al.*,[89] Bell *et al.*,[90] History of Insulin Web-site.[91]

Example Ex-5.3

Problem Statement and Approach

This refers to a more comprehensive and web-based search on combinatorial details pertinent to a topic of interest. The comprehensive mode of finding details is done by adding "limiting words" on the search. Relevant web-links would then allow surfing into a host of combinatorial set of details as needed and the "hits" get limited to specify just the added (limited) information being queried. An example of finding number of hits (such as on *Google* search) *versus* search entities progressively augmented with limiting descriptors on an enquiry on, for example, *human insulin* as in Table 5.5.

Problem P-5.1

As in the above examples, a web-based search can be performed on an extended topic specific to the protein structure of insulin using a keyword anchored on a suitable search engine.

Table 5.5 Application-specific search entities and number of hits.

Search entity: Descriptors of human insulin	Number of hits (e.g., on Google search)
Insulin	Highest count
Human + insulin	Decreasing values of counts of hits
Human + insulin + gene	↓
Human + insulin + gene + β-globin	
Human + insulin + gene + β-globin + protein	
Human + insulin + gene + β-globin + protein + diabetes	
Human + insulin + gene + β-globin + protein + diabetes +journals	
Human + insulin + gene + β-globin + protein + diabetes + publications	
Human + insulin + gene + β-globin + protein + diabetes + journals + Tutorials + MEDLINE + PubMed + etc.	Lower most count

Problem Statement

Perform a search on protein structure of insulin and gather details in making of the insulin from mRNA stage to eventual protein complex and summarize the underlying learning details on the map of *human β-globin gene versus protein folding.*

Problem P-5.2

Problem Statement

(i) Perform a blind annotation search to acquire basic details on fruit-fly (*D. melanogaster*) in the contexts of bioinformatics. From the search performed using a keyword anchored on a suitable search-engine, list the result with a precis writing on the following: (a) What is *D. melanogaster*? (b) Salient literature on this organism including book(s); (c) where do these species come from?; and (d) list the number of hits and the first ten priorities.

(ii) Similar to the above exercise, perform a blind annotation search to acquire details on zebra fish (*D. rerio* is a freshwater fish of *minnow* family, *Cyprinidae*). Perform a keyword anchored search and write a term paper on the results gathered on the following: its chromosomal details compared with those of human in formulating a relevant *Oxford-grid.*

Problem P-5.3

Problem Statement

Reference to a comprehensive web-based search on the combinatorial details pertinent to a given topic of interest, construct a table of hits *versus* the progressively increased search query on say, *D. melanogaster.* In each step, limit the detail added and write a note on the salient information searched with implied bioinformatic significance.

Example Ex-5.4

Problem Statement and Approach

This example illustrates an exercise on a database search (e.g., GenBank DNA sequence database at NCBI) performed to access a DNA sequence of a given organism. *Search query*: obtain the DNA sequence of *Bacillus subtilis* 2633[92] using "Search" function of Entrez at NCBI site.[93]

Tutorial T-5.1

This is a tutorial note on ORF finding. The outline thereof refers to DNA sequence of *Bacillus subtilis* 2633[92] of Example Ex-5.4. The ORF Finder searches for ORFs in the query DNA sequence entered. The program returns the range of each ORF along with its protein translation. (ORF Finder can also search the sequenced DNA for potential protein encoding segments, verify predicted protein using SMART BLAST or regular blastp.) The web version of the ORF finder is limited to the subrange of the query sequence up to 50 kb long.

Search procedure

Enter: query Sequence (of *Bacillus subtilis* 2633, α-amylase gene with Accession # X07796)[92]
Enter: accession number, gi, or sequence in FASTA format:
Submit ...
Result ...
Open Reading Frame Viewer
Sequence:
ORFs found: 31 Genetic code: 1 Start codon: "ATG" and alternative codons
ORFs were calculated on the interval from 1 to 2222 nt

Label	Strand	Frame	start	Stop	Length (nt*/aa**)
ORF4	+	2	599	2032	1434 | 477
ORF1	+	1	40	450	411 | 136
ORF3	+	2	17	412	396 | 131
ORF11	+	3	1420	1205	216 | 71

.......etc.

("nt")* means nucleotides and ("aa")** means amino acids

ORF4 (477 aa) Display ORF as ...
>lcl|ORF4
MFAKRFK ...CNTFFQ

Next:
Submit to: SMART BLAST (ORF4)
Result ... Summary

A concise summary of the three best matches in the sequence database together with the two best matches from well-studied reference species, showing phylogenetic relationships based on multiple sequence alignment and conserved protein domains.

Conserved domains for the query:

AmyAc_bacl_AmyA Aamy_C, etc.
Query: Unnamed protein product: *Chimeric alpha-amylase* precursor

Descriptions
Best hits: (Example)

Select... (Example)
alpha-amylase [*Bacillus subtilis*]

Max score	Total score	Query cover	E-value*	Ident	Accession #
987	987	100%	0.0	99%	ABW34932.1

Note: **E-value*: It is the estimate of how many times ("counts") one may expect a result (e.g., a score in a sequence comparison), at least as extreme as the one observed, occurring by chance. A value close to zero means that one can practically expect no unrelated sequence to score as high to the query sequence.

Suppose it is desired to get the six possible reading frames of a given query sequence. Relevantly, several sites on the web exist for the translation of an input sequence. For example, by clicking on the *Expasy link* cited as Website of EXPASY[94] opens the *Translate Tool**, which allows the eventual translation of a nucleotide (DNA/RNA) sequence to a protein sequence. By entering a DNA or RNA sequence in the box, the numbers and blanks in the query sequence are ignored by default and output is made available upon submission in a format such as Verbose ("Met," "Stop," spaces between residues).

Translating the DNA sequence and inferring the six reading frames is done by reading the nucleotide sequence three bases at a time and then looking at a table of the genetic code to arrive at an amino acid sequence. The test program examines the input sequence in all six possible frames (i.e., reading the sequence from 5′ to 3′ and from 3′ to 5′ starting with nt-1, nt-2, and nt-3).

Furthermore, by selecting one of the frames, it is possible to select the initiator and retrieve the amino acid sequence. Another package called *EMBOSS Six-pack* tool,[95] which can also read a DNA sequence and outputs the three forward and (optionally) three reverse translations in a visual manner.

Example Ex-5.5

Problem Statement and Approach

This exercise is relevant to BLAST search with a given amino acid query sequence and perform a database search so as to find the other amino acid sequences of corresponding proteins. Suppose the following is the given query sequence, perform a database search to find the amino acid sequences:

5′ ... MSTAVLENPGLGRKLSDFGQETSYIEDNCNQNGAISLIFSLKEEV
GALAKVLRLFEENDVNLTHIESRPSRLKKDEYEFFTHLDKRSLPALTN
IIKILRHDIGATVHELSRDKKKDTVPWFPRTIQELDRFANQILSYGA
ELDADHPGFKDPVYRARRKQFADIAYNYRHGQPIPRVEYMEEEKKT
WGTVFKTLKS ... 3′

Solution hint

Select "Standard Protein — protein BLAST [blastp]" under the "Protein BLAST" option

Continue the search procedure to reach the page that contains results.

Problem P-5.4

Problem Statement

Using the results of Example Ex-5.5, answer the following.

(a) *Why does an amino acid query sequence leads to a higher similarity than a nucleotide query sequence?*
(b) *What are the identifiable five database sequences that are very closely similar to the query sequence?*

For the identified five database sequences, list the following: name of the protein; source (organism/tissue) of the protein; amino acid identity; similarity scores; E-values; percentage of identity; and percentage positives (i.e., similar amino acids). Find the animals (apart from those five listed earlier) have this (query), protein with a similarity score in excess of 220 bits.

Problem P-5.5

Problem Statement

Given below is a BLAST search query DNA sequence.

Search for amino acid sequences that are similar to all six reading frames of the query DNA sequence supplied

5′ ... ATG TCC ATC GAG GAG GAA GAT ACA AAT AAG ATC ACA TGT ACG CAA GAC TCT TCA CCA ATA CTT TGT AAC TGA AAG GGT TAG CAT TCA ATT TGG GTT ATA ACA AGA CCG TAA AAA GGA TAA ATA AAG ATG AAT TTG AGG TAA CAT TAA GGC AGT AAA TTG TAT CAT GTC ATG GAC AAA CTA TCC TAA GCC TGG TAA AAC GAA CAG CTT CAA CGT ACC TCT TAA GCA ATT CCT TTA AGA AAT GCA ACA GTA TAA GGC AGT AAA TTG TAT CAG CTT CAA CGT ACC TCT TAA ... 3′

Solution hints

(i) Select "Nucleotide query — protein db [blastx]" under the "Translates BLAST searches" option

(ii) Search the "nr" database and remove the "√" (check) from the "low complexity" filter box

Example Ex-5.6

Problem Statement and Approach

This refers to a search exercise pertinent to *Saccharomyces* genome database. By opening the relevant link: *Saccharomyces Genome Database* (SGD), of Stanford-Cherry Lab[96] and clicking on "Gene/Seq Resources" the following DNA sequence can be copied and pasted into the box "to submit."

5′... AGGTCCATCG AGGGGGAAGA AACAAATAAG ATCACATGTA CGCAAGACTT TCTTCACCAA TACTTTGTAA CTGAAAGGGT TAGCATTCAA TTTGGGTTAA ATAACAAGAC CGTAAAAAGG ATAAATAAAG ATGAATTTGA TAAGGCAGTA AATTGTATCA

TGTCATGGAC AAACTATCCT AAGCCTGGGT TTAAACGAAC AGCTTCAACG TACCTCTTAA GCAATTCCTT TAAGAAATCT GCAACAGTAT ... 3′

After the submission is complete, a list of sequence analyses results is made available to choose and following questions can be posed thereof and results can be gathered.

How many hits were obtained with the DNA sequence upon submission? In addition to the Saccharomyces website given above, another tool can also be used: NCBI BLAST[97] yields more than six hits can be found.

Which of the hits can be regarded as: "Better"? Why?[98]

What gene is this DNA sequence a part of?[99]

Which part of the gene was the query sequence a part of (e.g., 5′ end, 3′ end or in the middle)?

What chromosome is the gene on?

Suppose the gene is selected by clicking it on the map. What is the systematic name for this gene?[100]

Where is this gene located with respect to the centromere (i.e., whether on the one on the left arm or on the right arm of the chromosome)?[101]

What is its mutant phenotype?[102,103]

What biological processes is the gene involved in?[104]

When compared, the gene to DNA from other organisms, what two other organisms are seen to have similar genes? Are these "homologies" biologically relevant? If so, why?[105]

Example Ex-5.7

Problem Statement and Approach

This example exercise is concerned with searching Medline details to find authors on the paper that outlines the work done to find query DNA sequences.

Query sequences on a pair of homologs can be selected following the order specified in Website on Kpol_1050p68,[106] and Website on pubmed.[107] Then, the following search exercises can be carried out.

By clicking on the accession number link for one of the two homologs, the 6-frame translation of the gene can be retrieved.

Next, an assigned sequence could be segmented, in a website, for example, Saccharomyces genome database official Site,[108] Saccharomyces–query sequences with NCBI taxnomy ID-H930;[108–111] and for each one of the segments, its 6 reading frames can be obtained.

A relevant set of example queries on the search is as follows.

What information is obtained using this function?

Which reading frame appears to contain the ORF?

Are there any introns observed in this gene? How is it decided?

How many DNA bases make up the ORF and how many amino acids does that encode?[111]

Example Ex-5.8

Problem Statement and Approach

Explained in this example is the method of accessing a link to find a specified organism and using Genscan to find the associated ORFs in its sequence. For example, suppose the sequence of a *yeast clone* is considered. Then, using "Vertebrate" as the specified organism, the solutions for the following questions are sought.

(i) *How many complete genes are there in the query sequence?*
(ii) *How many amino acids are present in ORF #3?*
(iii) *By copying the predicted protein sequence from ORF #3 and exercising an appropriate search with that sequence, find the identity of the protein and the gene that encodes it.*
(iv) *What is the name of the gene that encodes this protein?*

Solution hint...

The complete genes in the query sequence of a yeast clone:

Chosen clone NC_000074. It has 7864 bp as follows:

1 ttttattatt ttaagtatcc ... ... gatc 7864

Gene Name: HEPD0921_3_H11 Cacnbl and more ...

Problem P-5.6

Problem Statement

With reference to Example Ex-5.8, what is the molecular process the query gene supposedly involved (based on gene acronym and other information that might have already been found)?

Solution hint

By locating the DNA sequence of the gene, as in the example cited, the required solution can be attempted.

Example Ex-5.9

Problem Statement and Approach

Using relevant bioinformatic database search, this exercise is indicated to gain knowledge on sudden *acute respiratory syndrome* (SARS) virus and ways of achieving relevant antiviral efforts.

Suggested tasks

(a) Obtain the SARS virus genetic virus and gather info on proteins in that virus. Gather details on proteins made by this virus.
(b) Apply BLAST to compare SARS virus *versus Avian Bronchitis* virus.
(c) Write a note on the possibilities of developing vaccines for SARS virus.

Problem Objective

Obtain the SARS virus genetic virus and gather information on proteins in that virus. Furthermore, find details on proteins made by this virus. Apply BLAST to compare SARS virus *versus Avian Bronchitis* virus. Write about the possibilities of developing vaccines for SARS virus.

Descriptive Projects

Exercises outlined below are project-oriented problems, which can be prescribed, for example as term-papers in directed independent studies; and, some of them can even serve as seeds for advanced studies towards expanded academic thesis/dissertation efforts. However, such research tasks may need a prior knowledge and/or background detail in pertinent bioinformatic topics. As such, the project exercises presented here are written to forge details acquired by reading bioinformatic books (like the present one and/or peer-writings). A host of similar projects can be seen across the literature and readers are encouraged to access them to supplement the pedagogy associated with those outlined here.

Project Pr-5.1

This project is indicated to develop a feasible descriptive exercise so as to learn details on a specific topic, for example, on *highly expressed gene characterization of yeasts.* Yeasts exhibit biodiversity (depicting an abstract expression of all aspects of the variety of life, from bio-molecules to the variety of different species populations as well as communities of species). The common yeasts are: (i) Baker's yeast (*Saccharomyces cerevisiae*), (ii) a pathogen of humans *(Cryptococcus neoformans*), (iii) the dimorphic fungus (*Candida albicans),* which is another significant pathogen of humans, and (iv) common leaf-surface yeasts.

Problem Statement and Suggested Exercise/Tasks

Choosing any given yeast gene (listed above), list the conditions under which it can be regarded as being highly expressed. (Hint: Visit Stanford Microarray Database to learn and decide on any of the conditions represented in the Stanford database.)

Project Pr-5.2

Preliminary Details

The *hemoglobin beta, or HBB,* is the gene that gives instructions required in making of a protein known as *beta-globin.* This protein is a subunit of a larger protein called *hemoglobin* (which is found inside red blood cells or RBCs). The following sequence is given as the CDS of *HBB* (*hemoglobin beta*) gene in humans.

5′ ... acatttgctt ctgacacaac tgtgttcact agcaacctca aacagacacc atggtgcacc tgactcctga ggagaagtct gccgttactg ccctgtgggg caaggtgaac gtggatgaag ttggtggtga ggccctgggc aggctgctgg tggtctaccc ttggacccag aggttctttg agtcctttgg ggatctgtcc actcctgatg ctgttatggg caaccctaag gtgaaggctc atggcaagaa agtgctcggt gcctttagtg atggcctggc tcacctggac aacctcaagg gcacctttgc cacactgagt gagctgcact gtgacaagct gcacgtggat cctgagaact tcaggctcct gggcaacgtg ctggtctgtg tgctggccca tcactttggc aaagaattca ccccaccagt gcag-gctgcc tatcagaaag tggtggctgg tgtggctaat gccctggccc acaagtatca ctagctcgct ttcttgctgt ccaatttcta ttaaaggttc ctttgttccc taagtccaac tactaaactg ggggat-atta tgaagggcct tgagcatctg gattctgcct aataaaaaac atttattttc attgc ... 3′

Problem Statement and Suggested Exercise/Tasks

1. Suggest a database to obtain the above or similar sequence of *HBA* gene seen in relevant organisms.
2. Use the given HBB sequence to perform a BLAST search indicating the *HBB* gene from other organisms. (Do not close the window with the search results; it will be in need again.)
2. Find and add the DNA sequences from the *Northern elephant seal, orangutan, olive baboon,* and *crab-eating macaque.*
3. Now that you have five different sequence files, select them all and perform a CLUSTALW analysis (ignoring any warnings that may be given by default). Import the results of the CLUSTALW analysis and then use DRAWTREE to draw an unrooted phylogenetic tree. Print out the tree.
4. *What organisms are most related to be based on the HBB gene alignment? Is this expected?*
5. Now go back to the BLAST search results. Locate the same three non-human primate sequences, and this time, use the information in the Features section to determine and copy the coding region of the sequence.
6. Perform a CLUSTALW analysis on the coding regions for these five organisms, import the results, and use DRAWTREE again. Print out the tree and discuss its relevance *vis-à-vis* the analysis performed. *Is it meaningfully helpful? If not, why?*
7. Perform the pertinent exercises as above on the sequence of HBA gene.
8. Hence, find the organisms that are most related on the basis of HBA gene alignment.

Project Pr-5.3

Objective: This exercise is to perform datamining and analysis of the AIDS causing HIV. No programming is needed, but processing a vast data and rationalizing the results expected.

Problem Statement and Suggested Exercise/Tasks

(a) Obtain the genetic and protein data on HIV. Pick any complete sequence out of many strains. Gather info from protein databank and other protein/enzyme databases with regard to proteins made by this virus.
(b) Apply BLAST to align this sequence across other viruses.
(c) Detect similarity/grafted regions between HIV and Hepatitis virus.
(d) Create an evolutionary tree of the viruses and show the location of HIV.
(e) Another ancient virus-infecting residents across Australia's Northern Territory is *human T-cell leukemia virus type 1,* or HTLV-1. *Is it related to HIV?* Gather details and analyze its sequence details.

Project Pr-5.4

The scope of this exercise is to investigate the rice genome details and know about the growth productivity of *Oryza sativa,* commonly known as Asian rice, is the plant species most commonly referred to (in English) as rice. It is renowned for being easy to genetically modify and is a model organism for cereal biology.

Problem Statement and Suggested Exercise/Tasks

(a) Obtain rice genome data. Select a few (Public data source: Rice Genome program of Japan). Obtain some literature *via* PubMed.
(b) Prepare a list of rice genes and corresponding proteins.
(c) Recognize the genes and proteins involved in rice/paddy cultivation.
(d) Give some info on relevant enzyme reaction rates *versus* growth.
(e) *Wild rice* (genus *Zizania*) also known as Indian rice, water rice, or water oats, genus of four species of coarse grasses of the family *Poaceae,* the grain of which is sometimes grown as a delicacy. Perform the exercises similar to (a) through (d) on this species.
(f) There are other species like *Oryza meyeriana, Oryza ridleyi, Oryza schlechteri,* and *Oryza coarctata* complex that form the most primitive section of the genus. Also exists the *O. officinalis* complex with varying habitats and variations shown by their responses to pests and diseases. Develop a heuristic comparative study on such species similar to that indicated for *O. sativa.*

Project Pr-5.5

The objective of this exercise is to consider the *human genome project* (HGP) and relevant to any one of human chromosomes ascertaining gene-like patterns in the ORFs.

Problem Statement and Suggested Exercise/Tasks

(a) Obtain the genome data on a human chromosome at NCBI site. Get the physical and genetic atlas for these chromosomes that contain information on known genes. Identify ORF regions.
(b) Use gene-finding tools and look for gene-like patterns starting with ATG.
(c) Decide whether these patterns are declared in the literature as new genes.
(d) Learn on animal models used in studying genetic diseases; that is, transgenesis and induced mutation creating models of human genetic diseases with mice, flies, worms, and other animals. *But what do these models reveal about humans?* (Typically used such model organisms are *S. cerevisiae, Pisum sativum, D. melanogaster, C. elegans, D. rerio, Mus musculus,* and *Rattus norvegicus.*)

Project Pr-5.6

This exercise refers to considering the human genome and enumerating approximately the associated palindromic genes existing.

Preliminary Details

A palindromic sequence refers to sequence consisting of nucleic acids within double-helix of DNA and/or RNA that is the same when read from 5′ to 3′, on one strand, and 5′ to 3′, on the other, complementary, strand. It is also known as a *palindrome* or an *inverted-reverse sequence.* A suite of tools for genome sequence analysis called the Biological Language Modeling Toolkit allows identifying all palindromic sequences in the human genome, and it can be applied to study palindromes in whole genomes.

Problem Statement and Suggested Exercise/Tasks

(a) Obtain relevant genome data on a favorite human chromosome at NCBI site.
(b) Search for palindromic (at least approximate) patterns.
(c) Identify the proteins made by relevant genes.
(d) Write on the significance of palindromic sequences.
(e) Prepare a comprehensive note on human genome palindromes and related diseases.

Project Pr-5.7

Problem Statement

Objective: To learn the comparative genomics of human and a eukaryote (yeast): Suggested tasks/steps.

Obtain three yeast chromosome 3 sequence. List the genes and proteins in this chromosome.
Statistically estimate the similarities between these genes/proteins. Use CG cluster considerations.

Project Pr-5.8

Objective: To compare similarities and differences between two organisms at the genomic level; for example, mouse *versus* human or human *versus* zebra fish.

Problem Statement and Suggested Exercise/Tasks

1. Obtain the mouse and human genome data. Select one or two chromosomes considered similar.
2. Align the relevant amino acid sequences *via* BLAST.
3. Rationalize why humans are different from mouse at the genomic level highlighting the genes unique to humans.
4. Repeat the above on human *versus* zebra fish genome data.

Project Pr-5.9

Objective: To learn the comparative aspects of: (i) human and mouse at proteomic level and (ii) human *versus* medaka fish at genomic level.

Problem Statement and Suggested Exercise/Tasks

1. Obtain sequence data on human protein (HP).
2. Obtain sequence data on mouse protein (MP).
3. Blast align HP and MP.
4. Compare and contrast the proteomes of these two organisms.
5. Considering Medaka fish (*Oryzias latipes*), access the associated well-established resources for genetic and genomic resources and compare the human and medaka genomes to gain novel information on the sets of genes, repetitive elements, gene regulatory elements, and the evolutionary conserved elements at the nucleotide sequence level. Relevant genomic information implied chromosomal evolution in the vertebrate lineage and the use of medaka for medical science.

Project Pr-5.10

Objective: To compare and contrast at genomic level of two mycobacteria: *Mycobacterium tuberculosis* and *Leprae.*

Problem Statement and Suggested Exercise/Tasks

1. Obtain genome and proteome data on both bacteria.
2. Search for common genes and proteins.
3. Locate the organisms in the evolutionary tree and identify the common proteins in the organisms.
4. Learn relevant clinical informatics on the related diseases due to these organisms.

Project Pr-5.11

Objective: To create a knowledge base on model organisms.

Problem Statement and Suggested Exercise/Tasks

1. Write on the role of model organisms in drug discovery efforts.

2. Construct a knowledge base on: *Arabidopsis thaliana, D. melanogaster*, Zebrafish, *Chlamydomonas, Cyanobacteria, M. musculus* and *R. norvegicus, Macaca mulatta, Gorilla beringei, Pongo pygmaeus, Escherichia coli, S. cerevisiae*, and *C. elegans.*
3. Construct also a knowledge base on: *P. sativum, D. rerio, M. musculus, R. norvegicus*, and *Pan troglodytes.*
4. Prepare an annotated note on each of the aforesaid model organisms.

Project Pr-5.12

Objective: To learn the interactive map of the metabolism in human body.

Problem Statement and Suggested Exercise/Tasks

1. Prepare a flow chart of a few selected metabolic pathways (biochemical pathways) that include info on proteins, sequences, structure, substrates, chemical reactions, and related parameters.
2. Cross-link different pathways and show their inter-connectivity (Example: Insulin metabolism).
3. Write a pseudocode on tracking a metabolic pathway.

Project Pr-5.13

Objective: To gain knowledge on dengue virus and ways of achieving relevant antiviral efforts.

Problem Statement and Suggested Exercise/Tasks

1. Consider the four strains of dengue virus and gather info on proteins in one of them virus. Gather details on proteins made by this virus (see Chapter 8).
2. Compare and contrast the four viruses.
3. Write on the possibilities of developing vaccines for this virus.
4. Do similar studies on other possible viruses that have multiple strains.

Project Pr-5.14

Problem Statement

Objective: With reference to B19 virus, to find the palindromic sections in its ssDNA.

Suggested steps/tasks:

Obtain the ssDNA of B19 virus sequence
Identify the underlying palindromes
Write on the significance of such palindromic subspaces

Project Pr-5.15

Objective: To learn about human, bacterial, and YACs.

Problem Statement and Suggested Exercise/Tasks

1. Familiarize yourself with Entrez Genomes at NCBI or with TAIR.
2. Obtain an example of bacterial or human or YAC through appropriate information resource and run it through GENSCAN.
3. *How many "genes" are there on the artificial chromosome that you have chosen? Which putative gene is the largest (in terms of number of nucleotides)? How many nucleotides (ignore the promoter and the poly A + site) and how many introns are present?*
   (Example: *Via* the *Arabidopsis Information Resource,* the BAC is obtained through TAIR and running the *Arabidopsis* sequence (F7F1 BAC of *A. thaliana*) through GENSCAN, a text file can be created for saving.)
4. Perform a similar study in the context of a domestic animal like a cow.

Project Pr-5.16

Objective: To find the number of DNA molecules and base pairs in each case of simple members of eukaryota domain of life, namely, baker's yeast, fruit fly, and nematode worm.

Problem Statement and Suggested Exercise/Tasks

1. Learn using the search engine Entrez.
2. Use the link in Sequence of yeast clone #71020[112] to find the sequence of yeast clone #71020.
3. Find the sequence of *A. thaliana* chromosome 2 BAC F23H14 (Online at https:*//www.ebi.ac.uk/genomes/ath2.html*).

Project Pr-5.17

Objective: To familiarize with OMIM. Online Mendelian Inheritance in Man (OMIM) is a public domain database where one can obtain information regarding mutations in man and other mammals that cause inherited diseases.

Problem Statement and Suggested Exercise/Tasks

1. Go to NCBI > OMIM page> type cystic fibrosis and click on the first gene 602421.
2. Collect and report the details on relevant information of the gene, such as the clone, sequence ID, research team, laboratory, literature, gene function, sequence, etc.
3. Visit OMIM Entry — # 603903 — SICKLE CELL ANEMIA and collect/report the details as above.

Project Pr-5.18

Objective: To retrieve a protein sequence from the Web. For example, it is desired to retrieve the sequence of the protein performing the dUTPase function in *E. coli.*

Problem Statement and Suggested Exercise/Tasks

1. Perform the necessary retrieval and furnish the results by navigating the SWISS-PROT database home page.[113] Type dUTPase coli in the Quick Search Window.

   Perform Quick Search and summarize the results of the search and furnish the display hard copies.
2. Retrieve the protein sequence performing the dUTPase function in *S. Cerevisiae* and summarize the results.

Project Pr-5.19

Objective: To use Entrez search system and obtain the protein sequences for protein kinase.

Problem Statement and Suggested Exercise/Tasks

1. Visit NCBI > Entrez > protein. Type protein kinase.

   How many protein sequences are fetched?

2. Select FASTA and get details on the first sequence.
3. Gather details on autoregulatory region in protein kinase C depicting a pseudoanchoring site.

Project Pr-5.20

Objective: With reference to B19 virus to find the palindromic sections in its single-stranded DNA (ssDNA).

Problem Statement and Suggested Exercise/Tasks

1. Obtain the ssDNA of B19 virus sequence (see Chapter 8).
2. Identify the underlying palindromes.
3. Write on the significance of such palindromic subspaces.
4. Obtain the ssDNA of another virus sequence and do the exercises as above.

Project Pr-5.21

Objective: As in PROJECT PR-5.15, gather details on bacterial artificial chromosome (BAC) obtained through TAIR and running the *Arabidopsis* sequence *A. thaliana* chromosome 2 BAC F23H14.

Problem Statement and Suggested Exercise/Tasks

1. As in Project PR-5.15, familiarize yourself with Entrez Genomes at NCBI or with TAIR. *Via* pertinent *Arabidopsis* Information Resource, the BAC is obtained through TAIR and running the *Arabidopsis* sequence (*A. thaliana* chromosome 2 BAC F23H14 2 BAC F23H14 through GENSCAN, a text file can be created for saving).
2. Repeat the above exercise for: *A. thaliana chromosome* 2 clone F10A8 map rga, *A. thaliana chromosome 2 clone F2I9* map rga, and *A. thaliana chromosome 2 clone T23K3* map mi320.

Project Pr-5.22

Objective: To compare sequences of the same gene from three species: humans, common chimpanzees, and gorillas.

Problem Statement and Suggested Exercise/Tasks

1. Learn using the search engine *Entrez.*
2. Going to Entrez protein database, enter in the search box the name of protein and the name of one of the species: *Homo sapiens* (humans), *P. troglodytes* (common chimpanzee), or *Gorilla gorilla* (um, gorilla).
3. The gorilla sequence is an outgroup in assigning substitutions to the chimp and human lineages and a relative rate test can show whether one lineage has a faster rate of evolution than the other. Learn relevant details in phylogenetic perspectives (see Chapters 9 and 10).

Project Pr-5.23

Objective: Similar to Project PR-5.17 to obtain information regarding mutations in plant life by familiarizing with details.

Problem Statement and Suggested Exercises/Tasks

To learn about genes of disease-causing organisms in plants.

1. For example, in soy-bean, a seed-borne fungal pathogen called *Phomopsis longicolla* (*Diaporthe longicolla*) causes the so-called *Phomopsis seed decay* (PSD).
2. Relevant genetic base of fungal virulence factors allows understanding the mechanism of PSD disease development. Search for a database for *P. longicolla* that contains information about the associated genome assemblies (contigs), gene models, gene descriptions, and functional ontologies.
3. A web-based front-end to the database of such database will aid in the development of new control strategies for appropriate pathogens.

Project Pr-5.24

Objective: Similar to Project Pr-5.23 as above, to obtain information regarding disease-causing genes in farm animals.

Problem Statement and Suggested Exercise/Tasks

1. To learn about genes of disease-causing organisms in farm animals: for example, considering the cow, anthrax, brucellosis, cryptosporidiosis, dermatochalasis, *Escherichia coli*, giardiasis, leptospirosis, listeriotic, pseudocowpox, Q-fever, rabies, ringworm, salmonellosis, tuberculosis, and vesicular stomatitis are commonly known diseases. For example, anthrax is an infectious bacterial disease of animals, caused by the spore-forming bacteria *Bacillus anthracis*.
2. Search for a database for *Bacillus anthracis* that contains information about the pertinent genome assemblies (contigs), gene models, gene descriptions, and other functional ontologies.
3. A web-based front-end to the database of such a database will aid veterinarians in the development of new control strategies for appropriate pathogens.

5.8 Closure

Reviewed in this chapter are underlying details, considerations, and overviews on bioinformatic resources comprised of databases, networks, and tools. Identified and described are salient genomic and proteomic databases and other bioinformatic resources available worldwide. In short, the scope of this chapter allows the readers to familiarize with the state-of-the-art bioinformatic methodologies *via* the underlying resources, internetworking, and tools.

References

[1] D. Roy: *Bioinformatics*. Narosa Publishing House. New Delhi, India: 2009.

[2] NCBI Resource Coordinators: Database resources of the National Center for Biotechnology Information. *Nucleic Acids Research,* 2012, vol. 41 (Database issue), D8–D20.

[3] D. A. Benson, I. Karsch-Mizrachi, D. J. Lipman, J. Ostell and E. W. Sayers: GenBank. *Nucleic Acids Research,* 2011, vol. 39 (Database issue), D32–D37.

[4] D. A. Benson, M. Cavanaugh, K. Clark, I. Karsch-Mizrachi, D. J. Lipman, J. Ostell and E. W. Sayers: GenBank. *Nucleic Acids Research,* 2012, vol. 41 (Database issue), D36–D42.
[5] GenBank entry on table of features. Available online at: *www.ncbi.nlm.nih.gov/collab/FT/index.html.*
[6] NCBI-GenBank Flat File Release 215.0 Distribution Release Notes: GBREL.TXT Genetic Sequence Data Bank, August 15, 2016.
[7] S. P. Lapage, P. H. A. Sneath, E. F. Lessel, V. B. D. Skerman, H. P. R. Seeliger and W. A. Clark: *International Code of Nomenclature of Bacteria: Bacteriological Code — 1990 Revision.* American Society for Microbiology Press, Washington D. C., USA: 1992.
[8] D. A. Benson, I. Karsch-Mizrachi, D. J. Lipman, J. Ostell and D. L. Wheeler: GenBank. *Nucleic Acids Research,* 2007, vol. 35 (Suppl. 1), D21–D25.
[9] NCBI Homepage. Available online at: *www.ncbi.nlm.nih.gov.*
[10] The NCBI taxonomy homepage. Available online at: *www.ncbi.nlm.nih.gov/Taxonomy/taxonomyhome.html.*
[11] K. Clark, I-K. Mizrachi, D. J. Lipman, J. Ostell and E. W. Sayers: GenBank. *Nucleic Acids Research,* 2016, vol. 44 (Database issue), D67–D72.
[12] Sanger Institute history. Available online at: *Sanger Institute. UK: Archive.org.*
[13] Sanger campus. Available online at: *sanger.ac.uk/about/campus.*
[14] T. K. Attwood and D. J. Parry-Smith: *Introduction to Bioinformatics.* Pearson Education Ltd., England: 1999.
[15] DDBJ. Available online at: *http://www.ddbj.nig.ac.jp.*
[16] Y. Tateno, T. Imanishi, S. Miyazaki, K. Fukami-Kobayashi, N. Saitou, H. Sugawara and T. Gojobori: DNA Data Bank of Japan (DDBJ) for genome scale research in life science. *Nucleic Acids Research,* 2002, vol. 30(1), 27–30.
[17] Genome information broker (GIB). Available online at: *http://gib.genes.nig.ac.jp.*
[18] Human genome studio (HGS). Available online at: *http://studio.nig.ac.jp.*
[19] D. L. Wheeler, T. Barrett, D. A. Benson, *et al.*: Database resources of the national center for biotechnology information. *Nucleic Acids Research,* 2007, vol. 35 (Database issue): D5–D12.
[20] B. F. F. Ouellette and M. S. Boguski: Database divisions and homology search files: A guide for the perplexed. *Genome Research,* 1997, vol. 7(10), 952–955.
[21] E. Mardis, J. McPherson, R. Martienssen, R. Wilson and W. R. McCombie: What is finished, and why does it matter? *Genome Research,* 2002, vol. 12, 669–671.
[22] M. Olson, L. Hood, C. Cantor and D. Botstein: A common language for physical mapping of the human genome. *Science,* 1989, vol. 245, 1434–1435.
[23] S. T. Sherry, M. Ward and K. Sirotkin: dbSNP — database for single nucleotide polymorphisms and other classes of minor genetic variation. *Genome Research,* 1999, vol. 9(8), 677–679.

[24] SNP. Available online at: *http://www.ncbi.nlm.nih.gov/SNP/snp_summary.cgi.*

[25] D. Koboldt: The current state of dbSNP. Posted in MassGenomics. January 24, 2012. Available online at: *http://massgenomics.org/2012/01/the-current-state-of-dbsnp.html.*

[26] M. S. Boguski, T. M. J. Lowe and C. M. Tolstoshev: dbEST — database for "expressed sequence tag". *Nature Genetics,* 1993, vol. 4, 332–333

[27] R. J. Roberts: Restriction and modification enzymes and their recognition sequences. *Nucleic Acids Research,* 1982, vol. 10(5), r117.

[28] R. J. Roberts and D. Macelis: REBASE-restriction enzymes and methylases. *Nucleic Acids Research,* 1993, vol. 21(13), 3125–3137.

[29] The Sequence Read Archive (SRA). Available online at: *http://www.ncbi.nlm.nih.gov/sra.*

[30] M. J. Fullwood, C. L. Wei, E. T. Liu and Y. Ruan: Next-generation DNA sequencing of paired-end tags (PET) for transcriptome and genome analyses. *Genome Research.* 2009, vol. 19, 521–532.

[31] miRBase: The microRNA database. Available online at: *http://identifiers.org/mirbase/.*

[32] J. S. Griffiths: The micrRNA registry. *Nucleic Acids Research,* 2004, vol. 32 (Database issue), D109–D111. Available online at: *http://identifiers.org/pubmed/14681370.*

[33] J. S. Griffiths, R. J. Crocock, S. van Dongen, A. Bateman and A. J. Enright: MicrRBase-microRNA sequences, targets and gene nomenclature. *Nucleic Acids Research,* 2006, vol. 34 (Database issue), D140–D144. Available online at: *http://identifiers.org/doi/10.1093/nar/gkj112.*

[34] H. Berman, K. Henrick and H. Nakamura: Announcing the worldwide protein data bank. *Nature Structural Biology,* 2003, vol. 10(12), 980.

[35] Official website of PIR at Georgetown University. Available online at: *http://pir.georgetown.edu/.*

[36] C. Wu and D. W. Nebert: Update on genome completion and annotations: Protein information resource. *Human Genomics,* 2004, vol. 1(3), 229–233.

[37] D. C. George, W. C. Barker and L. T. Hunt: The protein identification resource. *Nucleic Acids Research,* 1986, vol. 14(1), 11–15.

[38] C. H. Wu, L. L. Yeh, H. Huang, L. Arminski, J. Castro-Alvear, *et al.*: The protein information resource. *Nucleic Acids Research,* 2003, vol. 31(1), 345–347.

[39] The munich information center for protein sequences (MIPS-GSF), Martinsried, Germany, MIPS: A database for genomes and protein sequences. Available online at: *http://www.mips.biochem.mpg.de/.*

[40] H. W. Mewes, K. Albermann, K. Heumann, S. Liebl and F. Pfeiffer: MIPS: A database for protein sequences, homology data and yeast genome information. *Nucleic Acids Research,* 1997, vol. 25(1), 28–30.

[41] W. C. Barker, J. S. Garavelli, P. B. McGarvey, C. R. Marzec, B. C. Orcutt, *et al.*: The PIR-international protein sequence database. *Nucleic Acids Research,* 1999, vol. 27(1), 39–43.

[42] A. Bairoch and R. Apweiler: The SWISS-PROT protein sequence data bank and its new supplement TREMBL. *Nucleic Acids Research,* 1996, vol. 24(1), 21–25.

[43] N. Pattabiraman, K. Namboodiri, A. Lowrey and B. P. Gaber: NRL-3D: A sequence-structure database derived from the protein data bank (PDB) and searchable within the PIR environment. *Protein Sequences & Data Analysis,* 1990, vol. 3(5), 387–405.

[44] N. Hulo, A. Bairoch, V. Bulliard, L. Ceruttil, E. De Castro, *et al.*: The PROSITE database. *Nucleic Acids Research,* 2006, vol. 934 (Database issue), D227–D230.

[45] N. Hulo, A. Bairoch, V. Bulliard, L. Cerutti, B. Cuche, *et al.*: The 20 years of PROSITE. *Nucleic Acids Research,* 2007, vol. 36(Database issue), D245–D249.

[46] C. J. A. Sigrist, C. Lorenzo, E. de Castro, P. S. Langendijk-Genevaux, V. Bulliard, A. Bairoch and N. Hulo: PROSITE, a protein domain database for functional characterization and annotation. *Nucleic Acids Research,* 2010, vol. 38 (Database issue), D161–D166.

[47] T. K. Attwood, P. Bradley, D. R. Flower, A. Gaulton, A. N. Maudling, *et al.*: PRINTS and its automatic supplement, prePRINTS. *Nucleic Acids Research,* 2003, vol. 31(1), 400–402.

[48] T. K. Attwood: The PRINTS database: A resource for identification of protein families. *Briefings in Bioinformatics,* 2002, vol. 3(3), 252–263.

[49] P. Scordis, D. R. Flower and T. K. Attwood: FingerPRINTScan: Intelligent searching of the PRINTS motif database. *Bioinformatics,* 1999, vol. 15(10), 799–806.

[50] T. K. Attwood, M. E. Beck, A. J. Bleasby and D. J. Parry-Smith: PRINTS — A database of protein motif fingerprints. *Nucleic Acids Research,* 1994, vol. 22(17), 3590–3596.

[51] R. D. Finn, J. Tate, J. Mistry, P. C. Coggill, S. J. Sammut, *et al.*: The pfam protein families database. *Nucleic Acids Research,* 2008, vol. 36 (Database issue), D281–D288.

[52] EMBL-EBI pfam help summary Pfam 30.0 (June 2016, 16306 families). Available online at: *http://pfam.xfam.org/help.*

[53] BLOCKS WWW Server. Available online at: *http://blocks.fhcrc.org/help.*

[54] S. Pietrokovski, J. G. Henikoff and S. Henikoff: The Blocks database — a system for protein classification. *Nucleic Acids Research,* 1996, vol. 24(1), 197–200.

[55] J. G. Henikoff and S. Henikoff: Blocks database and its applications. *Methods in Enzymology,* 1996, vol. 266, 88–105.

[56] C. G. Nevill-Manning, T. D. Wu and D. L. Brutlag: Highly specific protein sequence motifs for genome analysis. *Proceedings of the National Academy of Sciences* (*USA*), 1998, vol. 95, 5865–5871.
[57] Web site on motif. Available online at: *http://motif.stanford.edu/emotif.*
[58] J. Y. Huang and D. L. Brutlag: The EMOTIF database. *Nucleic Acids Research*, 2001, vol. 29(1), 202–204.
[59] C. G. Nevill-Manning, T. D. Wu and D. L. Brutlag: Highly specific protein sequence motifs for genome analysis. *Proceedings of the National Academy of Sciences (USA)*, 1998, vol. 95(11), 5865–5871.
[60] L. Lo Conte, B. Ailey, T. J. P. Hubbard, S. E. Brenner, A. G. Murzin and C. Chothia: SCOP: A structural classification of proteins database. *Nucleic Acids Research*, 2000, vol. 28(1), 257–259.
[61] Web-site on SCOP. Available online at: *http://scop.mrc-lmb.cam.ac.uk/scop.*
[62] C. A. Orengo, A. D. Michie, S. Jones, D. T. Jones, M. B. Swindells and J. M. Thornton: CATH — a hierarchic classification of protein domain structures. *Structure*, 1997, vol. 5(8), 1093–1109.
[63] Web-site on CATH. Available online at: *http://www.cathdb.info.*
[64] M. Knudsen and C. Wiuf: The CATH database. *Human Genomics*, 2010, vol. 4(3), 207–212.
[65] R. A. Laskowski: PDBsum: Summaries and analyses of PDB structures. *Nucleic Acids Research*, 2001, vol. 29(1), 221–222.
[66] T. A. de Beer, K. Berka, J. M. Thornton and R. A. Laskowski: PDBsum additions. *Nucleic Acids Research*, 2014, vol. 42 (Database issue), D292–D296.
[67] R. Sayle and E. J. Milner-White: RasMol: Biomolecular graphics for all. *Trends in Biochemical Sciences (TIBS)*, 1995, vol. 20(9), 374–376.
[68] A. Herráez: Biomolecules in the Computer. *Jmol to the Rescue. Biochemistry and Molecular Biology Education*, 2006, vol. 34(4), 255–261.
[69] A. J. Bleasby, D. Akrigg and T. K. Attwood: OWL — a non-redundant composite protein sequence database. *Nucleic Acids Research*, 1994, vol. 22(17), 3574–3577.
[70] B. K. Hou, J. S. Kim, J. H. Jun, D. Y. Lee, Y. W. Kim, *et al.*: BioSilico: An integrated metabolic database system. *Bioinformatics*, 2004, vol. 20(17), 3270–3272.
[71] J. D. Wren and A. Bateman: Databases, data tombs and dust in the wind. *Bioinformatics*, 2008, vol. 24(19), 2127–2128.
[72] R. L. Tatusov, M. Y. Galperin, D. A. Natale and E. V. Koonin: The COG database: a tool for genome-scale analysis of protein functions and evolution. *Nucleic Acids Research*, 2000, vol. 28(1), 33–36.
[73] P. S. Neelakanta: *A Textbook on ATM Telecommunication Engineering.* CRC Press, Inc., Boca Raton, FL, USA: 2000.
[74] P. S. Neelakanta and D. M. Baeza: *Next Generation Telecommunications and Internet Engineering.* Linus Publications Inc., Deer Park, NY, USA: 2009.

[75] MIT: IP addresses, host names, and domain names, information systems and technology. Available online at: *https://ist.mit.edu/network/*ip.

[76] J. Ian and W, Norman: Architecture of the World Wide Web, vol. 1. W3C Recommendation 15 December 2004. Available online at: *https://www.w3.org/TR/webarch/.*

[77] A. J. Gibbs and G. A McIntyre: The diagram, a method for comparing sequences. Its use with amino acid and nucleotide sequences. *European Journal of Biochemistry,* 1970, vol. 16, 1–11.

[78] S. B. Needleman and C. D. Wunsch: A general method applicable to the search for similarities in the amino acid sequence of two proteins. *Journal of Molecular Biology*, 1970, vol. 48(3), 443–453.

[79] P. H. Sellers: On the theory and computation of evolutionary distances. *SIAM Journal on Applied Mathematics*, 1974, vol. 26(4), 787–793.

[80] T. F. Smith and M. S. Waterman: Identification of common molecular subsequences. *Journal of Molecular Biology*, 1981, vol. 147, 195–197.

[81] W. R. Pearson and D. J. Lipman: Improved tools for biological sequence comparison. *Proceedings of the National Academy of Sciences USA*, 1988, vol. 85(8), 2444–2448.

[82] Website on biocorba project. Available online at: *http://biocorba.org/.*

[83] Website on bioperl. Available online at: *http://cvs.bioperl.org/.*

[84] Ensembl website. Available online at: *http://www.ensembl.org/.*

[85] Bioperl-db CVS directory. Available online at: *http://cvs.bioperl.org/cgi-bin/viewcvs/viewcvs.cgi/bioperl-db/?cvsroot=bioperl.*

[86] S. Raptis and G. Dimitriadis: Human insulin. *Clinical Physiology and Biochemistry*, 1985, vol. 3(1), 29–42.

[87] D. S. Nicole and L. F. Smith: Amino-acid sequence of human insulin. *Nature,* 1960, vol. 187, 483–485.

[88] A. Mandal: Insulin protein structure. Available online at: *http://www.news-medical.net/health/Insulin-Protein-Structure.aspx.*

[89] S. R. Joshi, R. M. Parikh and A. K. Das: Insulin — history, biochemistry, physiology and pharmacology. *Supplement of Journal of Association of Physicians (JAPI)*, 2007, vol. 55, 19–25.

[90] G. I. Bell, J. H. Karamt and W. J. Rutter: Polymorphic DNA region adjacent to the 5′ end of the human insulin gene. *Proceedings of the National Academy of Sciences USA*, 1981, vol. 78(9), 5759–5763.

[91] History of Insulin Web-site. Available online at: *http://www.med.uni-giessen.de/itr/history/inshist.html.*

[92] M. Emori and B. Maruo: Complete nucleotide sequence of an alpha-amylase gene from Bacillus subtilis 2633, an alpha-amylase extra hyper producing strain. *Nucleic Acids Research,* 1988, vol. 16(14B), 7178.

[93] Website of NCBI-Entrez. Available online at: *http://www.ncbi.nlm.nih.gov/Entrez/.*

[94] Website of EXPASY. Available online at: *http://web.expasy.org/cgi-bin/translate/dna_aa.*
[95] EMBOSS Six-pack tool. Available online at: *https://www.ebi.ac.uk/Tools/st/emboss_sixpack/.*
[96] Stanford website of Saccharomyces. Available online at: *http://genome-www.stanford.edu/Saccharomyces/.*
[97] NCBI-BLAST. Available online at: *http://blast.ncbi.nlm.nih.gov/Blast.cgi.*
[98] Sacharomyces-ORFmap. Available online at: *http://www.yeastgenome.org/cgibin/ORFMAP/ORFmap?chr=8&beg=412907&end=413156.*
[99] Saccharomyces Locus/YHR157W: Overview #sequence. Available online at: *http://www.yeastgenome.org/locus/YHR157W/overview#sequence.*
[100] Saccharomyces over view on locus/S000001200. Available online at: *http://www.yeastgenome.org/locus/S000001200/overview.*
[101] Saccharomyces: Nomenclature/name. Available onlie at: *http://www.yeastgenome.org/help/community/nomenclature-conventions#systematicName.*
[102] Saccharomyces-Locus/YHR157W: Overview #phenotype. Available online at: *http://www.yeastgenome.org/locus/YHR157W/overview#phenotype.*
[103] Saccharomyces-Locus/S000001200/phenotype. Available online at: *http://www.yeastgenome.org/locus/S000001200/phenotype.*
[104] Saccharomyces-Locus/S000001200/gene ontology. Available online at: *http://www.yeastgenome.org/locus/S000001200/go.*
[105] Website on NCBI. Available online at: *http://www.ncbi.nlm.nih.gov/gene.*
[106] Website on Kpol_1050p68. Available online at: *http://www.ncbi.nlm.nih.gov/pubmed/17494770/KAFRE0E02150.*
[107] Website on Pubmed. Available online at: *http://www.ncbi.nlm.nih.gov/pubmed?LinkName=gene_pubmed_pmc_nucleotide&from_uid=13883026.*
[108] Saccharomyces Genome Database — Official Site: Available online at: *https://www.ncbi.nlm.nih.gov/genome/?term=Saccharomyces+Genome+Database+official+site.*
[109] Global Catalogue of Microorganisms: Saccharomyces cerevisiae (strain ATCC 204508/S288c) (Baker's yeast). Available online at: http://gcm.wfcc.info/speciesPage.jsp?strain_name=Saccharomyces.
[110] Global Catalogue of Microorganisms: Saccharomyces bayanus (Yeast) (Saccharomyces uvarum). Available online at: http://gcm.wfcc.info/speciesPage.jsp?strain_name=Saccharomyces.
[111] NCBI Map Viewer: *Saccharomyces cerevisiae* genome data and search tips Available online at: *https://www.ncbi.nlm.nih.gov/projects/mapview/static/scerevisiaesearch.html.*
[112] Sequence of yeast clone #71020. Available online at: *http://genomewww4.stanford.edu/cgibin/SGD/getSeq?map=a3map&seq=71020&flankl=&flankr=&rev=.*
[113] SWISS-PROT database. Available online at: *http://www.expasy.org/sprot/.*

Chapter 6

Translational and Clinical Bioinformatics

Beyond the shores of omic perspectives lies in the horizon of bioinformatics, the so-called translational research informatics (TRI). It forms a unique subdomain of efforts dedicated to address transformational aspects of basic biosciences versus practical considerations usable in macroscopic landscape of biomedical, clinical, and public health sciences. Relevantly transforming the facts of basic sciences into details of health-related studies leads farther into conceivable heuristics of translational and clinical bioinformatics.

When wisdom is borne across translations, its charm is always gets lost ...; and, "...reading a poem in translation is like kissing a woman through a veil ..."

— Walter Benjamin and Annie Michaels

6.1 Introduction

Contrary to the caveat of quotation indicated above, the antonymous meaning of translation is borne rather objectively across the veil of details in biology, and the charm of disseminating the translated results of basic sciences into clinical aspects bioinformatics is seen uninhibitantly pursuable in health-related studies beneficial to invalids, and pertinent exercises of translating omic details of base-pairs into bedside pragmatics aim at accelerating patient-related needs *vis-à-vis* clinical pursuits (on diagnostic and therapeutic regimens) with no loss of any veiled details. Thus, the conceivable tasks of translational bioinformatics imply smartly borrowing the vast information from basic biosciences (such as microbiology, virology, epidemiology) and meaningfully translating them into useful informatics across societal health and clinical studies. That is, the facts of biosciences can be availed from their root and viably specified as omic details of microbiology; hence, the perceived details can be translated for widely applicable contexts of clinical tasks.

Furthermore, the associated data-centric biostatistics can be elegantly adopted in preventive medicine.

In bridging basic sciences and informatics of clinical details, the intervening translatory efforts also imply inevitable decision-theoretic logistics wherein "... sometimes you make the right decisions, sometimes you make the decision right ..." as observed by Phil McGraw (an American Psychologist). Hence, translational bioinformatics implies a carefully structured art of guided adjudications on health-specific decisions indicated for robust and reliable patient-related adoptions and bedside implementations.

Translational bioinformatics aims at establishing reliable and robust metrics on thresholds in such clinical decisions pertinent to diagnosis, progression, level or status of pathogeny, and prognosis of a disease. Specifiable items encountered thereof include the following: patient-related emergency criteria, viable treatment procedures, adequate therapy regimens, foreseeable post-recovery needs, and prescribing adequate preventive measures. It is imperative that all these tasks warrant a host of decision-theoretic heuristics and algorithms usable in clinical studies.

Thus, the concepts of basic sciences compiled and translated into a span of wider knowledge base in information science include grossly changing research traits involving a sizable data of statistics. Such traits and bioinformation data form perceivable gists of *translational bioinformatics* (abbreviated as TBI)[1] and the core of clinical informatics; hence, the *American Medical Informatics Association* (AMIA) has defined translational bioinformatics as, "... the development of storage, analytic, and interpretive methods to optimize the transformation of increasingly voluminous biomedical data into proactive, predictive, preventative, and participatory health ...."[2] Associated efforts include research studies on developing novel techniques toward the integration of biological and clinical data; hence, translational methodology and clinical informatics encompass the gamut of novel biological observations and related statistics and use them in enabling robust and bounded clinical decision-theoretics.

Today, TBI has become a vital part of biomedical research and bioinformatics geared toward *precision medicine*. It has emerged as a result of developments in high-throughput technologies and *electronic health records* (EHRs) posing a paradigm shift in both healthcare and biomedical research. The associated, increasingly voluminous and grossly disposed datasets of bio- and health sciences of medical interest have necessitated a unique class of tools and methods to convert (translate?) the underlying information into practical, pragmatic, as well as actionable knowledge.

This chapter is written to define and conceptualize the underlying aspects of TBI with deliberations on its past and present status plus foreseeing its futuristic scope. Hence, furnished are certain examples of medical interest.[3–5] In all, this newly indicated knowledge of TBI gained from integrative efforts of basic biology and applied medical sciences forms an end product, which can be disseminated to a variety of stakeholders (including biomedical scientists, clinicians, and patients). Evolving pertinent realms of TBI is outlined in this chapter as its first part (Part I) with appropriate details, examples, and exercises.

The clinical bioinformatics mentioned earlier refer to a more selective study. It is designed exclusively on the macroscopic aspects of bioinformatics beyond the heuristics of "omics" and it is extended to address the associated logistics *versus* decision-theoretic considerations exclusively in clinical contexts. Furthermore, it is also applied in specific patient management scenarios and in pertinent clinical studies (embracing related case studies) of patients. The scope and details of clinical bioinformatics are companion themes presented in the second part (Part II) of this chapter.

6.2 Part I: Translational Bioinformatics

As defined above, the emphasized integrative efforts of biology and applied medical sciences form the end product of TBI.[6,7] The associated tasks involve mining massive bioinformatic datasets and bridging them with the details on latest multimodal measurement technologies. Hence, the advocated directions on TBI-related studies are: (i) evolving integrative aspects of biology and clinical synergism; (ii) making computational tools to handle massive data set; (iii) performing a cohesive analysis of multimodal measurements; (iv) bringing the basic knowledge of "base-pairs to (patient) bedside"; and (v) detailing informatics of the biocomplex with elaborations of "big-data" perspectives.

Relevantly, the paradigm shifts in public policy, emergence of large datasets (gathered *via* multiple molecular level measurements) and emphasized use of *EHR*, plus "recent advances in natural language-processing, access to vast computing infrastructure, sophisticated ontologies, data-mining and machine learning tools" show a directed convergence of TBI *vis-à-vis* big datamining[6,7] considerations.

Classical scientific methods and statistical inferences across the spectrum of preclinical discovery applied in clinical practice aim at formulating "a hypothesis, test the hypothesis experimentally, analyze the data, and make informed decisions to accept, reject, or refine the hypothesis." The associated flow of drug-discovery suites and developments are iterative, collaborative, and multidisciplinary in nature. Such studies culminate in "pivotal clinical trials that provide evidence of efficacy and safety to support market access." Rapid identification of validated drug targets (e.g., pertinent to antiretroviral drugs for treating HIV and statins for managing high cholesterol in cardiovascular diseases) has led to uncovering many new medicines "that reduced morbidity and mortality and increased life expectancy." Thus, the emergence of TBI in the recent past shows the potentials to address associated challenges and it is nothing more than a validly refined bioinformatic aspects of thinking.

Furthermore, computational biology, informatic aspects of biocomplex, and bioengineering techniques of information exchange plus using *in silico* software platforms have enabled the emergence of TBI-specific multiple databases that permit comprehensive studies on, "structure–activity relationships, animal disease models, drug mechanisms of action, toxicology and human risk assessments, gene expression and genomic microarrays, sequence and functional annotations, and phenotypes of disease and drug response."

In addition, in the context of drug research,[8] TBI extracts actionable information on bioinformatic details available as large datasets. Hence, new understandings and novel knowledgebase are realized across various levels of "... spectrum of genes, subcellular compartments, intact cells, organs and tissues, signaling pathways, and drug response in populations and patient subgroups" in living systems; that is, TBI is a conceived route of bioinformatic knowledge on biological details of genes through whole-body systemics converged to frame bedside strategies toward preventing and managing wellness of living systems and conditions of serious chronic diseases.

While basic bioinformatics allow knowing all about genes in general, unique computational approaches are, however, required in understanding diseased genes. This specifies invoking a higher layer of TBI as detailed in Kann,[9] and Butte and Chen.[10] Relevantly, the annotated details of experiments (such as in NCBI *Gene Expression Omnibus*

[GEO]) are related to human diseases that can be correlated with PUBMED identifiers (representing publications on experiment). Hence, Butte comments[11] that the primary scientific efforts as above depict the "next step needed for the field of bioinformatics" with adunct provisioning of global communication and related networking infrastructure. A venues of relevant newer networking strategies (including *cloud computing)* could frame a conceivable key enabling the technology for future of omic research involving *high-throughput sequence* data analyses and related computation.[12] Hence, large-scale translational research in genomic medicine can be achieved.

6.3 TBI: Implementation

In modern context, implementing TBI needs unique algorithms; as such, it incorporates a gist of analytic disciplines including *artificial intelligence* (AI), *cognitive science, artificial neural networks* (ANNs), and *deep-learning techniques.* Transforming basic discoveries of physical and biological sciences into a practically applicable knowledge base (required in health sciences) leverages the massive range of biomedical data obtainable from traditional bioinformatic suites and *next-generation sequencing* (NGS). This concurrently leads to creating unlimited extent of electronic database (such as on healthcare records).[13] Primarily, TBI perspectives can be regarded as an off-shoot of the so-called *translational research informatics* (TRI), a subdomain of bioinformatic efforts dedicated to data management in biomedical, medical, and health sciences,[14] and it represents a multidisciplinary field of *health information technology* (HIT).

6.4 TBI: ITS Scope and Trends

Commensurate with the introductory details furnished above, the scope of TBI can be viewed in two perspectives: (i) in a broader sense, it is intended to evoke a diverse set of skills and experiences based on concepts, theories, and notions of bioinformatics and biomedical sciences and (ii) in a restricted sense, the purpose of associated pedagogy in TBI is to cultivate the ability to use informatics in bringing the results of clinical trials to practice, solving current healthcare challenges, and applying ethically, the policies related to computer technologies, information science, and medical/clinical practices. In short, the scope of TBI basically implies transition of data from "base-pairs to bedside," and it forms a branch of health informatics. In the perspectives of *National Library of Medicine* (NLM) and *National Institutes of Health* (NIH), health informatics is regarded as "the inter-disciplinary study of the design, development, adoption and application of information technology (IT)-based innovations in healthcare services delivery, management and planning." Thus, the scope of TBI expressed in terms of health informatics objectively includes basic and advanced biological sciences including genetics, fundamental and applied biomedical research, societal public health including preventive medicine and epidemic alerts, occupational therapy and industrial hygiene, medical practices covering nursing, clinical medicine, laboratory-based biochemical assays, instrumented diagnostic suites, regimes of therapeutics, avenues of dentistry, drug discovery and pharmaceutical research, allied health services (such as physical therapy, geriatric support), and alternative medicine including disciplines of nonallopathic indigenous approach.

6.5 Resources of TBI

In the contexts of TBI, the role of computational efforts is inevitable considering the enormous database details involved and accessing them as necessary for use worldwide. As such, a significant support of statistical methods and algorithmic suites is warranted toward gathering the data, analyzing the details, storing the information worldwide, sharing and dissemination of research pursuits and outcomes, and integrating the details of medical, biological, and health sciences for a cohesive data management. Constituents of pertinent computer-specific resources *vis-à-vis* TBI and associated details are as follows.

Electronic Medical Record

Healthcare management, especially of adults, to lead independent, long, and healthy lifestyle has been a topic of international interest and focused research and business (proprietary) efforts thereof have emerged in the recent past. Mostly, the underlying efforts are directed at (a) identifying a healthcare locale environment, (b) specifying relevant medical-related ambient (c) implementing patient-friendly diagnostic/therapeutic regimens, and (d) coordinating patient and healthcare/medical personal routinely to judge and proactively decide health management. Relevantly, digital patient-care information infrastructures known as *electronic medical record* (EMR) are traditionally designed in the context of hospital/clinical setting and in outpatient cases.[15,16]

Translational Research Informatics

This is described as "an integrated software solution to manage the: (a) logistics, (b) data integration, and (c) collaboration, required by translational investigators and their supporting institutions." Relevant class of informatic systems interoperates with *HIT* and EMR along with *clinical trial management system* (CTMS) in *clinical research informatics* (CRI), biotechnology, and pharmaceutical industries. Pertinent information systems depict a comprehensive resource that "maintains and manages planning, performing and reporting functions, along with participant contact information, tracking deadlines and milestones."[17]

In *de novo* perspectives of TRI, researchers actively engage in acquiring and integrating systems to enable the end-to-end TRI requirements. Furthermore, within the premise of CTMS, *eClinical* is a term used across biopharmaceutical industry with reference to *trial automation technology* and largely referring to electronic data capture, CTMSs, or *randomization and trial supply management* (RTSM) systems. The eClinical resources use applications, such as interactive voice response systems and electronic patient diaries.

In TRI contexts, *mobile health* or *mHealth* (also denoted as *m-health*) refers to the practice of medicine and public health supported *via* mobile devices.[18] It is regarded as the hybrid intersection between eHealth and smartphone technology. Using mobile communication devices, such as (smart) mobile phones, tablet computers, and PDAs in the medical applications and health services for communication and information processing, the mHealth field has emerged as a subsegment of *eHealth* which uses of *information and communication technology* (ICT) such as computers, mobile phones, communications satellite, patient monitors for health services and information with or without human-in-the-loop. These systems also need intense telecommunication support toward mHealth implementation including radio technologies, communication and cloud networks, and secure health-related applications and systems. Augmenting the percentage of health-related smartphone apps is in trend currently; however, there are some concerns about the accuracy, privacy, and unregulated status of health apps, but it is expected that such concerns would diminish in the near future.[19]

6.6 Systemic Aspects of TBI

Diagnostic plus therapeutic avenues can be conceived as viable suites of *eClinical solutions* in TBI contexts. An overview of underlying steps is summarized in Table 6.1.

6.7 TBI and Smart-Health Perspectives

As stated earlier, TBI is a rational way of bringing the base-pair concepts to bedside applications. Relevant scope can be seen further expanded in making of *smart healthcare systems* (or *s-health*).[20] In essence, realization of comfort, preference satisfaction, personalized services, and uncompromised wellbeing (tightly related to health with its continued care across all age groups) forms the concept of "smart healthcare" emerging as the pivotal norm in making of smart cities having such healthcare facility.

Focusing on *s-health*, Solanas *et al.*[20] introduced it as a novel concept complementarily to *mobile health* (m-health). The *m-health* is intended to improve healthcare services in terms of allowing access to services and knowledge, providing user-oriented features, and being personalized, and the s-health can provide context-specific services leveraging on the network and sensing infrastructure of smart cities. For example, Patsakis *et al.*[21] demonstrate how smart city sensors can be used to provide personalized medical services. Issues related to data interoperability in such smart healthcare information systems can be tackled as indicated in Liu *et al.*[22]

Smart healthcare thus implies pivotal system integration and interoperation in healthcare management. While modern technologies enable new health products, the smartness facilitates a change in how the healthcare is managed supporting a paradigm shift from a focus on cure towards a broader view of wellness management and healthy living. The international initiative on s-Health, therefore, includes the combination of m-Health and smart cities *via* ICT perspective *vis-à-vis* individuals and society.

6.8 TBI-Specific Patient-Care Systems

In the contexts of TBI, the details in the following subsection can be identified as feasible patient-care efforts based on global modeling of patient-care systems in terms of underlying informatic details in biology extended to medical/health sciences.

Table 6.1 Systemic management of translational bioinformatics efforts.

Systemic translational bioinformatics efforts	Descriptions
Studies on basic science	Theoretical models, analytical evaluations, and computational exercises
Remarks: Contexts: m-Health, eHealth, etc.	
Translational tasks	Investigations leading to data compilation, analyses, observational evaluations, and specific tasks, like biomarker validation
Remarks: Mostly, human-in-loop assisted by instrumented measurements	
Systems biology	Data compilation and management: locating and accessing databases with stored archival of raw instrument files and relevant content management
Remarks: Data pertinent to: Theoretical and wet-lab studies	
Electronic patient survey	Facilitating web-based proforma/forms to capture details of the participant subjects
Remarks: Exemplars are Electronically formatted patient questionnaires on patient's demographic data, pathogenic status, assertions on possible treatment regimens, and outcomes on diagnostic and prognostic pursuits	
Clinical information management	Collection and integration of clinical annotations archived from various information sources and databases
Remarks: Examples of information resources: HL7 EMR, Cancer Registries, Clinical Data Management Systems (CDMS), Clinical Data Warehouses (CDW), etc.	
Research avenues	Survey on available and potentially futuristic research avenues that lead to collaborative solutions across individual investigators and research teams so as to share project information, results data, and foreseeable insights
Remarks: Implementation of networking and software solutions for the collaboration	
Conceiving, biorepository	Management of biospecimens collected from study subjects, operating rooms, forensic ambient, etc.
Remarks: Strict confidentiality on details vis-à-vis patient privacy preservation	
Wet-lab/Bio-laboratory information systems	Management of laboratories of clinical, analytical, and life sciences core technology
Remarks: Laboratory studies relevant to Genomics, proteomics, basal metabolic panel (BMP) comprehensive metabolomics, molecular imaging, oncological precancer screening peptide synthesis, flow cytometry, blood chemistry test, etc.	

Example Ex-6.1

Problem statement: this exercise is to illustrate a conceivable smart, home-based diabetic patient-care management in line with the heuristics of TBI. This explorative study is indicated as a homecare facility in the context of adult diabetics management. (With no loss of generality, the associated heuristics can be pursued in the contexts of other diseases and related homecare management.)

Background details: the conceivable homecare facility in the context of adult diabetic (*diabetes mellitus* Type I and II: T1DM/T2DM) or in similar patient management can be conceived as a digital infrastructure based on TBI. It refers to a smart system, which can accommodate computer and communication units plus necessary sensors/monitors framing an overall *home-area network* (HAN) infrastructure. Relevant smart facility can gather a *cache* of data useful for self-management needs of patients for their robust wellbeing. In making of such a facility would, however, warrant developing models, algorithms, and computational methods for risk-related decisions on diagnostic/therapeutic regimens of the patients as well as developing necessary infrastructure hardware.

Objectively, in the context of adult diabetics management, this example (model) is conceived in TBI perspectives and its infrastructure includes relevant *EHR* and/or *EMR* systems designed for a systematized collection of patient details (*via* HAN) and electronically storing such health information in a digital format; further, these records can be availed and shared across different healthcare settings using the associated HAN facility on *ad hoc* basis.

The aforesaid smart healthcare functions can be achieved in the homecare ambient illustrated in Fig. 6.1. It obviates the need for expensive proprietary systems, and in relevant hardware implementations, the said facility includes HAN plus other patient-centric sensors/monitors, which can be deployed with ease of portability and implementation. Prototyping such a practical, operational unit is a technological viability achievable through TBI modeling. The smart home-based diabetic patient-care facility for adults shown in Fig. 6.1 has digital data organization with EMR-like functional units conceived *via* personal computer (PC) being the database (DB) core as illustrated in Fig. 6.2.

The subsystems marked as: Task (A), Task (B), and Target-base (C) in Fig. 6.1 are concerned with the tasks described below.

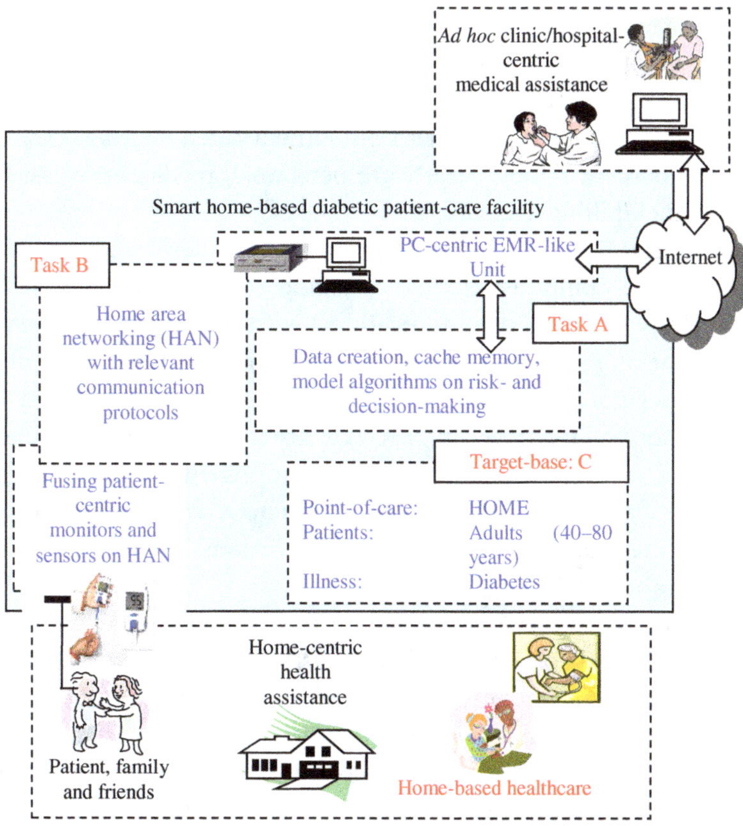

Fig. 6.1 Concept of translational bioinformatics-based diabetic patient-care facility for adults: home-centric healthcare ambient.

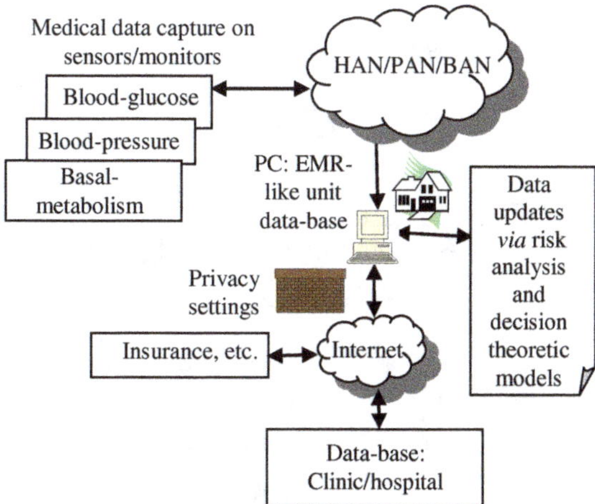

Fig. 6.2 Organization of EMR-like functional units conceived with personal computer as the database core at home.

Task (A): This refers to developing an EMR-like, personal computer (PC)-centric medical record system (similar to conventional EMR) framed as an appropriate *database* (DB) of digital infrastructure provisioned with secured privacy (with an authenticated sharing feasibility across the patient, family, healthcare personnel, medical professionals and insurance/billing agencies, etc.).

As illustrated in Fig. 6.2, the PC represents the core database (DB) at home and the computer is further connected to other peripherals, namely, HAN (for local communications) with the Internet (for remote communication). All communications are protected for privacy of the data. Furthermore, the HAN is augmented with personal-area and body-area networking (PAN and BAN) to accommodate patient-specific sensors and monitors as needed. This comprehensive infrastructure will allow a seamless information flow across communicating and storage entities specified in *Task-A.2* with various domains illustrated in Figs. 6.2 and 6.3.

Furthermore, *Task-A.3* is the subtask enabling data export and database-to-application program interactions. As shown in Fig. 6.2 a streamlined system configuration can be facilitated as a part of the EMR-like PC-based DB system being conceived. It can systematically configure the DB at home and similar such units remotely placed (as in a clinic) along with the other peripherals like patient-monitoring units, etc. All such data exports can be exercised at the physical layer level *via* wireline and/or wireless media with appropriate protocol and *media access control*

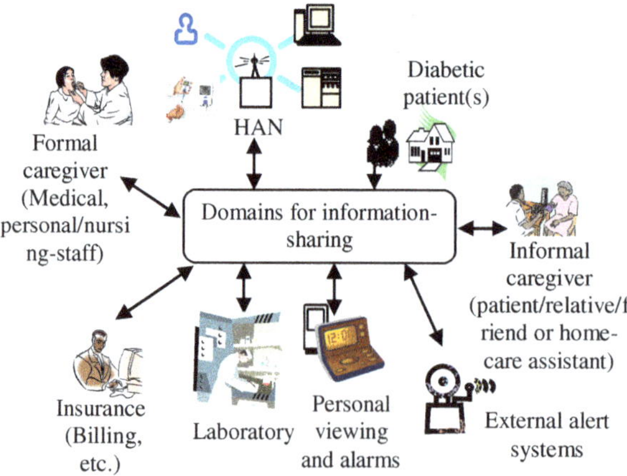

Fig. 6.3 Information sharing across facility constituents in the organization of EMR-like functional units.

(MAC); furthermore, both real-time (rt) data entry as well as nonreal time (nrt) data transports can be supported.[23,24] The associated information sharing across facility constituents is illustrated in Fig. 6.3.

Task (B): This task is to improvise the HAN interfaced with the digital infrastructure. The HAN is designed to contain a PAN and Internet connectivity with associated protocols framed across a fused set of electronic sensors/monitors in the network.

Task-B1: This task involves developing an exclusive HAN on wireline/wireless media so as to enable communication of patient-centric sensors with the entire facility and protocol as illustrated in Fig. 6.4.

Facilitating the HAN infrastructure could be a major task of interest and it can be evolved as an infrastructure of PAN subsets. Comprehensively, the HAN can be interfaced with the Internet for external communication. Connectivity of healthcare and well-being management, tele-monitoring, and developing wireless sensor network have been research interests.[25–27] However, considering simplicity and affordability as well as matching the needs of the

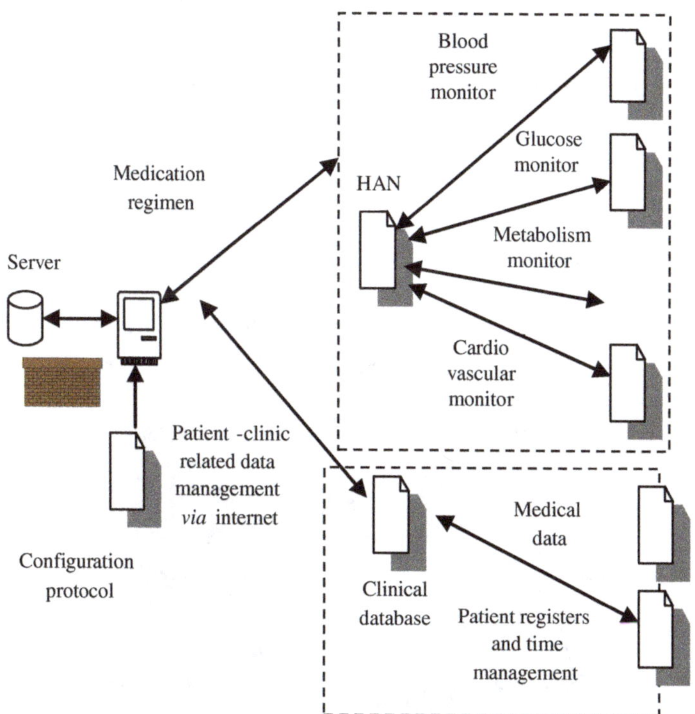

Fig. 6.4 Data exports/sharing in the patient-centric HAN.

exclusive objectives of T1DM/T2DM patient-management at home, a compatible system in TBI perspectives is rather imperative and imminent. Apart from monitoring sensor data, keeping track of diabetic patient's lifestyle (of diet and exercise) and other social habits (such as smoking and drinking) are health-specific prescriptions. Combining the cached data on patient-related details and information on tracked lifestyle statistics will be modeled toward timely risk-evaluation and decisions in the proposed endeavors. To simulate home-based patient-monitoring, a distributed set of electronic sensors/monitors are to be used and interfaced *via* wired and/or wireless media of the HAN (and PAN).

Modern communication strategies allow the use of advanced mobile telephony and related IT technologies[28] with touch-screen/iPAD-ready capability.

Task-B2: *Communication protocol development for the HAN*: The communication links and protocols can be established as per the standards of wireline and wireless local area networking as shown in Fig. 6.5. Conceiving HAN and the associated PAN implementation are consistent with IEEE 802.11, Bluetooth™, and ZigBee™. Accordingly, the electronic sensors and monitors can be interactively fused on to the network. Typically, the data rate required in the PAN contexts would correspond

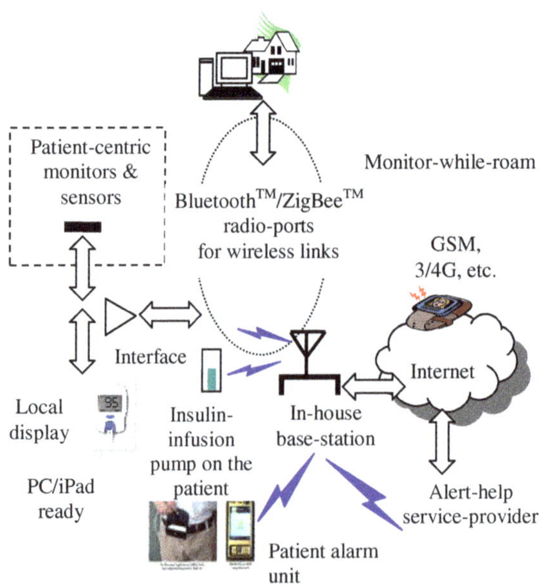

Fig. 6.5 HAN in the facility: IEEE 802.11 (WLAN), Bluetooth™, and ZigBee™ standards.

to low-rates inasmuch as, the patient-centric data are slowly varying function of time.

Therefore, most of the sensor-network fusion can be established *via* ZigBee^{TM} with its (IEEE 802.15.4) low-rate WPAN standards. As necessary, the Bluetooth^{TM} can also be adopted for necessary communications. Security and reliability should be kept in focus in forging HAN, and relevant specifications in building such communication networks (wireline/wireless) should aptly match the endeavors involved. Furthermore, interoperation and coexistence of multiple sensors in wireless ambient may cause mutual RF interference issues, which should be averted.

Task-B3: *Patient-centric sensor adoption: invasive and noninvasive types:* Considering patient-centric sensors and monitors in question in the case of diabetic management, essentially there are two vital items: (i) blood-glucose monitors and (ii) blood-pressure monitors. Commercially, both items are available at an affordable price. However, in the context of using them in the HAN, there are two issues: (a) the invasive versions (like finger-prick blood glucose monitors or traditional arm-cuff blood-pressure sphygmomanometer), have to be properly interfaced with the wireless system using Bluetooth^{TM} or ZigBee^{TM}. (b) Considering noninvasive versions, it will be more convenient to fuse them in the HAN system.

Task (C): Analytical and computational models on diabetic health conditions: In developing a system as described above, not only involves putting together the computer and communication entities with necessary protocols, it is necessary to develop certain analytical models, computer-specific algorithms, and software on *ad hoc* basis for incorporation in the EMR-like DB as outlined below.

Task-C1: *Modeling health conditions that are co-morbid with diabetes and quality-of-life*: this course of modeling involves the following efforts: (a) modeling quality-of-life of diabetics with depression; (b) modeling diabetics-related metabolic state: it refers to the *basal metabolic rate* (BMR) or energy liberated per unit of time by the patient; (c) modeling cardiovascular conditions in diabetic patients: adults with diabetes are two to four times more likely to have heart disease or a stroke than adults without diabetes. Related intensive glycemic control in evaluating the risk factor is also feasible.[28] (d) modeling pre-diabetic conditions on adult diabetics: information and modeling of metabolic disturbances associated with pre-diabetes is vital toward risk evaluation of cardiovascular diseases-cum-diabetes in adults. (e) modeling obesity conditions on adult diabetics: as a focused effort, obesity *versus* diabetic risks will be modeled. Compatible literature data as in Elhayany *et al.*[29] and Rodríguez-Villar[30] can be used in such studies. (f) glucose metabolism model: the modeling and simulation of glucose regulatory metabolism in human beings, for example, is reported in Neelakanta *et al.*,[31] Cobelli and Mari,[32] Cobelli and Mari,[33] Dupont,[34] and Sorensen,[35] where the underlying complexity of the processes involved are crucially addressed; furthermore, simulated results based on theoretical formulations validated against clinical data are available in the above references. (g) diabetics-related database for pathology-specific modeling: an exhaustive database on diabetics can be formed using National Diabetes Statistics of National Diabetes Information Clearinghouse (NDIC).

Task-C2: this is concerned with developing risk-evaluation and decision-theoretic models in pathogenic conditions toward patient-care. Relevant details are indicated as follows:

Typically, any prescriptions of risk-levels on pathogenic conditions should be specified with an error-bar having upper- and lower-bounds (UB and LB). For example, in the studies on glucose regulatory modeling described in Dupont,[34] the relative risk-factor (Rrf) due to T2DM *versus* patient's age is indicated with bounding-limits LBF-UB and LBF-LB as shown in Fig. 6.6, where LBF explicitly denotes the so-called *Langevin–Bernoulli function** (LBF)[36,39] (briefly outlined later).

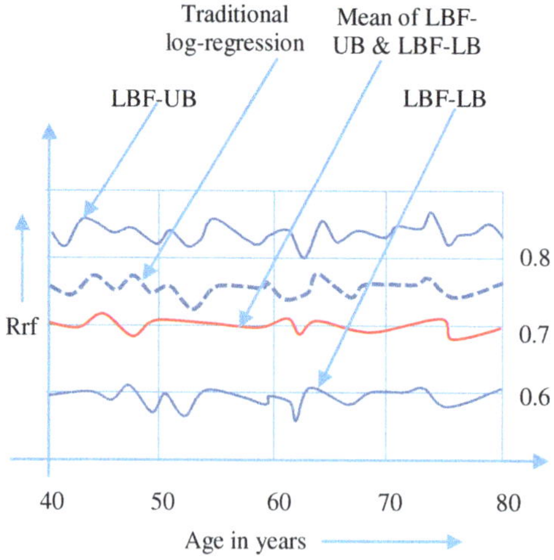

Fig. 6.6 Computed relative risk factor (Rrf) due to T2DM *versus* patient's age (in years) with bounding limits. LBF-UB and LBF-LB denote upper- and lower-bounds evaluated for the model described in [II.5.34]. (Explicitly, LBF denotes the so-called Langevin–Bernoulli function* outlined below)

In adult patient-care management, related clinical situations could conform to extremities such as terminal illness, state-of-health requiring an inevitably prolonged hospital/rehab-center-based care-taking (as in, Alzheimer's or Parkinson diseases) or simple outpatient care (as in chronic states of T1DM/T2DM). In such cases, appropriate TBI-based modeling should be conceived. For example, the chronic states of T1DM/T2DM are considered. The associated seriousness of the disease is evident from the statistics available in World health organization publication,[40] and American diabetes association diabetes statistics[41] implying a timely diagnostic and therapeutic regimens plus patients *via* counseling as needed; relevantly, an overall homecare patient management as illustrated in Fig. 6.1 would be suffice with the inclusion of smart communication links and networking improvised in the systems.[42–44] The logistics and rationale in conceiving such a homecare facility are as follows:

Typically, proprietary EMR units are improvised in clinical/hospital or rehab-center settings with a comprehensive electronic database (DB).[42,43] However, they are not affordable home-based care facility. As such, in situations like adult diabetic patient-care management, a conceivable translational bioinformatic project should be targeted toward enabling a "self-management style"[44] at home as illustrated in Fig. 6.1. Relevant cross-breeds of possible directives are as follows:

— Framing an in-house (home-based) digital health information infrastructure.
— Elucidating model-based algorithms for vis-à-vis the patient-specific diagnostics and prognostics (e.g., as required in blood glucose control in Type-I and/or -II diabetic conditions).[45]
— With available patient-centric medical data and features extracted and specified in terms of associated model parameters, decision-theoretic models can be evolved as regard to diagnosis and therapeutic needs. Pertinent classification algorithms (for risk evaluation and reaching proper decisions for adequate patient care) can be based on *linear discriminant analysis* (LDA), *hidden Markov model* (HMM), and ANN. Furthermore, such simulation exercises may involve *fuzzy data* considerations and analyses, as in glucose metabolism modeling described in Hanss and Nehls,[46] Hanss and Nehls,[47] and Mahfouf *et al.*[48]

*A Note on: Langevin–Bernoulli Function**

In the context of modeling the states of random entities in a stochastic system, pertinent nonlinear differential equation is indicated by the author Neelakanta[36] (and by, Neelakanta and De Groff[37]). It conforms to the family of Bernoulli–Riccati differential (stochastic) equations and relevant solution is specified by LBF. With reference to an arbitrary variable ξ, the LBF $L_q(\xi)$ is explicitly given by:

$$L_q(\xi) = \left(1+\frac{1}{q}\right) \times \coth\left[\left(1+\frac{1}{q}\right) \times \xi\right] - \left(\frac{1}{q}\right) \times \coth\left[\frac{1}{q} \times \xi\right] \quad \text{Eq. (6.1)}$$

with q depicting an *order-parameter* of the underlying features of a set of random variables or entities $\{\xi_i\}$ constituting a stochastic space. The LBF of Eq. (6.1) duly describes the random interactions of entities that decide the functional attributes of the variable, ξ. Furthermore, with $q \to 1/2$, the function $L_q(\xi) \to \tanh(\xi)$ depicts the resulting UB of the associated, totally disordered state/dispositions of the interactions and with $q \to \infty$, the function $L_q(\xi) \to \{\cot(\xi) - 1/(\xi)\}$, which refers to the LB of the associated stochasticity pertinent to the other extremity decided by the totally ordered state of the underlying entities. Thus, the range (1/2 to ∞) depicts the stretch from a totally anisotropic, disordered state to a totally isotropic ordered state with $q = 1/2$ depicting the UB and $q = \infty$ leads to the LB (in conformance to the theory of interaction in stochastic processes.[36–39] (Additional details and applications of LBF in the contexts of specifying bounds in risk factor specifications and in logistic regressions will be presented in Chapter 5.)

6.9 Review Section: Tutorials, Examples, and Exercises

This section is written to present some exemplar topics on TBI with necessary tutorials together with rider exercises indicated as problems for self-study learning and for developing research pursuits.

Tutorial T-6.1

This section outlines analytical details toward developing a clinically viable translational bioinformatic model of glucose regulatory metabolism in humans[31–35] based on underlying complex system heuristics:

The regulatory system in biological systems (such as glucose regulation unit in humans) can be assumed as an embodiment of a complex system made of interacting heterogeneous constituents having both statistical and deterministic functional attributes. Pertinent functional dynamics of such constituents can be specified in the spatiotemporal domain $\Omega(x, y, z; t)$, in terms of a *metric* of *complexity* (C) defined as[49]: $C(t) = \varphi\{n(t), N(x, y, z, t)\}$, where (x, y, z) are spatial coordinates, t is the time, and $n(t)$ denotes the time-dependent countable number of (large) constituent elements of the complex system, in question. Furthermore, N depicts the variety associated with the constituents and φ represents a function that has to be modeled, so that $C(t)$ denotes the extent (or measure) of complexity involved (consistent with empirical details of a "good-fit" on experimental observations). In general, the associated spatiotemporal dynamics can be regarded as a deterministic and/or stochastic process.

In a biological regulatory process (like glucose metabolism*), the complexity (C or φ) may depict a disorderly state (at some reference time, say t_o), and the regulatory process (enabled *via* feedback) strives to achieve the desired regulation (i.e., the associated dynamics of cybernetics seeks to reach an objective function (specifying the warranted goal) toward realizing an orderliness at a later instant of time, $t > t_o$, and possibly at a desired targeted location (x_R, y_R, z_R).[31–33] The process settling in a state of orderliness at $(x_R, y_R, z_R; t > t_o)$ implies a reduction in the system entropy (Chapter 2). (*Metabolism** signifies any (biochemical) reaction that takes place within the living system).

Glucose metabolism being considered here specifically refers to the metabolic reactions associated with glucose. The metabolized glucose is either stored or excreted. These locations of glucose metabolism are *Insulin-sensitive cells* (ISC), where glucose is metabolized and *noninsulin-sensitive cells* (NISC), which are other sites where glucose is metabolized and kidneys, where glucose is lost due to excretion *via* urine, as described in Neelakanta *et al.*[31] and Cobelli and Mari.[32] Maintenance of blood glucose level *via* antagonistic functional roles of insulin and glucagon hormones involves the architecture of the model depicting pertinent mass and control signal flows in the metabolic plant. Explicit variables (in the metabolic plant) are with reference to insulin, glucose, and glucagon subsystems that constitute the glucose regulatory process in human beings including related food intake considerations.[50]

A classical version of glucose regulation model is due to Cobelli and Mari[32,33] and others,[34,35] which consists of a metabolic plant made of glucose controller plus the pair of antagonistic hormonal controllers, namely, insulin and glucagon. Alternatively, a complex system model of glucose regulatory metabolism is described in Neelakanta *et al.*[31] Considering the glucose subsystem, it is a one-compartment model governing the associated flows of extracellular fluids. It illustrates the underlying maintenance process of blood glucose level. The whole gamut of associated regulatory metabolism, in essence, is decided by mass-flow considerations involving glucose, insulin, and glucagon at appropriate sites in the human body. Relevant dynamics of glucose regulation due to Cobelli and Mari[32,33] consists of a metabolic plant made of glucose controller plus the pair of antagonistic hormonal controllers, namely, insulin and glucagon.

The model in Neelakanta *et al.*[31] formulates an alternative complex system depiction of metabolic activity of physiological interest pertinent to glucose regulatory process (which is assumed to have deterministic and/or stochastic attributes). This model is aimed at predicting and inferring the temporal excursions of the variables of interest (in terms of mostly dormant subsets of related variables in the pertinent physioanatomical system). Such temporal excursions of blood-glucose concentration and the rate of appearance of glucose (RAG) in blood-plasma as regulated by the hormonal participants, namely insulin and glucagon bears statistical (upper and lower) bounds. The simulated results of the model in Neelakanta *et al.*[31] are validated and compared against available clinical data in Cobelli and Mari,[32] Cobelli and Mari,[33] and Sorensen.[35]

The modeling approach pursued[31] also facilitates modifying/tuning model components (within the statistical bounds established), and such modifications are useful, for example, to refine predictive control algorithms adopted in closed-loop glucose control regimens (administered *via* insulin infusion pumping) of maintaining normoglycemia in (Type I) diabetic patients.[51–53]

In the contexts of translational and clinical bioinformatics, modeling the required physiological activities (such as glucose metabolism) involves recognizing the underlying (model) components (depicting the associated physioanatomical parts)[54] and understanding the functional details (decided by the dynamics of the processes involved[55]). Hence, due attention is needed in archiving exhaustive details thereof and use them comprehensively in the model projected. The following illustrative examples are designed to outline relevant considerations on modeling.

Example Ex-6.2

Statement of the Exercise

Pursuant to the details in Tutorial T-6.1, this example is furnished to describe the two specific models of Neelakanta *et al.*,[31] Cobelli and Mari,[32] and Cobelli and Mari[33] indicated above with regard to the metabolic activity glucose regulatory process shown in Fig. 6.7. Considering the model in question, the glucose regulatory process involves mass

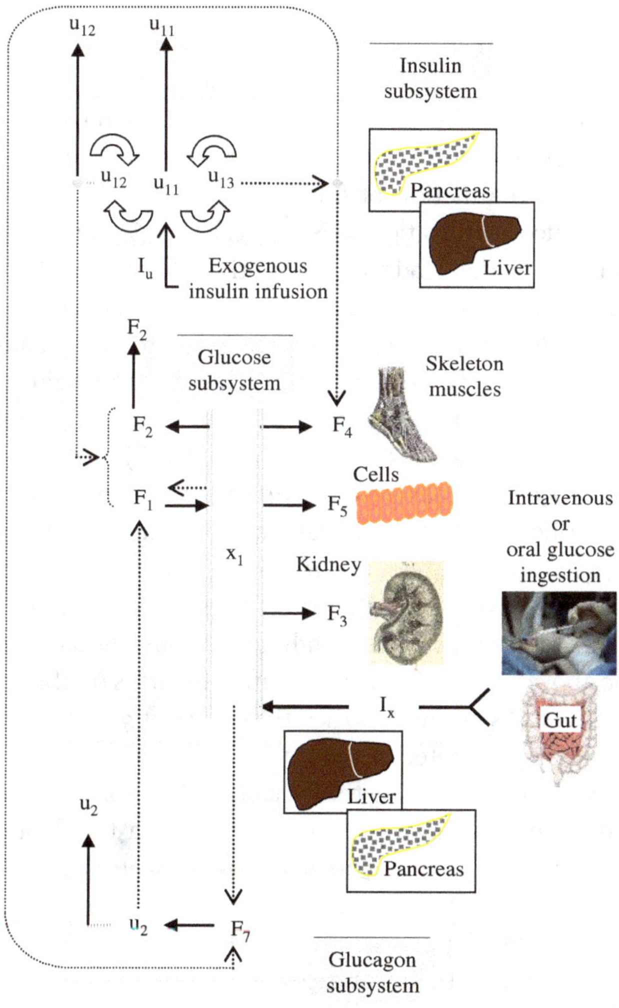

Fig. 6.7 Illustration of explicit variables involved in insulin, glucose, and glucagon subsystems of glucose regulatory process in human beings.[32,40]

and control signal flows in the metabolic plant, and there are explicit variables in the metabolic plant pertinent to insulin, glucose, and glucagon subsystems in human beings.[31,32]

The associated entities in the modeling are as follows: (x_1: Quantity of glucose in the plasma and extracellular fluids in mg); (u_{1p}: Quantity of pancreatic stored insulin in μU); (u_{2p}: Quantity of pancreatic, promptly releasable insulin in μU); (u_{11}: Quantity of insulin in plasma in μU); (u_{12}: Quantity of insulin in the liver in μU); (u_{13}: Quantity of insulin in the interstitial fluids in μU); and (u_2: Quantity of glucagon in the plasma and interstitial fluids in pg).

In addition, the following valid functional details as regard to the model can be specified: (F_1: Amount of liver glucose production ⇒ Nonlinear function of $\{x_1, u_{12}, u_2\}$); (F_2: Amount of liver glucose uptake ⇒ Nonlinear function of $\{x_1, u_{12}\}$); (F_3: Amount of renal excretion ⇒ Nonlinear function of x_1); F_4: Peripheral insulin dependent glucose utilized ⇒ Nonlinear function of $\{x_1, u_{13}\}$); and (F_5 = Peripheral insulin independent glucose utilization ⇒ Nonlinear function of (x_1). The mass-flow considerations that lead to deciding the rate of flow of x_1 (i.e., dx_1/dt pertinent to the quantity of glucose in the plasma and extracellular fluids) are detailed in the classical approach due to Cobelli and Mari.[32] It is conceived in a deterministic (but, interactive) suite. However, glucose metabolism dynamics can be specified more comprehensively in a complex system domain (as in Neelakanta *et al.*[31]) by duly taking into account of stochastic plus interactive considerations. The approach thereof prescribes a set of nonlinear (logistic) functions for the set of variables, $\{x_1, u_{1p}, u_{2p}, u_{11}, u_{12}, u_{13}, u_2\}$. Such functions are expressed in Neelakanta *et al.*[31] as general solutions of a Bernoulli–Riccati differential equations,[36–39] and in the corresponding solutions, the associated functions of the abovesaid variables are specified in terms of LBF, $L_{qS}(.)$ (and/or its derivatives). As outlined earlier, the subscript q_S in the LBF depicts the spatial-order parameter of the underlying stochastic process (with the subscript, S (in q_S) depicting the variable G, H, F_1, or F_4). Furthermore, q_S is specified in the range (1/2 to ∞) depicting a totally anisotropic, disordered state to a totally isotropic ordered state, respectively.[36–39] In the studies on glucose regulatory process (modeled in Neelakanta *et al.*,[31] Cobelli and Mari,[32] Cobelli and Mari,[33] Dupont,[34] Sorensen[35]), it is observed that the nonlinear aspect of the rate of change of associated regulatory variables is significant for higher levels of enzymatic doses present in the system during the early regimes of temporal discourse. However, at a later stage (i.e., towards terminal dynamics of the process), such variations tend to cease depicting more or less an invariant state of flow of the participant fluxes, (namely, glucose, glucagon, and insulin in glucose metabolism).

Example Ex-6.3

Problem Statement

This exercise exemplifies the details gathered from computations performed on model validation of analytical considerations involved in the dynamics of glucose and insulin flows in glucose metabolism as in the previous example (Example Ex-6.2) in terms of clinically ascertained results.

With reference to assessing the glucose metabolism in humans, relevant clinical studies involve collecting data on glucose and insulin levels present in the blood of a nondiabetic adult subject (typically of about a normal weight of 70 kg). Example details are available in Cobelli and Mari,[33] Dupont,[34] and Sorensen.[35]

Relevant clinical data is obtained (from a nondiabetic adult subject) by a test known as *oral glucose tolerance test* (o.g.t.t.). It refers to the following: after an overnight fast, the adult subject is made to drink a solution containing a known amount (100*g*) of glucose. Blood sample is obtained before the subject drinks the glucose solution and tested for blood glucose level. Subsequently, blood is drawn from the subject every 30–60 minutes (for up to 3 hours) and glucose level in each sample is assayed and recorded; hence, the *RAG* $|(dx_1/dt)|$ is plotted as in Fig. 6.8.

It is shown as a dotted-line graph. Furthermore, simulated results based on complexity model[31] are obtained and presented as a cluster of data obtained *via* an ensemble set of values of q_s, randomly chosen (with uniform distribution) between ½ and a large value (>1000). As indicated earlier, the range $(1/2<q_s<\infty)$ depicts UB and LB of the estimations framing the error-bars in Fig. 6.8. The test data and results as above,[32] enable ascertaining details on the following: (i)Time-course of glucose and insulin plasma concentration for a long oral glucose load; (ii) shape and integral data on the glucose rate of appearance in plasma; and (iii) disposal of the glucose load among various tissues (i.e., in liver and in peripheral tissues).

The analytical model described in Neelakanta *et al.*[31] refers to a stochastic differential equation on the aforesaid temporal dynamics of glucose (and insulin) in o.g.t.t. It is solved numerically considering a

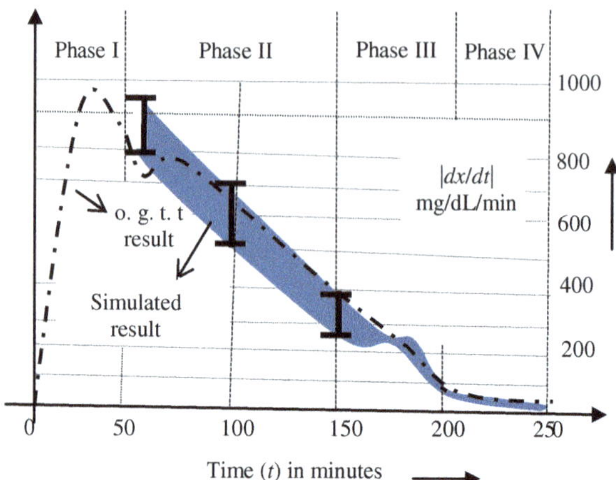

Fig. 6.8 Clinical results of o.g.t.t performed are shown in dotted line.[32] It refers to *rate of appearance of glucose* $|(dx_1/dt)|$ (RAG) *versus* the passage of time in the (100 *g*) in an o.g.t.t regimen on a normal (70 *kg*) adult. Simulated results (based on complexity model[31]) are presented as a cluster of data. The cluster of details shown are obtained *via* ensemble set of values of q_S, randomly chosen (with uniform distribution) between ½ and a representative, large value (≈1000). As indicated in the text, the range ($1/2 < q_S < \infty$) leads to LBF-based estimations across upper- and lower-bounds (UB and LB) framing the error bar as shown.

differential span of time, Δt_i over the different phases of the time-interval (between initial and terminal events) and the corresponding differential levels, Δx_i (of glucose or insulin) are noted. An ensemble of computations is performed with the value of q_s picked randomly from a uniformly distributed set of numbers starting from 1/2 to a large value, theoretically tending to ∞, and the ensemble of such computational runs of simulations yield unbiased stochastical variations in the time course of the functions involved. Typically, such random variations could justifiably prevail in the interactive domain of the complex metabolism under consideration with the range of disorder specified within the selected bounds of $q_s = (\frac{1}{2}$ to $\infty)$. The computed details as above results as available in Cobelli and Mari[32] are presented in Fig. 6.8 with error-bars depicting the associated data spread. Relevant results illustrate model validation details within the estimated LB and UB of the computed outcomes in Neelakanta *et al.*[31]

A typical use of the model narrated above (as in Neelakanta *et al.*[31]) is, for example, in refining adaptively, the so-called *insulin infusion pump algorithms*. In a technique known as *insulin pump therapy*,[51–53] such insulin infusion pump algorithms can be adopted toward a controlled-delivery of insulin by means of a compatible, software-activated release pump.

6.10 TBI Versus Pathogenic Considerations

Another art of applied bioinformatics with an extended scope refers to a translational framework that fits into the clinical contexts of certain common pathological states, for example, allergy-related conditions[56] in human beings. The associated translational informatic considerations enable building a knowledge base pertinent to clinical aspects of allergenic reactions (that mostly result from "maladaptive immune response in predisposed subjects to otherwise harmless molecules"[57]). Relevant allergy and allergic reactions form morbidity of global concern[58–67] and need remedial counter measures, which are designed *via* translational/clinical informatics. Classification of allergies as *IgE*-mediated* and *non-IgE mediated* versions details on *antigenic determinants* (epitopes**) and related process of *sensitization* are topics of interest in allegy bioinformatics.

(The *immunoglobulin E* (IgE*) are antibodies produced by the immune system, and an allergy refers to the immune system overreacting to an allergen by producing the antibodies, IgE. These antibodies can proliferate across the cells that release chemicals, causing an allergic reaction. An *epitope** is a part of an antigen that is recognized by the immune system, specifically by antibodies, B cells, or T cells).

Adequate listing of documented allergens exist in the IUIS allergen nomenclature database[64] and in allergome database.[65] Many IgE-binding epitopes have been identified as *sequential epitopes,* although for many, this does not represent the full epitope. Sequence comparisons can be exercised to a database of known IgE-binding epitopes. The class of allergy bioinformatic came into being with an established procedure (e.g., *via* FASTA or BLAST) and formalizing a threshold greater than 35% identity in 80 or greater amino acids so as to identify potential allergenic cross-reactivity of transgene encoded proteins in genetically enhanced crops.[66]

Aalberse[58] pointed out that proteins sharing less than 50% sequence identity are rarely cross-reactive. In contrast, proteins that share at least 70% identity often show cross-reactivity. Sequence comparisons can be exercised *via* databases of known IgE-binding epitopes.[67]

6.11 Review Section (Part I: TBI-Related Exercises)

Presented in this section is a set of exercises on topics related to TBI as discussed in Part I of this chapter. The notions of TBI also portray a gist of feasible studies that can be extended into thesis/dissertation pursuits in clinical informatics.

Problem P-6.1

Problem Statement

Considered here are certain heuristics on TBI for diagnostic and prognostic prediction of cancer. In the past few decades, high-throughput microarray profiling had been considered to track complex molecular aberrations in carcinogenesis. For example, performed in Chen *et al.*[68] is a comprehensive search in the *GEO* for the array-based profiles in human *Prostate cancer* (PCa). The retrieved GEO series seems to fall into five categories: Gene expression profiling, noncoding RNA profiling, genome binding/occupancy profiling, genome methylation profiling, and genome variation profiling. As an example, the discovery of PCa biomarkers has been boosted by the advent of NGS technologies. More details on GEO data types and technologies can be found in Barrett.[69]

Evaluating the cancer biomarkers by high-throughput sequencing technologies is a viably projected TBI (*via* the potential use of cloud computing in NGS data processing) is the exercise in hand. There are NGS-specific strategies in cancer research with a focus on deciding the biomarker candidates useful in the diagnosis and prognosis of cancer. Relevantly, the feasible futuristic perspectives of TBI and cloud computation refer to improving cancer management.

Following the concepts and approach indicated in Gómez-López and Valencia,[70] the proposed exercise is to develop a similar heuristic study as regard another cancer type (other than PCa) cancer treatment (such as liver cancer diagnosis, prognosis and treatment of cholangiocarcinoma).

Problem P-6.2

Problem Statement

TBI approach can be applied in the contexts of clinical bioinformatics tailored to address *personalized medicine*. In the clinical framework of modern medicine, "personalized medicine" starts even before a disease is manifested in an individual, many times at a point when the disease or condition is preventable." In general, data that could be gathered in personalized clinical contexts are useful in developing individualized (disease) preventive models and therapeutic regimens, for example, of negative breast cancer depicting "a particular form of cancer for which there are currently not many targeted therapeutics."[71] Personalized patient rehab *via* stem-cell therapy* denote state-of-the art considerations posing a broad futuristic scope in personalized medicine of clinical informatics. (*Stem-cell therapy** refers to using stem cells to treat or prevent a disease or symptomic conditions. Bone marrow transplant method is widely used in stem-cell therapy, but other versions based on umbilical cord blood are also in vogue).

Problem P-6.3

Problem Statement

In the case of allergic bioinformatics described earlier, assessment of personalized allergic pathogeny can be attempted *via* TBI approach in elucidating allergic symptoms, eliciting exposure to a structurally diverse group of proteins (depicting, the allergens). A pertinent example is the art of understanding allergenicity at the molecular level (with a wide application in food safety and in treating allergic diseases).

A related topic of interest in TBI platform is to ascertain what makes a protein allergic. Advanced computational methods based on relevant heuristics could lead to tools of bioinformatics that can be applied to address the above problem. It has been found that allergens are usually foreign proteins with few/no bacterial homolog. They are clustered into few protein families (associated with a limited number of protein domains) opposing the idea that any protein can be allergenic.

Problem P-6.4

Problem Statement

This is concerned with an exclusive application of TBI in forensic data "translation" into informatic details for robust analyses in medical jurisprudence.

With reference to details gathered in forensic scenes, a new class of *forensic bioinformatics* can be conceived. It involves reviewing cases pertinent to forensic DNA testing and automated analysis systemic data.

Relevant TBI research allows better understanding and defining the issues on forensic DNA testing and fast interpretations of gathered, valuable information about forensic determinants in the jurisprudence efforts of: (i) crime-related scenarios and (ii) person identification tasks. "Forensic bioinformatician aims (in essence) to solve mysteries of biomarker studies."[72–74]

Typically, *short tandem repeats* (STR), single nucleotide polymorphisms, and whole mitochondrial analyses are three classes of markers often considered in *forensic DNA typing. Massively parallel sequencing* (MSP*) platforms could be useful in such forensic science so as to reveal new information (such as details on the complexity and variability of the markers that were previously unknown), also relevant methods enable handling amounts of data too immense for analyses by manual means. Pertinent sequencing chemistry of bioinformatic methods are viably considered in processing and interpreting new and extensive data associated with forensic DNA analyses.

Forensic applications of bioinformatics, however, need more developments as well as standardization of efficient, favorable tools for each stage of data processing so as to carry out faster, more accurate/robust methods of forensic analysis. Relevantly, developing a set of optimal pipeline of tools is a possible theme of clinical/translational bioinformatic techniques, and supplying such tools to forensic laboratories for sequencer software schedules can make the wet-lab analyses more robust and convenient. Such efforts depict novel research effort pursuable in forensic bioinformatics and making tools required thereof can enable rapidly foreseeable forensic markers sequenced on MPS platforms.

Problem P-6.5

Problem Statement

This exercise refers to TBI realization using fuzzy clinical/health data sets. It is often an open-question in research pursuits in clinical and other health science-related contexts to evolve strategies to handle "fuzzy" data, translate them into clinically viable decision-theoretic details so as to achieve robust analyses and reliable results. Such efforts warrant deducing defuzzified risk parameters in diagnostic suites and finding *imminence-level indicators* in therapeutic regimens.

Concerned models and asserted details can be formulated with necessary bounds *via* logistic considerations plus superposition of fuzzy logic heuristics of Hanss and Nehls,[46] Hanss and Nehls,[47] Mahfouf *et al.*[48] Allen and Sefton,[53] Scalon and Sanders,[54] and Khoo[55], and other logistic functional details (including LBF) outlined earlier may be required on *ad hoc* basis. A pertinent exercise is described in Lee *et al.*[75] in modeling clinical decision-making in diagnostic aspects of pulmonary infection. Similar or more advanced research topics can also be formulated toward the proliferation dynamics of cancer[76] and in metabolic processes (where the associated variables could be largely fuzzy). Furthermore, pertinent concepts of fuzzy control can be adopted in modeling metabolic feedbacks observed in physiological systems and in designing microinfusion pumps. These are plausible topics in TBI conceivable as thesis/dissertation efforts.

6.12 Part II: Clinical Bioinformatics

The theme of this part (Part II) of this chapter is to present a gist of salient details on clinical bioinformatics, which depicts a subset of informatic basics of translational informatics. Considering the incentivized regulatory efforts impressed on hospitals and clinics, the underlying quality of patient care is now a focal topic and perused strategy. As such, translating fundamental aspects of microbiological bioinformatics into *ad hoc* applications on patients' bedside needs warrant a devoted study on explicit concerns in establishing reliable and robust metrics on thresholds in clinical decisions. Elucidating such decisive details *vis-à-vis* patient wellness assessed in the clinical and/or at the bedside sites can be conceived and applied in the tasks pertinent to diagnosis, progression, level or status of the pathogeny, prognosis of a disease, and therapeutic regimens. Furthermore, deciding on emergency criteria, hospital admission protocols, treatment procedures, therapy regimens, post-recovery needs, and preventive measures also warrant a host of decision-theoretic algorithms, and their applications in clinical practice (and in health sciences) require an exclusive art of bioinformatic efforts — dubbed here as clinical informatics.

Clinical bioinformatics thus implies a subset art of TBI and it is conceived to address the global areas of healthcare sciences and specific aspects of medical practice directed at developing requisite methods for acquiring, storing, organizing and analyzing biological data, clinical details, and healthcare information that support the delivery of adequate patient care. The clinical bioinformatics is centered on analytical efforts and computer science including making use of software tools useful in conceiving biological knowledge base (especially with the associated "big data" framework). Thus, the tasks of clinical bioinformatics are concerned with assessing the informatic aspects of biological and medical details accessed in individualized healthcare applications in clinical and/or hospital settings.

Traditional genomic and proteomics strategy tools in bioinformatics focus on similarity searches among genomic sequences and structure prediction toward protein modeling. The associated DNA/RNA level microscopic details of biology, however, need to be translated and expanded for patient-related macroscopic contexts for clinical/hospital applications. That is, in clinical bioinformatics, the genomic and proteomic data are meaningfully useful only when they are integrated with clinical data.

Clinical bioinformatics includes elaborate studies of bioinformatics tools and various facets of proteomics related to drug target identification and clinical validation in terms of pharmacogenomics. In all, clinical bioinformatics could provide benefits toward improving healthcare strategies, disease prevention methods, and establishing robust health maintenance *vis-á-vis* personalized medicine[77,78] and hospital/clinic management efforts.

Trends in Clinical Bioinformatics

The recent pursuits in clinical bioinformatics refer to framing the associated scope as a conceivable study geared to address post-genomic and proteomic technologies in the efforts of clinical research, practices of medicine, and in the preventive tasks of societal health sciences. Such exercises are expected to provide necessary technical infrastructure and expanded knowledge in allowing individualized healthcare with relevant sources of medical information and applications of bioinformatic concepts. Pertinent medical information focuses on the acquisition, storage, and usage of underlying information in exercising positive healthcare in diagnostic and therapeutic methods being conceived in clinical informatics as outlined below.

Microarray technology and clinical bioinfomatics: Microarray technology has facilitated global analyses of gene expressions of thousands of genes in a single assay, and pertinent experiments have produced enormous amounts of data with positive leads toward comprehensive microarray analyses, "big" data-mining and elaborate data-processing approaches using various statistical methods and decision-theoretic methods. Hence, a host of microarray analysis software packages have been developed by commercial enterprises as well as by academic researchers. The associated computational efforts on statistical assays have been significantly assumed in clinical usage and medical practice, and they are useful tools in the expanding art of clinical informatics.

Art of proteomics translation into clinical bioinformatics: methods of omics have expanded to handle large extents of heterogeneous datasets, which have strengthened the knowledge discovery processes in clinical bioinformatics. The associated high-throughput acquisition of proteome data is a modern asset; furthermore, recent proteomics aspects of bioinformatics platforms have emerged as data management systems and knowledge bases in proteomics. Pertinent tools allow exploring the protein complex in detail in clinical bioinformatic contexts.

Wet-lab studies in clinical bioinformatic contexts: The *two-dimensional electrophoresis* (2-DE) gel studies have established itself as the *de facto* approach in isolating proteins from cell and tissue samples, and the role of clinical bioinformatics in such 2-DE gel and in applying the 3D protein structure analyses in patient-related studies are actively pursued topics in modern patient-care strategies enclaving clinical informatics.

Another growing study refers to clinical bioinformatic aspects of using pharmacology details conceived in drug-*target identification* methods, clinical-trial strategies, studies on *biomarkers,* and in realizing appropriate *toxicogenomic* and *pharmacogenomic* tools. Pertinent computational programs facilitating *in silico* pharmacology fuse the details of genomes, transcriptomics, and proteomes with cytogenetic pathophysiology, and they are indicated for viable use in clinical bioinformatics.

Furthermore, in oncologic pursuits of medicine (that deals with prevention, diagnosis, and treatment of cancer), clinical bioinformatics is considered in involving high-throughput genome technology and massive quantities of cancer-related data. Such data are comprised of genome sequences, SNPs and microarray-specific gene expressions. The gene expression details in the contexts of clinical bioinformatics lead to identification of sequence motifs warranted in cancer research, and relevantly, early cancer detection and management, risk identification, risk reduction, and cancer prevention measures frame a comprehensive scope for the underlying clinical bioinformatic studies.

Heuristics of "Omics" in Clinical Informatics

In the context of microbiology, the term "omics" refers to technologies that are primarily aimed at the universal detection of genes (genomics), mRNA (*transcriptomics*), proteins (proteomics), and metabolites (*metabolomics*) present in a specific biological sample, and as indicated in other chapters, bioinformatics in bioengineering sense implies applying computational methods to store, organize, archive, analyze, and visualize microbiological details. The associated technology in practice provides accessible interfaces that allow searching databases and assess the whole genomes. Furthermore, bioinformatic analysis leads to extracting information from databases on genomes and proteomes, which can be used, for example, drug companies who need genome details for drug targets.

Resource Searching in Clinical Informatics

Clinical bioinformatics also involves expanded strategies that enable online database searching on query genomic sequences, related proteins, associated mutations, annotated scientific literature, models of gene regulatory networks, and metabolic pathways. In short, when clinical considerations are viewed in omics sense, the emerging study refers to clinical bioinformatics. The data models evolved thereof (such as GenBank files), the big genomic databases (such as NCBI, EMBL, Nucleotide Sequence Database, and DDBJ), as well as other retrieval systems that provide users to access the sequences, mapping, taxonomy, and structural data are supportive aspects of clinical bioinformatics.

In all, clinical bioinformatics is a devoted branch of bioinformatics pertinent to clinical applications of informatics concerning omics considerations in biosciences, developments of relevant tools for medical and health practices, and expanding the informatic avenues and approaches for clinical research [II.5.78] and translator methodologies. Thus, development of personalized healthcare, medication, and therapies frame the core thematic scope of clinical bioinformatics. Relevantly, analysis of human microarray and other omics data *via* details of human databases aims at establishing a scientific approach to translate bioinformatic concepts to clinical and medical application so as to extend molecular and cellular mechanisms and therapies for human diseases.

Applications of Clinical Bioinformatics: A Review

The conceivable applications of clinical bioinformatics stem from basics of the following microbiological paradigms and related informatic details:

(a) *Knowing the prospects of "omics" technology*: As stated earlier, the term "omics" refers to a new concept implied in molecular research that encompasses the gene-related set {genomics, transcriptomics, proteomics, and metabolomics}.
(b) *Making clinical informatic decisions on metabolic and signaling pathways*: The metabolic pathways refer to any biosequence representation of biochemical reactions, catalyzed by a host of enzymes that occur in all living cells in the contexts of exchanges of energy and chemicals, and signaling pathways describe a group of molecules in a cell that work cohesively to control one or more cell functions, such as cell division or cell demise.
(c) *Discovering and development of biomarkers of interest in clinical bioinformatics*: Biomarker refers to an epoch of interaction between a biological system and a potential hazard, which may be chemical, physical, or biological. For personalized treatment regimens targeted in clinical research, identifying novel biomarkers for diagnosis, patient stratification, and extending personalized treatment options have sprouted grossly, for example, in cancer research.[79]
(d) *Performing computational biology tasks (pertinent to the omics) relevant to clinical bioinformatics*: Classical system biology morphing into its modern traits includes computational aspects of system biology (in terms of *in vitro* and *in silico* omics) toward understanding the behavior of cells, tissues, organs, and whole organisms.[80]
(e) *Exercising high-throughput image analysis in clinical bioinformatic contexts*: Imaging and bio-image informatics, including the whole span of microscopic and biomedical image acquisition methods and applications.
(f) *Evaluating human molecular genetics in clinical bioinformatics*: Human molecular genetics is a subject developed to include a broad spectrum of topics covering molecular basis of human genetic diseases, developmental genetics, cancer genetics, neurogenetics, chromosome and genome structure and functions, gene therapy, mouse and other models of human diseases, functional genomics, and computational genomics.[81]
(g) *Making human tissue banks and clinical bioinformatic considerations*: A tissue bank is an establishment that collects and recovers human cadaver tissue for the purposes of medical research, education, and allograft transplantation. With potential biomarkers identified *via* clinical informatics and experimental designs thereof

are in place, for the tissue banks play the next role.[82] Tailored efforts in clinical bioinformatics *via* IT could help in the management of biobanks personal data with highly secure access.

(h) *Formulating mathematical medicine and biology in clinical bioinformatics*: mathematical medicine and biology is a subject with a profuse extent of mathematical content, analytical methods, and computational techniques addressing topics in medicine and biology. Relevant areas of life sciences include: cell biology, developmental biology, ecology, epidemiology, immunology, infectious diseases, neuroscience, pharmacology, physiology, and population biology.

(i) *Performing protein expression and profiling through systems biology extended into clinical bioinformatics*: clinical bioinformatics can enable discovery and developments in new diagnostics and therapies for diseases. Relevant studies imply the analysis and visualization of complex medical datasets on patient complaints, history, therapies, clinical symptoms and signs, results of physician's examinations, outcomes of biochemical analyses, imaging profiles, and measured extents of pathologies involved.

(j) *Integrating the efforts in clinical bioinformatics*: Mostly the integrated relationship between clinical observations *versus* the underlying microbiological (molecular) mechanisms in clinical contexts may not be explicit. Therefore, fresh methods of combining clinical measurements and their global functions within a particular omics category may be required to decide explicitly the underlying relationships, and they can be conceived as advances in clinical and TBI.

6.13 Modeling in Clinical Bioinformatics

This section indicates some avenues of modeling that can be considered as topics that overlap the concepts of biomedical heuristics, TBI, and clinical informatics. Relevantly, formulating newer concepts and models in clinical bioinformatics are presented in the following subsections addressing: (a) models of basic biology of clinical relevance; (b) disease-specific, clinical study models; and (c) patient-related risk evaluation paradigms and models.

Models of Basic Biology with Clinical Relevance

These are unique models that could be developed on innumerable systems pertinent to a variety of biological details. Such details enclave the structures that make up the living systems and the associated small or big various biological functions. Inasmuch as biological systems encompass a vast, diverse array of topics and problems, innumerable models can be proposed; for example, in the book by Haefner,[83] some essential aspects of modeling are furnished on a few biological systems and their subdisciplines with necessary analytics and computer simulations. Relevant examples of models in mathematical biology as in Haefner[83] include a span of classical and state-of-the-art topics, such as *cellular automata* and *artificial life*. Hence, unique models of basic biology can be posed in a translated perspective for adoptions in clinical bioinformatics.

Models of Disease-specific Clinical Details

Unique models are needed in clinical applications with annotated details on the states of diagnosed pathogeny. Notwithstanding, however, almost countless diseases in existence, modeling strategies can be trimmed to address a few (within the scope of clinical informatics), for example, *Type-1* and *Type-2 diabetes mellitus* (T1DM and T2DM)-related biological considerations. Hence, the associated modeling considerations involve the following clinical informatic details.

(a) *Modeling metabolic state under diabetic/nondiabetic conditions* refers to elucidating the *basal metabolic rate* (BMR) or energy liberated per unit of time by the patient in relation to the underlying glucose metabolism.
(b) *Modeling cardiovascular conditions coexisting in diabetic patients*: this effort is to ascertain the risk associated with cardiovascular disease (which is generally regarded as a preventable complication of T2DM).[84–86] Details available in the literature concerning underlying risks could form the benchmark data for necessary modeling in question.[87–89]
(c) *Modeling prediabetic conditions on adult diabetics*: information and modeling of metabolic disturbances associated with prediabetes is vital toward risk evaluation of cardiovascular diseases-cum-diabetes in adults.[90]
(d) *Modeling obesity conditions on adult diabetics*: as a focused effort, obesity *versus* diabetic risks can be modeled. Compatible literature data can be availed from Elhayany *et al.*[91] and Rodríguez-Villar *et al.*[92]
(e) *Modeling glucose metabolism*: this model refers to simulation of glucose regulatory metabolism in human beings, for example, as

reported by the author (with others) in Neelakanta *et al.*,[31] Cobelli and Mari.[32] Simulated results based on theoretical formulations of the model in question can be validated against clinical data.

(f) *Modeling diabetics-related database for pathology-specific ambient*: an exhaustive database on diabetics and the associated risk factors can be availed from *National Diabetes Statistics of National Diabetes Information Clearinghouse* (NDIC) and *National Diabetes Statistics, NIH Publication No 11-3892, 2011*; hence, model perspectives can be appropriately included in the database for comprehensive risk and decision evaluations (e.g., in home-cared patient management described earlier).

Patient-Centric Decision-Theoretics and Risk-Evaluation Models in Clinical Bioinformatics

Modeling in clinical informatics could include medically specified decision-theoretic considerations in patient evaluations (*via* appropriate analytical methods). Associated schemes include statistical data analyses, prescribing metrics on patient-care, finding statistically specified levels toward deciding whether a subject need to be an in- or out-patient, identification of high-risk patients, administering procedures for robust diagnostics, and extending optimal therapy regimens.

Modeling risk-analysis and decision-theoretics in clinical efforts of bioinformatics are based on a host of statistical models and classification algorithms (such as *LDA*[93] plus *Kalman filtering*[94]) toward risk-tracking and HMM.[94–96] Furthermore, ANN[36,37,97] can also be adopted in modeling efforts on adequate patient-care[23–25] strategies.

Modeling risk analysis and establishing decision-theoretics in clinical bioinformatics based on statistical details could warrant regression analysis, and most often, the outcome of such regression analysis on medical data may require prescribing statistical UB and LB on the predicted value (e.g., as indicated earlier in Fig. 6.1, a sample set of results on risk aspects of T2DM *versus* age[34]), the traditional risk-factor *versus* modeled version of bounds shows a significant difference).

Decision Paradigms for Hospital-Related Issues and Patient-care Management Procedures: Clinical Bioinformatic Modeling

Another targeted effort in clinical bioinformatics is to identify methods and means of imparting quality and adequate care to patients. Such facilitation is also governed by government-legislated regulations. Hence, indentifiable modeling can include salient perspectives of patient admission protocols (on placing or admitting/readmitting patients in hospitals) and patient-care management procedures as exemplified below.

Example Ex-6.4

Statement of the Problem

This example outlines hospital-related patient-care issues *vis-à-vis* procedural aspects of placing or admitting/readmitting the patients. The underlying considerations are as follows: it has become necessary in modern times to identify and classify patients on statistical basis so as to ascertain their hospital admission/ readmission needs. Correct identification of high-risk patients could allow hospitals and other medical facilities to accommodate them with logistically justified prioritizations based on the extent of risk involved and medical urgency required; it also implies smartly utilizing the limited resources viewed in terms of space (beds), equipment, ambulatory, medical-assistance personnel (at all levels), nursing staff, physicians, and the availability of *emergency room* (ER) support while mitigating hospital readmissions. However, existing and traditional methods are mostly subjective and are based on rule-of-thumb decisions adopted on *ad hoc* basis in most of the clinical settings. Furthermore, readmission systems in vogue mostly lack scientific metrics on the underlying protocols and procedures and require a review on the basis of clinical informatics. A practical approach need to be framed to adopt clinical natural language processing in decision-theoretics on hospital readmissions.

Often binary-tree-based readmission classification is considered for patients readmitted for varying reasons with no due concerns on predicting the primary cause of readmission. Also, often exist scores of "co-occurring evidence discovery of clinical terms with primary diagnosis" that can be hardly useful in a clear decision protocols on readmission requirements.

Any unplanned hospital readmissions could augment healthcare management shifting the underlying focus from quantity-of-care to *quality-of-care* (QoC) paradigm without placing undue burden on resources, cost, and patient inconvenience. As such, a modern approach required in reanalyzing the patient care is to evolve relevant metrics and policy details toward care quality *versus* hospital readmission procedures.[98]

Problem P-6.6

Statement of the Problem

Consistent with patient admission contexts, this problem is indicated as regard to ER support and reviewing the *triage* provisioning and deciding on acuity-level involved. The underlying question is: *what are the issues involved in patient intake into ER?*

When patient admission/readmission pursuits are sought, the first level of patient entry into a hospital could be the ER facility where immediate patient-care needs and decisions on admission or readmission requirements are evaluated at the triage level. The triage prioritizes the patient-care based on the severity of their chief health issues and complaints. The purpose of triage is to determine whether the patient is in need of immediate care so as to initiate necessary treatment and stabilize the patient. In an ER (or *emergency department*, ED), an individual can walk in as a nonemergency case when he/she is not brought in by an ambulance (or in ambulated state), and the triage is responsible to determine the patient's condition on the severity and urgency required with pertinent attention to be exercised by a physician and/or ER medical personnel. Thus, the triage nurse is responsible with the task of assigning a patient with a "triage acuity level" by assessing the severity of the patient's main complaint. In a comprehensive suite, the triage nurse takes down notes on the complete history, checks vital signs, and asks about any allergies of the patient, and a focused triage is also practiced wherein, more defined attention to injuries and/or minor illnesses as well as the history limited to that specific complaint are ascertained. Typically, the triage acuity is stratified into five levels with level-1 being the most severe with the patient in need of immediate and urgent care and level-5 refers to the least severe and the patient can wait to be seen.

The pursuable flow as regard to how a patient's level of condition (in an ED) can be assessed involves the following steps: (a) emergency department flow: patient mode of arrival, (walk-in or EMS). (b) triage nurse involvement: taking history, assessing vital signs and deciding the acuity level. (c) *ad hoc* bed placement. (d) further nurse's assessment: confirming chief complaint and making a brief note on past medical history. (e) physician assistant and/or physician examination.

Unplanned protocols and procedures, if pursued in the ER, could create patient-life endangerment and backup of patients, (known as *logjam* that are waiting to be seen by the triage nurse) and would implicate hospital admission/readmission status. Patients visiting the ER and requiring hospitalization should not only be treated for immediate symptoms but helped with the necessary steps to prevent unnecessary future hospital visits. Alternative to being hospital in-patient, suggestions can be provided at out-patient facility and/or at ER informing the patients with the possibility of engaging a home healthcare professional (so as to avoid hospital admission or readmission). Patients under constant watch by such a home healthcare professional would be unlikely to require hospital visits. Unnecessary hospital visits and/or admission or readmission steps could mean poor usage of resources and prohibitively expensive burden to patients.

Appropriate predictive analytics could offer potential decisions for hospitals to classify patients, who most likely need admission/readmission. Relevant predictive models in practice, however, have many shortcomings. Many such systems are created from clinical experience and lack statistical foundation. Systems built from statistical learning approaches often use general purpose techniques with little concern for the properties of clinical data. This has led to poor model performance.

Example Ex-6.5

Problem Statement

This example outlines details on clinical informatics pursuits in hospital readmission system. As observed earlier, hospital admission could mean a first-time entry of a patient for a given illness, etc.; alternatively, it refers to a hospital readmission being sought. That is, hospital readmission system refers to an episode of a patient who had been discharged earlier from a hospital and is getting admitted again within a specified time period. Such readmission rates have been increasingly used as an outcome measure in health services and often regarded as a quality benchmark for health systems.

However, considering limited resources and costs involved, revised readmission needs are being considered. "Hospital readmission rates are formally included in reimbursement decisions for the *Centers for Medicare and Medicaid Services* (CMS) as part of the *Patient Protection and Affordable Care Act* (ACA) of 2010, which penalizes health systems with higher than expected readmission rates (ERRs) through the Hospital Readmission Reduction Program." Since the inception of this penalty, there have been other programs that are practiced with the aim to decrease hospital readmission so as to meet the limited resources and economic reasons.[98–100]

The design and implementation of the so-called *patient hospital readmission system* (PHRS) could be based on discharge summaries. Such methods on PHRS procedure require finding co-occurring evidence of diseases, symptoms, and medications. Existing systems largely use co-morbidity indexes (like *Charlson comorbidity index** built on correlations). However, finding co-occurring terms with potentially similar root causes could allow developing better intelligent comorbidity indexes. Furthermore, a big-data approach can be introduced to allow national and international clinical datasets to be used in formulating PHRS schemes. (*The *Charlson comorbidity index* developed in 1987 by Mary Charlson and colleagues) refers to a weighted index that predicts the risk of death within one year of hospitalization for patients with specific comorbid conditions. There are 19 such conditions indicated with each condition assigned a weight coefficient from 1 to 6, based on the estimated 1-year mortality hazard ratio from a Cox proportional hazards model, and these weights are summed to produce the *Charlson comorbidity score*.[101] Modeling a PHRS prediction can depend on the ERR defined as follows: [(Predicted readmission rate/ERR) – 1], and the readmission systems toward predictive modeling should not ignore misclassification costs, and the associated *false positive* (FP) and *false negative* (FN) misclassifications should be weighted with appropriate costs *via* cost-sensitive *confusion matrix* written, for example, in terms of: (Fraction of actual true cost *versus* false predicted cost: λ) and (fraction of actual false cost *versus* true predicted cost: μ).

The underlying decision costs could rely on: *test diagnostic group* (DG), new patients in the DG, risk-coefficient assignment for each patient in the DG, actual number of readmissions done in a fiscal year, ERR and probability of patients considered for intervention through homecare. Appropriate models can be prescribed to deuce μ and λ in the cost-matrix. Hence, a cost function can be formulated and optimized toward, for example, (a constrained) minimum ERR. Various methods, such as naïve Bayes' rule, annealing technique, ANN can be pursued toward the underlying optimization.

6.14 Review Section: Clinical Bioinformatics

This section reviews the subject of clinical informatics with additional details, related examples, and rider problems (posed as self-study exercises). Tutorials as needed are also included.

Example Ex-6.6

Problem Statement

Further to adult healthcare facility briefly considered in an earlier section, this example is indicated to outline on growing need for home-based healthcare, specifically the aged persons. This is a matter of societal and governmental concern that spins across the globe to facilitate and manage effectively the health of ageing population. As reported in World Health Organization publication,[40] in 2000, one in ten, or about 600 million, people were 60 Dudley years or older, and this figure is expected to reach 1.2 billion people by 2025, and in 2050 around 1.9 billion. An example on home-based adult healthcare for diabetics as a TBI-specific task was explained in an earlier section emphasizing high prevalence of diabetes across growing geriatric population.[40,41]

A modeling effort with translational notions in clinical bioinformatics needs first a base of target patients (such as aging subjects having chronic states like T1DM/T2DM) as described earlier, and relevant exercise can be conceived in terms of acutely needed considerations, which cross-breed the associated concerns and possible directives on: (a) framing an in-house (home-based) digital health information infrastructure; (b) compiling routine data on patient information and translating the data into a knowledge base toward risk evaluations and decision details; (c) making context-aware man–machine interfaces in the healthcare infrastructure, toward home-based "self-management"; (d) integrating computer-to-patient (and caregivers) with necessary information-sharing; enabling "self-management" in question *via* a tailored self-monitoring set of routines supporting education and counseling; and (e) constructing a database to hold a cache of socioeconomic profile and day-to-day life style information.

The logistics and rationale in conceiving the facility as above can be identified as follows: (i) provisioning proprietary EMR units in clinical/hospital or rehab center settings to facilitate a comprehensive *electronic database* (DB) may not be economically viable for adoption in a home-care facility[102,103]; (ii) sparsely existing practical and affordable home-based care facility (for adult patients) targeted toward enabling a "self-management style" at home; and (iii) supporting incumbent patients in getting educated/counseled concurrently so as to lead a comfortable home-based lifestyle with normal societal interests, exercise routines, dietary habits within their means of economics/cost-of-living availing insurance, and/or governmental support.[42]

Example Ex-6.7

Problem Statement

Indicated in this example are details on clinical/patient-related data collection, data grouping *via* logistic regression and data analyses toward risk predictions, and elucidating UB and LB on the evaluated risk. Hence, more examples and exercises on case-studies (and/or clinical details) relevant to the tasks of clinical bioinformatics are presented and specified as extensions to the context of this chapter.

Risk-evaluation methods in clinical bioinformatics, an overview: Described in this subsection are methods of applying risk-evaluation techniques useful in clinical diagnostic studies. Heuristics on pertinent algorithmic pursuits (that follow logistic-regressions of medical diagnostic test data enabling nonlinear risk-predictive technique) are presented.

1. *Medical/health suites of risk-evaluations (required in patient-care management) motivations*: Clinical decisions (with UB and LB) can be regarded as motivated objectives based on underlying biostatistics, medical/health suites, and clinical bioinformatics. The associated stochastic considerations determine the probability of occurrence (p) of an event (depicting a clinical epoch or test/diagnostic outcome), and this probability value can be inferred empirically by fitting a dataset to a (*logit function*) logistic curve with UB and LB on the basis of stochastic heuristics.
2. *Model development using the available (clinical) data:* the aforesaid task (of risk-evaluation with bounds) implies conceiving *ad hoc* models using the available (clinical) data so as to illustrate the bounds-limited performance of the (risk) prediction.

The art of medical healthcare statistics as an adjunct topic of clinical datamining, and it is regarded as a knowledge discovery effort in databases related to bio-/medical sciences.[42–44,103] For example, it is often required to distinguish between two entities (such as healthy and pathogenic subjects) in diagnostic pursuits. This is done in medical situations on the basis of clinical decisions made with a gamut of data acquired *via* laboratory tests and/or from clinical diagnostics.

Regression analysis: Normally, the collected clinical datasets are subjected to biostatistical analysis (in bioinformatic sense) by applying the so-called regression predictive techniques. Such regression methods are empirically conceived to describe the relational attributes between a *dependent* or *response variable* on a set of *independent* or *predictive* or *explanatory variable.*

The risk factors denote a set of *categorical variables,* meaning that such variables may take two or more possible values. Therefore, in biological and/or medical models, it is necessary to describe a reasonably best-fitting and most parsimonious empirical relationship between an outcome (*dependent* or *response variable*) on a set of independent (*predictive* or *explanatory variable*)[104,105]; that is, the model that "fits" the data should yield a "*fair goodness-of-fit*" and lead to interpretation of results with ease and adequacy.[106] In the context of biological sciences, biomedical research, bioengineering, bioinformatics, and clinical analysis, the response sought could be largely of binary nature, 0 or 1, (with the associated Bernoulli statistics).

Given a set of regressors depicting clinical observations, a function should be evolved to predict the mean patient response. Such a regression can be done by means of the so-called *logistic-regression model.* Relevantly, in pursuing logistic regression toward analyzing and evaluating clinical or biological data statistics, the underlying objectives can be enumerated as follows:

- Conceiving a systematic procedure toward diagnostic decisions based on the statistics of availed clinical data
- Reviewing the literature on the state-of-the-art aspects of logistic regression in the context of applying biostatistics and clinical information for the intended diagnostic decisions
- Prescribing a set of bounds (upper and lower) on the predictions (responses), such as the risk factors *via* logit function decisions
- Justifying the underlying stochastic considerations in prescribing the bounds on the response predicted
- Imparting meaningful coefficients of regression and interpreting the associated logistic regression in terms of the risk factor (specified within its bounds) *vis-à-vis* real-world clinical data available in literature
- Identifying the efficacy of the pursuit, ascertaining the merits and demerits so as to look for possible future scope

A set of feasible avenues specific to logistic regression in the field of clinical bioinformatics can be listed as follows:

- Overlaying the well-known logistic regression efforts (adopted in biostatistics, medical informatics, and other areas[24]) with an explicit prescription of definitive UB and LB on the regressed output

- Evaluating such bounding limits of logistic regression can be done by resorting to the LBF mentioned in an earlier section so as to account for the underlying stochastic considerations[36,37]
- Testing the applicability of LBF in logistic regressions by using clinical data available in literature
- Comparing the results on risk-factors evaluated for a set of clinical data with those deduced *via* traditional (non-LBF logistic regressions)
- Identifying, relevant items of open-questions for plausible studies in clinical bioinformatics

The scope of regression analysis of data and corresponding analytical as well as computational considerations can be applied conveniently in clinical contexts. For example, the epidemiological aspects of associating second-hand smoke *versus* heart ailment and evaluating the risk are described in Walpole *et al.*,[106] Glantz and Parmley[107] using relevant regression analysis.

Such regression analyses bring out the relation between two sets of variables, namely: (i) the independent (observed) variables $\{x\}$ and (ii) the dependent (predicted) variables $\{y\}$, and it is attempted to find a reasonably good-fit on the causal link between $\{x\}$ and $\{y\}$ *via* the regression analysis studies.[24,108,109]

In summary, while performing regression analysis with clinical/health-related data (or otherwise) and hence evaluating the risk, the underlying procedure in practice involves the following steps: (i) data collection, presentation, and applying enumerative statistics; (ii) selection of compatible regression curve functions; (iii) adopting multiple regression analysis and correlation analytics as needed; (iv) prescribing dummy variables; and (v) performing forecasting *via* regression curves.

6.15 Logistic-Regression in Clinical Bioinformatic Contexts

In clinical contexts, the logistic-regression procedure is specific to patient-related, (mostly) binary (dependent) responses deduced from health-related details and datasets. Though medical and health-related biostatistics recognize such logistic-regression models, relevant models are sparsely used in clinical trials. Nevertheless use of such models is favored in recent years as can be seen across some salient efforts published in the archival of literature. The following are citable efforts:

(i) using logistic regression suites in deducing the "risk of determining the risk" with multivariable models in medical contexts has been indicated by Concato *et al.*[110] and proportional hazard elucidation is described in Katz and Hauck[111]; (ii) a cohesive interpretation of health research data *via* statistical considerations and logistic regression is available in Hirsch and Riegelman,[112] Armitage and Berry,[113] Harrell *et al.*,[114] and Khan *et al.*[115]; (iii) logistic regression models for prognostic predictions are discussed in Khan *et al.*[115]; and (iv) exclusive used of logistic-regression models in obstetrics and gynecology is reviewed in Khan *et al.*[115]. Relevantly, some exemplary clinical case studies are described below with the application of appropriate logistic regression procedure.

Example Ex-6.8

Problem Statement

This example illustrates a case study on infantile mortality statistics where logistic regression is pursued. The details are as follows: infantile mortality statistics pertinent to South Carolina, USA, is presented in Yoon[116] with respect to live birth/infant death cohort, and statistical inference on infant mortality rate thereof is discussed and it is concluded that infant mortality rate could be exponentially higher among the (infants of) lowest birth weights.

In the underlying exercise, logistic regression is appropriately adopted in deriving conclusive inferences on regressors such as birth weight, maternal weight, extent of prenatal care (as per *Kessner–Kotelchuck index**), etc.[117–119] The **Kessner–Kotelchuck index* (KI) refers to an *Adequacy of Prenatal Care Utilization* (APNCU) index, which is based on certain pregnancy-related crucial details. This index primarily implies a way to classify the adequacy of prenatal welfare on the basis of received types of clinical services and number of prenatal visits (*versus* the expected number of visits for the period between when care began and the delivery date). The expected number of visits is based on the *American College of Obstetricians and Gynecologists* prenatal care standards for uncomplicated pregnancies, and it is adjusted for the gestational age when care began and for the gestational age at delivery.[118,119] The aforesaid KI is thus a composite index consisting of three categories, namely, adequate, intermediate, and inadequate. It is defined as per a combination of values decided by (i) time of entry into PNC; (ii) number of prenatal visits; and (iii) gestational age (in weeks) of activity. Scoring for KI can be availed from Bloch *et al.*,[117] Kessner *et al.*,[118] Kotelchuck,[119] and Singer *et al.*[120]

In all, maternal wellness characteristics can be specified by a host of variables and corresponding parameters specified as number values.[117] Relevant characteristics of states of maternity and possibility of infantile mortality are dependent on a number of societal as well as patient-specific details that can be availed in clinical contexts and assessed for the risk involved *via* compatible risk-analyses methods including logistic regression approach. Maternal characteristics implicating infantile mortality are: maternal age, maternal education, marital status, maternal race, KI of *prenatal care* (PNC), use of tobacco during pregnancy, use of alcohol during pregnancy, multiple births, gestational period, and birth-weight group (classified as very low, moderately low, and normal).

Model development: Considering the characteristics that decide infantile mortality *versus* the states of maternity as indicated above, a logistic regression model can be framed to assess the associated risk-factor on infantile mortality. The procedure thereof first involves stipulating an outcome variable, z, in terms of a predictor variable (regressor variable) set $\{x_i\}$ with appropriate prescription of a set of coefficients (regressor coefficients), $\{\beta_i\}$ (for each corresponding predictor variable indexed by i). Concerning the present example of infantile mortality being modeled, the predictor variables refer to the following: maternal age (x_1), maternal education (x_2), maternal race (x_3), marital status (x_4), KI of PNC (x_5), tobacco use during pregnancy (x_6), alcohol use during pregnancy (x_7), number of previous children (x_8), birth-weight group (x_9), and gestational period (x_{10}). Categories of adequacy/inadequacy of (KI) depend on gestational weeks at delivery, number of PNC visits, and trimester of first visit. Each of the regressor variables identified in Table 6.2 is a predictor variable that decides the infantile mortality and the descriptive observations presented statistically decide the eventual infant mortality.[116]

Using the set of regressor variables characterized as above, the outcome variable z can be modeled as follows:

$$z = b_0 + b_1x_1 + b_2x_2 + \cdots + b_{10}x_{10} \qquad \text{Eq. (6.2)}$$

where the β-coefficients can be quantified *via* statistical details gathered on the regressor variables described above.

A possible set of exemplar β-coefficients scored toward the variables involved in the example being considered are presented in Table 6.3.

Table 6.2 Predictor variables that decide infantile mortality.

Parameter: x_i	Significance in infantile mortality
Maternal age (x_1)	No significance in infantile deaths
Maternal education (x_2)	Significant, regardless of race of the mother
	Black mother *versus* white mother: Infantile
Race of the mother (x_3)	mortality rate is almost doubled
	Significant across all ethnicities
	Intermediate and inadequate prenatal care
Marital status (x_4)	lead to risk of infant deaths
Prenatal care (x_5)	Significant risk to the infant
Smoking during	
Pregnancy (x_6)	No significant risk factor
Drinking during	
Pregnancy (x_7)	Significant risk can be foreseen
Multiple births (x_8)	VLW infants' mortality rate is higher
Birth-weight group (x_9)	Than that of MLW infants. Mortality rate
	decreases as birth-weight increases
Gestational period (x_{10})	Short gestation is a leading cause in infantile deaths

Table 6.3 Scoring for β-coefficients: an example set.

β	Parameter type	Score	β	Parameter type	Score
β_0	Expected value	—	β_6	—	1.5
β_1	—	0	β_7	—	0.5
β_2	—	1.0	β_8	—	2.0
β_3	White	1.0	β_9	NBW	1.0
				MLBW	2.0
				VLBW	3.0
	Hispanic	2.0			
	Black	3.0			
β_4	—	1.0	β_{10}	3 trimesters	1.0
				2+ trimesters	2.0
β_5	Adequate	1.0			
	Intermediate	2.0			
	Inadequate	3.0			

The numerical values of the coefficients shown depict clinically viable, relative influence of each predictor variable involved. (However, the number values indicated are subjective, but are clinically significant. They approximately reflect proportionate aspects of implications on the outcome of the predictions.)

With reference to the collective variable, z of Eq. (6.2), its logistic regression can be specified traditionally by a functional relation: $f(z) = 1/\{1+\exp(-z)\}$. That is, given a variable, z depicting the net influence of regressors (plotted along the along the *abscissae*), its corresponding dependent variable, $f(z)$ can be presented along the vertical (y) axis. Such a plot of logistic-regression, however, conforms to a response constrained to one single value of $f(z)$ for each value of z. But, in view of statistical dispersions of clinical variables, a smear of variation of the computed outcome $f(z)$ is appropriate, when the regressing influences refer to an ensemble of random variations (as in clinical findings). That is, the response ascertained using logistic-regression should not be constrained to one single value, but it should be stretched across an ensemble values, and this stretch represents an *error-bar* on the computed outcome of $f(z)$.

Thus, for each logistic regression output, an error-bar could exist with an UB and LB. Hence, a modified expression for $f(z)$ is necessary to account for the stochastic aspects of the details on $f(z)$ arising from possible variations in β-values. The underlying considerations in formulating a modified logistic-regression and finding the logit predictions of bounds in clinical decisions are outlined in a tutorial presented later.

In the present example on infantile mortality, the logistic regression refers to establishing the *risk-factor of mortality* (RFM) *versus* birth weight of infants. The computed details are plotted (in normalized form) in Fig. 6.9, where the curves shown refer to the following computation: with the data on β-values randomly changed within conceivable statistical limits (that account for the associated randomness of the variables above), the plotted details conform to an ensemble of ten computations for each birth weight, 1000 g through 3000 g (with an increment of 100 g). The theoretical prediction of the associated UB and LB on the results will be projected after considering the heuristics of modified and stochastically justified prescription of $f(z)$.

The profile of the maximum values of RFM is indicated in Fig. 6.9 by an upper curve and, the minimum values are shown by the envelope of the bottom curve. The mean value of upper and lower plots is also indicated in Fig. 6.9. That is, with reference to the above case study,

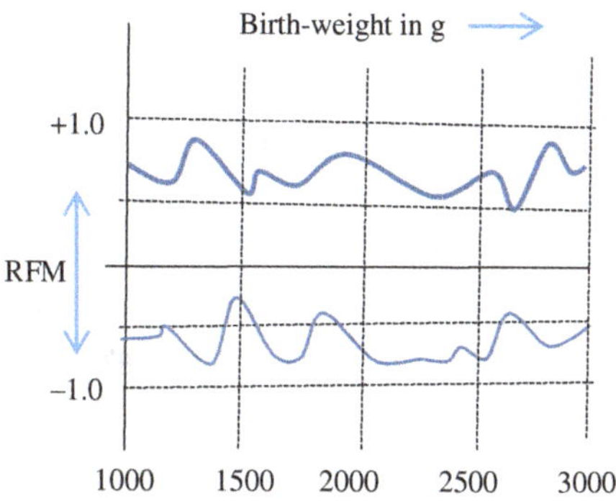

Fig. 6.9 Envelopes of upper- (top-curve) and lower-bounds (bottom-curve) on *risk-factor of mortality* (RFM) *versus* birth-weight of infants computed *via* logistic regression: $f(z) = 1/\{1 + \exp(-z)\}$, with z constituted by a random ensemble of maternity-related regressor variables for each birth-weight value and $f(z)$ is the corresponding RFNM estimated.

computations on pertinent logistic regression show that the risk of mortality has a range bounded by an upper- and a lower-limit. Elucidating these bounds theoretically will be described later in terms of a modified logistic regression function.

Tutorial T-6.2

This is a tutorial note on logistic regression, which as indicated above is a technique useful in decision-making problems having multivariate details plus associated coefficients, such as those encountered in clinical and epidemiological pursuits. As stated earlier, the regression approach *via* logit function, $f(z) = [1/\{1 + \exp(-z)\}]$ is viable and widely used in practice with $z = \Sigma_i \beta_i \times x_i$, where $\{x_i\}$ is a set of regressor variables, and each variable is weighted by a coefficient, β_i. The resulting log-regressed details yield a single (deterministic) value for the risk factor. However, the risk factor involved in a conceived diagnosis or therapy regimen, it implies inferring a probabilistic output of some specific value (between 0 and 1). That is, the plot of $f(z)$ *versus* z should imply a smear of variation of the computed outcome across a designated *error bar* specifying an UB and LB on estimated, $f(z)$.[34]

Mathematically, the traditional logit function $f(z) = y = 1/\{1 + \exp(-z)\}$ indicated above can be written as $\{(1/2)+(1/2) \times \tanh(z/2)\}$, and it can be rewritten as equal to: $\{(1/2)+(1/2) \times L_q(z/2)\}$, where $L_q(.)$ denotes the LBF, described earlier. Relevant logistic-regression in terms of LBF gets its UB specified by the order function, $q = ½$, and correspondingly, $q \rightarrow \infty$ gives the LB, and no empirical details nor any "curve-fitting" strategy (applied to empirical data) are required in deciding these bounds; as such, it is very compatible in clinical assessments of risk factors. Furthermore, as characterized by LBF, the algorithm for $f(z)$ in terms of $L_q(.)$ becomes stochastically justifiable. Pertinent illustrative clinical case studies wherein the risk factor is decided *via* LBF are presented in the following examples.

Example Ex-6.9

Problem Statement

This example is an exercise on deciding the risk factor *via* logistic regression in the case of infantile mortality statistics considered earlier.

Now, using the same dataset, birth weight, maternal weight, etc. as in Tables 6.2 and 6.3 are considered (in the earlier example), the LBF-based modified log regression is now evaluated with its UB and LB, and the computed results are presented in Fig. 6.10. The associated risk factor as decided by traditional logistic regression, $f(z) = 1/\{1 + \exp(-z)\}$ is also presented in Fig. 6.10, computed using an ensemble of uniformly distributed random values of associated regressor variables. Details of LBF-based logistic regressions presented in Fig. 6.10 show the ability to determine the corresponding UB and LB of the risk factor without the need for any empirical simulations with random ensemble dataset.

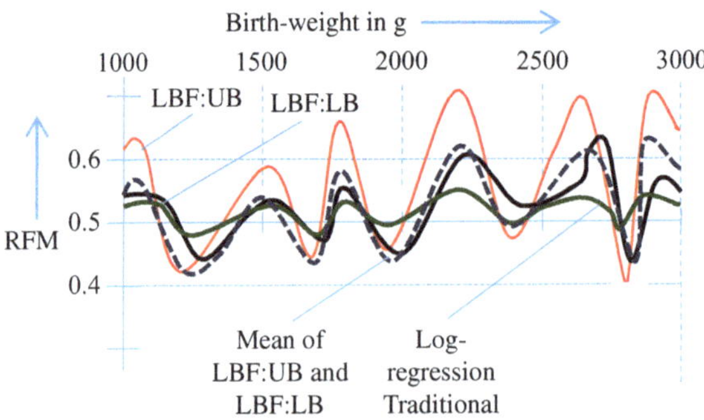

Fig. 6.10 Risk factor of mortality (RFM) *versus* birth weight of the infants: Results deduced *via* logistic regression using LBF.

Example Ex-6.10

Problem Statement

This is another example of logistic regression applied to the clinical context of deciding the risk factor in the treatment of urinary tract infections (UTI). A clinical protocol toward managing acute urinary symptoms in healthy adult woman is indicated in Williams.[121] It refers to clinical/diagnostic evaluation of bacterial infection of the bladder, (known as *cystitis* or *urinary tract infection,* or UTI).

Based on clinical statistics as in Williams,[121] the following are prediction model parameter/values for the UTI: (i) dysuria (pain or burning on urination) denoted by symbol, *D*; (ii) urinary frequency: denoted by symbol, *F*; (iii) and vaginal symptoms (discharge or itch): denoted by symbol, VAG SX. In addition, the urine analysis assesses the presence and amount of a number of chemicals in the urine, and it also reflects details about the health of the kidneys and types of the cells that may be present in the urine. Such information provides certain predictive details as regard to: (a) bacterial content being present in urine sediment (as depicted by the microscopic examination of a centrifuged portion of urine) to a level and (b) the presence of more *white blood cells* (WBC).

A decision-tree is formulated in Williams[121] as regard to the beginning of UTI symptoms with a set of track/route parameters toward clinical decisions on UTI, where the set $\{P_1, P_2, P_3, P_4, P_5\}$ is probability values estimated for the decision on intended diagnostics of the UTI. Suppose the clinical observations on UTI conform to the following set: dysuria (painful urination) (*D*) — present; urinal frequency (*F*) — absent; vaginal Sx (VAG SX) — absent; bacteria in sediment (*B*) — scaled as $\geq 2+$; and WBC scaled as ≥ 10. Hence, P_1 depicts the initial probability of UTI prescribed on the basis of aforesaid symptoms *D*, *F*, and VAG SX plus the two urine analysis data, and P_2 denotes the modified probability (of UTI) by tracing a decision-tree. Furthermore, the set $\{P_3, P_4, P_5\}$ can be traced *via* algorithmic routes with the stipulation of symptomatic conditions and urine analysis data as follows: P_2 [($B + W$+)/(D, F, VAG SX)]; P_3 [($B + W$−)/(D, F, VAG SX)]; P_4[($B - W$+)/(D, F, VAG SX)]; and P_5 ($B - W$−)/(D, F, VAG SX)].[121]

A typical exemplification of the probabilities as above refers to data in Williams[121] with the presence and absence of causative factors denoted by binary coefficients 1 and 0, respectively, and the associated heuristics yields results on probabilistic strategies on "best" adoptable decisions specified in three categories: "Treat now," "trade-off range", or "wait and treat later." Relevant empirical details[121] give a pair of output variable index as follows:

For "treat now" strategy:

$$z_1 = (D + F + V) + (P_2 - 0.30) + (P_3 - 0.24) + (P_4 - 0.36) \quad \text{Eq. (6.3a)}$$

For "wait" strategy:

$$z_2 = (D + F + V) + (P_3 - 0.03) + (P_4 - 0.06) + (P_5 - 0.02) \quad \text{Eq. (6.3b)}$$

Hence, the corresponding traditional logistic regressions are stipulated as follows:

For "treat now" strategy: $p(z_1) = 1/(1+\exp(-z_1))$ Eq. (6.4a)
For "wait" strategy: $p(z_2) = 1/(1 + \exp(-z_2))$ Eq. (6.4b)

The estimated probability set $\{P_2, P_3, P_4, P_5\}$ on UTI example toward best strategies on the diagnostic decisions, for "treat now," "treat-later," "wait for the treatment," and "trade-of-region" are tabulated in Williams,[121] and relevantly, the case-study data on UTI subjected to logistic regression analysis enable deciding two treatment regimens: "treat-now" (say, with antibiotics) and "treat-later" (consistent with statistical set of regressors on clinical information and patient conditions). The gray area in depicting the divergent decision would possibly call for second expert opinion. Should the gray area be narrow, then the affirmative action on the treatment becomes more critical, and the extent of decision boundary is considered as a good indication on the severity of treatment regimen to be pursued. Results obtained by traditional logistic regression depicting (relative) risk factor of UTI *versus* varying extent of clinical observations (deciding the logistic variable, z) show a trade-off range as illustrated in Fig. 6.11(A). Furthermore, distinctively identifiable UB and LB of the "treat-now" (TN) and "treat-later" (TL) regimens required in patient management are obtained by adopting the LBF-based logistic-regression as shown in Fig. 6.11(B).

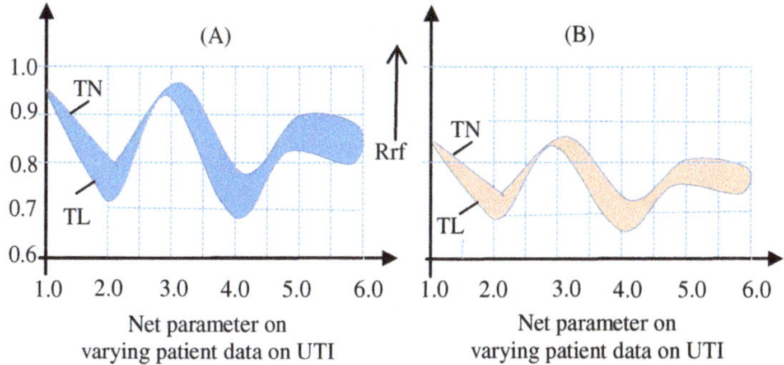

Fig. 6.11 (A) Relative risk factor (Rrf) of UTI (deduced by computations based on traditional logistic regression: $f(z) = [1/(1 + \exp(-z)])$ *versus* net parameter (z) obtained by varying regressors on clinical conditions/patient data. (TN: Treat now and TL: Treat later regimens) (B) Relative risk factor (Rrf) of UTI (deduced by computations based on LBF-based logistic regression: $f(z) = [(1/2)+ (1/2) \times L_q(z/2)])$ *versus* net parameter (z) obtained by varying regressors on clinical conditions/patient data. (TN: Treat now and TL: Treat later regimens)

Example Ex-6.11

Problem Statement

This is another example of decision-making in the clinical contexts of Type 2 *diabetes mellitus* (T2DM) condition *via* logistic regression characterized by following regressor variables depicting major signs and symptoms of T2DM: polyuria (frequent urination), polydipsia (increased thirst), polyphagia (increase hunger), fatigue, erectile dysfunction (ED), weight-loss and increased risk of cognitive dysfunction, and dementia.[122–127] In addition, the following are typical blood-test related risk-level indication parameters (which are identified as low-, borderline, and high-risk levels): LDL cholesterol, HDL cholesterol, triglycerides, total cholesterol, cholesterol/HDLC, hemoglobin A1C, glucose level (with fasting).[60,61]

Decision-making in patients with diabetes: model development: the risk-factor assessment and primary care decisions on diabetes management can also be done *via* logistic-regressions so as to explore the associated determinants of response. For example, in Agardh *et al.*,[127] American Diabetes Association,[128] and D'Arrigo,[129] relative risks are calculated with reference to socioeconomic position at certain stages of life in middle-aged T2DM patients using multiple logistic-regression analyses.

Likewise, multivariate logistic regression analyses can be done to identify the extent of risk of T2DM in women facing work stress and low emotional support. In general, the screening process of identifying those individuals who are sufficiently high risk of a specific disorder (like T2DM), so as to warrant further investigation or direct actions, can be exercised *via* logistic-regression approaches.[130]

A model development for T2DM toward using logistic-regression is as follows: As indicated earlier, relevant set of β-coefficients that define z are first identified so as to formulate the logistic-regression outcome, $f(z)$. Hence, pertinent regressor dataset, $\{x_i\}$ and β-coefficients are gathered on clinical characteristics of T2DM patients and their symptomatic details as listed below (Table 6.4). The β-coefficients assigned as indicated for the regressor parameters are mostly subjective, but clinically justifiable consistent with medical expertise.

Hence, computed details ascertaining the risk factor (*via* the logistic function $f(z)$ with $z = \Sigma_i \beta_i \times x_i$) are presented earlier in Fig. 6.6. The plots therein depict the risk factor due to T2DM *versus* patient's age obtained by $f(z) = [(1/2) + (1/2) \times L_g\ (z/2)]$ as per model prescription detailed in Dupont.[34] As mentioned earlier, $Lq(.)$ denotes the LBF with q being its stochastic order parameter. Further the computed data, as shown in Fig. 6.6, has bounding limits LBF-UB and LBF-LB determined by the function LBF in $f(z)$ with (q = ½) and ($q \rightarrow \infty$) depicting the UB and LB, respectively; these limits explicitly provide an error-bar at each evaluated value of the risk factor. The β-coefficients for z-function for T2DM are as in Table 6.4.

Example Ex-6.12

Problem Statement

This exercise is yet another example of using logistic regression in clinical contexts. It is concerned with *retinopathy of prematurity* (ROP), a pathological state which refers to a disease of the eye affecting primarily premature infants and is regarded as the leading cause of blindness in newborn babies.[106]

Relevant to the incidence of ROP, the associated statistical data and risk-factor considerations are reported in Walpole *et al.*,[106] where the study enumerates 29 weighted predictor variables $\{\beta_i \times x_i\}$ amenable for logistic regression *via* an output variable, and the corresponding risk-factor is specified in terms of $f(z)$. The study emphasizes risk-factors that correlate the most with the incidence of ROP. Such factors are clinically important since they can alert the physicians to the neonates most likely to contract ROP and help them to assess the condition of the patient.

Table 6.4 Assignment of β-coefficients for *z*-function: T2DM

PS	N	P-β	CTP	N	P-β
Column notations as above are expaned in the bottom of this table					
Gender (male/female)	G	Male: 1.0 Female: 1.0	LDL (mg/dL)	L	≤100:1.0 ≥130:3.0
Age (in years)	A	>50:2.0 <50:1.0	HDL (mg/dL)	H	Men ≥45:1.0 ≤35:3.0 Women ≥55:1.0 ≤45:3.0
BMI (in kg/m^3)	B_m	>27:2 <24:1.0	Hemoglobin H1C	H_c	≤5.7:1.0 ≥6.1:3.0
Smoking habit (yes/no)	S	Yes: 1.5 No: 0	Triglycerides (mg/dL)	T	<200:1.0 ≥400:3.0
High/moderate Alcohol consumption (yes/no)	A_l	Yes: 1.5 No: 0	Total cholesterol (mg/dL)	C_T	125–150:1.0
Physical exercise (yes/no)	P	Yes: 1.5 No: 2.5	Fasting C-peptide (nmol/L)	F_C	1.09:2.0
Polyuria (yes/no)	P_U	Yes: 1.5 No: 1.0	Fasting plasma-glucose (nmol/L)	F_{P1}	<5.2:1.0 >8.4:2.5
Polydipsia (yes/no)	P_D	Yes: 1.5 No: 1.0	2h-Postprandial plasma glucose (nmol/L)	F_{P2}	6:1.0 >15:2.5
Polyphagia (yes/no)	P_P	Yes: 1.5 No: 1.0	Weight-loss (yes/no)	W	Yes:1.5 No:1.0
Fatigue (yes/no)	F	Yes: 1.5 No: 0.5	Noticeable cognitive dysfunction or dementia (yes/no)	C	Yes:1.5 No:0.5
Erectile dysfunction in men (yes/no)	E_d	Yes: 2.0 No: 0.5	Regimented diet (yes/no)	R	Yes:0.5 No:2.5
Pregnancy in women (yes/no)	P_g	Yes: 1.5 No: 0.5	PS: Patient details and symptom characteristics N: Notation P-β: Prescription of β-coefficients CTP: Clinical test parameters		

Gestational age and the child's weight are regarded as being primarily significant *vis-à-vis* ROP, and breast milk consumption is considered as a protective factor against ROP. Referring to the details in Walpole *et al.*,[106] the set $\{x_i\}$ denoting the discrete ensemble of predictor variables

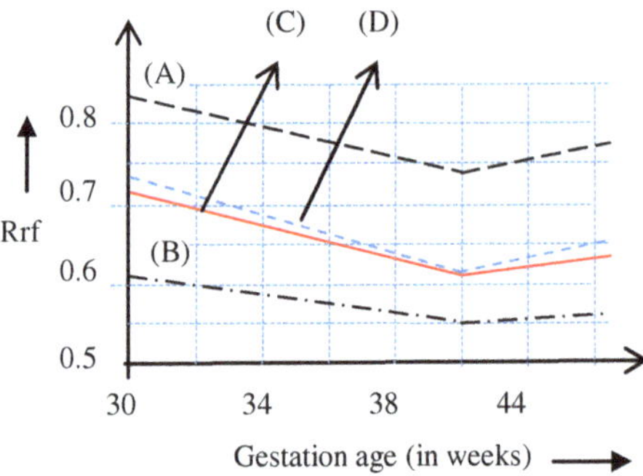

Fig. 6.12 Risk-factor (Rrf: Relative risk-factor) of ROP *versus* gestational age. (A): LBF-UB; (B): LBF-LB; (C): Mean of LBF-UB and LBF-LB; and (D): Traditional log-regression.

and corresponding weighting coefficients $\{\beta_i\}$ are listed. Hence, the ROP-related risk factor of the case study is evaluated as a function of gestation age, and the corresponding logistic regression curves obtained are illustrated in Fig. 6.12. The computed details presented in Fig. 6.12 are based on both traditional functional relation: $f(z) = 1/[1 + \exp(-z)]$ and LBF-based logistic-regressions, namely, $f(z) = [(1/2) + (1/2) \times L_q(z/2)]$. The results in Fig. 6.12 illustrate the ability of LBF in providing explicit details on the limits of the risk involved *via* LB and UB, and the traditional $f(z)$ yields results that conform approximately to the mean values of the deduced UB and LB as shown.

Problem P-6.7

Problem Statement

This problem is a rider exercise on logistic-regression-based decision-theoretics on heart ailment risks. Pertinent details of the model are as follows: A case study assessing the risk of fatality associated with heart ailment is considered. In such cases of heart-related pathology, the associated risk could result from a number of etiological input factors (or predictor variables, $\{x_i\}$) namely: age (x_1), gender (x_2: 0, for male and 1, for female), blood cholesterol level (x_3), and body mass index, or BMI (x_4). Hence, for the intended assessment of heart ailment risk, an ensemble set of random data of $\{x_i\}$ can be adopted to deduce the function, $f(z)$ as narrated below.

As described earlier (and elaborated in Dupont[34]), relevant clinical study details suggest that the total contribution (z) toward the risk of heart attack can be specified by: $z = (\beta_0 + \beta_1 x_1 + \beta_2 x_2 + \beta_3 x_3 + \beta_4 x_4)$, where the set $\{\beta_{i\,=\,1,\,2,\,3,\,4}\}$ refers to regression coefficients (per year), and the risk (of death of the patient) thereof is implied in terms of an estimated number of years of survival. It is given by $f(z)$ denoting the logistic function. The clinical trial details indicated in Dupont,[34] suggest the following intercept and regression coefficients (per year) in the expression for z: ($\beta_0 = -5$; $\beta_1 = +2$; $\beta_2 = -1$; $\beta_3 = +1.2$; $\beta_4 = +0.1$), corresponding to: age (x_1) >50; sex (x_2) is male (0) or female (1); cholesterol level (x_3) >5 mmol/L and BMI (x_4) >25. Furthermore, based on clinical observations, the risk in question presumably increases with age, with z going up by 2 every year above the age of 50; female has a decreased risk with z going down by 1; cholesterol level would increase the risk inasmuch as, z goes up by 1.2 for each mmol/L; and lastly, when the BMI value above 25 increases the risk involved.

As stated earlier, the risk assessment should also conform to stochastic considerations with the associated ensemble of predictors depicting unbiased random variables (RVs). Relevant RVS can be heuristically considered in terms of random samples of subjects (male or female) across an age group of, say 50–55, as well as the subjects having random BMI values (in excess of a nominal value of 25) plus cholesterol levels fluctuating above 5 mmol/L.

The aforesaid statistical profile can be simulated by generating an ensemble of pseudoreplicates of the regressing parameters, $\{x_i\}$ each specified over a (clinically justifiable) range. The pseudoreplicates can be generated using *statistical bootstrapping* as detailed in Rush.[105] Hence, for each pseudoreplicate set simulated, $z = (\beta_0 + \beta_1 x_1 + \beta_2 x_2 + \beta_3 x_3 + \beta_4 x_4)$ is obtained and the result on the logit-functional plot of age *versus* the risk-factor, $f(z) = 1/\{1+\exp(-z)\}$ is constructed. The cluster of such curves obtained is thus pertinent to the ensemble of pseudoreplicates (of multivariates) leading to a pair of empirical statistical limits of the risk involved. That is, the cluster of curves constructed using dataset as above conform to a set of multivariable subjected to a logistic-regression, $f(z)$. And, the results decide/predict the trend on the outcome of the collective implications arising from the multiple input set.

However, for the same problem as above, use of LBF leads to results illustrated in Fig. 6.13, which explicitly provide the UB and LB, respectively, by using the LBF logit-regression, namely, $f(z) = [1/2+\{1/2 \times L_q(z/2)\}]$ with $q \to 1/2$ and $q \to \infty$, respectively; here, relevant

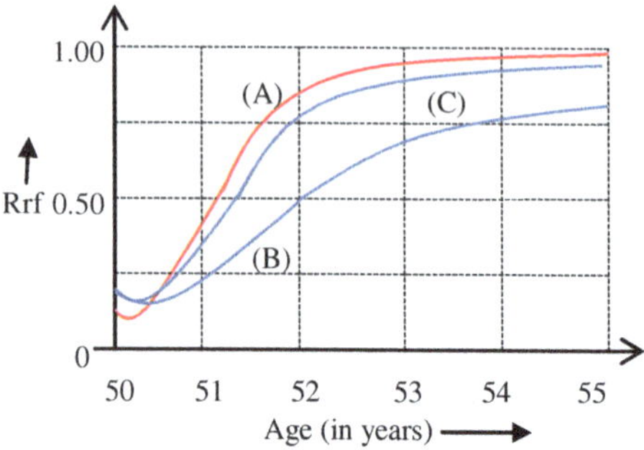

Fig. 6.13 Age *versus* (normalized/relative) risk-factor (Rrf) versus age of the patient with heart ailment. Results: (A): LBF-based, UB (with $q \to ½$); (B) LBF-based, LB (with $q \to \infty$, say, $\approx$1000); and (C) plot of risk-factor deduced from traditional log-regression. It depicts average values of Rrf obtained from the cluster of details inferred using pseudoreplicated regressors.

computations involve no empiricism in arriving at decision bounds on the risk-factor calculated.

6.16 Clinical Bioinformatics: A Summary

In summary, this section (Part II) of the chapter identifies some thematic details on clinical bioinformatics. The associated details refer to a set of perusal of topics introduced in Part I on translational informatics. That is, in translating the concepts of microbiological bioinformatics into *ad hoc* applications on patients' bedside, diagnostic and therapeutic needs, clinical bioinformatics is devised. In prescribing decision-theoretic bounds of clinical assessment of risks, as explained in Neelakanta,[36] Neelakanta and De Groff,[37] Neelakanta *et al.*,[38] and Neelakanta *et al.*,[39] the LBF has inherent mathematical characteristics of specifying stochastic bounds implying asymptotic limits on the estimates of a statistical experiment. Use of LBF in the contexts of logistic-regression is a neoteric avenue in clinical bioinformatics sparsely seen in literature. It enables analytical prescription of LB and UB on regressed data without resorting to empirical curve fitting (using, for example, the Richards' curve[139]). Furthermore, blending Bayesian binarization (such as *Neyman–Pearson decisions*)[131–134] and logistic-regression decision methods could be another useful tool in biostatistics extended to clinical informatics of decision theoretics.[135]

There are other bounds (e.g., the so-called *Cramér–Rao bounds)* that can be adopted to specify the accuracy of unbiased estimations of statistical parameters[131–134]) in clinical contexts.

Furthermore, statistical bounds presumably depict a confidence region on the uncertainty of an estimated parameter in the universe of statistical details. Determinants of different types are adopted in practice to determine the statistical bounds. Both nonrandom parameters as well as random entities are used in deciding the bounds as described in Cramer[133]

Inasmuch as, statistical decision-making involves a large extent of statistical variables, which are mostly described only in linguistic norms and overlaps; the deduced decisions could invariably be fuzzy. Use of fuzzy logistic-regression is still an unexplored strategy[136] in medical/biostatistics, and introducing fuzzy considerations in clinical bioinformatics is a viable open question for relevant studies.

Automated trend analysis toward evolving risk and/or health improvement in a clinical setting is a desirable project in modern healthcare engineering. Developing a biostatistical effort thereof would involve decision-theoretic methods using logistic-regression methods addressed in this chapter. That is, developing healthcare assessment *via* clinical bioinformatics is still in infantile level, though it is much warranted in clinical bioinformatic directions on patient-care management strategies.[135–137] A gist of such efforts is studied and elaborated in Dupont.[34]

Associating multivariate regression analysis in decision-making *vis-à-vis* microbiological data (e.g., as in asserting sequence information) opens a broad scope for future research[138] in clinical/TBI contexts. Furthermore, studies on building and applying logistic-regression models and including bias in *odds-ratios* in relevant clinical decision modeling (as seen in Agresti,[139] Peduzzi *et al.*,[140] and Nemes[141]) worth further considerations.

Decision theoretics in health management is viably useful in medical insurance actuary efforts toward inferring the extent of risk specified by pathological/clinical conditions.

In all, clinical bioinformatics has provisions to embrace both microbiological studies on sequence signature analysis of model organisms as well as it includes macroscopic perspectives of realizing diagnostic models based on personalized analysis of trends.[142,143]

6.17 Closure

A devoted, medically implied and health science-related bioinformatic efforts depict exemplary directions in translational and clinical bioinformatics. Translating informatic prospects of microbiology and the associated molecular omic details enables comprehending clinical aspects of bedside-related usable information (as described in Part I). Strategies thereof in extending genomic, transgenomic, and proteomic information into clinically viable diagnostic and therapeutic applications frames the arts of clinical bioinformatics.

In summary, the translational bioinformatic details conceived (and detailed in Part I) aim at "enhancing the adoption of best practices (in health sciences) in the community." Subsidiary trends of translational bioinformatics lead to clinical bioinformatics described in Part II. Relevant considerations imply assessing stochastically bounded decision-theoretic results in diagnostic and therapeutic contexts, and the associated analyses enable framing the scope for a constricted, but a more comprehensive study on pathology-specific informatics.

References

[1] J. D. Tenenbaum, N. H. Shah, and R. B. Altman: Translational bioinformatics. In: E. H. Shortliffe and J. J. Cimino (Eds.). *Biomedical Informatics*. Springer-Verlag, London, UK: 2014, pp. 721–754.

[2] A. J. Butte: Translational bioinformatics: Coming of age. *Journal of the American Medical Informatics Association*, 2008, vol. 15(6), 709–714.

[3] J. D. Tenenbaum: Translational bioinformatics: Past, present, and future. *Genomic Proteomic Bioinformatics*, 2016, vol. 1491, 31–41.

[4] F. J. Azuaje, M. Heymann, A. Ternes, A. Wienecke-Baldacchino, D. Struck, D. Moes, and R. Schneider. Bioinformatics as a driver, not a passenger, of translational biomedical research: Perspectives from the 6th Benelux Bioinformatics Conference. *Journal of Clinical Bioinformatics*, 2012, vol. 2(7), 1–3.

[5] N. H. Shah, C. Jonquet, Y. A. Lussier, P. Tarzy-Hornoch, and L. Ohno-Machado: Ontology-driven indexing of public datasets for translational bioinformatics. *BMC Bioinformatics*, 2009, vol. 10(2), S1.

[6] E. A. Mendonca: Selected proceedings of the 2010 summit on translational bioinformatics. *BMC Bioinformatics*, 2010, vol. 11(9), 1–4.

[7] I. N. Sarkar, A. J. Butte, Y. A. Lussier, P. Tarczy-Hornoch, and L. Ohno-Machado: Translational bioinformatics: Linking knowledge across biological and clinical realms. *Journal of American Medical Information Association*, 2011, vol. 18, 345–357.

[8] L. J. Lesko: Drug research and translational bioinformatics. *Clinical Pharmacology & Therapeutics*, 2012, 91(6), 960–962.

[9] M. G. Kann: Advances in translational bioinformatics: Computational approaches for the hunting of disease genes. *Briefings in Bioinformatics*, 2010, 11(1), 96–110.

[10] A. J. Butte and R. Chen: Finding disease-related genomic experiments within an international repository: First steps in translational bioinformatics. *Proceeding Archives of AMIA Annual Symposium,* 2006, vol. 2006, pp. 106–110.

[11] A. J. Butte: Translational bioinformatics applications in genome medicine. *Genome Medicine,* 2009, vol. 1(6), I64.

[12] J. T. Dudley: Translational bioinformatics in the cloud: An affordable alternative. *Genome Medicine,* 2010, vol. 2(8), 51.

[13] Q. Yan: Translational bioinformatics and systems biology approaches for personalized medicine. *Methods in Molecular Biology,* 2010, vol. 662, 167–178.

[14] P. Payne, D. Ervin, P. Dhaval, T. Borlawsky, and A. Lai: TRIAD: The translational research informatics and data management grid. *Applied Clinical Informatics,* 2011, vol. 293, 331–344.

[15] M. Sherling: EMR meets modern technology, *Practical Dermatology,* November 2010, 27–28.

[16] T. D. Gunter and N. P. Terry: The emergence of national electronic health record architectures in the United States and Australia: Models, costs, and questions. *Journal of Medical Internet Research,* 2005, vol. 7(1), e3. Published online: 2005 March 14. doi: 10.2196/jmir.7.1. e3.

[17] H. Leroux, S. McBride, and S. Gibson: On selecting a clinical trial management system for large scale, multi-centre, multi-modal clinical research study. *Studies in Health Technology and Informatic,* 2011, vol. 168, 89–95.

[18] S. Adibi. (Ed.) *Mobile Health: A Technology Road Map.* Springer international publishing AG, Switzerland: 2015.

[19] J. A. Wolf, J. F. Moreau, O. Akilov, T. Patton, J. C. English, J. Ho, and L. K. Ferns: Diagnostic inaccuracy of smartphone applications for melanoma detection. *JAMA Dermatology,* 2013, vol. 149(4), 422–426.

[20] A. Solanas, C. Patsakis, M. Conti, I. S. Vlachos, V. Ramos, F. Falcone, O. Postolache, P. A. Pérez-Martínez, R. Di Pietro, D. N. Perrea, and A. Martínez-Ballesté: Smart health: A context-aware health paradigm within smart cities. *IEEE Communications Magazine,* 2014, vol. 52(8), 74–81.

[21] C. Patsakis, R. Venanzio, P. Bellavista, A. Solanas, and M. Bouroche: Personalized medical services using smart cities' infrastructures. *Proceedings of IEEE International Symposium on Medical Measurements and Applications (MeMeA 2014),* (Lisboa, Portugal, 11–12 June, 2014), pp. 1–5.

[22] S. Liu, W. Li, and K. Liu: Pragmatic oriented data interoperability for smart healthcare information systems. *Proceedings of 14th IEEE/ACM International Symposium on Cluster, Cloud and Grid Computing (CCGrid 2014),* (Chicago, IL, 26–29, May, 2014), pp. 811–818.

[23] P. S. Neelakanta: *A Text-book on ATM Telecommunications: Principles & Implementation*, CRC Press, Boca Raton, FL, USA: 2000.

[24] P. S. Neelakanta and D. M. Baeza: *Next-generation Telecommunications & Internet Engineering*, Linus Publications, New York, NY: USA 2009.

[25] M. N. K. Boulos, R. C. Lou, A. Anastasiou, C. D. Nugent, J. Alexandersson, G. Zimmermann, U. Cortes, and R. Casas: Connectivity for healthcare and well-being management: Examples from six European projects. *International Journal of Environmental Research and Public Health*, 2006, vol. 6, 1947–1971.

[26] V. Rialle, J. B. Lamy, N. Noury, and L. Bajolle: Telemonitoring of patients at home: A software agent approach. *Computer Methods and Programs in Biomedicine*, 2003, vol. 72, 257–268.

[27] H. Alemdar and C. Ersoy: Wireless sensor networks for healthcare: A survey. *Computer Networks*, 2010, vol. 54, 2688–2710.

[28] J. B. Buse, H. N. Ginsberg, G. L. Bakris, N. G. Clark, F. Costa, R. Eckel, V. Fonseca, H. C. Gerstein, S. Grundy, R. W. Nesto, M. P. Pignone, J. Plutzky, D. Porte, R. Redberg, K. F. Stitzel, and N. J. Stone: Primary prevention of cardiovascular diseases in people with diabetes mellitus: A scientific statement from the American Heart Association and the American Diabetes Association. *Circulation*, 2007, vol. 115(1), 114–126.

[29] A. Elhayany, A. Lustman, R. Abel, J. Attal-Singer, and S. Vinker: A low carbohydrate mediterranean diet improves cardiovascular risk-factors and diabetes control among overweight patients with type 2 diabetes mellitus: A 1-year prospective randomized intervention study. *Diabetes Obesity and Metab*olism, 2010, vol. 12(3), 204–209.

[30] C. Rodríguez-Villar, A. Pérez-Heras, I. Mercadé, E. Casals, and E. Ros: Comparison of a high-carbohydrate and a high-monounsaturated fat, olive oil-rich diet on the susceptibility of LDL to oxidative modification in subjects with Type 2 diabetes mellitus. *Diabetic Med*icine, 2004, vol. 21(2), 142–149.

[31] P. S. Neelakanta, M. Leesirikul, Z. Roth, and S. Morgera: A complex system model of glucose regulatory metabolism, *Complex Systems*, 2006, vol. 16, 343–367.

[32] C. Cobelli and A. Mari: Control of diabetes with artificial systems for insulin delivery: Algorithm independent limitations revealed by a modeling study", *IEEE Transactions on Biomedical Engineering*, 1985, vol. BME-32, 840–845.

[33] C. Cobelli and A. Mari: Validation of mathematical models of complex endocrine-metabolic systems. A case study on a model of glucose regulation," *Medical and Biological Engineering and Computing*, 1983, vol. 21, 390–399.

[34] A. N. Dupont: Risk evaluation in clinical diagnostic studies: Ascertaining statistical bounds *via* logistic regression of medical informatic data,

M S Thesis, Florida Atlantic University, Boca Raton, Florida, USA: 2011.

[35] J. T. Sorensen: A physiologic model of glucose metabolism in man and its use to design and assess improved insulin therapies for diabetes. *Doctor of Science Thesis,* Department of Chemical Engineering, Massachusetts Institute of Technology, Boston, USA: 1985.

[36] P. S. Neelakanta. (ed.), *Information-theoretic Aspects of Neural Networks.* CRC Press, Boca Raton, FL, USA: 1999.

[37] P. S. Neelakanta and D. De Groff: *Neural Network Modeling: Statistical Mechanics and Cybernetic Principles,* CRC Press, Boca Raton, FL, USA: 1994.

[38] P. S. Neelakanta, S. Abusalah, D. De Groff, and R. Sudhakar: Logistic model of nonlinear activity in a neural complex: Representation *via* Riccati differential equation. *Cybernetica,* XXXIX (1996), 15–30.

[39] P. S. Neelakanta, R. Sudhakar, and D. De Groff: Langevin machine: A neural network based on stochastically justifiable sigmoidal function. *Biological Cybernetics,* 1991, vol. 65, 331–338.

[40] World Health Organization publication. Available on-line at: *http://www.who.int/ageing/publications/Hutton_report_small.pdf.*

[41] American Diabetes Association diabetes statistics: Available on-line at: *http://www.diabetes.org/diabetes-basics/diabetes-statistics.*

[42] S. G. Mougiakakou, C. S. Bartsocas, E. Bozas, N. Chaniotakis, D. Iliopoulou, I. Kouris, S. Pavlopoulos, A. Prountzou, M. Skevofilakas, A. Tsoukalis, K. Varotsis, A. Vazeou, K, Zarkogianni, and K. S. Nikita: SMARTDIAB: A communication and information technology approach for the intelligent monitoring, management and follow-up of Type 1 diabetes patients. *IEEE Transactions on Information Technology in Biomedicine,* 2010, vol. 14(3), 622–632.

[43] Health News: *Palm Beach Post,* January 25th, 2012.

[44] C. Chen: Web-based remote human pulse monitoring system with intelligent data analysis for home health care. *Expert System Applications,* 2011, vol. 38, 2011–2019.

[45] R. S. Parker, F. J. Doyle III, and N. A. Peppas: A model-based algorithm for blood glucose control in Type I diabetic patients. *IEEE Transactions on Biomedical Engineering,* 1999, vol. 46, 148–157.

[46] M. Hanss and O. Nehls: Simulation of human glucose metabolism using fuzzy arithmetic. *Proceedings of the Nineteenth International Conference of the North American Fuzzy Information Processing Society* (NFIPS, 13–15 July 2000), pp. 201–205.

[47] M. Hanss and O. Nehls: Enhanced parameter identification for complex biomedical models on the basis of fuzzy arithmetic. *Proceedings of the Joint IFSA World Congress and Twentieth NAFIPS International Conference* (25–28 July, 2001), vol. 3, 1631–1636.

[48] M. Mahfouf, M. F. Abbod, and D. A. Linkens: A survey of fuzzy logic monitoring and control utilisation in medicine. *Aritificial Intelligence in Medicine*, 2001, vol. 21, 27–42.

[49] P. S. Neelakanta, J. C. Park, and D. De Groff: Complexity parameter *vis-à-vis* interaction systems: Applications to neurocybernetics, *Cybernetica*, XL (1997), 243–253.

[50] D. Manojlovic, V. Todorovic, M. Pavlovic, V. Ristic, and M. Ristic: The effect of high-fiber diet upon glucoregulation and lipidaemia in diabetic patients. *Casopis Lékaru Ceských*, 1986, vol. 125(47), 1437.

[51] R. B. Northrop and E. A. Woodruff: Regulation of a physiological parameter or in vivo drug concentration by integral pulse frequency modulated bolus drug injections, *IEEE Transactions on Biomedical Engineering*, 1986, vol. BME 33, 1010–1020.

[52] C. Deeney: Insulin therapy by continuous subcutaneous infusion, *The Pharmaceutical Journal*, 2003, vol. 271, 206–208.

[53] D. G. Allen and M. V. Sefton: A model of insulin delivery by a controlled release micropump. *Annals of Biomedical Engineering*, 1986, vol. 14. 257–276.

[54] V. C. Scalon and T. Sanders: *Essentials of Anatomy and Physiology*. F. A. Davis Company, Philadelphia, PA, USA: 2003.

[55] M. C. K. Khoo: *Physiological Control Systems: Analysis Simulation and Estimation*. IEEE Press, Piscataway, NJ, USA: 1999.

[56] A. Tao and E. Raz (Eds.), *Allergy Bioinformatics*. Springer Scientific + Business Media, Dordrecht, Germany, 2005.

[57] D. Ghosh and S. G. Bhattacharya: Allergen bioinformatics: Recent trends and developments. Selected work in bioinformatics. Available on-line at: *http://cdn.intechopen.com/pdfs-wm/21897.pdf*.

[58] R. C. Aalberse: Structural biology of allergens. *Journal of Allergy and Clinical Immunology*, 2000, vol. 106(2), 228–38.

[59] B. T. K. Lee and V. Brusic: Allergen Bioinformatics. In C. Schönbach, S. Ranganathan, and V. Brusic (Eds.), *Immunoinformatics*, Springer, New York, NY, USA: 2008. pp. 91–107.

[60] A. Mari, E. Scala, P. Palazzo, S. Ridolfi, D. Zennaro, and G. Carabella: Bioinformatics applied to allergy: Allergen databases, from collecting sequences information to data integration. The Allergome platform as a model. *Cellular Immunology*, 2006, vol. 244(2), 97–100.

[61] S. C. Bischoff: Role of mast cells in allergic and non-allergic immune responses: Comparison of human and murine data. *Nature Reviews Immunology*, 2007, vol. 7, 93–104.

[62] D. Ghosh and S. G. Bhattacharya: Allergen bioinformatics: Recent trends and developments. Selected work in bioinformatics. Available on-line at: *http://cdn.intechopen.com/pdfs-wm/21897.pdf*.

[63] Universal protein knowledge base: Available on-line at: *http://www.uniprot.org.*

[64] Official site for the systematic allergen nomenclature that is approved by the WHO and International Union of Immunological Societies (WHO/IUIS) allergen nomenclature sub-committee: Available on-line at: *www.allergen.org.*

[65] Allergome–an on-line resource on allegens: Available on-line at: *www.allergome.org.*

[66] R. E. Hileman, A. Silvanovich, R. E. Goodman, E. A. Rice, G. Holleschak, J. D. Astwood, and S. L. Hefle: Bioinformatic methods for allergenicity assessment using a comprehensive allergen database. *International Archives of Allergy and Immunology*, 2002, vol. 128(4), 280–291.

[67] P. Chatchatee, K. M. Jarvinen, L. Bardina, K. Beyer, and H. A. Sampson: Identification of IgE and IgG-binding epitopes on a_{s1}-casein: Differences in patients with persistent and transient cow's milk allergy. *Journal of Allergy and Clinical Immunology*, 2001, vol. 107(2), 379–383.

[68] J. Chen, D. Zhang , W. Yan, D. Yang, and B. Shen: Translational bioinformatics for diagnostic and prognostic prediction of prostate cancer in the next-generation sequencing era. *BioMed Research International*, 2013, vol. 2013, Article ID 901578, 13 pages. Available on-line at: *http://dx.doi.org/10.1155/2013/901578.*

[69] T. Barrett: Gene expression omnibus (GEO). *The NCBI Handbook* (Internet/2nd Ed./2013). Available on-line at: *https://www.ncbi.nlm.nih.gov/books/NBK159736/.*

[70] G. Gómez-López and A. Valencia: Bioinformatics and cancer research: Building bridges for translational research. *Clinical and Translational Oncology*, 2008, vol. 10(2), 85–95.

[71] A. J. Butte and L. Ohno-Machado: Making it personal: Translational bioinformatics. *Journal of the American Medical Informatics Association*, 2013, vol. 20(4), 595–596.

[72] Y. Y. Liu and S. A. Harbison: A review of bioinformatic methods for forensic DNA analyses. *Forensic Science International: Genetics*, 2018, vol. 33, 117–128.

[73] P. Datta, S. Sood, P. Rastogi and M. Yadav: DNA profiling in forensic dentistry. *Journal of Indian Academy of Forensic Medicine*, 2012, vol. 34(2), 156–159.

[74] A. Shcherbina, N. Chiu, E. Schwoebel, J. Harper, M. Petrovick, T. Boettcher, C. Zook, J. Bobrow, and E. Wack: Sherlock's toolkit: A forensic DNA analysis system. *2015 IEEE International Symposium on Technologies for Homeland Security* (HST, 14–16 April 2015, Waltham, Massachusetts, USA).

[75] I. K. Lee, H. S. Kim, and H. Cho: Design of activation functions for inference of fuzzy cognitive maps: Application to clinical decision making in diagnosis of pulmonary infection. *Healthcare Informatics Research,* 2012, vol. 18(2), 105–114.

[76] I. Ramis-Conde, M. A. J. Chaplain, and A. R. A. Anderson: Mathematical modelling of cancer cell invasion of tissue. *Mathematical and Computer Modelling,* 2008, vol. 47(5–6), 533–545.

[77] P. L. Chang: Clinical bioinformatics. *Chang Gung Medical Journal,* 2005, vol. 28(4), 201–211.

[78] X. Wang and L. Liotta: Clinical bioinformatics: A new emerging science, *Journal of Clinical Bioinformatics,* 2011, vol. 1(1), 1–3.

[79] J. Griss, Y. Perez-Riverol, H. Hermjakob, and J. A. Vizcaino: Identifying novel biomarkers through data mining a realistic scenario? *Proteomics Clinical Applications,* 2015, vol. 9(3–4), 437–443.

[80] S. L. Ward: Omics, bioinformatics, computational biology. Available on-line at: *http://alttox.org/mapp/emerging-technologies/omics bioinformatics-computational-biology/.*

[81] T. Strachan and A. Read: *Human Molecular Genetics.* Garland Science/ Taylor and Francis Group, New York, NY, USA: 2010.

[82] K. S. Suh, S. Sarojini, M. Youssif, K. Nalley, N. Milinovikj, F. Elloumi, S. Russell, A. Pecora, E. Schecter, and A. Goy: Tissue banking, bioinformatics, and electronic medical records: The front-end requirements for personalized medicine. *Journal of Oncology,* 2013, vol. 2013, Article ID 368751, 12 pages.

[83] J. W. Haefner: *Modeling Biological Systems: Principles and Applications.* Springer Science + Business Media Inc., New York, NY, USA: 2005.

[84] J. Zeber and M. L. Parchman: Cardiovascular disease in type 2 diabetes: Attributable risk due to modifiable risk factors. *Can Family Physician,* 2010, 56(8), e302–7.

[85] J. B. Saaddine, B. Cadwell, B. E. W. Gregg, M. M. Engelgau, F. Vinicor, G. Imperatore, and K. M. Narayan: Improvements in diabetes processes of care and intermediate outcomes: United States, 1988–2002. *Annals of Internal Medicine,* 2006, vol. 144(7), 465–474.

[86] S. J. Beaton, S. S. Nag, M. J. Gunter, J. M. Gleeson, S. S. Sajjan, and C. M. Alexander: Adequacy of glycemic, lipid, and blood pressure management for patients with diabetes in a managed care setting. *Diabetes Care.* 2004, vol. 27(3), 694–698

[87] Y. Ohkubo, H. Kishikawa, E. Araki, T. Miyata, S. Isami, S. Motoyoshi, Y. Kojima, N. Furuyoshi, and M. Shichiri: Intensive insulin therapy prevents the progression of diabetic microvascular complications in Japanese patients with non-insulin-dependent diabetes mellitus: A

randomized prospective 6-year study. *Diabetes Research and Clinical Practice*, 1995, vol. 28(2), 103–117.

[88] J. S. Skyler, R. Bergenstal, R. O. Bonow, J. Buse, P. Deedwania, E. A. Gale, B. V. Howard, M. S. Kirkman, M. Kosiborod, P. Reaven, and R. S. Sherwin: Intensive glycemic control and the prevention of cardiovascular events: Implications of the ACCORD, ADVANCE, and VA diabetes trials: A position statement of the American diabetes association and a scientific statement of the American college of cardiology foundation and the American heart association. *Circulation*, 2009, vol. 119(2), 351–357.

[89] J. B. Buse, H. N. Ginsberg, G. L. Bakris, N. G. Clark, F. Costa, R. Eckel, V. Fonseca, H. C. Gerstein, S. Grundy, R. W. Nesto, M. P. Pignone, J. Plutzky, D. Porte, R. Redberg, K. F. Stitzel, and N. J. Stone: Primary prevention of cardiovascular diseases in people with diabetes mellitus: A scientific statement from the American heart association and the American diabetes association. *Circulation,* 2007, vol. 115(1), 114–126.

[90] J. B. Meigs: Epidemiology of type 2 diabetes and cardiovascular disease: Translation from population to prevention: The Kelly West award lecture 2009. *Diabetes Care,* 2010, vol. 33(8), 1865–1871.

[91] A. Elhayany, A. Lustman, R. Abel, J. Attal-Singer, and S. Vinker: A low carbohydrate mediterranean diet improves cardiovascular risk factors and diabetes control among overweight patients with type 2 diabetes mellitus: A 1-year prospective randomized intervention study. *Diabetes Obesity and Metabolism,* 2010, vol. 12(3), 204–209.

[92] C. Rodríguez-Villar, A. Pérez-Heras, I. Mercadé, E. Casals, and E. Ros: Comparison of a high-carbohydrate and a high-monounsaturated fat, olive oil-rich diet on the susceptibility of LDL to oxidative modification in subjects with Type 2 diabetes mellitus. *Diabetic Medicine*, 2004, vol. 21(2), 142–149.

[93] F. L. Lewis: *Optimal Estimation: With an Introduction to Stochastic Control Theory,* John Wiley & Sons, New York, NY, USA: 1986.

[94] U. A. Khan, and J. M. F. Moura: Distributing the kalman filter for large-scale systems, *IEEE Transactions on. Signal Processing,* 2008, vol. 56(10), 4919–4935.

[95] R. Bhar and S. Hamori: *Hidden Markov Models.* Kluwer academic publishers, dordrecht, The Netherlands: 2004.

[96] W. Zucchini and I. L. MacDonald: *Hidden Markov Series for Time Series.* Chapman & Hall/CRC Press, Boca Raton, FL, USA: 2009.

[97] P. S. Neelakanta, M. A. Dabbas, and D. De Groff: Constructive ANN with dynamically set sigmoid: A simulation tool for technoeconomic

forecasting. *International Journal of Latest Trends in Computing*, 2012, vol. 3(2), 30–37.

[98] R. N. Axon and M. V Williams: Hospital readmission as an accountability measure. *Journal of American Medical Association*, 2011, vol. 305(50), 504–505.

[99] Centers for medicare and medicaid services: Readmissions reduction program, 2014: Available on-line at: *http://go.cms.gov/1gLbnoa.*

[100] J. Hoffman: Overview of CMS readmissions penalties for 2016. Available on-line at: *http://www.besler.com/2016-readmissions-penalties/.*

[101] NCI Comorbidity index overview. Available on-line at: *https://healthcaredelivery.cancer.gov/seermedicare/considerations/comorbidity.html.*

[102] S. Kirmiz and J. E. Weinreb: Diabetes in the elderly, chapter 5 in: J. E. Morley, L. van den Berg (Eds.), *Endocrinology of Aging.* Humana Press, Totowa, NJ, USA: 2011.

[103] P. McGowan: Patient self-management. *Background Paper to the New Perspectives*: Presented in: *International Conference on Patient Self-Management*, (Victoria, BC, Canada. September 2005).

[104] D. W. Hosmer and S. Lemeshow: *Applied Logistic Regression*, John Wiley Inc., New York, NY, USA: 1989.

[105] S. Rush: Logistic regression: The standard method of analysis in medical research, 2001. Available on-line at: *http://ramanujan.math.trinity.edu/tumath/research/studpapes/s3.pdf.*

[106] R. E. Walpole, R. H. Myers, S. L. Myers, and K. Ye: *Probability & Statistics for Engineers & Scientists.* Pearson/Prentice Hall, Upper Saddle River, NJ, USA: 2007.

[107] S. A. Glantz and W. W. Parmley: Passive smoking and heart disease: Epidemiology, physiology, and biochemistry. *Circulation*, 1991, vol. 83, 1–12.

[108] S. A. Glantz and W. W. Parmley: Passive smoking and heart disease: Mechanisms and risk, *Journal of American Medical Association*, 1995, vol. 273, 1047–1053.

[109] S. A. Glantz: *Primer of Biostatistics*, McGraw-Hill, New York, NY, USA, 1981.

[110] J. Concato, A. R. Feinstein, and T. R. Holford: The risk of determining risk with multivariable models, *Annals of Internal Medicine*, 1993, vol. 118, 201–210.

[111] S. M. H. Katz, and W. W. Hauck: Proportional hazards (Cox) regression, *Journal of General Internal Med*icine, 1993, vol. 8, 702–711.

[112] P. R. P. Hirsch, and R. K. Riegelman: *Statistical First Aid. An Interpretation of Health Research Data.* Blackwell Scientific Publications, Boston, MA, USA: 1992.

[113] P. P. Armitage, G. Berry: *Statistical Methods in Medical Research.* Blackwell Scientific, London: UK: 1994.

[114] F. E. Harrell, K. L. Lee, D. B. Matchar, and T. A. Reichert: *Regression models for prognostic prediction: Advantages, problems, and suggested solutions, Cancer Treatment Rep*orts, 1985, vol. 69, 1071–1077.

[115] K. S. Khan, P. F. W. Chien, and L.S. Dwarakanath: Logistic regression models in obstetrics and gynecology literature. *Obstetrics and Gynecol*ogy, 1999, vol. 93, 1014–1020.

[116] J. Yoon: South Carolina infant mortality statistics. *Public Health Statistics and Information Reports*, 2005, vol. 1, 1.

[117] J. R. Bloch, K. Dawley, and P. D. Suplee: Application of the kessner and kotelchuck prenatal care adequacy indices in a preterm birth population. *Public Health Nursing*, 2009, vol. 26(5), 449–459.

[118] D. M. Kessner, C. E. Kalk, and J . Singer: Assessing health quality–the case for tracers. *New England Journal Medicine*, 1973, vol. 288(4), 189–94.

[119] M. Kotelchuck: An evaluation of the Kessner adequacy of prenatal-care index and a proposed adequacy of prenatal-care utilization index, *American Journal of Public Health*, 1994, vol. 4, 1414–1420.

[120] D. Singer, C. Kalk, and E. Schlesinger: Infant death: An analysis by maternal risk and health care. 1973, Institute of Medicine, Washington, D. C, USA.

[121] B. T. Williams: *Computer Aids to Clinical Decisions*, vol. I, CRC Press, Boca Raton, FL, USA: 1982.

[122] C. D. Saudek, R. R. Rubin, and C. S. Shump: *The John Hopkins' Guide to Diabetes*, The John Hopkins University Press, Baltimore, MD, USA: 1997.

[123] M. Greenwood-Robinson: *Control Diabetes*, St. Martin's Press, New York, NY, USA: 2002.

[124] V. Kumar, N. Fausto, A. K. Abbas, S. Ramzi, and S. L. Robbins: *Robbins and Cofran Pathologic Basis of Disease*, Saunders Publications, Philadelphia, PA, USA: 2005, pp. 1194–1195.

[125] A. Rignell-Hydbom, J. Lidfeldt, H. Kiviranta, P. Rantakokko, G. Samsioe, C. Agarda, and L. Rylander: Exposure to p, p'-DDE: A risk-factor for Type 2 diabetes, *PLos ONE*, 2009, vol. 4(10), 1–6.

[126] M. Norberg, H. Stenlund, B. Lindahl, C. Anderson, T. W. Eriksson, and L. Weinehall: Work stress and low emotional support is associated with increased risk of future type 2 diabetes in women. *Diabetes Research and Clinical Practice*, 2007, vol. 76, 368–377.

[127] E. E. Agardh, A. Ahlbom, T. Andersson, S. Efendic, V. Grill, J. Hallquist, and C. G. Ostenson: Socioeconomic position at three points in life in association with type 2 diabetes and impaired glucose tolerance in middle-aged swedish men and women. *International Journal of Epidemiology*, 2007, vol. 36, 84–92.

[128] American diabetes association: Standards of medical care in diabetes-2010, *Diabetes Care*, 2011, vol. 3, suppl. I, (5–1)–(5–61).

[129] D'Arrigo: Cholesterol: The good, the bad, and the ugly. *Diabetes Forecast*, 1999, 54–58.

[130] Repose of a WHO and international diabetes federation meeting: Screening for Type 2 diabetes. *WHO/NMH/MNC/03.1, WHO, department of noncommunicable disease management*, Geneva, 2003.

[131] B. Efron: Bootstrap methods: Another look at the jackknife. *Annals of Statistics*, 1979, vol. 7(1), 1–26.

[132] C. R. Rao: Information and the accuracy attainable in the estimation of statistical parameters, *Bulletin of the Calcutta Mathematical Society*, 1945, vol. 37, 81–89.

[133] H. Cramer. *Mathematical of Statistics*, Princeton University Press, Princeton, NJ, USA: 1946.

[134] M. D. Srinath, P. K. Rajasekaran, and R. Viswanathan, *Introduction to Statistical Signal Processing with Applications*, Prentice Hall, Englewood Cliffs, NJ, USA: 1996.

[135] A. Meloni, A. Ripoli, V. Positano, and L. Landini: Mutual information preconditioning improves structure learning of Bayesian networks from medical databases. *IEEE Transactions on Information Technology in Biomedicine*, 2009, vol. 13, 984–989.

[136] T. Deutsch, E. Carson, and E. Ludwig: *Dealing with Medical Knowledge: Computers in Clinical Decision Making*, Plenum Press, New York, NY, USA: 1994.

[137] H. Ying, F. Lin, R. D. MacArthur, J. A. Cohn, D. C. Barth-Jones, H. Ye, and L. R. Crane: A fuzzy discrete event system approach to determining optimal HIV/AIDS treatment regimens. *IEEE Transactions on Information Technology in Biomedicine*, 2006, vol. 10, 663–676.

[138] F. J. Richards: A flexible growth function for empirical use. *Journal of Experimental Botany*, 1959, vol. 10, 290–300.

[139] A. Agresti: "Building and applying logistic regression models" In *An Introduction to Categorical Data Analysis*, John Wiley and Sons, Hoboken, NJ, USA: 2007, pp. 138.

[140] P. Peduzzi, J. Concato, E. Kemper, T. R. Holford, and A. R. Feinstein: A simulation study of the number of events per variable in logistic regression analysis. *Journal of Clinical Epidemiology*, 1996, vol. 49, 12.

[141] S. Nemes, J. M. Jonasson, A. Genell, and G. Steineck: Bias in odds ratios by logistic regression modeling and sample size. *BMC Medical Research Methodology*, 2009, vol. 9, 56.

[142] P. Stenberg, F. Pettersson, A. O. Saura: A. Berglund, and J. Larsson, Sequence signature analysis of chromosome identity in three *Drosophila* species. *BMC Bioinformatics*, 2005, vol. 6, 1–17.

[143] D. L. Hudson and M. E. Cohen: Diagnostic models based on personalized analysis of trends (PAT), *IEEE Transactions on Information Technology in Biomedicine*, 2010, vol. 14, 941–948.

Section IV
Bioinformatics of Cytogenetic Complex and Viral Omic Landscape

In the realm of bioinformatics, two other vital topics of interest that are covered in this section: (1) *cytogenetic informatics,* a branch of bioinformatics pertinent to entropy perspectives of genetic biology in ascertaining informatic aspects of structures and functions of a biological cell. Furthermore, such *organelle* level informatic studies include chromosomal synteny details of a cellular complex and (2) *viral informatics* on ascertaining exclusive omic details framed by viral species, and the associated bioinformatics implies thematic efforts on multiple sequences, for example, of a viral serovar that yield results on identifying the associated motifs useful in rational vaccine designs. Hence, the landscape of this section thus includes the following two chapters:

Chapter 7: Bioinformatics of Cellular Complex and Chromosomal Synteny
Chapter 8: Bioinformatics of Viral Omics

Chapter 7

Bioinformatics of Cellular Complex and Chromosomal Synteny

Part I of this chapter describes paradigms of cytogenetic bioinformatics. Relevant entropy details of chromosomal informatics sparsely seen in peer writings, as well as new avenues of bioinformatics at organelle level of the cellular complex.

Addressed in Part II of this chapter are cytogenetic details adopted to ascertain synteny details (appropriately defined via entropic- or information-theoretic metrics) depicting the negentropy features of blocks that represent contiguous genes located within the same chromosome.

> *"... our bodies are made up of trillions of cells, each governed by diverse 'software' processes. You and I are walking around with outdated software running in our bodies, which evolved in a very different era ..." (Ray Kurzweil)*

7.1 Introduction

The universe of living organisms represents a complex system infested with stochastic features of underlying omics, functional and structural aspects of cells and genetic statistics of whole species[1]; hence, an ambiguous set of studies is typically formalized to depict random instances of software processes of diverse informatics pertinent to omic, organelle, and whole-life constituents of living systems. However, the topics of cellular (organelle) informatics are sparsely seen in the peer writings of general bioinformatics; as such, relevant considerations form the focused themes elaborated in this chapter.

Organization of the cellular complex represents a unique stochastic framework constituted by random entities. Hence, pertinent details at organelle level mostly bear entropy features. Therefore, ascertaining the exclusive footprint of pedagogy on the underlying entropy-theoretics is imminent within the global perspectives of cytogenetic bioinformatics.

It includes genetic considerations associated with structural and functional aspects of biological cells possessing chromosomal inclusions along with other biological entities and elucidating associated informatic details arising from perceivable stochastic variations in the functional diversity and structural artifacts of the cellular body (such as *chromosomal aberrations* or CAs) is an open question, and it dictates a perusal toward unique studies in the contexts of cytogenetics as conceived here (in Part I of this chapter). Such studies are viably of interest and practically usable in medical/health-sciences-related bioinformatics.

The organelle informatics addressing the entropy features of chromosomal constituents (within the cells) holds a body of details commencing at the omics levels of the associated DNA strings. Thus, the cell in its cytogenetic framework depicts an entropy-related universe (in Shannon's sense), wherein exclusive informative details could prevail beyond the central dogmatic perspectives posed at omics level (elaborated earlier in Chapter 2). Furthermore, could prevail within the cellular infrastructure, atypical versions of cells (or entities having structural abnormality in one or more chromosomes) as well. They denote CAs implying certain observable (genetic) pathogeny (generally known as *genetic disorders*) at the species level.

Thus, while framing *cellular bioinformatics*, presence of both normal and abnormal chromosomal sets existing (in varying proportions) within the cytogenetic complex should be duly considered, and the heuristics of stochastic mixture theoretics can be evoked in ascertaining the associated *information redundancy* as a function of fractional abnormal chromosome population being present. This bioinformatic measure of redundancy can then serve as a track-parameter toward the progression of genetic functions as well as disorders (e.g., the growth of cancer).

7.2 Part I: Cytogenetic Informatics

As stated earlier, this chapter is comprised two parts: Part I describes entropic details of cytogenetics and Part II is devoted to outline the informatics of synteny heuristics and related correlation considerations.

As regard to Part I, relevant cytogenetic concepts implied refers to a branch of genetic biology that correlates the structure, number, and behavior of chromosomes in the cells (which primarily depict a set of DNA strings seen) portraying hereditary as well as abnormal (diseased) states. Cytogenetics, in short, is concerned with organized and/or disorganized aspects of cellular/chromosomal framework[1–4] with its constitutive entities as follows.

Chromosome and Its Structure

Normal human body consists of about 50 trillion cells. In the nucleus of relevant eukaryotic cell, the genetic material is well-organized with compactly packaged DNA and *histone proteins**. The plethora of such structures is known as chromosomes mentioned earlier. The DNA is packaged and statistically ordered into structural units called *nucleosomes* as indicated in Fig. 7.1. They are the main protein components of chromatin "acting as spools" around which the DNA winds and performs the role in gene-regulation. (*Histones** are highly alkaline proteins found in eukaryotic cell nuclei that package and order the DNA into structural units called nucleosomes. Histones play a viable role in gene regulation.)

As described in Chapter 1, a cellular constituent refers to a collection of original DNA in transcriptomic form specified as chromosomes shown in Fig. 7.1. With reference to chromosomal constituent structure, its distinct parts are further identified in Fig. 7.2. Described in O'Connor[4] are pertinent details on the parts of a chromosome. Centromere shown in Figs. 7.1 and 7.2 refers to a DNA region found in the vicinity of the middle section of chromosomes, where the "two sister chromatids" are proximally set close in contact. Furthermore, each chromosome has a constriction point known as *centromere* that divides the chromosome into two sections or arms. The short-arm of the chromosome is labeled as *p-arm* and the long-arm of the chromosome is labeled as *q-arm* as

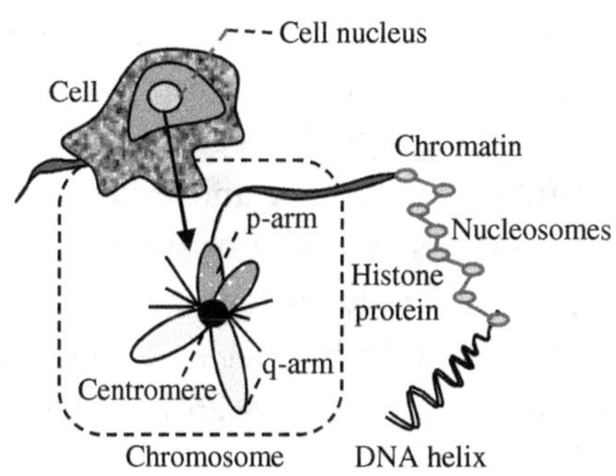

Fig. 7.1 Constituents of a chromosomal structure along with DNA and histones.

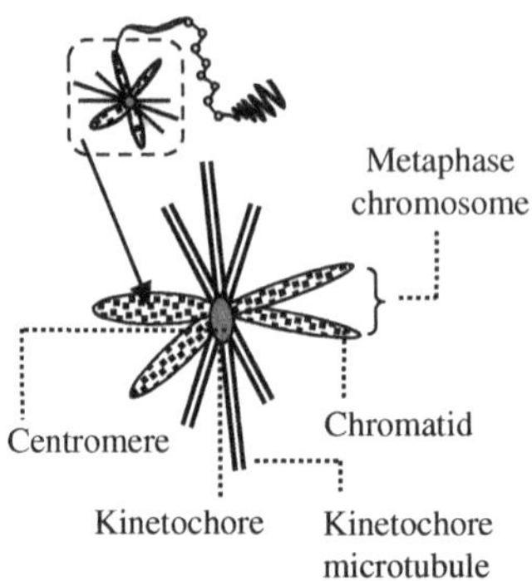

Fig. 7.2 Expanded representation of a centromere of a chromosomal unit.

shown in Fig. 7.1. The location of the centromere gives the chromosome its characteristic shape. The chromosomes are essential units responsible for cellular division, and they must be replicated, divided, and passed successfully to their daughter cells, so as to ensure genetic diversity and survival of the progeny. Moreover, an ordered organization of the genetic material at molecular level is required by the cell for its normal functionality across all living systems.

In the illustrations of Figs. 7.1 and 7.2, it can be seen that a chromosome has two arms, *p*-(shorter) and *q*-(longer) *arms*. Furthermore, considering the location of the centromere and connection of the arms, the chromosomes can be divided into six types, namely, *metacentric, submetacentric, acrocentric, telocentric, subtelocentric,* and *holocentric.* (However, human chromosomes are classified into only four types, namely, *metacentric, submetacentric, acrocentric,* and *telocentric.*) Each of these centromere versions is defined and described in 7.4.

Structurally, the DNA of a centromere is bendable (or curved) so as to support the associated tight-winding of DNA around protein binding sites (in prokaryotes and eukaryotes[5]). For example, the budding yeast (*Saccharomyces cerevisiae*) has a very simply organized centromeric DNA of only 125–200 bp long, and the region of 125 bp locus (known as *CEN*) plays a significant role in mitotic and meiotic chromosome segregation. The CEN sequence is typically organized into three domains, consisting of two highly conserved protein-binding sites (termed as *CDE I* and *CDE III*) flanking a 78–86 bp high ($A + T$) central sequence (*CDE II*). Mutational analyses point out that the 25 bp CDE III binding site is absolutely essential for centromere functions.[6]

The functional attributes in centromere are conserved in all eukaryotes and relevant features (such as in yeast to humans) and can be listed as follows: (i) centromere is the site of kinetochore assembly (a protein complex that drives chromosome segregation); they are formed at only one site on each chromosome. Furthermore, centromere depicts the last region where sister chromatids remain tethered by cohesion until the *anaphase* depicting the third stage of mitosis, where the daughter chromosome moves toward opposite poles; (ii) centromere incorporates a sensor, known as the *spindle checkpoint* that monitors the attachment of sister kinetochore to microtubules from both poles; hence, the spindle checkpoint oversees the tension across sister centromeres; and (iii) kinetochore-associated motor proteins are responsible for the movement of chromosomes along microtubules toward the spindle poles.

Human centromeres: These are relatively large, made of several million base pairs (bps) framing a bp sequence (known *α-satellite DNA*) and constituted by a number of *centromeric proteins* (CENPs), such as *CENP-A, CENP-B, CENP-C,* and *CENP-G* of protein family.[7,8] CENPs help in the regulation of *heterochromatic modification**, later followed by cell division. (**Heterochromatic modification* refers to certain changes observed in a tightly packed form of DNA (or *condensed DNA)*, which may exist in multiple varieties). Human centromere is recognized by the components of the kinetochore in cell division phases, and the centromeric DNA is normally in a condensed heterochromatin state.

There are two major classes of centromeres, namely, (i) *regional centromeres* wherein the DNA sequences are not defined and their functions are not known. These centromeres contain large amounts of DNA and are often packaged into heterochromatin. (ii) *Point centromeres,* which denote entities wherein the DNA sequences are both necessary and sufficient to specify the centromere identity as well as the functions in organisms. They are smaller and more compact and contain highly conserved DNA sequences that serve as binding sites for essential kinetochore proteins.[7–10] The so-called *somatic cells* of humans make up the internal organs, bones, blood, connective tissues, and skin. They consist of 46 chromosomes organized into 23 pairs. In each pair, one chromosome comes from maternal and the other from paternal source. There are 22 pairs of *autosomes* that determine the genetic traits and one pair of *allosomes* that specifies the sex, typically indicated as XX for female and XY for male.[2–4]

7.3 Chromosomal Aberrations

CAs reflect the abnormality aspects of the chromosome *vis-à-vis* the number or structure. They usually occur whenever there is an error in cell-division following meiosis or mitosis. Some factors that influence *CAs* are as follows: higher maternal age during accouchement, abiotic ambient, ionizing irradiations, autoimmunity, viral infections, and chemical toxins, etc.

Numerical aberrations in chromosomes occur when a default is encountered in the usual number of chromosomal pairs. They depict whole chromosomal aneuploidy representing a significant proportion of chromosomal changes found in humans. These aberrations can occur as a consequence of chromosome segregation defects during cell division. When an individual is missing a pair of chromosomes, a condition known as *monosomy*, as in *Turners syndrome* 45X occurs or when an individual has more than two chromosomes in a pair, a genetic disorder occurs as observed in *Down's syndrome 21 Trisomy*.

Structural aberrations in chromosomes imply the state that individual chromosomal structure is altered. Relevant structural variations can be classified into the following five categories: (i) *deletions* depicting a portion of chromosome missing or deleted; (ii) *duplications* where a part of chromosome is duplicated resulting in extra genetic material; (iii) *translocations* implying a part of chromosome transferred to another chromosome; (iv) *inversions* denoting a part of chromosome broken and again getting attached upside down rendering, the genetic material inverted, and lastly (v) the formation of *rings*, with a portion of chromosome broken off and assuming a shape of ring (with or without the loss of genetic material).[3]

7.4 Disorganization in a Cytogenetic Framework

When normal chromosomes are spliced and incorrectly repaired, then chromosomes with absence of centromere or addition of multiple centromeres could be formed. Relevant cytogenetic framework depicts a disorganized biocomplex of stochastic attributes. Such aberrant structures are undesirable inasmuch as they do not segregate properly and may often get lost from the dividing cell, and disorganization in cells implies errors in centromere proteins resulting in autoimmune diseases.[11] Furthermore, a perceivable dimension of cellular complexity could arise from *chromatin** constituents. (*Chromatin** refers to a complex material constituted by DNA and protein in eukaryotic cells.) Its primary function is packaging very long DNA molecules into a more compact, denser shape

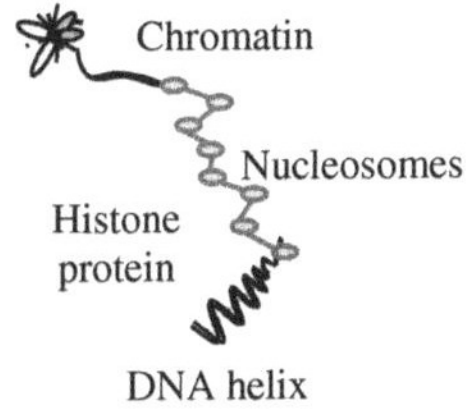

Fig. 7.3 Structure of chromatin.

which prevents the strands from becoming tangled and plays important roles in reinforcing the DNA during cell division.[12]) Details of cellular complexity can be observed in human cytogenetic features as follows: human beings have 23 pairs of chromosomes per cell. That is, a total of 6 billion bps of DNA exists per cell, and a normal human body may consist of about 50 trillion cells.

In general, considering a eukaryotic cell, the genetic material is well organized by compactly packed DNA (*via* electrostatic forces) with histone proteins in the nucleus achieving the status of chromatin. (Chromatin is not found in prokaryotes, like bacteria, that lack nucleus.) Furthermore, the phosphate groups of the DNA composition are fundamental structural units of chromatin. It is an assemblage of DNA by ester bonds. The simplified structure of chromatin is illustrated in Fig. 7.3. Mostly the associated protein consists of multiple copies of five versions of histones. (Chromatin may also consist of nonhistone proteins, such as *transcriptional factors* (TF) that are transiently present in very small amounts).

Histone Proteins

Histones keep the DNA organized and also they help to regulate expression of genes. Chemically, they are highly alkaline and their function is to order the DNA into structural units of nucleosomes. In all, the histones being major protein components of chromatin act as spools around which the DNA helically winds.

Nucleosome represents a subunit of chromatin composed of a short-length DNA wrapped around a core of histone proteins. The human genome contains 23 chromosomes; that is, approximately 3 billion nucleotide pairs are formed into a compact organization of nucleosome. Each such nucleosome is about 11 nm in diameter. The DNA double-helix wraps around a central core of eight histone proteins to form a single nucleosome and a second histone protein (H1) fastens the DNA to nucleosome core.[12,13]

Chromatin Organization

Based on microscopic observations, two-levels of chromatin organization exist: (i) *heterochromatin* and (ii) *euchromatin*. They refer to states of compaction (DNA and histones) and their transcriptional potential. Heterochromatin denotes a tightly coiled form of DNA into a condensed compaction. It is genetically inactive but controls metabolism, transcription, and cell division (interphase) activities. Genes present in heterochromatin are generally inactive (with no transcription) and an increased methylation of cytosine could be observed in CpG islands of the promoter in the gene. In *euchromatin DNA*, histones are loosely or lightly packed and are transcriptionally active. Chromatin in this state stains lightly in *karyograms*.[14] Heterochromatin and euchromatin differ in their biophysical conformations and in metabolic expression of their genes but same in their basic structure of DNA arranged within chromosomes.[14]

Prenucleosomes

A chromatin particle in prenucleosomes depicts a precursor of nucleosome made up of intermediate DNA–histone complex.[15] The nucleosome is the basic repeating unit of chromatin and prenucleosomes get converted into nucleosomes by motor proteins that use the energy molecule, Adenosine Triphosphate (ATP). Packing of DNA with histone proteins in forming the chromatin allows stabilizing the chromosomes and enables regulation of genes, mainly in DNA replication. The prenucleosome could inform how genetic material is duplicated and used.[15]

7.5 Chromosomal Aberrations

Entropy-inspired cytogenetic analyses can be indicated to determine quantitatively the extent of normal and abnormal features of chromosomes. Any atypical features in chromosomes imply *chromosomal abnormality* (*CA*) or *chromosomal aberrations*, and there are two regimes wherein the *CA* can be observed and assessed. They correspond to: (i) aberrations at nucleotide level and (ii) those observed at cytogenetic level. Pertinent details are described as follows:

CA at nucleotide level: The chromosomal variations at nucleotide level account for about 10%–12% of human genomic DNA; such genetic variations include deletion, duplication, or inversion occurring in the range of 1 kb to several mb in size; in contrast, *single nucleotide polymorphisms* (SNPs) that differ *copy number variations* (CNVs)* on a single nucleotide base. (The CNV* is a type of structural variation depicting genome sections wherein genomes are repeated, and the number of such repeats could vary between individuals across the domain of human population. Typically, CNVs conform to duplication or deletion events affecting a considerable number of base pairs.) In essence, the CNVs play a role in evolutionary adaption in humans as well as in other mammals.[16–19]

CA at cytogenetic level: CNVs can be seen in one or more sections of the DNA of the chromosomes in the cells. Such CNVs include large regions of genome that are insertions, deletion, duplications, and translocations of base pairs.

Variation details in cytogenetic CNVs are caused by inheritance or by *de nova mutations** (A *de nova mutation** or *de novo variant* implies a new mutation, or a new variant, which is neither parent-possessed nor transmitted, and it denotes a genetic alteration that prevails for the first time in one family member due to a variant (or mutation) in a germ cell (egg or sperm) of one of the parents or a variant that could stem from a fertilized egg itself during early embryogenesis).

CNV detection: The classical method used in detecting, analysis, and/or validation DNA CNV was based on *bacterial artificial chromosome* (BAC) arrays, and today CNVs are detected by other types of tests as well, such as *fluorescence in situ hybridization, comparative genomic hybridization, high-resolution array-based tests via n-array comparative genomic hybridization*, and *quantitative PCR-based technique,* and the most efficient method used in detecting CNVs is called array-based method or *virtual karyotype.*

SNPs and CNV

As observed in earlier chapters, a genome represents all the genetic material within the chromosomes. In such genomic contexts, the *SNPs* (as contrasted with CNVs earlier) represent a DNA sequence variation that occurs at a single nucleotide. The SNPs occur throughout the individual DNA, mostly in noncoding regions in the *coding sequence* (CDS) part of the genome. In the suite of central dogma of microbiology, the DNA complex is transcripted into RNA, and relevantly the *transcriptome* is defined as a set of all RNA molecules in one cell or a population of cells.[20]

Dysregulation of normal genes can cause the conversion of normal cells into cancerous cells. Typically, overexpression of gene is observed in the dysregulation. Furthermore, mutations in human DNA repair genes can also deregulate normal cellular processes; as well as, active genes when transposed could result in activation of silent genes.[21]

Clonal or Nonclonal Chromosomes

Cytogenetic characterization and classification of CA involve elucidating normal karyotypic picture of chromosomes and identifying major recurrent CAs. Cells that accumulate mutations correspond to a somatic evolution and implicate natural processes such as aging and development of diseases, like cancer. The survival of the cell in the somatic evolution with acquired mutation is decided by the increased fitness of the cells.

Cells in neoplasmic (tumorous type of abnormal and excessive growth) conditions try to increase their fitness with the resources like oxygen, glucose, and space. Furthermore, they generate more daughter cells competitively as compared to cells that are devoid of mutation. Such mutant cells (called *clones*) expand in the neoplasm with the available resources. Relevant clonal expansions form the signature of natural selection in cancer; and correspondingly, the population of aberrated cells (that promote the neoplasmic, cancerous conditions) are called *clonal chromosomes*.

The biological significance of clonal, karyotypic abnormalities in neoplasms indicated above has been understood to a fair extent. However, there are CAs that are not consistent within the cytogenetic definition of a clone, but do exist as an artifact manifesting in random losses of cells.[22,23]

Contamination Karyotypes

In cytogenetic contexts, molecular karyotyping involves tissue culture, which shows the presence of contaminations that may limit the efficacy of karyotype evaluations. For example, details pertinent to karyotyping products of conception with maternal cell contamination (MCC) are useful in assessing the fetal welfare. That is, a cross-contamination of cytogenetic entities of two systems (such as maternal and fetal) implies contamination type karyotypes. Males with so-called 46 *XX* karyotype are referred to as *XX male syndrome* or *de la Chapelle syndrome*. They are phenotypically male, but they are genetically classified as female (clinically termed as *XX* male syndrome). Such incidences are however rare (4-in-10^6). Similarly, an individual having 46 *XY* and not representing male characters represent *XY complete gonadal dysgenesis* (CGD) or *Swyer syndrome,* although in genetically in such male fetuses, the *Y*-chromosome is present.

CA and Genetic Disorders

Genetic disorder is a diseased-state caused by abnormalities in genes or chromosomes. Usually they could be present from or before, the birth. Sometimes genetic disorders are passed down from parent's genes. Alternatively, they could be caused by new mutation changes in the DNA. For example, same type of cancer can be caused by an inherited genetic condition or by mutation due to nongenetic causes.

A few thousands of human diseases are caused by single-gene disorder arising from a mutated gene. Based on chromosomal location, genetic disorders are classified as *autosomal* and *X-linked* types*. (**Autosomal* and *X-linked* types of genetic disorders depict two versions of inheritance patterns that describe the inheritance of a particular genetic trait from one generation to the next.)

CA and Oncological Relevance

Accumulation of genetic alterations in the cells causes human cancer. Nowell and Hungerford in 1960[24] discovered the CA associated with cancer (as in chronic myeloid leukemia, which implies an unrestricted growth of myeloid cells in the bone marrow). Relevantly, a gene that has the potential to cause cancer can be specified. It is known as an *oncogene.* Oncogenes can be classified into five groups based on the functional and biochemical properties of protein products and their normal counterparts (proto-oncogenes).[25,26] In tumor cells, oncogenes are often mutated or expressed at high levels.

The model depicting occurrence of genetic abnormalities can be divided into two types namely, the *linear structure model* and *oncogenetic tree-based structure model* as identified below:

Linear model: The genetic profile of individual tumors varies widely because no single mutation is present in all tumors, and certain genetic changes tend to occur early in the development and others relatively late stages.[27]

Oncogenetic tree model: Oncogenetic trees show multiple possible pathways and parallel progression along several pathways in the same tumor. Relevant model was first studied by Desper *et al.*[28] It is subdivided into branching trees and distance-based trees.

7.6 Evaluation Of CA

Traditional evaluation that measures cell contamination is based on representing cell contents by simple fractions of normal and abnormal chromosomes present. For example, indicated in Nikitana *et al.*[29] is an estimation method on the incidence of chromosomal abnormalities (for a given male/female sex ratio of conception) *versus* spontaneously observed abortions masked by *MCC*. Pertinent results obtained in Nikitana *et al.*[29] are reproduced in Table 7.1. The pursued analysis in Nikitana *et al.*[29] refers to determining the frequency (or relative percentage) of spontaneous abortion *versus* MCC under the conditions, such as those indicated in Table 7.1.

Essentially, the traditional approach (as in Nikitana *et al.*[29]) of analyzing a mixture of normal cellular contents with contaminations like

Table 7.1 Details on MCC in two groups of "46, *XX*" spontaneous abortions, one with and the other without *Y*-chromosome.[29]

Exogenous characteristics	Symbol	"*Y*+" spontaneous abortions (n = 18): With *Y* chromosome	"*Y*−" spontaneous abortions (n = 94): Without *Y* chromosome
Gestation	X_1	8.880 ± 2.600	8.860 ± 2.760
Maternal age	X_2	26.00 ± 6.540	25.54 ± 5.270
Paternal age	X_3	27.21 ± 6.360	28.28 ± 5.580
Diagnosis: blighted ovum	X_4	4	20
Diagnosis: missed abortion	X_5	12	62
Diagnosis: others	X_6	2	12
Tissue type for culture initiation: *extra-embryonic mesoderm*	X_7	14	64
Tissue type for culture initiation: chorionic villi	X_8	3	30
Tissue type for culture initiation: duration of long-term culture	X_9	35.08 ± 16.88	24.05 ± 13.86

abnormal chromosomes relies on specifying each content in terms of their relative percentage; that is, in terms of simple prorated values. Suppose the normal chromosomes content is N_c% and the percentage of aberrated chromosome is N_{ac}%, the total ($N_c + N_{ac}$) is equal to 100%. Then, the data acquired clinically or otherwise would denote the underlying inferential statistics on the associated variances of N_c and N_{ac}.

The results in Table 7.1 on MCC (relevant to two groups of "46, *XX*" spontaneous abortions, one with and the other without *Y*-chromosome) are deterministically specified with no error bar on the observed inferences; that is, the analysis does not account for any possible statistical spread on the associated variables (enlisted as X_1 through X_9). Therefore, the model in Nikitana *et al.*[29] can be improved to include risk statistics posing an error bar, allowing an explicit statistical spread on the results. Hence, a logistic regression on the exogenous entities can be prescribed as described in Chapter 6 and upper- and lower-bounds on the underlying risk statistics of the MCC can be deduced using Langevin–Bernoulli function (LBF)-based logit function. Defining a variable, $z = (X_1 + X_2 ... + X_9)$, the risk-factor (RF) associated (e.g., with the aforesaid set of variables, $\{X_1, X_2, ... X_9\}$) can be deduced using a modified the logit-function: $f(z) = \{1/2 + (1/2) \times L_q(z/2)\}$ as indicated by DuPont,[30] where $L_q(.)$ denotes the LBF. Relevantly, as elaborated in Chapter 6, the entity q associated with LBF depicts two extremities of stochastic orders, corresponding to ($q \rightarrow$ ½) and ($q \rightarrow \infty$), and they prescribe a pair of limits, namely, the upper- and lower-bounds (UB and LB) on the outcomes of $f(z)$ obtained by a set of ensemble of random trials on $\{X_i\}$ (depicting, for example, the MCC-specific details as above). The UB and LB on the underlying risk statistics of the MCC can be deduced using the LBF-based logit function. The following example illustrates the approach as above.

Example: Ex-7.1

Statement of the Problem

This example refers to the problem of assessing the RF of net contamination details using the clinical data on MCC[29] presented in Table 7.1, and the log-regressed details are enforced with UB and LB limits, *via* the stochastic orders, ($q \rightarrow$ ½) and ($q \rightarrow \infty$), respectively. The following cytogenetic mixture model and statistical aspects of MCC are applied to the case-study being modeled.

Considering the MCC to depict the presence of certain undesired cellular entity (that could lead to spontaneous abortions), a simple *proportionality model* is indicated in Nikitana *et al.*[29] on the incidence of MCC *versus* chromosomal abnormality. Relevant modeling enables the estimation of the risk involved (as a result of MCC) as decided by the variables of the data in Table 7.1 (that correspond to clinical samples acquired from 97 patients with aborted embryos).

The RF deduced thereof in Nikitana *et al.*[29] however conforms to yielding a single value and it ignores various other associative considerations of stochastic nature of regressing parameters characterized by the set, gestational age, maternal and paternal ages, diagnostic aspects of blighted ovum and missed abortion details, tissue type used for culture initiation, and duration of the long-term culture. All such parameters, however, possess characteristics that are mostly random (and partly deterministic), but they all fall within a pair of specified upper- and lower-bounds. As such, the parameters designated by an ensemble set of exogenic, (random) variables $\{X_i\}$ is compatible for log-regression *via* the function $f(z)$ with $z = \sum_i X_i$; hence, the risk factor due to MCC can be assessed.

Model preview: with reference to MCC, the analytical model formulated in Nikitana *et al.*[29] refers to estimating the incidence of chromosomal abnormalities and a set of proportional modeling of cytogenetic factors of prenatal selection corrected for cell contamination, and empirical expressions thereof on observed and expected values of cytogenetic factors (namely, 46 *XX* frequency, frequency of female and male spontaneous abortions with chromosomal abnormalities, frequency of chromosomal abnormalities in total sample, frequency of chromosomal abnormalities in the "46 *XX*" group, and sex ratio in spontaneous abortions with normal karyotype) are presented.

A high rate of MCC may distort the state of chromosomal abnormality encountered in spontaneous abortions across the first trimester of pregnancy. In estimating such MCC-based implications at cytogenetic level of prenatal situations *vis-à-vis* abortions, the model due to Nikitana *et al.*[29] considers N number of samples of spontaneous abortions observed with four major versions of chromosomal constituents, namely, (A) 46, *XX*; (B) female spontaneous abortions with chromosomal abnormalities; (C) 46, *XY*; and (D) male spontaneous abortions with chromosomal abnormalities.

Hence, N taken as the sum of (A), (B), (C), and (D), where the set $\{A, B, C, D\}$ depicting the karyotypes, is explicitly specified as follows:

(A): 46, XX, denotes chromosomes in maternal cells and it is subdivided into four types with reference to the sex-ratio of the embryo: (A_{fn}: XX-female normal); (A_{fa}: female abnormal having spontaneous abortions); (A_{mn}: XY-male normal); and (A_{ma}: male abnormal having spontaneous abortions). As such, it follows that $A = (A_{fn} + A_{fa} + A_{mn} + A_{ma})$. For a representative sample of N, the relative proportions of B, C, and D and A_{fn} and A_{fa} are indicated in Nikitana *et al.*[29] The mathematical model in Nikitana *et al.*[29] is based on the structure of cytogenetic factors expressed in terms of proportions of various chromosomal entities identified earlier as A, B, C, and D; hence, a pertinent factor is indicated as $k = (C + D)/(C + D + B)$. It defines the probability of male embryo detection in the A group. The influence of MCC as a function of k is then estimated on the spontaneous abortions observed. For a specific set of details presented in Nikitana *et al.*[29] and considering the karyotypes (46, XX), relevant representative values on B, C, and D are, respectively, as follows: B (abnormal female = 139), D (abnormal male = 94), and C (46, XY = 86). The numerical values indicated in Nikitana *et al.*[29] are typical sample sizes of the observed numbers in the model of MCC.

With reference to Table 7.1 and (Y+) status, the sum total of exogenous characteristics that decide the underlying risk, namely, $z = \Sigma_i X_{i=1,2,\ldots 9}$ is determined for a nominal set of $\{X_i\}$. Hence, $f(z) = [(1/2) + (1/2) \times L_q(z/2)]$ is evaluated. Inasmuch as Table 7.1 suggests, the existence of a span of deviation with respect to the nominal values of each X_i, a number of uniformly distributed random trials on each X_i is specified, and corresponding $f(z)$ is evaluated for each ensemble set of $\{X_i\}$. (For example, say, with $X_1 = 8.88$ taken as a nominal value, it is varied randomly in each trial with the deviation, ± 2.60 indicated in Table 7.1). Hence, the risk factor is deduced *via* $f(z) = [½ + ½ \times L_q(z/2)]$ over an ensemble of several statistical trials. The evaluated result is plotted in Fig. 7.4a; furthermore, with $q = ½$ (upper-bound) and $q \to \infty$ (lower-bound), corresponding data computed are also shown in Fig. 7.4a. In addition, the *infimum* and *supremum* pair is deduced on the bounds assuming a statistical quantile of 48% about the mean value.

With reference to MCC, there are two groups of "46, XX" specific spontaneous abortions, with and without Y chromosomes identified, respectively, as follows: *Case-1a — "Y+" spontaneous abortions* and *Case-1b — "Y−" spontaneous abortions.* The results on these case studies show the following observations: (i) with no MCC being present, the associated risk is as low as 20% and (ii) statistical variations implied on various factors of

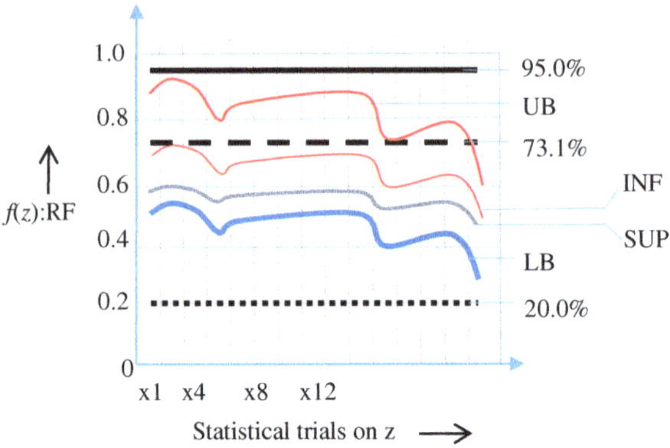

Fig. 7. 4a Risk-factor (RF) *versus* ensemble of trials (T_1, T_2...,T_{14}) of data presented in Table 7.1 for (*Y*+) state: Case 1a. (UB: Upper-bound; LB: Lower-bound; INF: *Infimum*; SUP: *Supremum*. 20% RF implies no contamination).

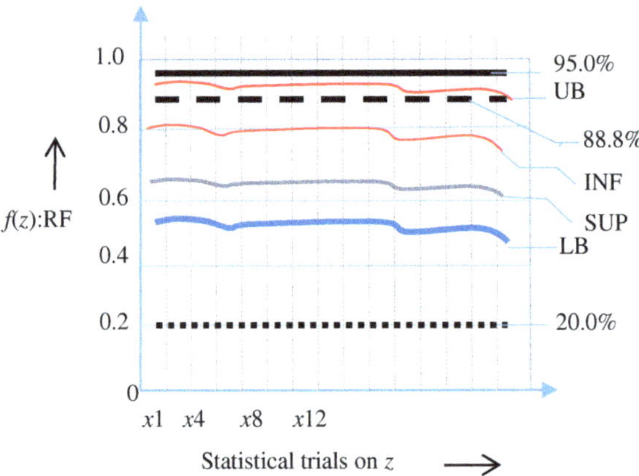

Fig. 7.4b Risk factor *versus* an ensemble of trials (T_1, T_2...T_{14}) of data presented in Table 7.1 for (*Y*–) case: Case b). (UB: Upper-bound; LB: Lower-bound; INF: *Infimum*; SUP: *Supremum*. 20% RF implies no contamination).

the set, $\{X_i\}$ of the contaminated state lead to the prediction of the risk *via* logistic-regression as mentioned before. Correspondingly, an upper- and a lower-bound of the RF can be deduced along with *infimum* and *supremum* limits as illustrated in Figs. 7.4a and 7.4b.

In summary, the model presented in Nikitana *et al.*[29] signifies the extent of possible spontaneous abortion that could result from MCC due to various prenatal selection factors, and the relevant study is based on proportional mixture considerations on contaminated and uncontaminated entities in the cultured cells. For a given set of *A*, *B*, *C*, *D*, *N*,

k, and male–female ratio (given in Table 7.1), the associated risk value as per traditional logistic regression for (Y+) case is 0.731, and the corresponding value for (Y–) case is 0.888. In short, the work reported in Nikitana *et al.*[29] gives on a single RF value. However, an extended study based on LBF logistic-regression provides a range on contamination-related risk factor of possible abortion (specified with an error span between upper- and lower-bounds, LB and UB, depicting a statistically plausible result).[31]

Tutorial: T-7.1

This section outlines stochastic mixture modeling of a cytogenetic complex containing an admixture of normal and aberrated chromosomes. The following details are first indicated thereof on relevant considerations:

A randomly mixed state of two or more entities can be generally specified as a simple, prorated ratio of each existing population, as in Example Ex 7.1. However, such an arithmetic proportion could only be a limiting case of a "truly statistical mixture" as observed by Lichtenecker and Rother[32] (and by Neelakanta *et al.*[33]). Though not specifically applied in biological contexts, such statistical mixture formulations of Lichtenecker and Rother,[32] Neelakanta,[33] and Neelakanta[34] can be considered as judicious candidates, for example, to quantify the state of chromosomal mixture constituents in a cytogenic complex. Hence, considered here is a way to model the quantitative extents of normal and aberrated chromosomes that may exist as a statistical mixture in a receptacle. Hence, the following can be stated toward intended modeling: identifying statistically mixed constituents at cytogenetic level *via* statistical mixture theory, availing literature details on typical and pathogenic cytogenetic mixture constituents, modeling the dynamics of growth (or decay) of a specific constituent (such as aberrated chromosomes), determining the extent of CAs expressed *via* statistical mixture theory and studying the competitive growth and decay dynamics of normal and abnormal chromosomes

Proportional-content theoretics: revisiting the details in Example Ex 7.1, the cytogenetic state of the associated chromosomal constituents is viewed in terms of fractional population of normal chromosomes *versus* other cell contaminations. Hence, a mathematical model depicting the MCC estimated in terms of various embryonic chromosomal attributes is illustrated and the fraction of cell contamination implied is termed as a *cytogenetic factor*, and the net effect of such contamination is addressed *via* traditional statistics of expected average of the contents involved, and the pertinent study in Nikitana *et al.*[29] follows the classical approach using statistical χ^2-method. Hence, the significance of such analysis is indicated toward prenatal selections in spontaneous abortions caused by abnormal chromosomal contents. Thus, the simple proportion model on normal and abnormal chromosomal contents at embryonic level enables deducing the gross abortion risks in subjects at clinical levels.

In terms of informatic statistics of chromosomes, an efficiency factor related to the constituents of the cytogenetic complex (possessing aberrant chromosomes) and the extent of abnormality can be optimally deduced *via cytogenetic information efficiency factor* (η) defined as follows: suppose the universe (Ω) represents the cellular complex system of mixture entities containing normal and abnormal chromosomes. Denoting the total chromosomal population as N and a constant c_i is assigned as a *cost-factor* to each chromosome (indexed as $i = 1, 2\ldots, N$) with an occurrence probability, $P_{\ell 1}$, the said information efficiency (η) factor of the cytogenetic complex depicts the ratio of average information (per chromosome) of the ensemble to the maximum possible (average) information (per chromosome). It is specified as follows[35]: $\eta = H(x)/[\bar{C} \times \ln(N)]$, where $\bar{C} = \Sigma_i^N c_i \times P_{\ell i}$ is the average (first moment) of C, and concurrently, $(1 - \eta)$ can be regarded as a *redundancy factor* (R). It refers to the reduction in information content of an ensemble from the maximum possible value, and it is given by $R = (1 - H/H_{max})$. Furthermore, H denotes the entropy functional depicting any statistical divergence (or distance) metric, such as Jensen–Shannon (JS), Kullback–Leibler measure, etc.[31] as elaborated in Chapter 2.

Mixture attributes of cytogenetic complexity: the aforesaid redundancy factor (R) can be used to quantify the mixture inclusion attributes of a cytogenetic complex *via* statistical mixture theoretics due to Lichtenecker and Rother.[32,36] The underlying heuristics specifies a weighted probability, r that describes the effective statistical attribute of the mixture proportioned by θ and $(1 - \theta)$ depicting the fractions of binary inclusions. That is, with reference to a binary mixture of two constituents (1 and 2) depicting populations (n_1 and n_2) placed in a receptacle, θ and $(1 - \theta)$ are defined as follows: $\theta = [n_1/(n_1 + n_2)]$ and $(1 - \theta) = [n_2/(n_1 + n_2)]$, and the associated weighted probability r can be written as follows: $r(\theta) = P_{1\ell}^{\theta} \times P_{2\ell}^{1-\theta}$, which is valid, within the statistical upper- and lower-bounds, namely ($r_{min} \leq r \leq r_{max}$). Explicitly, r_{min} and r_{max} are given by $r_{max} = [\theta \times P_{1\ell}] + [(1-\theta) \times P_{2\ell}]$ and $(1/r_{min}) = [\theta / P_{1\ell}] + [(1-\theta)/P_{2\ell}]$.

The statistical bounds indicated above conform to the extreme arithmetic and geometrical-mean statistics of the constituents. With reference to the set $\{r\}$, the corresponding Shannon measure of entropy (negentropy) can be written as a function of r as follows: $I(r) = -H(r) = [(-r) \times \ln(r)]$ and its maximum and minimum limits[37] can be obtained by replacing r, respectively, by the set: $\{r \rightarrow r_{min}$ and $r_{max}\}$.

Suppose one of the constituent entities of the statistical mixture, say, the one with a population n_2 has a uniform distribution implying that the occurrences of its elements (in the statistical-mixture space) are equally likely. That is ($P_{21} = P_{22} = P_{23} = \cdots = P_{2n2} = 1/n_2$), so that $[P_{21} + P_{22} + P_{23} + \cdots + P_{2n2} = 1]$. In contrast, the other constitutive entity (having a population n_1) is presumed to be of elements each bearing a distinct probability of occurrence. That is, ($P_{11} \neq P_{12} \neq P_{13} \neq \cdots \neq P_{1n}$) and $[P_{11} + P_{12} + P_{13} + \cdots + P_{1n} = 1]$.

In the context of cytogenetic complex, the population of aberrated chromosomes can be regarded as one with no information; as such, it belongs to a subset of cardinality, n_2 having a probability distribution of equally likely occurrences. Such uniformly distributed entities are consistent with the so-called Laplacian hypothesis on probability of equally likely epochs. The aberrated chromosomes coexisting with normal chromosomes in a mixed state within a cellular system (with population n_2) are noninformative as regard to the regular functions of the cellular complex. On the contrary, presence of such aberrated chromosomes implies a state of maximum entropy resulting from the uniformly distributed statistical attributes. However, pertinent to each normal chromosome, it prevails with a distinct (unequal) occurrence probability as decided by its designated structural and functional characteristics. Hence, such normal chromosomes can be regarded as informative (negentropic) entities of the cytogenetic complex (and these informative chromosomes could be redundantly present).

In view of general considerations on cytogenetic complexity expressed in terms of statistically mixed state of normal and abnormal chromosomes as outlined above, relevant scope can be extended to address the following specific tasks on certain clinically observed pathogenic conditions (such as, cancer growth) *versus* the chromosomal abnormality: (i) *copy-number alterations/variations* (CNVs) *versus* their involvement in tumor growth; (ii) *ploidy* and *aneuploidy* involvement in oncological contexts; (iii) mixture-state model of cytogenetic complex in terms of *clonal* and *nonclonal alterations* implicating cancerous growths; and (iv) spatiotemporal dynamics of cancer growth.

7.7 Review Section: Tutorials, Examples, and Exercises

Tutorial: T-7.2

This tutorial is intended to provide necessary informatic details toward developing a cytogenetic-specific, pathogenic models. One such model refers to *copy number alterations* (or CNVs, depicting sections of a genome characterized with repeated residue details). The number of such repeats could vary between individuals in the human population. Furthermore, CNV could imply a type of structural variation involving duplication or deletion of an event that may affect a considerable number of base-pairs on a genomic sequence. In short, copy number and its alterations depicting changes in DNA of a genome could eventually result in the cell having abnormal number of copies of one or more sections of the DNA.[38] Aberrations of this type may affect functions of a gene. Understanding (as well as quantifying such aberrations) is essential in comprehending the associated disease etiology, and it could possibly help developing targeted therapies for gene-related pathogenic states.

Developed in Taylor *et al.*,[38] is a multicomponent scoring model for CNV/CNA behind genetic defects. The presence of such undesirable CNA entities is modeled as "noise" and the associated tumor heterogeneity is described in terms of related "noisy impurities" commonly designated as *stromal admixture constituents*. Typically, four scoring parameters are identified in Taylor *et al.*[38] in this context to quantify the CNA. These are: (i) *single-copy gain* (A_0), (ii) *amplification* (A_1), (iii) *hemi-zygomatic loss* (D_0), and (iv) *homozygous deletion* (D_1). More related definitions on CNV-related considerations are as follows:

Ploidy: This depicts the number of complete sets of chromosomes in a cell on possible alleles for *autosomal* and *pseudoautosomal* genes (*Autosomal gene* a gene located on a chromosome other than the sex chromosomes and a *pseudoautosomal gene* depicts a gene that occurs on both sex-determining heteromorphic chromosomes).

Aneuploidy: It refers to the presence of abnormal chromosomes in a cell and it is an indication of chromosomal abnormality. Aneuploidy could imply either missing or the presence of extra chromosome states within the cytogenetic complex causing genetic disorders with relevance to some forms of cancer.

Somatic structural variation (SV): This includes rearrangements seen in the cellular body — involving CNV manifesting as deletions and amplifications (usually above 1 kb), intrachromosomal inversions, interchromosomal rearrangements and the insertion of transposable elements (TEs), and other exogenous sequences like viral DNA and mitochondrial DNA. Such somatically acquired structure variations (SVs) as well as CNVs can induce genetic changes that are directly related to tumor genesis. Somatic SV/CNV detection using *next-generation sequencing* (NGS) data faces major challenges resulting from tumor sample characteristics, such as ploidy, heterogeneity, and purity.

The fraction of ploidy and aneuploidy states, for example, may estimate the extent of tumor growth as described in Taylor *et al.*[38] with reference to breast carcinoma. The infiltration aneuploidy into the nonaberrant cells would result in tumor conditions across different extents of ploidy, from haploid to polyploidy levels. Given a ploidy, the tumor is observed when a corresponding aberrant cell fraction exists. Typically for low ploidy, high fractions are indicated to confirm tumor conditions. With higher ploidy, however, tumor can be seen even at lower aberrant cell fraction. An exercise called *allele-specific copy number analysis of tumors* (ASCAT) first determines the ploidy of tumor cells and specifies the fraction of aberrant cells.

Example: Ex 7.2

Statement of the Exercise

This example illustrates developing a model that represents the dynamics of cancer evolution resulting from patterns of CAs.[39] In cytogenetic contexts, the abnormal number of chromosomes (or chromosome abnormality) can be specified in terms of *clonal chromosomal aberrations* (CCA) and *nonclonal chromosomal aberrations* (NCCA). The definitions of CCA and NCCA are as follows: the NCCAs include numerical changes (aneuploidy and structural aberrations). When the aberration observed due to translocation is less than 20%, then it is considered as NCCA; otherwise, it is CCA.

The genesis and growth of cancer can be specified by a kinetic model in terms of proportions of NCCA and CCA. Presented in Pierdomenico *et al.*[40] is such a kinetic model of cancer evolution and the cancer progression is elucidated in terms of certain entities and their related pathway considerations. The interacting dynamics of normal cell (A), NCCA (N), and CCA (C) plus the associated pathways of interaction are illustrated in Fig. 7.5.

In Fig. 7.5, there are interacting coefficients denoted as k_1, k_2 ..., k_6. Furthermore, the set $\{k_1, k_2\}$ imply balancing trend imposed toward depletion of A, if balancing trend is positive; otherwise, A is subjected to repair and application; likewise, k_3 and k_4 represent the balance between the depletion of CCA (C) and NCCA (N), respectively, if positive, and if negative, it implies the repair and replication of CCA and NCCA, and the set k_5 and k_6 denotes the balance between the demise of NCCA and CCA, respectively, if positive, and when negative, it implies the repair and replication of NCCA and CCA. Furthermore, the four quantities A, N, C, and D (denoting the concentration of nonfunctional chromosomes) can be rendered as dynamic variables changing with respect to time (τ), and at any instant (τ), $A(\tau) + N(\tau) + C(\tau) = 1$.

It is assumed in Pierdomenico *et al.*[40] that A, N, and C are growth functions (e.g., exponential functions of time with a time constant) and relevantly, the resulting cancer growth function is indicated in terms of the functional attributes of A, T, and C with respect to time in the context of tumor proliferation. A complex system model of the statistical

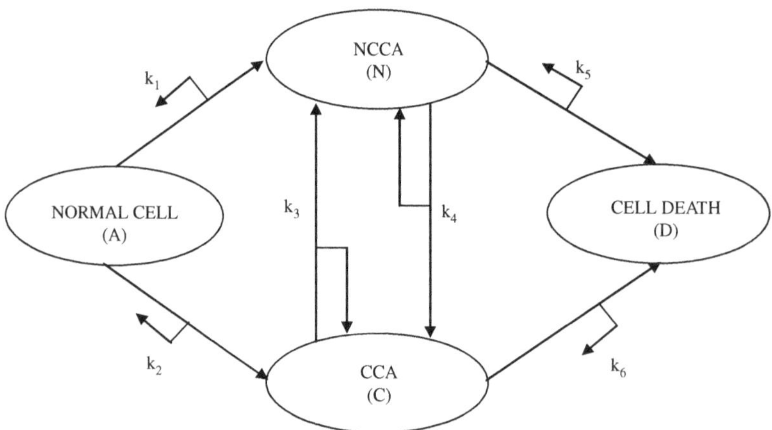

Fig. 7.5. Proposed pathway for cancer evolution and progression when k_n is equal to the pseudo reaction rate constant.[40]

mixture (of CCA and NCCA entities) can be ascribed to specify the associated spatiotemporal dynamics implying the growth or decay of constituent populations.

Tutorial: T-7.3

Narrated in this tutorial are pertinent details toward quantifying the stochastic mixture attributes of the contents in the cytogenetic complexity in terms of redundancy attributes.[35]

Furthermore, the maximum entropy concept[41–43] can be applied to each attribute group of the complex system universe, **Ω**: *X*, (which depicts, in essence, a mosaic of statistical mixture constituted by a pair, or multiple, subsystems of compositional domains). The model is defined in terms of the large set of disordered entities (constituting the complex system), and a complexity metric S is defined as a dichotomy of two regimes: (i) $0 \leq S < 1$ and (ii) $(1 < S < \infty)$. When S is very small $(S \to 0)$, the system is regarded as "simple", and as $S \to \infty$, the system becomes totally complex. (Hence, the value of $S = 1$ is a transition that bifurcates the system of being simple or complex when viewed in terms of the entropy involved.)[44–47]

If the cellular contents at the cytogenetic level are expressed in terms of the quantitative extents of the mixture constituted by normal and aberrated chromosomes, pertinent details can be adopted in deducing certain genetic-based diseases such as *carcinogenesis.* As such, mathematical models have been developed to model the cancer growth depicting the neoplasmic evolution with respect to time (t).[48] Modeling of tumor development, in general, can help understanding carcinogenesis and the related dynamics of pathogenic conditions. Hence, basic linear models were classically developed, for example, as in Vogelstein *et al.*[49] for colorectal tumors. In addition, a biologically comprehensive neoplastic development should be viewed in a stochastic framework (rather in terms of deterministic variables) consistent with the complexity of the cellular system and its contents having spatiotemporal randomness.

An effort toward addressing balanced/unbalanced states of cytogenetic contents, which takes into relevant interaction considerations alone, has been exercised by a stepped progression of the *number of imbalances of the tumor* (NIBT).[50] It denotes a discrete version of nonlinear evolution of cancer growth. More rigorously, a continuous nonlinear mathematical model in the context of interaction between tumor cells and oncolytic viruses has been developed in Agarwal and Bhadauria.[51] The stochastic aspect of tumor invasion in the surrounding space degrading the extracellular matrix is studied *via* two classical efforts due to Anderson and Chaplin[52] and Othmer and Stevens.[53]

More comprehensively, developed in Kubo[54] is an asymptotic profile of the solution to parabolic ordinary differential equations pertinent to tumor angiogenesis dynamics. It resolves thereof certain solvability issues of ODEs seen in Anderson and Chaplin,[52] Othmer and Stevens[53] Yet another model on early tumor growth and invasion has been developed *via* cellular automaton considerations by Patel *et al.*[55] Essentially, a hybrid cellular automaton model of early tumor growth that describes the activity of individual cells and continuous evolution and their microenvironment forms the main theme in Patel *et al.*[55]

Tutorial: T-7.4

Stochastic profile of chromosomal aberrations: an important query of interest in modern context is to know whether the CAs are random events or came from an internal (endogenous) deterministic mechanism. Discussed in Castro *et al.*[56] is the stochastic nature of CAs in solid tumors and several related Shannon information functions are evaluated thereof to describe the disorderliness present inside a tumor. Hence, suggested in Castro *et al.*[56] is that, in the context of quantifying the spread of aberrations, the generating process is neither deterministic nor totally random, but it produces variations that can be specified between two extrema. The study in Castro *et al.*[56] is fortified with relevant data on 79 different kinds of solid tumors having 30 or more karyotypes retrieved from Mitelman.[57]

The mixture state of contents expressed in terms of the stains at DNA level in the chromosomal structures is featured and expressed in terms of autosomal genetic markers in Fukshansky and Bar.[58] Specifically, the combinations of alleles at different locales (*loci*) on the chromosome that are transmitted together are considered. This one fold, single, and simple allele (known as *haplotype*) forms the general framework to test the hypothesis on mixed stain analysis advocated in Fukshansky and Bar.[58] The implication of neoplastic transformational dynamics concerning the induction of CA through direct and bi-stander mechanism has been addressed *via* a state-vector model.

Tutorial: T-7.5

Notwithstanding the existence of several growth models of tumors caused by CAs, yet another method to predict the dynamics of evolution of a tumor in the presence of associated disordered entities is based on ascribing stochastic features to the temporal dynamics (*via* Bernoulli–Riccati equation) of tumor growth *versus* time. Relevant model emphasizes stochastic aspects of the cytogenetic complex with the associated entities constituting a stochastic mixture of normal and abnormal chromosomes. Such chromosomes could be present in the receptacle matrix of cellular body, where the tumor is observed, and the spatiotemporal framework of these disordered set of entities is assumed as a complex system. Hence, a stochastic time evolution of tumor growth can be indicated by a function in closed form, and this growth function is verified against results obtained by other models. Relevantly, the following example is presented to illustrate the underlying considerations.

Example: Ex-7.3

Statement of the Problem

This example is furnished to elaborate the aforesaid stochastic differential calculus pertinent to the temporal growth/decay dynamics of the constituents present (as a stochastic mixture) in the cytogenetic framework. In general, the temporal evolution of a variable, $y(t)$ can be ascertained by a differential equation with specified coefficients, $p_{o,}$ and r_o denoting the attributes of (nonlinear) activity present, and assuming the extent of nonlinearity limited to second-order influence, a Riccati equation, namely, $y'(t) + r_o(t)y^2(t) + p_o(t) = 0$ can be specified, and it is indicated in Neelakanta[42] that the relevant solution Riccati equation can be written explicitly as follows[42]: $y(t) = L_q(t)$. As indicated in Chapter 6, $L_q(.)$ denotes the LBF with the parameter q deciding the extent of disorderliness associated with the system. Eq. (7.19)

The temporal dynamics of $y(t)$ in terms of $L_q(t)$ can be applied to CAs *versus* progression of cancer as described in Pierdomenico *et al.*[40] Relevant model leads to an insight into cancer initiation (and progression) useful in diagnostic and therapeutic regimens. The modeling data adopted thereof[40] refer to experimental studies representing stages of visual snapshots of biological process for a specific cancer progression. In relevant empirical depictions, the regressed growth (and decay) functions depict exponential functions having appropriate coefficients and reaction rate constants for the growth or decay. These coefficients are taken as a normalized set of rate constants, $0 \leq \{k_1, k_2, ..., k_6\} \geq 1$. Hence, a set of differential equations for A, C, and N is specified and relevant dynamics are established.

In lieu of the empirical approach indicated above, nonempirical formulations in depicting the temporal variations of clonal (C) and nonclonal (N) constituents in the cellular complex are indicated in Neelakanta and Yassin,[59] Neelakanta and Sardenberg[60] by prescribing logistic growth/decay function models for C and N as functions of time (τ). Hence, the LBF is shown as a feasible, logistically implied temporal growth (or decay) profile of $C(\tau)$ or $N(\tau)$. That is, the LBF $L_q(t)$ is proposed to denote the temporal growth or decay profiles of a given variable x (representing the population of C or N, both taken in normalized form), and the corresponding decay function of x is $\{1 - L_q(t)\}$.

Furthermore, the parameter q concurrently decides the initial rate (M_o) of such growth (or decay) with the corresponding order parameter (θ) of the system specified by $\theta = \{3/2 \times (M_o) - 1/2\}$, and, in terms of q, the initial growth rate, $M_o = (1/3 + 1/3q)$ yielding, $q = 2\theta$ as shown by Neelakanta and De Groff.[61]

The order parameter θ (and hence, q) are specific to *equipartition* considerations,[61] and it represents one-third of the constituent population, namely, N^*, C^*, and CH^* , where the asterisk denotes explicitly the normalized values and CH denotes the total chromosomal content assumed as 100% of clonal aberrations plus nonclonal aberrations and normal chromosomes. The normalization as above is done with respect to the total sum of C, N, and CH contents. With reference to the temporal growth stages of cancer, pertinent results are computed using the data furnished in Pierdomenico *et al.*[40] in terms of normalized entities: $C^* = C/S$, $N^* = N/S$, $CH^* = CH/S$ and $S = (C + N + CH)$ *versus* stepped temporal stages ($\tau = 0.5, 1.5, 2.5, 3.5, 4.5, 5.5$, and 6.5). Thus, $(\theta = N^* + C^* + CH^*)/3$ is determined, and hence, $q = 2\theta$ is ascertained and the functions $L_q(\tau)$ and $\{1 - L_q\ (\tau)\}$ are determined. The computed details as above are furnished along with experimental data and empirical evaluations available in Pierdomenico *et al.*[40] in Figs. 7.7–7.9. For comparison, illustrated further are plots of $C(\tau)$, $N(\tau)$, and $CH(\tau)$ relevant to: (i) experimental results,[40] (ii) empirical results,[40] and (iii) using the Langevin–Bernoulli algorithms in Figs. 7.6 and 7.8.

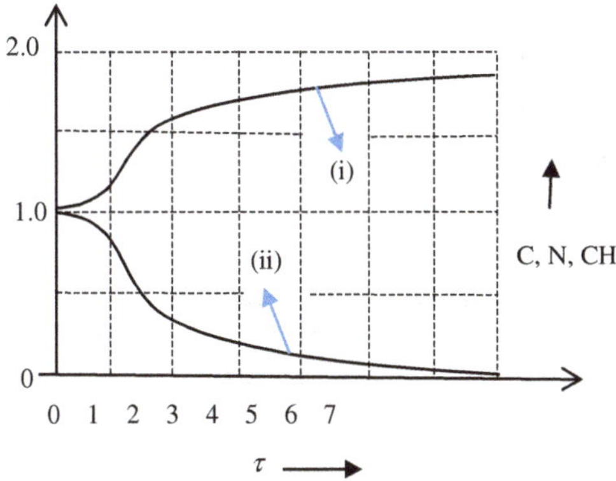

Fig. 7.6 Plots of: (i) growth curve, C: $L_q(\tau)$ and (ii) decay curve, *N*: $\{1 - L_q(\tau)\}$.

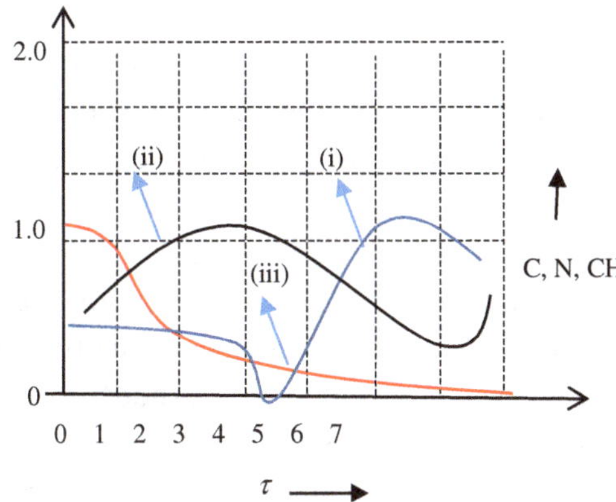

Fig. 7.7 Experimental data in normalized form of $C(\tau)$ – Curve (i) $N(\tau)$-Curve (ii) and CH (τ)-Curve (iii) *versus* τ denotes the stage of cancer growth as per the data in Pierdomenico *et al.*[40]

Example: Ex-7.4

Linking cancer and viruses *via* mathematical considerations and associating cell responses thereof are studied in genetic engineering with relevance to interaction between tumor cells and oncolytic viruses. For example, as analyzed and presented in Agarwal and Bhadauria,[51] the study is concerned with mathematical modeling on pertinent interaction dynamics, and the related stability considerations are also obtained by analyses on both infected and uninfected tumor cells being present; hence, the possibility of tumor load getting eliminated with time using virus therapy is considered.

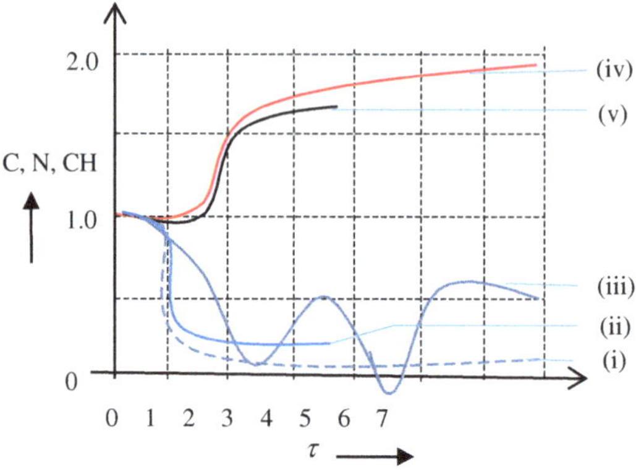

Fig. 7.8 Plots of $C(\tau)$, $N(\tau)$, and $CH(\tau)$: (i) Decay-using present method, $[1-L_q(\tau)]$); (ii) decay[40]; (iii) decay[40]; (iv) growth-using present method, $L_q(t)$; and (v) growth.[40]

In the model of Agarwal and Bhadauria,[51] the tumor cell population is represented as X and the virus-infected tumor cell population by Y. Furthermore, they are assumed to grow in a logistic fashion with the following assumptions in the model proposed: the tumor cells get infected with invading oncolytic viruses (*Oncolytic viruses** represent viruses that preferentially infect and lyse/break down the cancer cells. They enter the tumor cells and replicate subsequently, and the infected tumor cells subsequently cause infection in other tumor cells).

The dynamics of infection is modeled in Agarwal and Bhadauria[51] as follows: the initial conditions on X and Y are $X(0) = X_o$ (>0) and $Y(0) = Y_o$ (>0). Corresponding set of bounded solutions on cancer growth is obtained *via* differential equations formulated using the growth logistics. Pertinent solutions are indicated in Agarwal and Bhadauria[51] *via* numerical procedure on differential equations using Runge–Kutta method. The following parameters are used thereof in the computed results of Agarwal and Bhadauria:[51]

r_1 =	Maximum growth rate of uninfected cells	(= 40.00)
r_2 =	Maximum growth rate of infected cells	(= 2.000)
K =	Holding/carrying capacity of the cellular media	(= 100.0)
a =	Measure of the immune response of individual virus preventing it from destroying the cancer	(= 0.050)
b =	Transmission rate of viral dispersions	(= 0.020)
α =	Rate of killing infected cells by viruses	(= 0.003).

Corresponding to the presumed data as above, the computed densities of tumor cells as functions of time are presented in Agarwal and

Bhadauria.[51] For the values of r_1 assumed as above, the initial growth rate (m) obtained in Agarwal and Bhadauria[51] is in the range of 0.35–0.4. With the value of b changed to 0.06, this initial slope m becomes approximately, 0.7.

Alternative to the solution suites of Agarwal and Bhadauria,[51] another closed form solution for the associated growth function can be attempted as follows: assuming a stochastic framework of growth/decay dynamics and with the knowledge of initial slope, m (and hence q) as considered earlier, the functions $L_q(t)$ and $[1 - L_q(t)]$ to depict the growth and decay profiles, respectively, and the q-value related to m by the relation: $q = [1/(3m - 1)]$ implies an order-function (or disorder factor). Hence, for example, when $m \approx 0.7$, $q \to 1$ and when $m \approx 0.35$, $q \to 20$. The data corresponding to $b = 0.06$ leading to $(q \to 1)$ implies an upper-bound on the underlying growth dynamics; furthermore, data for $b = 0.02$ specifies $(q \to 20)$ depicting the lower-bound of the growth dynamics involved.

Shown in Figs. 7.9a and 7.9b are computed results obtained with the $L_q(.)$ function specified on growth and decay profiles. The results in Agarwal and Bhadauria[51] are also shown in these figures for comparison and the following inferences can be gathered from these illustrations: considering the results of Agarwal and Bhadauria[51] presented in Figs. 7.10a and 7.10b show the growth/decay dynamics that reside in the regime outside the upper/lower-bound constraints posed by $q \to 0.5$ or less. For example, with $q = 0.1$, the results of Agarwal and Bhadauria[51] are closer to the proposed model with $L_q(.)$ function.

Similar observations can be made with growth/decay curves of Figs. 7.9b, Figs. 7.10a and 7.10b. It implies that in the initial stage, the constituent of the cellular medium, namely, the infected and noninfected parts,

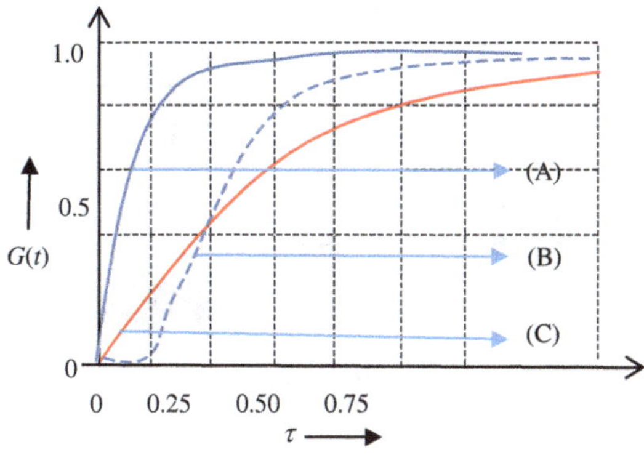

Fig.7.9a Growth function curves: (A) Reference[51]; (B): $q = 0.1$; and (C): $q = 0.5$.

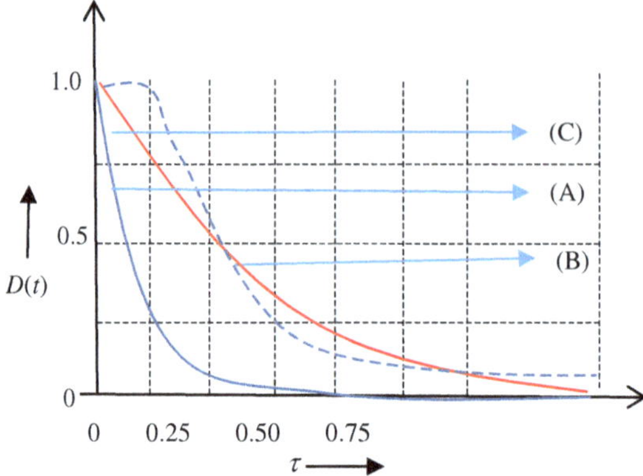

Fig. 7.9b Decay function curves: (A): reference [7.51]; (B): $q = 0.1$; and (C): $q = 0.5$.

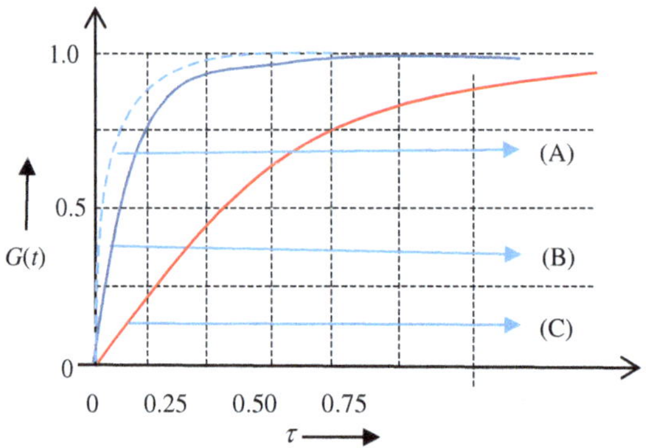

Fig. 7.10a Growth function curves: (A): Reference[51]; (B): $q = 0.1$; and (C): $q = 0.5$.

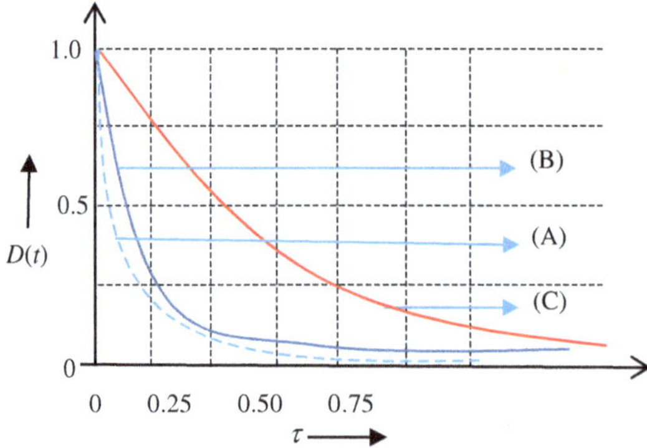

Fig. 7.10b Decay function curves: (A): Reference[51]; (B): $q = 0.1$; and (C): $q = 0.5$.

interact with the oncolytic viruses with a dispersion rate of $b = 0.02$. After a certain growth/decay time, the interaction ceases. When the b value, namely, the transmission rate of viral dispersions changes to 0.06, the decay and growth curves are shown in Figs. 7.10a and 7.10b. In this case, the states of growth and decay conform to non-interaction of included constituents in the cells.

Tutorial: T-7.6

This tutorial is on *NCCA* generated by genomic instability. Commensurate with the observation in Heng *et al.*[62] that the genomic instability is mainly responsible for cancer progression, a related problem is to model the dynamic relationship between NCCAs and *CCAs*, so as to explain the mechanism of chromosomal-based cancer evolution; hence, certain open questions on related issues pertinent to broad clinical applications can be conceived. The following note outlines the avenue of details thereof:

Cancer initiation and progression are identified with specific genes and pathways, and cancer is regarded as due to stepwise accumulation of genetic aberrations. However, some clinical findings show that tumors display high levels of genetic heterogeneity and distinctive karyotypes. Furthermore, chromosomal instability is regarded as being primarily responsible in generating stochastic karyotypic changes that lead to random progression of cancer. By tracing karyotypic patterns of individual cells that contain either defective genes responsible for genome integrity or subjected to challenges by oncoproteins or carcinogens destabilizing the genomes cancer progress can be studied. Pertinent analysis could include tracing the patterns of karyotypic evolution during different stages of cellular immortalization, and underlying results may reveal NCCAs (of both aneuploidy and structural aberrations) and nonrecurrent CCAs directly influencing genomic instability and karyotypic evolution.

Problem: P-7.1

Statement of the Problem

This problem is posed to identify a self-study exercise to model the dynamic relationship between NCCAs and CCAs explaining the mechanism of chromosomal-based cancer evolution and decide on open questions pertinent to broad clinical applications.

Having said that, Part I of the chapter is a debut attempt in fusing the concepts of bioinformatics, biostatistics, and datamining principles in the context of cytogenetics. The background details are narrated as textual topics and tutorial, and they are supplemented by a few examples and pertinent self-study problems. Though not exhaustive, the contents of this chapter offer new avenues of thinking for further studies in bioinformatics cast in the exclusive frameworks of cytogenetics and cellular functions.[63]

7.8 Part II: Synteny Concepts in Cytogenetics

This second part (Part II) of this chapter is devoted to describe information-theoretics and entropy features of chromosomal genomic loci *vis-á-vis* the so-called "synteny" details. Here, the implied synteny corresponds to a block representing a set of contiguous genes located within the same chromosome, and such a block is seen as a conserved entity between a pair of, or among, various species depicting a shared synteny. Organisms of relatively recent divergence may show similar "blocks of genes" or synteny in the same relative positions in the genome.

An objectively conceivable task in cytogenetics is to ascertain the chromosomes, the maximum plausible extent of shared-synteny that could possibly be encountered across an exhaustively searched, large number of orthologous genomes pertinent to a given pair of (test) species being compared. Determining such maximum plausible extent of shared-synteny in a sample space is described here. The search is done first across a limited number of underlying orthologs of test species gathering a sparse set of synteny data, and the associated entropy details are estimated. By judiciously extending such entropy details, the maximum synteny that could be expected when an ensemble of an infinitely large number of ortholog pairs is exhaustively searched is determined. That is, by appropriate choice of information-theoretic metrics, the required maximum extent of plausible shared-synteny is decided. In other words, by assessing (information-theoretics-based) entropy features of chromosomal genomic loci of a given pair of orthologs, the associated synteny correlations are estimated; hence, the maximum plausible extent of shared synteny (in chromosomes), when an infinitely large number of orthologous genomes of a given pair of (test) species that is exhaustively searched is determined.

Suppose when a search on similar "blocks of genes" or synteny on a sample space having a limited number of orthologs of the test species is done. Then the associated statistical data (on synteny correlation) so gathered could be sparsely sized; hence, the underlying entropy details ascertained could be small and incomplete; as such, by judiciously extending such limited entropy details, the theoretical maximum synteny that could be encountered if an ensemble of infinitely large number of ortholog pairs is exhaustively searched can be estimated (in terms of appropriately defined metrics). The search exercises as above are done with necessary constraint posed by the fixed number of chromosomes in each test species being considered.

Hence, the theme of the details addressed below are particulars on proposing relevant metrics, so as to eventually estimate the maximum extent of plausible shared synteny as a function of the number of ortholog pairs compared, also indicated are upper- and lower- (stochastic) bounds (designated, respectively, as UB and LB) on the estimations done. Test analyses performed thereof refer to the following pairs of orthologous species: mouse *versus* human, Medaka fish *versus* human, and Medaka fish *versus* Zebra fish. In all, described here is a novel strategy to find the maximum plausible extent of shared synteny between a pair of extensive test species. The results obtained are consistent with the stochastic basis of Shannon information, entropy-theoretics, and Schur convexity considerations.[64]

7.9 Diversity of Life at Cytogenetic Level

Genomic rearrangement events are inevitable in long-term evolution processes of living systems, and tracking such events allows understanding the diversity of life emphasized at cytogenetic levels. For example, relevantly known as *genetic speciation,* it denotes an evolutionary framework wherein new biological species could arise, and the associated genetic drift renders the formation of three classes of genes, namely, (i) the *homologs* depicting genes related to a second gene by descent from a *common ancestral globin root*; (ii) from the root, the globin families (like α- and β-globin families) are formed due to *gene duplication;* (iii) From each such family, *speciation* would occur leading to the sets of *species* (such as α_1, α_2, α_3 ... and β_1, β_2, β_3 ... etc.); (vi) the sets {α_1, α_2, α_3 ...) and {β_1, β_2, β_3 ...) so formed depict the *orthologs* denoting the lineage of genes in different species that have evolved from a common ancestral gene *via speciation;* and (*v*) lastly, the set of *paralogs* representing genes produced via gene duplication within a genome can also be realized. That is, paralogs represent the lineage of crossing between the sets of species, {α_1, α_2, α_3 ...) and {β_1, β_2, β_3 ...), and the evolutionary status of orthologs between the species could also be seen across more than two species.

In summary, the narration as above outlines the evolution of differential orthologs and paralogs in the lineage of gene duplication in an evolutionary scenario of a set of species.[64–66] Relevant events of gene evolution include *vertical descent* (also known as speciation), gene duplication, possible gene losses, horizontal gene transfer (HGT), and rearrangement of the structure. Correspondingly, the following versions of genes exist in nature: *Homologs* are genes that share a common ancestral root

> *Orthologs* are genes derived from a single ancestral gene in the *last common ancestor* (LCA) of the compared species
> *Paralogs* genes are related *via* duplication

In addition to the kinds of genes seen in the evolutionary lineages as above, suppose one or more genes in one lineage that are orthologous to one or more genes in another lineage (in lineage-specific duplications), the result of this finer classification can be stated as lineage-specific gene-duplication termed as *co-orthologs.* Relevantly, *symparalogs* (also known as *in-papralogs*) can be defined as a class of paralogous genes stemming from a lineage-specific duplications subsequent to a given speciation event.

Also, the so-called *out-paralogs* (or *alloparalogs*) imply another class of paralogous genes resulting from a single or more duplications preceding a given speciation event. Furthermore, whenever lineage-specific gene losses are encountered, genes that are paralogs may appear as orthologs caused by differential lineage-specific gene losses. Furthermore, in the context of evolutionary process, the so-called *HGT* can take place implying a *lateral gene transfer* (LGT). It refers to a movement of genetic material between unicellular and/or multicellular organisms (in contrast to the vertical transmission of DNA from parent-to-offspring implying a *reproduction*).

Furthermore, referring to Fig. 7.11, gene-Bh (of human chromosome #1) is defined as an orthologous gene-Bm (in mouse chromosome #1), and a transitive relation may also exist, namely, the gene-Bm (of mouse chromosome #1) also being orthologous to the gene-Bg in the chromosome #1 of a third species say, gorilla.

Genomic Similarity and Synteny

A study pertinent to evolution process of living systems is concerned with genomic sequencing and mapping wherein exclusive efforts are envisaged

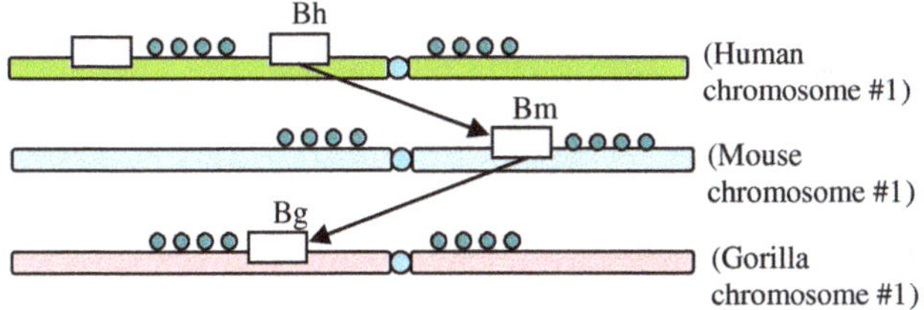

Fig. 7.11 An example of orthologs relative to three different species.

so as to enable comparing genomic structures of different species. As mentioned earlier, organisms of relatively recent divergence may show similar blocks of genes in the same relative positions in their genomes. This situation as indicated earlier refers to a state of synteny,[66,67] a term that translated roughly into "a state of possessing common chromosome sequences or features." For example, many of the genes of humans are syntenic with those of other mammals, not only apes but also cows, mice, and so on. Study of synteny can show how the genome is "cut and pasted" in the course of evolution, so that the observed commonness across chromosome of different species is feasible. That is, in the perspectives of genetics observed at cytogenetic level, there could be *physical co-localization of genetic loci* on the same chromosome within an individual or other species. Biologists refer such co-localization as the synteny, which refers to a conservation of blocks of order within two sets of chromosomes being compared. This state emphasizes a *shared synteny* being present.

Thus, in scientific perspectives, a synteny block represents a set of contiguous genes located within the same chromosome and such a block could be well-conserved across a pair of species or among various species. Knowledge on the distribution of conserved gene blocks facilitates tracing the diversity of life in evolutionary studies. It also helps in orthologous gene detection and gene annotation across the genomic era.

7.10 The Oxford Grid

The synteny correlation that may prevail across orthologs of two species can be mapped in a matrix, known as the *Oxford grid*.[67–71] It is a tool that helps to obtain a comparative mapping of conserved synteny between two species. Relevant synteny comparison is deduced by searching genes of a test pair of species mapped on the grid, and the comparative details enable historical inferences relating to the co-ancestry in phylogenetic contexts. Furthermore, syntenic comparison corresponds to assessing the so-called *persistent mutual information* (PMI) or *persistent cross-entropy* (PCE) that prevails within the ensemble of histories behind the test orthologs being searched.[72] That is, the underlying PMI/PCE depicts as well as quantifies the strong emergence of features conserved along the evolutionary phylogenetic tree, from its ancestral root to the taxonomic leaf.

In estimating synteny conservation between two species, the Oxford-grid representing a matrix can be used. The matrix is framed by the number of ortholog genes on each of the chromosome (along the row 1 through r) of the first species, which is also simultaneously present on each of the chromosomes (across the column 1 through c) of the second species. That is, the Oxford-grid illustrates a display of homologues loci from two species in terms of a matrix, $[m] = [r \times c]$. The synteny (of ortholog genes) mapped on the Oxford-grid are not just randomly seen scattered across m grids, (each grid depicting a node of an intersecting row and column of a pair of chromosomes). Some of the grid cells of chromosome pairs may contain many orthologs posing synteny and some may contain none.

Thus, shown in Fig. 7.12 is an Oxford-grid wherein the chromosome set $\{x_i\}$ of the species X and chromosome set Y: $\{y_j\}$ of the species Y make a matrix framework. Each grid-cell contains statistically

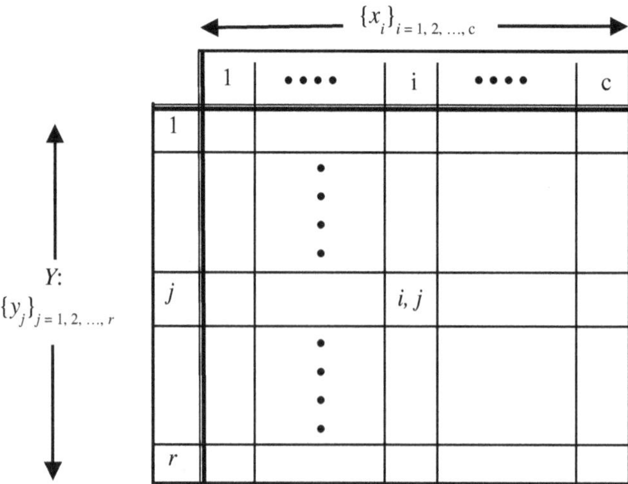

Fig. 7.12 An Oxford-grid matrix of the chromosome-set $\{x_i\}$ of the species X and chromosome-set $\{y_j\}$ of the species Y. Each grid-cell contains statistically enumerated counts (of finite or null values) on the state of homologous loci deduced from comparing the chromosomes of two species, X and Y on synteny correlations.

enumerated count details on the state of homologous loci deduced from comparing chromosomes of the two species, *X* and *Y* on synteny correlations, and relevant number count gathered *via* statistical enumeration is posted on each grid cell. The value so posted could either be finite- or of null-value.

7.11 Oxford-Grid of a Pair of Species: Examples

With the reference to the study indicated here, pairs of test species considered and used to illustrate an Oxford-grid are as follows: (i) mouse (*Mus musculus*) *versus* human (*Homo sapiens*), (ii) human (*H. sapiens*) *versus* medaka (*Oryzias latipes*), and (iii) medaka (*O. latipes*) *versus* zebra fish (*Danio rerio*). Relevant evolutionary relationship can be examined by applying the associated phylogenic distance measures. The Oxford-grid details of the aforesaid three pairs of species are reported in:

I: Human *versus* mouse[66,70]
II: Human *versus* medaka fish[73]
III: Medaka *versus* zebra fish.[66,70,73]

The phylogeny of test pairs of species considered above is shown in Fig. 7.13. The test pairs (I, II, and III) associated in the vertebrate topology (used at UCSC genome browser) depict, respectively, the following: (I) mouse (*M. musculus*) *versus* human (*H. sapiens*), (II) human (*H. sapiens*) *versus* medaka (*O. latipes*), and (III) medaka (*O. latipes*) *versus* zebra fish (*D. rerio*).

The test pairs (I, II, and III) of species, the underlying synteny heuristics can be objectively adopted to address the following: (i) to review the genetic aspects of comparative synteny between the pair of test species so as to know the historical inferences relating the co-ancestry in phylogenetic contexts and (ii) to apply the concepts of information-theoretics in order to determine the extent of entropy-based syntenic correlation that could exit across an exhaustive set of orthologs of the compared pair of species. Biological details of the species as above are available in Naruse *et al.*,[73] Introduction to scientific names for humans,[74] Mouse,[75] *Oryzias latipes*,[76] Japanese rice fish,[77] Zebrafish,[78] and *Danio rerio*,[79] and an outline of the zoological details of the four test species is as follows:

The homologous loci across a set of chromosomes of orthologs (between a pair of test species) indicated above can be tracked to infer

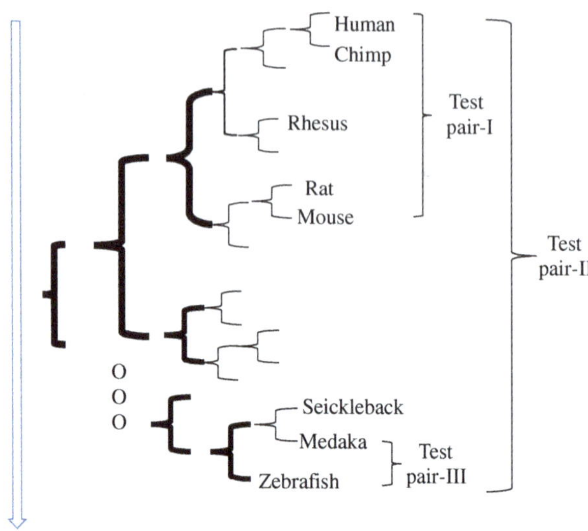

Fig. 7.13 Evolutionary relation between test species (pairs) seen across the phylogenic framework.

Table 7.2 Zoological details of the test species.

	Human (1)[74]	Mouse (2)[69,75]	Medaka (3)[73,76,77]	Zebrafish (4)[78,79]
Kingdom	ANIMALIA	ANIMALIA	ANIMALIA	ANIMALIA
Phylum	CHORDATA	CHORDATA	CHORDATA	CHORDATA
Class	MAMMALIA	MAMMALIA	ACTINOPTERY GII	ACTINOPTERY GII
Order	PRIMATES	RODENTAI	BELONIFORMES	CYPRINIFORMES
Family	HOMINIDAE	MURIDAE	ADRIANICHTHYI-DATE	CYPRINIDAE
Sub/super family	HOMININAE Sub-family	MUROIDEA Super-family MURINAE Sub-family	ORIZIINAE Sub-family	—
Tribe	HOMININI	—	—	—
Genus	HOMO	MUS	ORYZIAS	DANIO
Species	*HOMO SAPIENS*	*MUS MUSCULUS*	*ORIZIAS LATIPES*	*DANIO RERIO*

the extent of maximum synteny correlation that may prevail when an exhaustive set of orthologs are searched and their genes are mapped *via* the Oxford-grid — a tool as mentioned earlier that helps to obtain a comparative mapping of conserved synteny between two test species being compared.[80,81] Relevantly, a proposal toward seeking an answer is indicated for a query posed by Postlethwait in Houseworth,[66] Edwards,[67] Bengtsson *et al.*,[68] Mouse Genome Database (MGB),[69] Schena,[70] Murphy *et al.*,[71] Ball *et al.*,[72] Naruse *et al.*,[73] Introduction to scientific names for humans,[74] Mouse,[75] *Oryzias latipes*,[76] Japanese rice fish,[77] Zebrafish,[78] *Danio rerio*,[79] and Houseworth and Postlethwait[80] as described below.

7.12 Postlethwait Query

Suppose an Oxford-grid of a given pair of test species is mapped to depict grid-cells formed by a matrix: [r, rows (of chromosomes) $\times$ c, (column of chromosomes)]. That is, the Oxford-grid mapping contains $m = [r \times c]$ grid-cells and each cell is marked with statistically enumerated synteny counts (finite or null value) inferred from a compared set of chromosomes of n ortholog pairs. That is, the number of orthologs pairs searched corresponds to a limited search space of n ortholog pairs. Assuming that the k number of cells in the matrix show a finite synteny counts, then the number of cells that show null synteny count is $(m-k)$.

Next, suppose a new search space of the test species pair is considered to ascertain the extended statistics on synteny correlation between an infinitely large extent of orthologous genomes by rendering n to assume a large value, N_{max} ($\rightarrow \infty$). That is, by considering the number of ortholog pairs in an extended phase of exhaustive search space with $n \rightarrow N_{max}$ ($\rightarrow \infty$), Postlethwait indicated a problem[66,80] to determine the number of ℓ cells (replacing $k < \ell$) that would show finite synteny counts in the new search space with n $\rightarrow N_{max}$ ($\rightarrow \infty$); hence, the corresponding number of remaining cells showing null synteny count will be: $(m - \ell)$. Deducing this unknown statistics toward ascertaining ℓ cells (replacing k) in an exhaustive search space with $n \rightarrow N_{max}$ ($\rightarrow \infty$) implies solving the so-called *Postlethwait problem* as detailed by the author in Neelakanta and Sharma.[64]

In summary, suppose k depicts the number of grid cells in the Oxford-grid (containing a total of $m = [r \times c]$ cells) observed with in a finite count of conserved synteny relevant to a correlation mapping of n orthologs of two species. Then, the Postlethwait problem is to ascertain the increased number ($\ell > k$) of such grid-cells marked with (finite) number of conserved synteny after "all the orthologs" shared by both species are exhaustively searched with $n \rightarrow N_{max}(\rightarrow \infty)$ and their genes are compared so as to elucidate the existing synteny across the mapping exercised. Hence, the computation required is to infer the value of ℓ (> k), when n tends toward N_{max} ($\rightarrow \infty$) denoting an exhaustive exercise on the number of ortholog pairs being compared in the test search space.

In search of an answer for the aforesaid Postlethwait query, necessary algorithms are proposed in the sections that follow and pertinent simulations are elaborated in Neelakanta and Sharma.[64] Computational exercises are carried out thereof use Oxford-grid data on (n; k) gathered from the literature in Houseworth,[66] Edwards,[67] Bengtsson *et al.*,[68] Mouse Genome Database (MGB),[69] Schena,[70] and Naruse *et al.*[73] Pertinent chromosome details are concerned with the following pairs of species (mentioned earlier): (I) human *versus* mouse; (II) human *versus* medaka and (III) medaka *versus* zebra fish, and each dataset of the test pair is availed in a format framed as an Oxford-grid.[81]

In summary, the Oxford-grid details of a test pair (yielding a finite synteny count posted in k cells) are deduced *via* a comparative statistical search on a limited number of n orthologs of the test pair; relevant synteny comparison implicitly corresponds to assessing the PMI details[9] prevailing within the ensemble of phylogenetic histories of evolution behind such orthologs. That is, the underlying PMI depicts as well as quantifies the strong emergence of features that are conserved along the phylogenetic tree. When the search is expanded to an exhaustive space with $n \rightarrow N_{max}$ ($\rightarrow \infty$), then k will increase to assume a new value, ℓ (> k). Finding the maximum of ℓ (with $n \rightarrow \infty$) implies the Postlethwait query under discussion.

Solution to Postlethwait Query

The pedagogy of this section refers to seeking a possible approach to answer the Postlethwait question. The associated exercise is to quantify the extended heuristics on the emergence of genetic epochs at cytogenetic level (in line with PMI observed in the past with fewer [limited] search space of orthologs of n species). In tracking such orthologs by establishing the associated synteny correlation is done *via* Oxford-grid statistics. Relevantly, the Postlethwait query refers to imply whether it is possible to use and expand the statistical data (n; k), so as to infer the extent of synteny correlations that may likely to prevail when the comparative search on chromosomal pair (of the test species) is exercised exhaustively on their orthologs by allowing the number n to approach an (infinitely) large value; that is, by rendering, $n \rightarrow (N_{max} \rightarrow \infty)$; hence, the resulting fresh value of $k \rightarrow \ell$ is ascertained.

Concurrently, it may also be of interest to find whether there is a limiting maximum value of ℓ when the comparison of the test pair of species is infinitely increased; that is, when $n \rightarrow N_{max}$ ($\rightarrow \infty$). In this expanded search, it should be recognized that the search space of chromosomal comparison (in finding the homologous loci between two species toward inferring the syntenic correlation) is a domain, which could be mostly stochastic, largely nondeterministic and partly deterministic; as such, the underlying process is a stochastic exercise in a complex system designated by the filled-in statistics of the framework of m = [r-rows × c-columns] Oxford-grid cells. Any assertion in accruing synteny details derivable from this expanded matrix should possess statistical upper- and lower-bounds (LB and UB) on the value of ℓ estimated when $n \rightarrow (N_{max} \rightarrow \infty)$. That is, as $n \rightarrow (N_{max} \rightarrow \infty)$ relevant trend of $k \rightarrow \ell$ would grow monotonically and reach a value ($\ell_{max} < m$) at ($n \rightarrow N_{max}$), and any further synteny search on the test pair with $n > N_{max}$, the enumerative exercise may not yield a solution for ℓ larger than ($\ell_{max} < m$). Hence, relevant to Postlethwait query, it is proposed that there is a maximum value on the synteny specified by $\ell \rightarrow (\ell_{max} < m)$ *vis-à-vis* the corresponding maximum number of orthologs n tending to N_{max} (beyond which, the synteny search can be regarded as being futile).

In pursuing the synteny correlation *via* a statistical search as outlined above, necessary background details on genomic features of the diversity of life across the evolution process seen at the cytogenetic level refer to knowing the heuristics of similarity that may exist across the genes of test orthologs. An overview on underlying synteny considerations is outlined below.

7.13 Chromosomal Syntenic Similarity

As stated before, organisms of relatively recent divergence may show similar blocks of genes in the same relative positions in the genome. This synteny consideration corresponds to a block representing a set of contiguous genes located within the same chromosome or such a block is seen as a conserved entity between a pair of species or among various species, depicting a shared synteny. In practice, the extent of such synteny details is ascertained from a search space pertinent to a limited set of samples of orthologous genomes of a given pair of (test) species. However, Houseworth and Postlethwait in Houseworth and Postlethwait[80] have suggested schemes to estimate the "true" number of conserved synteny between the two test species by duly accounting the dependency of the number of orthologs searched toward chromosome pairing. Relevantly, proposed here is yet an enhanced and a computationally feasible strategy that accommodates the dependency of the number of orthologs involved when a large number (with a hypothetical limit tending to infinity) of orthologs are searched toward chromosome pairing.

That is, the query, "When we look at maps of conserved synteny between two species, we see that the orthologous genes are not randomly scattered between the pairs of chromosomes. Some chromosome pairs contain many orthologs and some contain none. If we have mapped some (n) of the orthologs between two species and observe k conserved syntenies, can we estimate the number of conserved synteny ℓ that will exist after we have mapped all the orthologs shared by both species?" was posed by Postlethwait as outlined in Houseworth,[66] Edwards,[67] Bengtsson *et al.*,[68] Mouse Genome Database (MGB),[69] Schena,[70] Murphy *et al.*,[71] Ball *et al.*,[72] Naruse *et al.*,[73] Introduction to scientific names for humans,[74] Mouse,[75] *Oryzias latipes*,[76] Japanese rice fish,[77] Zebrafish,[78] *Danio rerio*,[79] and Houseworth and Postlethwait.[80]

With reference to knowing the synteny conservation between two test species (i.e., in ascertaining the shared synteny), the required data (as explained earlier) are first represented as a matrix depicting the Oxford-grid mapping framed by "r" rows of chromosomes of one species *versus* and "c" columns of chromosomes of the other species as illustrated in Fig. 7.15, where the matrix $[r \times c]$ summarizes the map of the chromosome set $\{x_i\}$, (with $i = 1$ through c) of the first species X and the chromosome set $\{y_j\}$ (with $j = 1$ through r) of the second species Y.

Correspondingly, suppose the genes on each of the chromosomes (i: 1 through r) are concurrently present in the chromosomes (j: 1 through c) and specified in the matrix cells implying the existence of shared synteny. Hence, the matrix framed with a total number of cells, $m = (r \times c)$ as indicated in the Oxford-grid[67,68] of Fig. 7.14, where each grid-cell contains statistically enumerated count details on the state of homologous loci deduced from comparing the chromosomes of two species, X and Y on synteny correlations. Such number counts gathered *via* statistical-enumeration and posted on each grid-cell could be of either finite- or null-values.

By extending the analyses on Oxford-grid data, a solution to Postlethwait query can be pursued. Illustrated (partially) in Fig. 7.17 is an exemplar Oxford-grid availed from Coriell institute for medical research[82] for an orthologous pair of (test) species, namely, X: mouse

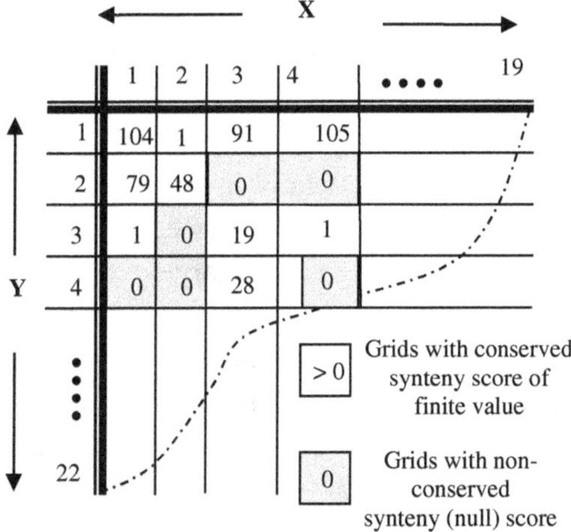

X
1
2
3
4
....
19
Y
1
2
3
4
22
104
1
91
105
79
48
0
0
1
0
19
1
0
0
28
0
>0
Grids with conserved synteny score of finite value
0
Grids with non-conserved synteny (null) score

Fig. 7.14 Oxford-grid layout (partial) with entry of synteny scores in the grids: an example for the test species mouse *versus* human. (Chromosome set: X: $\{x_i\}$ of mouse *versus* chromosome set: Y: $\{y_j\}$ of human). Complete Oxford-grid is available in OXGRID–The Oxford grid project,[81] Coriell institute for medical research.[82]

(*M. musculus*) *versus* Y: human (*H. sapiens*). Associated details indicate that there are [$i = 1, 2, \ldots, (c = 19)$, vertical columns] by [$j = 1, 2, \ldots, (r = 22)$, horizontal rows] so that the total number of Oxford-grid cells is equal to $m = (r \times c) = 418$, and among these m cells exists a random dispersion of k (out of m) cells, each posted with a finite value of counts on conserved synteny. Correspondingly, there are $(m-k)$ cells, each registering a null value on the synteny count. (Full details of Fig. 7.14 are available, for example, in Fig. 12.2 of Coriell institute for medical research.[82])

Furthermore, as regard to the Oxford-grid details of Fig. 7.14, the total (limited) number of orthologous genomes searched is $n = 3512$. and, as indicated above, the number of cells having finite number of conserved synteny scores marked out of a total number of m $(= r \times c) = (22 \times 19) = 418$, matrix cells. Using the matrix details as above, it can be attempted to find a solution to estimate the maximum plausible conserved synteny between species pairs observable in an exhaustive search space of orthologous genomes (consistent with the query posed by Postlethwait.[66,80] Hence, the required step-by-step algorithmic pursuit can be summarized as follows:

Step I: relevant to the limited sample space of n orthologous genomes searched pertinent to a test species pair and mapped on the Oxford-grid (as illustrated in Fig. 7.14, the synteny count details are summarized in

Table 7.3, and the underlying stochastic details imply probabilistic norms of the data *versus* the associated entropy (H) or *persisting mutual information* (PMI) (in Shannon's sense).

Step II: with reference to the details in Table 7.7, the solution being sought on Postlethwait's query involves ascertaining the new value for $(k \leq m) \rightarrow \ell$, when n is increased to a large extent, $n \rightarrow (N_{max} \rightarrow \infty)$. Correspondingly, the entropy details (H) pertinent to the random dispersion of $(k \leq m)$ matrix cells in the test Oxford-grid will attain a maximum value, H_{max}. That is, when the search space entity $(n \rightarrow n_{uk})$ is monotonically increased, the trend of the associated entropy (of the Oxford-grid complex) will be $(H \rightarrow H_{max})$, leading to ascertaining the

Table 7.3 Details of the Oxford-grid mapping (specified for a given pair of test species) depicting the number of matrix-cells containing finite and null synteny counts.

Case (A): synteny count details of a limited sample space of orthologs.[71,82]
Case (B): synteny count details of a new search space pertinent to an infinitely large extent of orthologous genomes (as warranted in the context of solving the Postlethwait problem.[66,80])

Case (*A*) Synteny count details of a limited sample space of orthologs.[82] Limited sample space statistics available on the existing Oxford-grid pertinent to a pair of test species	Given a pair of test-species, details on Oxford-grid mapping framed by a total number of cells on the matrix: $m = [r \times c]$		
	Number of ortholog pairs searched in a limited search space	Number of matrix cells showing finite synteny counts	Number of cells showing null-synteny count
	n	K	$(m-k)$
	Postlethwait problem		
Case (*B*) Synteny count details of a new search space pertinent to an infinitely large extent of orthologous genomes (as warranted in the context of solving the Postlethwait problem)	Number of ortholog pairs searched in an extended phase of exhaustive search space	Resulting cells showing finite number of synteny counts in the new search space	Corresponding number of cells showing null synteny count
	$n \rightarrow (N_{max} \rightarrow \infty)$	$\ell = ?$	$(m - \ell) = ?$

eventual extent of $(k \rightarrow \ell)$ under the constraint that the total number of cells $m = (r \times c)$ remains invariant.

Step III: the analytical pursuit and computational details pertinent to the steps is specified in Table 7.7. It involves procedural efforts as regard to the sparse entropy, $\{H: n; k, (m-k)\}$.

The analytical pursuit in tracking the sparse entropy, $\{H: n; k, (m - k)\}$ toward its exhaustiveness specified by H_{max} such that $\{H: n; k, (m-k)\} \rightarrow \{H_{max}: (N_{max} > n_{uk}); (\ell \geq k), (m - \ell)\}$. The procedure is outlined below.

Compilation of the Data of the Test Oxford-Grid

Considering an example of test species (say, mouse *versus* human), with Oxford-grid details as in Coriell institute for medical research[82] show that the total number of orthologous genomes searched is $n = 3512$. Other pertinent data observed in the Oxford-grid test are $(k = 154)$ denoting the number of cells having finite number of conserved synteny scores marked out of a total number of $m (= r \times c) = [22 \times 19] = 418$, matrix cells. Relevantly, the following ratios are first defined and calculated:

> Ratio of the number of cells marked with a finite conserved synteny score to the total number of orthologous pairs searched: $D = (k /n) = (154/3512)$
>
> Ratio of the number of cells marked with a finite conserved synteny score to the total number of cells in the Oxford-grid: $E = [k/(m = r \times c)] = (154/418)$
>
> Ratio of the number of cells marked with a null score (implying nonconserved synteny status) to the total number of cells in the Oxford-grid: $F = [(m-k)/m] = [(418 - 154)/418]$

7.14 Statistical Aspects of Oxford-Grid

As stated before, the Oxford-grid mapping is a representation of randomly mixed grid cells. That is, the Oxford-grid framework can be regarded as a receptacle dispersed with weighted proportions of randomly mixed entities, namely, k and $(m-k)$ cells, and this mixed state of such a binary set of cells depicts a universe of random space. This stochastic space holds an entropy profile of uncertainty as a result of its randomly dispersed binary contents of k and $(m-k)$ cells.

Considering such a random binary mixture, its effective probabilistic attribute can be quantitatively specified in terms of the so-called *Lichtenecker–Rother* (LR) *formulation,*[32,83] and in the present context, this LR formulation can be specified in terms of random-mix ratios of normalized entities, namely, $E = k/m$ and $F = (m-k)/m$. That is, the effective statistics of mixed entities k and $(m-k)$ can be specified by a parameter G that can be deduced explicitly using the set, $\{D, E, F\}$ as follows: $G = \{D^E \times D^F\} = D^{(E+F)}$; or, $\log_e(G) = [E \times \log_e(D) + F \times \log_e(D)] = [(E + F) \times \log_e(D)]$.

Known also as the *logarithmic law of mixing,* the LR formulation indicated above provides a true and effective description of randomness of the underlying statistical mixture. However, the value of G evaluated as above is constrained by Neelakanta,[31] Lichtenecker and Rother,[32] Neelakanta,[33] Neelakanta *et al.,*[34] Neelakanta *et al.,*[35] Neelakanta,[36] Kapur and Kesavan,[37] Taylor *et al.,*[38] Loo *et al.,*[39] Di Pierdomenico *et al.,*[40] Ferdinand,[41] Neelakanta,[42] Bendett and Neelakanta,[43] Jaynes,[44] Cover and Thomas,[45] Kullback,[46] Lin,[47] Xiao-bo,[48] Vogelstein *et al.,*[49] Hoglund,[50] Agarwal and Bhadauria,[51] Anderson and Chaplin,[52] Othmer and Stevens,[53] Kubo,[54] Patel *et al.,*[55] Castro *et al.,*[56] Mitelman,[57] Fukshansky and Bar,[58] Neelakanta and Yassin,[59] Neelakanta and Sardenberg,[60] Neelakanta and De Groff,[61] Heng *et al.,*[62] Karri,[63] Neelakanta and Sharma,[64] Koonin,[65] Houseworth,[66] Edwards,[67] Bengtsson *et al.,*[68] Mouse Genome Database (MGB),[69] Schena,[70] Murphy *et al.,*[71] Ball *et al.,*[72] Naruse *et al.,*[73] Introduction to scientific names for humans,[74] Mouse,[75] *Oryzias latipes,*[76] Japanese rice fish,[77] Zebrafish,[78] *Danio rerio,*[79] Houseworth and Postlethwait,[80] OXGRID — The Oxford grid project,[81] Coriell institute for medical research,[82] and Neelakanta[83]: $[(D/E) + (D + F)]^{-1} \leq (G = D^E \times D^F) \leq [(D \times E) + (D \times F)]$. (These constraints are known as *Wiener inequalities*), and the parameter G depicts a statistically justifiable value of effective probability ascribed to the randomness of the stochastic framework constituted by the set $\{D, E, F\}$, and the associated entropy details (in Shannon's sense) of this stochastic space can be deduced as described below.

Entropy Details of the Oxford-Grid

Inasmuch as Oxford-grid mapping depicts a stochastic phase of randomly mixed grid cells (having finite and null scores), an entropy attribute can be prescribed to it; relevant considerations thereof, can be elucidated as follows: using the stochastic parameter G pertinent to the Oxford-grid, corresponding Shannon entropy of the randomly mixed grid-cells can be denoted as SH-E, which can be deduced in terms of Shannon entropy as follows[42]: *SH-E* = – [$G \times \log_e(G)$] nats. It represents the uncertainty feature of the test Oxford-grid constituted by sparsely known synteny data availed from a limited search exercised on synteny correlation of a test pair of species (specified by the set {n; k, $(m-k)$} indicated in Table 7.3.

In order to decide on maximum plausible value of H_{max} denoting an entropy functional pertinent to the set {ℓ, $(m - \ell)$}, the status of sparse entropy can be extended with: {ℓ, $(m - \ell)$} → {ℓ_{max}, $(m - \ell_{max})$} that would eventually prevail in the Oxford-grid, when ($N_{max} \to \infty$). This corresponds to the grid matrix having a total count of m cells containing ℓ_{max} cells with the finite enumerated synteny counts and $(m - \ell_{max})$ cells posted with null counts when the search space, $n \to (N_{max} \to \infty)$ is subject to the constraint ($\ell_{max} \leq m$).

To determine this new value of the entropy (H_{max}), first, the extent of search space of the orthologous pairs, namely n, is specified as a variable, n_{uk}, and it is rendered to increase say, linearly from a minimum to a maximum value. Next, for each value of ($n \to n_{uk}$) linearly changed, and seeking the corresponding unknown value of ($k \to k'$), a set of uniformly distributed random numbers can be assumed for the unknown k' within the constraining limit, ($k \leq k' \leq m$). The choice of uniformly distributed random numbers is commensurate with the Laplacian hypothesis of unknown statistics on unknown k' being prescribed. Hence, the ratios defined earlier can be calculated for each set, {n_{uk}; k', m} and the associated value of G as well as; the corresponding estimate of entropy (SH-E) are determined.

Thus, the resulting entropy, SH-E *versus* $n \to [n_{uk} \to (N_{max})]$ is plotted as illustrated in Fig. 7.15 using Oxford-grid data of mouse *versus* human.[70] The patch of results on SH-E seen along the ordinate *versus* the abscissae data on linearly varying ($n \to n_{uk}$) denote details obtained with an ensemble of randomly chosen (and uniformly distributed) set of values assigned to k' such that (k = 154) ≤ k' ≤ m (= 418), for each linearly incremented value of n_{uk} in the range (3,512 ≤ n_{uk} ≤ 50,000). Now, for the clustered patch of computed details as seen in Fig. 7.15, a regressed, isotonic trendline is obtained with coefficients (of regression), α_1 and α_2 and expressed as follows: SH-E[($n \to n_{uk}$)] = [– [$\alpha_1 \times \log_e[(n \to n_{uk})] + \alpha_2$]. For the test species (mouse *versus* human) being considered, this trendline so estimated has the coefficients, α_1 = – 0.0523 and α_2 = 0.5846 as shown in Fig. 7.15. Hence, for any presumed (and unknown) value of ($n \to n_{uk}$) searched in the range (3,512 ≤ n_{uk} ≤ 50,000 → ∞), the

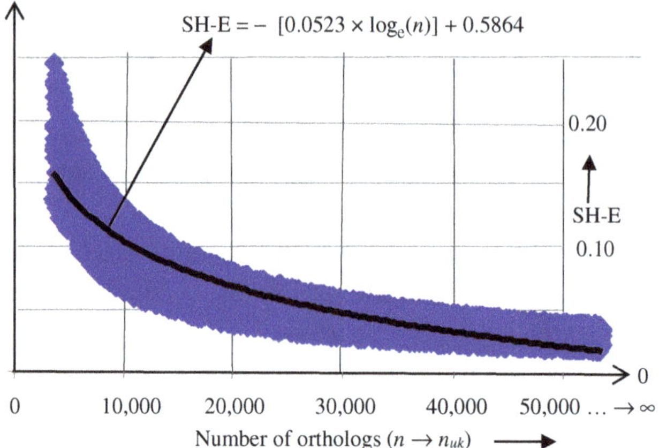

Fig. 7.15 Shannon entropy (SH-E) *versus* number of orthologous genomes ($n \rightarrow n_{uk}$), linearly changing in the search space of test species, mouse *versus* human. *SH-E* is estimated as indicated earlier using the Oxford-grid data of Schena.[70] Furthermore, m (= 418) and the ranges: $154 \leq (k \rightarrow k' \leq (m = 418)$ and $3{,}512 \leq n_{uk} \leq (50{,}000 \rightarrow \infty)$ are adopted in the computations leading to an empirical curve fit on SH-E *versus* n data. ($SH\text{-}E = -[0.0523 \times \log_e(n)] + 0.5864$).

associated entropy (SH-E) is estimated. For example, relevant computation yields, SH/E = 0.08322 for an assumed value of $(n \rightarrow n_{uk}) = 15{,}000$.

Expanded Search on Orthologs

Considering a value of $\ell > (k \leq m)$, the next step involves finding the value of G corresponding to a known (estimated) entropy value SH-E. For example, considering SH-E (for n_{uk} = 15,000) = 0.08322 as assessed above, an inverse solution of the entropy relation, SH-E(n_{uk}) = $-$ [$G_{uk} \times \log_e(G_{uk})$] can be adopted to find the corresponding value of G_{uk}. That is, considering the Shannon entropy equation in its most general form: $v = f(u) = [-u \times \log_e(u)]$, relevant inverse relation is given by: $u = f^{-1}(v)$ or explicitly, $(1/u)^u = \exp(v)$, with (v: SH-E and u: G). However, the relation, $(1/u)^u = \exp(v)$ depicts a transcendental equation. As such, a simple analytical solution for u for a given value of v is not straightforward. Hence, a graphical solution can be pursued as illustrated in Fig. 7.16.

The entities (v: SH-E) *versus* (u: G) are plotted in Fig. 7.16 with u incremented linearly from an initial (small) value (say, for example, 0.0001, 0.0002, etc.), and the corresponding Shannon-entropy, namely, v: SH-E is computed and plotted as a dependent variable as shown in Fig. 7.16. Now, a value G_{uk} can be identified and marked on (u *versus* v) graph, such that it corresponds to the intersection of the computed, (*u*: *G*) *versus* (*v*: SH-E) curve and the known value of *v*: SH/E = 0.08322 (corresponding to n_{uk} = 15,000) deduced earlier. Thus, the solution for the *G*-value being sought (and denoted as G_{uk}) is found to be 0.02175 as shown in Fig. 7.16. Likewise, for every value of n_{uk} in the range, $(3{,}512 \leq (n \to n_{uk}) \leq 50{,}000)$, the corresponding value of G_{uk} can be determined.

Using the deduced value of G_{uk} as above, the procedure toward elucidating $k' \to (\ell \leq m)$, with the ortholog search space of $n_{uk} \to N_{\max}$ (depicting a large value tending toward infinity) can be attempted as follows: suppose the variable k' is increased (say, in the incremental steps of 5) from its base-value to the maximum value equal to $m = (r \times c)$. In each case of k' so chosen, the Oxford-grid denotes a domain with a new stochastic state invoked by the presence of a random mix of entities, namely, k' and $(m-k')$ cells. Thus, a set of metrics implicitly depicting the *G* value of LR heuristics can be defined, and their statistical limits of upper- and lower-bounds (LB and UB) can be ascertained as summarized follows: (i) statistical random mixture *LR-formulation-based metric*[32,83]: $S_1 = [(k'/n_{uk})^{(k'/m)}] \times [(k'/n_{uk})^{(1-k'/m)}]$; (ii) statistical metric that yields the lower-bound (LB) *via Neelakanta's susceptibility formula*[83]: $S_2 = 1 - [(4 - k'/n_{uk})^{(k'/m)} \times (1-k'/m)^{(k'/m)}]$. Relevantly, the following other versions of UB and/or LB limits can also be specified: (a) statistical *Wiener upper-bound* (UB) metric (*version 1*): $S_3 = [(k'/n_1) \times (k'/m)]/[(1- k'/m) \times (k'/m)]$ and (b) statistical *Wiener upper-bound* (UB) metric (*version 2*), known also as *Beer measure*[83]): $S_4 = [(k'/n_1) \times (k'/m)]$.

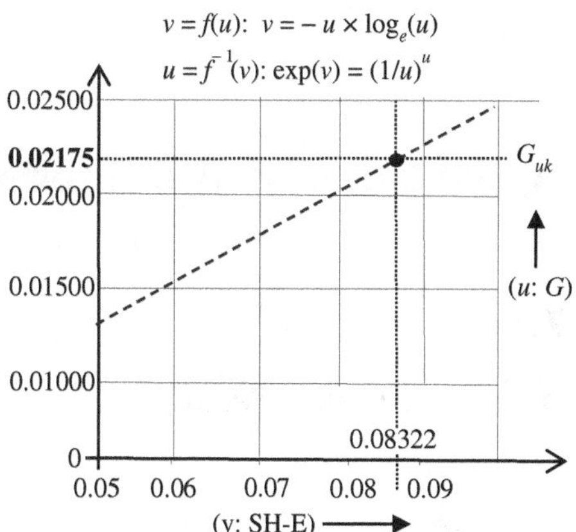

Fig. 7.16 Graphical solution to find the G-value (u) for a known entropy, SH-E (v) using the inverse entropy relation of $[v = f(u) = -u \times \log_e(u)]$, namely, $[u = f^{-1}(v): \exp(v) = (1/u)^u]$.

In essence, the metrics specified as above denote variates of LR formulation depicting the G value consistent with stochastic mixture theoretics. Specifically, the metric S_1 exactly refers to the LR formulation and the other metrics, namely, S_2, (S_3 and S_4), respectively, denote the lower- and upper-bounds of S_1 as per the so-called Wiener limits.

In all, for a given set of {n; k, (m–k)}, and a chosen value of $n \rightarrow n_{uk}$, each of the metrics S_1, S_2, S_3, and S_4 can be computed as a function of the variable, $k \rightarrow k'$. Correspondingly, the particular value of k' that yields the assessment of the metric in question matching G_{uk} (deduced as explained earlier) would denote the maximum count k namely, ℓ ($\leq m$). It depicts the value of ℓ ($\leq m$) sought for the search space of n_{uk} of the pair of test species. For example, as described in Neelakanta and Sharma,[64] considering the dataset, {n_{uk} = 15,000, G_{uk} = 0.02175} of mouse *versus* human, the assessed variation of S_1 *versus* k' shows the value of the count k ascertained with the metric, [S_1 = {$(k'/n_{uk})^{(k'/m)}$} × {$(k'/n_{uk})^{(1-k'/m)}$}] (LR formulation) as $\ell\,|_{S1}$ equal to: (304 + 309)/2 = 307. *In lieu* of the metric S_1, other measures namely, S_2, S_3, and S_4 can also be prescribed in the procedure narrated above. Corresponding results of the computations are denoted, respectively, as: $\ell|_{S2}$, $\ell|_{S3}$, and $\ell|_{S4}$ using the same dataset adopted in the case of $\ell|_{S1}$ assessment with the metric: S_1. Relevant computational details in Neelakanta and Sharma[64] offer explicit results on the following pertinent to the test Oxford-grid of Mouse *versus* Human[70] dataset as above. With [S_2 = 1–{$(1 - k'/n_{uk})^{(k'/m)}$ × $(1- k'/m)^{(k'/m)}$] (Neelakanta's susceptibility formula), $\ell|_{S2}$ is equal to: (259 and 264) ≈ 262; with [S_3 = {$(k'/n_1) \times (k'/m$}]/{$(1-k'/m) \times (k'/m)$}], (Wiener UB version 1), $\ell|_{S3}$ is equal to (364 and 369) ≈ 367 and, with [S_4 = {$(k'/n_1) \times (k'/m)$}/{$(1-k'/m) \times (k'/m)$} (Wiener UB version 1), $\ell|_{S4}$ is equal to: (364 and 369) ≈ 367.

7.15 Exploring the Synteny with Tutorials and Examples

Example: Ex 7.5

Example-pairs of Orthologs and Results on Their Synteny Correlations

In essence, the analyses and computations described so far show how by stretching the parameter n (depicting the number of ortholog pairs searched) across a stretch ranging from a low-value to a large-value of, n

→ N_{max} (→ ∞), the maximum plausible count on k-cells (containing finite values of synteny counts) in the Oxford-grid tending toward ℓ (= l_{max}), can be obtained (for each value of n), and the result on ℓ is expressed in terms of its set $\{\ell|_{S1}, \ell|_{S2}, \ell|_{S3}, \ell|_{S4}$ depicting the limiting bounds. Hence, by considering n across a range from a low value, say, 500 to N_{max} = 50,000, the maximum plausible count on k is deduced (and expressed in terms of the set $\{\ell|_{S1}, \ell|_{S2}, \ell|_{S3}, \ell|_{S4}\}$).

The results are presented in Figs. 7.17a–7.17c, as plots of ($k \to k'$) *versus* ($n \to n_{uk}$), respectively, for the test pairs: mouse (*M. musculus*) *versus* human (*H. sapiens*); medaka fish (*O. latipes*) *versus* human (*H. sapiens*); and, medaka fish (*O. latipes*) *versus* zebra fish (*Donio rerio*). Hence, the value of estimated maximum count-k encountered in each case is indicated as ℓ_{max} at the corresponding value of $n = n_{max}$.

In each case of the species pairs considered thereof, the results in the graphs of Figs. 7.17a–7.17c show an initial monotonically increasing trend of k' *versus* ($n \to n_{uk}$), and subsequently, the plots assume a convex form with a maximum value of $k' \to \ell$ at a certain value of $n_{uk} = n_{max}$. Figs. 7.17a–7.17c also include details on statistical lower- and

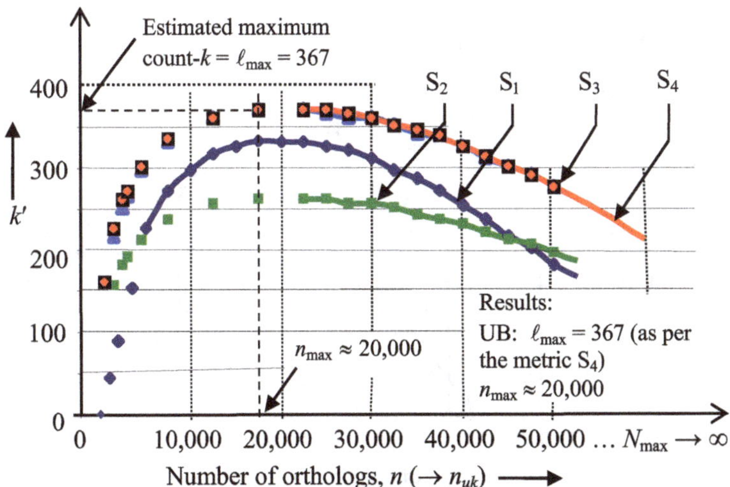

Fig. 7.17a Plot of simulation results on the variable k' (denoting the synteny count-k) *versus* number of orthologs, $n_{uk} \to$ 50,000 (= N_{max}) → ∞ searched. The variable k', ($0 \leq (k' \to \ell) \leq m$) denotes the number of cells on the Oxford-grid having a finite value of observed syntenic correlation at random cells, (i, j).

Results: deduced for the Oxford-grid of the test species pair[70]: mouse (*M. musculus*) *versus* human (*H. sapiens*).

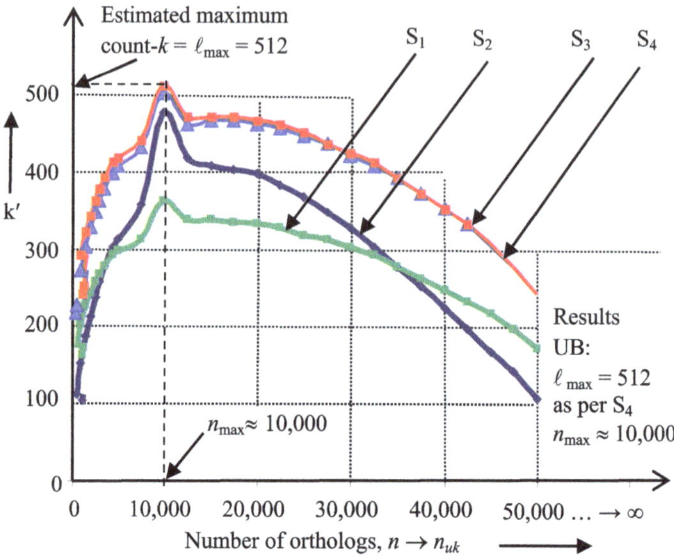

Fig. 7.17b Plot of simulation results on the variable k' (denoting the synteny count k) *versus* number of orthologs, ($n_{uk} \rightarrow N_{max} = 50{,}000$) searched. The variable k', $(0 \le (k' \rightarrow \ell) \le m$ denotes the number of cells on the Oxford-grid having a finite value of observed syntenic correlation at random cells, (i, j): Oxford-grid of the test species pair used: medaka fish (*O. latipes*) *versus* human (*H. sapiens*).

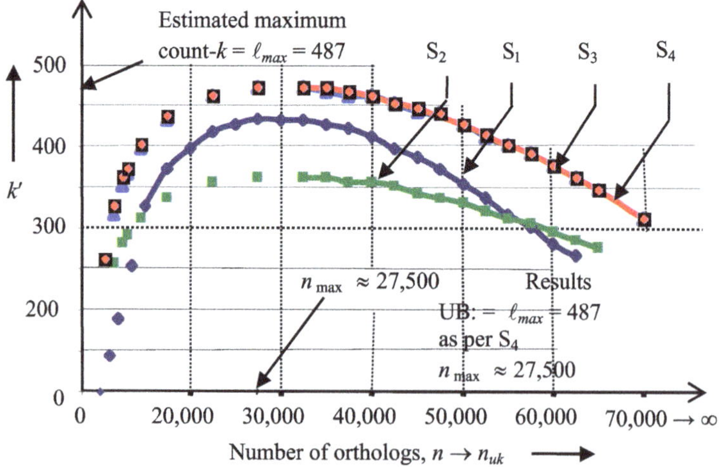

Fig. 7.17c Plot of simulation results on the variable k' (denoting the synteny count k) versus number of orthologs, ($n_{uk} \rightarrow N_{max} = 50{,}000$) searched. The variable k', $(0 \le (k' \rightarrow \ell) \le m$ denotes the number of cells on the Oxford-grid having a finite value of observed syntenic correlation at random cells, (i, j): deduced for the Oxford-grid of the test species pair: medaka fish (*O. latipes*) *versus* zebra fish (*Donio rerio*).[73,84]

upper-bounds on S_1 as decided, respectively, by the Wiener limits, S_2 and (S_3, S_4). In each case of the test species pairs studied, a summary of the pertinent results is presented in Table 7.4.

Having said that, while seeking an answer on the maximum plausible extent of syntenic correlation that could prevail between chromosomes of a pair of test species, if their orthologous genomes are searched exhaustively, the strategy proposed hypothesizes that with limited statistics of k (out of m entities) filled in the Oxford-grid (of the compared species) would dictate a sparse value of underlying (Shannon's) entropy features (stemming from statistically mixed pair of cell entities, k and (m – k) in the Oxford-grid), and they implicitly depict the information

Table 7.4 Computed data and a summary of results relevant to Postlethwait query.[64,66,80]

<table>
<tr><td rowspan="5">Test species pairs</td><td colspan="3" rowspan="3">Available Oxford-grid data of the test species pairs: {k, n}, m and (m–k)}</td><td colspan="5">Estimated values: (via present method) $k \to (\ell \le m)$ at n when $n \to (N_{max} =$ 50,000) keeping the value of m constant as specified by the Oxford-grid</td></tr>
<tr><td colspan="4">Proposed metrics</td><td rowspan="4">$(n \to n)$ $n_\ell \approx$:</td></tr>
<tr><td rowspan="2">S_1</td><td>LB</td><td colspan="2">UB</td></tr>
<tr><td rowspan="2">{k, n}</td><td rowspan="2">m = (r × c)</td><td rowspan="2">(m–k)</td><td>S_2</td><td>S_3</td><td>S_4</td></tr>
<tr><td colspan="4">Estimated $k \to (\ell \le m)$
ℓ_{S1} | ℓ_{S2} | ℓ_{S3} | ℓ_{S4}</td></tr>
<tr><td>Mouse versus human</td><td>{154, 3512}</td><td>418 (22 × 19)</td><td>304</td><td>307</td><td>262</td><td>367</td><td>369</td><td>20,000</td></tr>
<tr><td>Medaka fish versus human</td><td>{104, 818}</td><td>552 (24 × 23)</td><td>448</td><td>477</td><td>362</td><td>502</td><td>512</td><td>10,000</td></tr>
<tr><td>Medaka fish versus zebra fish</td><td>{125,255}</td><td>600 (24 × 25)</td><td>475</td><td>397</td><td>347</td><td>483</td><td>487</td><td>27,500</td></tr>
<tr><td>Mouse versus Human</td><td colspan="3">[154 and 3512]</td><td rowspan="3">⇔</td><td colspan="4">[369 at 20,000]</td></tr>
<tr><td>Medaka fish versus Human</td><td colspan="3">[104 and 818]</td><td colspan="4">[512 at 10,000]</td></tr>
<tr><td>Medaka fish versus Zebra-fish</td><td colspan="3">[125 and 255]</td><td colspan="4">[487 at 27,500]</td></tr>
</table>

on the patterns of weighted probability that describe the effective stochastic attributes of the associated mixture constituents, as indicated by Neelakanta *et al.*[35]

Correspondingly, the proposed metric S_1 measures the effective statistical attributes (of the mixture constituents in the test Oxford-grid), and it is bounded by the statistical Wiener limits, given by: $[S_2 < S_1 < (S_3, S_4)]$. Furthermore, the assessed sparse-entropy plots in Figs. 7.17a–7.17c show a convexity profile, when the search is extended on more and more number of pair of orthologs with n tending to: $(n_{uk} \rightarrow N_{max} \rightarrow \infty)$.

Relevantly, the simulated results conform to Schur convex functional attributes consistent with maximum entropy heuristics.[37] This leads to finding an answer for the Postlethwait's query concerning syntenic correlation attributes *vis-à-vis* the posterior (*ex ante*) status of shared synteny that can be inferred in an exhaustive search made on test orthologous pairs, by projecting the sparse data (availed *ex post* from the test Oxford-grid).

The aforesaid methodology implies a stochastic modeling of the associated Oxford-grid parameters and applying entropy theoretic considerations. Efficacy of the proposed approach is implicitly confirmed by the Schur convexity of entropy-based results obtained in conformance with the analytics of maximum entropy optimization principles as indicated above.[37,85]

Tutorial: T-7.7

Syntenic correlation measures: this tutorial review is concerned with measures of conserved synteny correlation evaluations. Measures of conserved synteny are important for estimating the relative rates of chromosomal evolution in various lineages. A natural way to view the synteny conservation between two species is to represent the comparative details on an Oxford-grid denoting a matrix with $m = (r \times c)$ grid cells, wherein the number of orthologous genes on each chromosome of the first species *versus* on each of the chromosome of the second species. That is, the Oxford-grid, in essence, illustrates the display of homologous loci from two species in terms the matrix, $[m] = [r \times c]$.

The number of chromosomal segments conserved during the evolution of two species denotes also a measure of their genomic distance. In general, the number of conserved segments containing homologous genes can be estimated by comparing synteny relations within and between the two genomes. However, there are three possible sources of underestimation stemming from the following: (i) conserved segments in which genes are yet to be identified in one or both species; (ii) repeated translocations or transpositions resulting in not just one, but several conserved segments from a chromosome in one species being located on a single chromosome in the other; and (iii) due to intrachromosomal rearrangements like, inversion.

Suppose a restricted number (n) of the orthologs of a given pair of species are mapped between the two species with k grid cells out of $m = (r \times c)$ show finite entries on the conserved synteny. Then, Postlethwait poses the following question: instead of the limited extent of prevailing (n) orthologs, if an exhaustive search with $n \rightarrow (N_{max} \rightarrow \infty)$ is performed. Then, what will be the increased extent of $k \rightarrow \ell$?

In essence, the interest that lies in the present study conforms to developing a method that quantifies the extent to which the future (*ex ante*) evolution of the dynamics concerning the synteny would prevail between orthologs of a given pair of species.[86] Resorting to information-theoretic principles, the entropy of such a unit can be specified in terms of Shannon information and entropy heuristics.[27]

Relevant to the present study concerning syntenic correlation between a pair of species, first considered is the entropy profile of an Oxford-grid wherein the extent of synteny conservation between the orthologs (of the two test species) for a set of n orthologs is known *a priori,* and using this known (*ex post*) data, suppose the synteny search is extended to an exhaustively large number of orthologs (of the test species pair), then an algorithm is sought to deduce the *ex ante* details on the corresponding extent of syntenic correlation.

At chromosomal level, the synteny consideration of interest implies deducing the extent of identical genes being present in a pair of chromosome of two species. The statistical correlation is a viable parameter for such comparison. In addition, an association has also been considered to access the synteny between chromosomes belonging to a pair of species. The concept of statistical association was proposed by Guttman in 1941.[85–87] When applied to chromosomes, the association parameter measures the proportion of error made relationship, whereas the term correlation strictly implies a linear relationship in question is not necessarily causal. Statistical ε-lambda and Cramer's-V,[88] tetrachoric correlation and odd's ratio.[89]

Cramer's V-measure[88] is also referred to as Cramer's phi (ϕ). It measures the association between two nominal variables giving a result as (0

$\leq \lambda \leq 1$). It is based on Pearson's χ^2-statistics. Considering two discrete variables, ϕ_c denotes the inter correlation measures between them. It is symmetric measure with 0 denoting nil association between the variables and 1 denotes a total association. ϕ_c is calculated by taking the square root of the χ^2 statistic dividing by the sample size and the length of minimum dimension, that is, the minimum dimension refers to k denoting the smaller of the number of a rows, r or column, c of a matrix $[r \times c]$ written in terms of the two sets of compared variables.

Goodman and Kruskal's lambda (GK-λ)[85]: this measure of association denotes the proportion reduction in error in cross-tabulation analysis. With reference to a sample composed of a nominal independent variable and a dependent variable GK-λ; denotes the extent to which the modal categories and frequencies for each value of the independent variable differ from the overall modal category and frequency for all the values of independent variable together, that is, $\lambda = (\varepsilon_1 - \varepsilon_2)/\varepsilon_1$, where ε_1 is overall nonmodal frequency and ε_2 is the sum of the nonmodal frequencies for the each value of the independent variable, and λ lies between 0 and 1 both inclusive.

In information-theoretic perspective (as described in Chapter 2), the statistical features of comparison can be viewed in terms of entropy measures (such as Kullback–Leibler metric and Jenson and Shannon measure and a host of other Csiszár measures,[42,90]) and these measures implicitly denote the underlying relational association between compared entities in entropy framework.[85,86] Details thereof can be stated as follows:

The syntenic correlation (ρ) as detailed in this chapter refers to a measure viewed as a natural statistical entity that compares syntenic conservation across pairs of species. Genome evolution in a multichromosome organism may involve: (i) translocation of genes between chromosomes, rearrangement of genes on chromosomes, splitting and fusion of chromosomes, and gene (genome) duplication events. The underlying details on the measures of rearrangement distances between a pair of chromosomes denote complex information profile. Mostly, measures on such synteny are restricted to gene arrangements in highly conserved orthologs.

Considering various lineage of species, measures of conserved synteny signify the need to estimate the relative rates of chromosomal evolution. This conserved synteny between two species as indicated earlier is presented in Oxford-grid format that conforms to a matrix,

$m = [r \times c]$, which summarizes the number of orthologs genes on each of the chromosomes (1 through r rows of the first species) that corresponds to each chromosome (1 through c column of the second species).

Originally, synteny was defined with reference to genes on the same chromosome, and conserved synteny between two species refers to the present of two or more orthologous synteny in each of the two species. Subsequent studies identify the conserved synteny by the presence of one or more markers or orthologs. As such, a conserved synteny track refers to the presence of one (or more orthologs) on a pair of chromosomes (considering one chromosome rom each species). Correspondingly, a synteny plot can be drawn, which indicates the relative position of each orthologous gene-pair on the chromosome of each species, for example, as considered in this chapter: human and mouse. The width and the length of each box wherein the dots are placed in the synteny plot denote the relative position proportional to the length of the chromosomes. This version of syntenic plot is due to Murphy *et al.*[71] Furthermore, a syntenic dot plot can also be drawn to depict a kind of scatter plot. Here, each axis denotes a sequence laid end-to-end, and each dot in the scatter plot represents a putative homologous match between two sequences. Often, dot plots are used for whole genome comparison within the same genome or across two genomes from different taxa in order to identify synteny.

Notwithstanding syntenic correlation measures comparing genomic distances across many pairs of species (proposed by Houseworth and Postlethwait[66]), similar measures also exist due to Bengtsson *et al.*[68] and due to Zakharov and Valeev.[91,92] Pertinent results give a measure with nonzero lower-bound (LB) that depends on the probability that a pair of gene will be syntenic in each species. The measure due to Houseworth and Postlethwait,[66] on the other hand, provides a correlation measure between (0 and 1). Its value is 1, whenever the two genomes have identical syntenic groups and zero, if the orthologs genes are randomly scattered between the two genomes. Relevantly, the estimated syntenic correlation (ρ) is explicitly given in Houseworth[66] in terms of the aforesaid entities. Relevant measure of syntenic correlation as above implies an implicit a measure of association. Similar degree of association *via* a matrix of comparison has also been indicated in other contexts of comparisons using statistical data by Fisher,[93] Goodman and Kruskal,[94] and Guttman.[87]

The genomic-map information gathered often refers only to a restricted knowledge on orthologous chromosome number, for a given gene in a genome. An alternative approach due to Kumar *et al.*[95] takes into an account, the distribution of the number of markers in contiguous sets of autosomal markers, in order to estimate the number of conserved segments. Comprehensive details on synteny-related considerations can further be seen in the host of literature.[96–104]

Problem: P-7.2

Statement of the Problem

This problem is designed to formulate a relational algorithm between the synteny correlation and statistical divergence between the compared chromosomes. Relevant to the tutorial details furnished above, a rider problem can be designed as presented below. It can be first attempted as a directed independent study/term-paper, and it can be expanded into thesis/dissertation project/exercise.

Using the statistical distribution (such as β-distribution) approach, the synteny comparison as in Waddington *et al.*[104] of the numbers of conserved loci (per linkage group *versus* linkage group length) across two test species, relevant problem can be addressed in terms of the mutual/divergent information (or cross-entropy such as Csiszár metrics[42,90]) of the statistical data of the test species. Furthermore, given a statistical distribution (like β-distribution), the associated entropy can be estimated as described in Estimating the entropy.[105]

Problem: P-7.3

Statement of the Problem

In comparative genome analysis, determining reliable measures of synteny conservation that are nonindependent of comparisons enable identifying conserved and disrupted syntenies and finding redundant rearrangements (without systematic errors tending to overestimate the degree of conservation) is needed. In Hannenhalli *et al.*[98] certain methods to estimate the total number of conserved syntenies within the genome, counting both those that have already been described and those that remain to be discovered are indicated, and results on relative occurrence events of interchromosome *versus* intrachromosome rearrangements for certain lineages (such as those leading to humans and mice) are presented.

This problem is designed to formulate an algorithm to formulate synteny disruption for selected lineages by applying posentropy concept to "disruptions" involved. Probabilistic occurrence of interchromosome *versus* intrachromosome rearrangements for certain lineages (such as humans and mice) can be modeled thereof in deducing the entropy measure. This problem can be first devised as a directed independent study/term-paper and then expanded into a thesis/dissertation project/exercise.

Problem P-7.4

Statement of the Problem

In Kumar *et al.*,[95] it is indicated that genomic divergence between species can be quantified in terms of the number of chromosomal rearrangements that have occurred in the respective genomes following their divergence from a common ancestor. Relevant rearrangements often disrupt the structural similarity between genomes, with each rearrangement producing additional, *albeit* shorter, conserved segments. Hence, a simple statistical approach is proposed in Kumar *et al.*[95] using the distribution of the number of markers in contiguous sets of autosomal markers to estimate the number of conserved segments. Such an identification requires information on the relative locations of orthologous markers in one genome and only the chromosome number on which each marker resides in the other genome. A mathematical model devised thereof accounts for the effect of the nonuniformity of the breakpoints and markers on the observed distribution of the number of markers in different conserved segments.

Following the heuristics of Kumar *et al.*,[95] explain the context of molecular clocks *versus* divergence times of species and their relevance in synteny conservation. Hence, propose a possible mathematical model to account for the presence of nonuniformity of the breakpoints and markers on the observable distribution of the number of markers in different conserved segments.

Problem P-7.5

Statement of the Problem

Comparison of paralogs using Oxford-grid data as indicated in Neelakanta and Sharma[64] and elaborated in this chapter is indicated for a set of three species pairs. As discussed earlier, relevant genomic comparisons provide invaluable information about evolutionary relationships, morphology, and biogenesis. As well-known, homologous sequences between various species point out a common ancestor and comparison of similar sequences determines conserved sequences that may serve more important functions. Oxford-grids are adopted to decide on the number of homologous genes between the chromosomes of compared species. More genome-sequence comparisons have been pursued to identify possible common sequences variants and existence of similar sequences among species. The future application of Oxford grids can give enhanced insight into the phylogenic relationship among species. Useful databases exist for sequenced genomes of many organisms and research on human–mouse, medaka–human, medaka–zebra fish is the theme in Neelakanta and Sharma.[64] Following such considerations, the following problem is indicated as a pertinent exercise.

Following the analysis of Neelakanta and Sharma,[64] develop syntenic correlation and elucidate the underlying synteny conservation details of the following pairs of species: mouse/human–cat, human–cattle/sheep, human–dog, human–pig, and human–horse. Some useful data are as follows.[106–112]

For human–cattle pair: humans have 22 chromosomes (On y-axis: Number of rows: 22), and cattle have 29 chromosomes (On x-axis: Number of columns: 29). Hence, $m = 22 \times 29 = 638$; $k = 86$, $(m - k) = (638 - 86) = 552$, and $n = 1229$.

For human–pig pair: humans have 22 chromosomes (On y-axis: number of rows: 22), and pigs have 18 chromosomes (On x-axis: number of columns: 18). Hence, $m = m = 396$, $k = 72$, $(m - k) = (396 - 72) = 324$, and $n = 1012$.

For human–sheep pair: humans have 22 chromosomes (On y-axis: number of rows: 22), and sheep has 26 chromosomes (On x-axis: number of columns: 26). Hence, $m = 22 \times 26 = 572$, $k = 60$, $(m - k) = (572 - 60) = 512$, and $n = 480$.

For horse–human pair: horses have 31 chromosomes (On y-axis: number of rows: 31), and human has 22 chromosomes (On x-axis: number of columns: 22). Hence, $m = 31 \times 22 = 682$, $k = 50$, $(m - k) = (682 - 50) = 632$, and $n = 143$.

Synteny Evaluation: Pseudocode

With reference to elucidating the synteny details as considered in this chapter (while searching for an answer for the Postlethwait query), a set of relevant pseudocodes can be indicated toward evaluating the number of orthologs $n \rightarrow (n_{uk} \rightarrow N_{\max} \rightarrow \infty)$ *versus* the count-k placed in the Oxford-cell grids. Pertinent to test species, namely, human *versus* mouse, human *versus* medaka fish, medaka fish *versus* zebra fish,[70,73,80] pseudocode(s) on evaluating the extent of synteny *versus* the number of orthologs-pair being searched are available in Sharma.[107]

Synteny Evaluation: Clinical Usage

The heuristics of this chapter, for example, can be pursued to verify (or screen) any prevailing cytogenetic aberrations (CAs). For example, with reference to mouse (representing a model organism), relevant synteny correlation against a human subject (having no CA) could be different from that of humans with CA being present.[48]

7.16 Closure

Synteny as indicated defines two or more genomic regions that are derived from a common ancestral genomic region. Presence of synteny refers to identifiable set of homologous genes in two genomes having a collinear arrangement. When such a pattern of gene-order conservation is discovered, the most parsimonious explanation is that the two regions are related through a common ancestor.[47] This chapter is conceived to indicate the bioinformatic prospects of synteny analysis and determine (using Oxford-grid), the ultimate limit on maximum extend of countable syntenic correlation when the search space of orthologs being compared is exhaustively increased.

References

[1] D. J. Bolding, M. Bishop, and C. Canning (Eds.): *Handbook of Statistical Genetics*. John Wiley and Sons Ltd., West Sussex, England: 2007.

[2] A. T. Annunziato: DNA packaging: Nucleosomes and chromatin. *Nature Education*. 2008, vol. 1(1): 28.

[3] A. J. F. Griffiths, W. M. Gelbart, J. H. Miller, and R. C. Lewontin: *Modern Genetic Analysis*, W. H. Freeman and Company, New York, NY, USA: 1999.

[4] C. O'Connor: Chromosomal segregation in mitosis: The role of centromeres. *Nature Education*, 2008, vol. 1(1).,

[5] C. Hoischen, M. Bussiek, A. Derome, F. Hayes, and S. Diekmann: Bacterial centromeres and kinetochore complexes. Available online at:

http://www.fli-leibniz.de/images/groups/diekmann/hoischen/ch_kinetochore.pdf.

[6] K. S. Bloom and J. Carbon: Yeast centromere DNA is in a unique and highly ordered structure in chromosomes and small circular mini chromosomes. *Cell*, 1982, vol. 29(2), 305–317.

[7] *Mapping protein/DNA interactions by cross-linking, PMID: 21413366,* Institut national de la santé et de la recherche médicale (INSERM), Paris, 2001. Available on-line at: *http://www.ncbi.nlm.nih.gov/books/NBK7107/.*

[8] G. M. Copper, *The Cell: A Molecular Approach*, Sinauer Associates Publishers, Sunderland, MA, USA: 2000.

[9] A. F. Pluta, A. M. Mackay, A. M. Ainsztein, I. G. Goldberg, and W. C. Earnshaw: The Centromere: Hub of chromosomal activities. *Science*, 1995, vol. 270, 1591–1594.

[10] J. C. Lamb, J. Theuri, and J. A. Birchler: What's in a centromere? *Genome Biology*, 2004, vol. 5(9), 239–245.

[11] D. J. Amor and K. H. A. Choo: Neocentromeres: Role in human disease, evolution and centromere study. *American Journal of Human Genetics*, 2002, vol. 71, 695–714.

[12] S. K. Knutson: *In Vivo* Characterization of the Role of Histone Deacetylase 3 in Metabolic and Transcriptional Regulation. *Ph. D. Dissertation*, Graduate School of Vanderbilt University, Nashville, TN, USA: 2008. Available online at: *http://etd.library.vanderbilt.edu/available/etd-05222008-164304/unrestricted/.*

[13] T. Kouzarides: Chromatin modifications and their cell function. *Cell*, 2007, vol. 128, 693–705.

[14] C. L. Woodcock and R. P. Ghosh: Chromatin higher-order structure and dynamics. *Cold Spring Harbor Perspectives in Bio*logy, 2010, 2: a000596, 1–25.

[15] S. E. Torigoe, D. L. Urwin, H. Ishii, D. E. Smith, and J. T. Kadonaga: Identification of a rapidly formed non-nucleosomal histone-DNA intermediate that is converted into chromatin by ACF. *Molecular Cell*, 2011, vol. 43, 638–648.

[16] V. L. Chandler, Paramutation: From maize to mice. *Cell*, 2007, vol. 128, 641–645.

[17] S. Sharma, T. K. Kelly, and P. A. Jones: Epigenetics in cancer. *Oxford Journals*, vol. 31(1), 2009, 27–36.

[18] P. Stankiewicz and J. R. Lupski: Structural variation in the human genome and its role in disease. *Annual Review of Medicine*, 2010, vol. 61, 437–455.

[19] G. H. Perry, N. J. Dominy, K. G. Claw, A. S. Lee, H. Fiegler, R. Redon, J. Werner, F. A. Villanea, J. L. Mountain, R. Misra, N. P. Carter, C. Lee,

and A. C. Stone: Diet and the evolution of human amylase gene copy number variation, *Nature Genetics.* 2007, vol. 39, 1256–1260.

[20] A. R. Varsale, A. S. Wadnerkar, and R. H. Mandage: Cancer investigation: A genome perspective. *Academic Journals,* 2010, vol. 5(5), 79–86.

[21] R. Bishop: Applications of fluorescence in situ hybridization (FISH) in detecting genetic aberrations of medical significance. *Oxford Journals,* 2010, vol. 3(1), 85–95.

[22] H. H. Heng, J. B. Stevens, G. Liu, S. W. Bremer, K. J. Ye, P. V. Reddy, G. S. Wu, Y. A. Wang, M. A. Tainsky, and C. J. Ye: Stochastical cancer progression driven by non-clonal chromosome aberrations. *Journal of Cell Physiology,* 2006, vol. 208(2), 461–472.

[23] D. Barh: A normal 46 XX karyotype does not always represent female phenotype. *IIOAB Journal,* 2012, vol. 3(3), 49–50. Available online at: *https://iioab.org/Vol3(3)2012/3(3)49-50.pdf.*

[24] I. Lobo: Chromosome abnormalities and cancer cytogenetics. *Nature Education,* vol. 1(1), 2008.

[25] A. I. Baba and C. Catoi: *Comparative Oncology,* The Publishing House of the Romanian Academy, Bucharest, Romania: 2007.

[26] P. U. Devi: Basics of carcinogenesis. *Health Administrator,* 2005, vol. XVII(1), 16–24.

[27] B. Vogelstein, E. Fearon, S. Hamilton, S. Kern, A. Preisinger, M. Leppert, Y. Nakamura, R. White, A. Smits, and J. Bos: Genetic alterations during colorectal-tumor development. *New England Journal of Med*icine, 1988, vol. 319(9), 525–532.

[28] F. Jiang, R. Desper, C. H. Papadimitriou, A. A. Schaffer, O. P. Kallioniemi, J. Richter, P. Schraml, G. Sauter, M. J. Mihatsch, and H. Moch: Construction of evolutionary tree models for renal cell carcinoma from comparative genomic hybridization data. *Cancer Research,* 2000, vol. 60(22), 6503–6509.

[29] T. V. Nikitana, I. N. Lebedev, N. N. Sukhanova, E. A. Sazhenova, and S. A. Nazarenko: A mathematical model for evaluation of maternal cell contamination in cultured cells from spontaneous abortions: Significance for cytogenetic analysis of prenatal selection factors. *Fertility and Sterility,* 2005, vol. 83(4), 964–972.

[30] A. N. DuPont: Risk-Evaluation in Clinical Diagnostic Studies: Ascertaining Statistical Bounds *via* Logistic Regression of Medical Informatics Data. *M.S. Thesis,* College of Engineering and Computer Science, Florida Atlantic University, Boca Raton, FL, USA: 2011.

[31] P. S. Neelakanta: Application of logarithmic law of mixing to electric susceptibility. *Electronics Letters,* vol. 25, 1989, 800–802.

[32] K. Lichtenecker and K. Rother: Die herleitung des logarithmischen mischungsgesetzes aus allegemeinen prinzipien der stationären strömung. *Physik Zeitschrift,* 1938, vol. 32, 255–260.

[33] P. S. Neelakanta: Complex permittivity of a conductor-loaded dielectric. *Journal of Physics*, 1990, vol. 2, 4935–4947.

[34] P. S. Neelakanta, R. I. Turkman, and T. K. Sarkar: Complex permittivity of a dielectric-mixture correlated version of logarithmic law of mixing. *Electronic Letters*, 1985, vol. 21(7), 270–271.

[35] P. S. Neelakanta, T. V. Arredondo, and D. De Groff: Redundancy attributes of a complex system: Application to bioinformatics. *Complex Systems*, 2003, vol. 14, 215–233.

[36] P. S. Neelakanta: Permittivity of dielectric-conductor mixture: Application of logarithmic law of mixing to electrical susceptibility. *Electronics Letters*, 1989, vol. 25, 800–8022.

[37] J. N. Kapur and H. K. Kesavan: *Entropy Optimization Principles with Applications.* Academic Press/Harcourt Brace Jovanovich publishers, Boston, MA, USA: 1992.

[38] B. S. Taylor, Jordi, Barretina, N. D. Socci, P. DeCarolis, M. Ladanyl, M. Meyerson, S. Singer, and C. Sander: Functional copy-number alterations in cancer. *PLoS ONE*, 2008, vol. 3(9), e3179.

[39] P. V. Loo, S. H. Nordgard, O. C. Lingjaerde, H. G. Russnes, I. H. Rye, W. Sun, V. J. Weigman, P. Marynen, A. Zetterberg, B. Naume, C. M. Perou, A. B. Dale, and V. N. Kristensen: Allele-specific copy number analysis of tumors. *Proceedings of the National Academy of Sciences (USA)*, 2010, vol. 107, 39.

[40] J. Di Pierdomenico, H. Ying, F. Lin, and H. H. Q. Heng: A mathematical model relating chromosome aberrations to cancer progression. *Proceedings of the 28th IEEE EMBS Annual International Conference* (New York City, USA, Aug 30–Sept 3, 2006), 1-4244-0033-3, 2006, pp. 2028–2031.

[41] A. E. Ferdinand: The theory of system complexity. *International Journal of General Systems*, 1994, vol. 1, 190–235.

[42] P. S. Neelakanta: *Information Theoretic Aspects of Neural Networks*, CRC Press, Boca Raton, FL, USA: 1999.

[43] R. M. Bendett and P. S. Neelakanta: A relative complexity metric for decision-theoretic applications in complex systems. *Complex Systems*, 2000, vol. 12, 281–295.

[44] E. Jaynes: On the rationale of maximum entropy methods. *Proceedings of IEEE*, 1982, vol. 70, 939–952.

[45] T. M. Cover and J. Thomas: *Elements of Information Theory*, John Wiley and Sons Inc., New York, NY, USA: 1991.

[46] S. Kullback: *Information Theory and Statistics*, Wiley Interscience Publications, New York, NY, USA: 1959.

[47] J. Lin: Divergence measure based on the Shannon entropy. *IEEE Transactions on Information Theory*, 1991, vol. 37, 145–151.

[48] L. Xiao-bo: Mathematical modeling of carcinogenesis based on chromosome aberration data. *The Chinese Journal of Cancer Research,* 2009, vol. 21(3), 240–246.

[49] B. Vogelstein, E. R. Feason, S. R. Hamilton, S. E. Kern, A. C. Preisinger, M. Leppert, A. M. M. Smits, and J. L. Bos: Genetic alterations during colorectal-tumor development. *New England Journal of Medicine,* 1988, vol. 319(9), 525–532.

[50] M. Hoglund, A. Frigyesi, T. Sall, D. Gisselsson, and F. Mitelman: Statistical behavior of complex cancer karyotypes. *Journal Genes, Chromosomes and Cancer,* 2005, vol. 42, 327–341.

[51] M. Agarwal and A. S. Bhadauria: Mathematical modeling and analysis of tumor therapy with oncolytic virus. *Journal Applied Mathematics,* 2011, vol. 2, 131–140.

[52] A. R. A. Anderson and M. A. J. Chaplin: Continuous and discrete mathematical models of tumour-induced angiogenesis. *Bulletin of Mathematical Biology,* 1998, vol. 60, 857–900.

[53] H. G. Othmer and A. Stevens: Aggregation blowup and collapse: The ABC's of taxis in reinforced random walks. *SIAM Journal of Applied. Mathematics,* 1997, vol. 57(4), 1044–1081.

[54] A. Kubo: Mathematical analysis of some models of tumour growth and simulations, tumour angiogenesis dynamics. *WSEAJ Transactions on Biology and Biomedicine,* 2010, vol. 7(2), 31–40.

[55] A. A. Patel, E. T. Gawlinskis, S. K. Lemieux, and R. A. Gaten: A cellular automaton model of early tumor growth and invasion; The effects of native tissue vascularity and increased anaerobic tumor metabolism. *Journal of Theoretical. Biology,* 2001, vol. 213(3), 315–331.

[56] M. A. A. Castro, T. G. H. Onsten, J. C. F. Moreira, and R. M. C. de Almeida: Chromosomal aberrations in solid tumors have a stochastical nature. *Mutation Research/Fundamental and Molecular Mechanisms of Mutagenesis,* 2006, vol. 600(1–2), 150–164.

[57] F. Mitelman: Recurrent chromosome aberrations in cancer. *Mutation Research/Reviews in Mutation Research,* 2000, vol. 462(2–3), 247–253.

[58] N. Fukshansky and W. Bar: DNA mixtures: Biostatistics for mixed stains with haplotypic genetic makers. *International Journal of Legal Medicine,* 2005, vol. 119, 285–290.

[59] P. S. Neelakanta and R. Yassin: A co-evolution model of competitive mobile platforms: Technoeconomic perspective. *Journal of Theoretical and Application Electronic Commerce Research,* 2011, vol. 6(2), 31–49.

[60] P. S. Neelakanta and R. C. T. Sardenberg: Consumer benefit versus price elasticity of demand: A nonlinear complex system model of pricing Internet services on QoS-centric architecture. *Netnomics,* 2011, vol. 2(1), 31–60.

[61] P. S. Neelakanta and D. F. De Groff: *Neural Network Modeling: Statistical Mechanics and Cybernetics Perspectives,* CRC Press, Boca Raton, FL, USA: 1994.

[62] H. H. Q. Heng, J. B. Stevens, G. Liu, S. W. Bremer, K. J. Ye, P.-V. Reddy, G. S. Wu, Y. A. Wang, M. A. Tainsky, and C. J. Ye: Stochastic cancer progression driven by non-clonal chromosome aberrations. *Journal of Cellular Physiology,* 2006, vol. 208(2), 461–472.

[63] J. Karri: Cytogenetic Bioinformatics of Chromosomal Aberrations and Genetic Disorders: Data-mining of Relevant Biostatistical Features. *MS Thesis,* College of Engineering and Computer Science, Florida Atlantic University, Boca Raton, FL, USA: 2012.

[64] P. S. Neelakanta and S. Sharma: Estimating maximum plausible conserved synteny between orthologous genomes of a species pair compared in an exhaustive search-space. *European Journal of Bioinformatics,* 2016, vol. 3, 01–09.

[65] V. Koonin: Orthologs, paralogs and evolutionary genomics. *Genetics,* 2005, vol. 39, 309–338.

[66] E. A. Houseworth: Measure of conserved synteny. Abstract for IMA: RECOMB Satellite Workshop on Comparative Genomics (Oct. 20–24, 2003, University of Minnesota). Available online at: *http://www.ima.umn.edu/2003-2004/W10.20 24.03/activities/Housworth Elizabeth/housworth.pdf.*

[67] J. H. Edwards: The Oxford-grid. *Annals of Human Genetics,* 1991, vol. 55, 17–31.

[68] B. O. Bengtsson, K. K. Levan, and G. Levan: Measuring genome reorganization from synteny data. *Cytogenetics and Cell Genetics,* 1993, vol. 64(3–4), 198–200.

[69] Mouse Genome Database (MGB): Mouse genome informatics website, The Jackson laboratory, Bar Horbor, ME, USA. Available online at: *http://www.informatics.jax.org/.*

[70] M. Schena: *Microarray Analysis,* Wiley-Liss, (A John Wiley & Sons, Inc.), Wilmington, DE, USA: 2003.

[71] W. J. Murphy, S. Sun, Z.-Q. Chen, N. Yuhki, D. Hirschmann, M. Menotti-Raymond, and S. J. O'Brien: A radiation hybrid map of the cat genome: Implications of comparative mapping. *Genome Research,* 2000, vol. 10, 691–702.

[72] R. C. Ball, M. Diakonova, and R. S. Mackay: Quantifying emergence in terms of persistent mutual information. *Advances in Complex System,* 2010, vol. 13, 327–338.

[73] K. Naruse, M. Tanaka, K. Mita, and A. Shima: A Medaka gene map: The trace of ancestral vertebrate proto-chromosomes revealed by comparative gene mapping. *Genome,* 2004, vol. 14, 820–828.

[74] Introduction to scientific names for humans. Available online at: *http://www.tutorvista.com/science/scientific-names-for-humans.*

[75] Mouse: Wikipedia, the free encyclopaedia. Available online at: *http://en.wikipedia.org/wiki/mouse.*

[76] *Oryzias latipes*: Available online at: *http://www.uniprot.org/taxonomy/8090.*

[77] Japanese rice fish: Wikipedia, the free encyclopedia. Available online at: *http://en.wikipedia.org/wiki/japanese_rice_fish.*

[78] Zebrafish: Wikipedia, the free encyclopedia. Available online at: *http://en.wikipedia.org/wiki/Zebrafish.*

[79] *Danio rerio*: Available online at: *http://www.uniprot.org/taxonomy/7955.*

[80] E. A. Houseworth and J. Postlethwait: Measure of syntenic conservation between species pair. *Genetics*, 2002, vol. 162, 441–448.

[81] OXGRID–The Oxford grid project. Available online at: *http:// oxgrid.angis.org.au/.*

[82] Coriell institute for medical research: Cytogenetics: Available online at: *http://www.coriell.org/research-services/cytogenetic/what-is-cytogenetics.*

[83] P. S. Neelakanta: *Handbook of Electromagnetic Materials.* CRC Press, Boca Raton, FL, USA: 1995, p. 166.

[84] C. Y. Lin, C. Y. Chang, and H. J. Tsai: Zebrafish and Medaka: New model organisms for modern biomedical research. *Journal of Biomedical Science*, 2016, vol. 23, p. 11.

[85] J. Yee, M. S. Kwon, T. Park, and M. Park: A modified entropy-based approach for identification gene-gene interaction in case-control study. *PLosONE*, 2013, vol. 8(7), 1–8.

[86] M. de Andrade, and X. Wang: Entropy based genetic association tests and gene-gene interaction tests. *Statistical Applications in Genetics and Molecular Biology*, 2011, vol. 10(I), Article 38.

[87] L. Guttman: An outline of the statistical theory of prediction. *The Prediction of Personal Adjustments*, Social Science Research Council, New York. 253–318.

[88] Cramèr's V: From Wikipedia, the free encyclopedia. Available online at: *http://en.wikipedia.org/wiki/Cram%C3%A9r%27s_V.*

[89] Tetrachoric correlation: Introduction to the tetrachoric and polychoric correlation coefficient. Available on-line at: *http://john-uebersax.com/stat/tetra.htm.*

[90] I. Csiszàr: A class of measure of informatively of observation channels. *Periodica Mathematica Hungarica*, 1972, vol. (2), 191–213

[91] I. A. Zakharov and A. K. Valeeve: Quantitative analysis of evolution of mammalian genomes by comparison of genetic map. *Proceedings of the Academy of Science (USSR)*, 1988, vol. 301, 1213–1218.

[92] I. A. Zakharov and A. K. Valeeve: Quantitative analysis of evolution of mammalian genomes by comparison of genetic map. *Genetika,* 1988, vol. 28, 1988, 77–81.

[93] R. A. Fisher: *Statistical Methods for Research Workers.* Oliver and Boyd, Edinburgh, UK: 1938.

[94] L. A. Goodman and W. H. Kruskal: Measures of association for cross classifications, IV: Simplification of Asymptotic Variances. *Journal of the American Statistical Association,* 1972, vol. 67(338), 121–146.

[95] S. Kumar, S. Gadakar, A. Filipski, and X. Gu: Determination of the number of conserved chromosomal segments between species. *Genetics,* 2001, vol. 157, 1387–1395.

[96] J. Ehrlich, D. Sankoff, and J. H. Nadeau: Synteny conservation and chromosome rearrangements during mammalian evolution. *Genetics,* 1997, vol. 147, 289–296.

[97] D. Graur and W. H. Li: *Fundamentals of Molecular Evolution.* Sinauer Associates, Sunderland, MA, USA: 2000.

[98] S. Hannenhalli, C. Chappey, E. V. Koonin, and P. A. Pevzner: Genome sequence comparison and scenarios for gene rearrangements: A test case. *Genomics,* 1995, vol. 30, 299–311.

[99] S. Kumar and S. B. Hedges: A molecular timescale for vertebrate evolution. *Nature,* 1999, vol. 392, 917–920.

[100] J. H. Nadeau and D. Sankoff: Counting on comparative maps. *Trends Genetics,* 1998, vol. 14, 495–501.

[101] D. Sankoff and J. H. Nadeau: Conserved synteny as a measure of genome rearrangement. *Discrete Applied Mathematics,* 1996, vol. 71, 247–257.

[102] D. Sankoff, G. Leduc, N. Antoine, B. Paquin, B. F. Lang, *et al.*: Gene order comparisons for phylogenetic inference: Evolution of the mitochondrial genome. *Proceedings of National Academy of Science* (*USA*), 1992, vol. 89, 6575–6579.

[103] A. Apostolico, J. Hein, D. Sankoff, M. N. Parent, I. Marchland, and V. Ferretti: On the Nadeau-Taylor theory of conserved chromosome segments. *Proceedings of Eighth Annual Symposium on Combinatorial Pattern Matching,* (CPM 97: Aarhus, Denmark, June 30–July 2, 1997). pp. 262–274.

[104] D. Waddington, A. Springbett, and D. W. Burt: A chromosome-based model for estimating the number of conserved segments between pairs of species from comparative genetic maps. *Genetics,* 1999, vol. 154, 323–332.

[105] Estimating the entropy: Available on-line at: *https://mathematica.stackexchange.com/questions/35472.*

[106] C. Kemkemer, M. Kohn, D. N Cooper, Lutz, F. J. Högel, Hameister, and H. Kehrer-Sawatzki: Gene synteny comparisons between different vertebrates provide new insights into breakage and fusion events during

mammalian karyotype evolution. *BMC Evolutionary Biology*, 2009, vol. 9, p. 24.

[107] S. Sharma: Cytogenetic Bioinformatics of Chromosomal Synteny Evaluation: Application, Towards Screening of Chromosomal Aberrations/ Genetic Disorder. *MS. Thesis*, College of Engineering and Computer Science, Florida Atlantic University, Boca Raton, FL, USA: 2014.

[108] C. G. Elsik, R. L. Tellam, and K. C. Worley: The Genome Sequence of Taurine Cattle: A window to ruminant biology and evolution. Available on-line at: *https://www.ncbi.nlm.nih.gov/pmc/articles/PMC2943200/ Published in final edited form as: Science. 2009 Apr 24, 324*(5926), 522–528. doi:[10.1126/science.1169588]

[109] L. Goodstadt and C. P. Ponting: Phylogenetic reconstruction of orthology, paralogy and conserved synteny for dog and human. *PLoS Computational Biology*, 2006, vol. 29, 2(9), e133.

[110] A. R. Caetano, Y. L. Shiue, L. A. Lyons, S. J. O'Brien, T. F. Laughlin, A. T. Bowling, and J. D. Murray: A Comparative Gene Map of the Horse (*Equus caballus*). *Genome Research*, 1999, vol. 9(12). 1239–1249.

[111] MGI Oxford-grid: Human + cat: Available on-line at: *https://images.search.yahoo.com/yhs/search;_ylt=AwrCwLU7hhFcr3cA0jQPxQt.;_ylu=X3oDMTByMjB0aG5zBGNvbG8DYmYxBHBvcwMxBHZ0aWQDBHNlYwNzYw?p=human+cat+synteny+oxford+grid&fr=yhsptypty_maps&hspart=pty&hsimp=yhspty_maps#id=0&iurl=https%3A%2F%2Fimage3.slideserve.com%2F5499841%2Fmgi-oxford-gridn.jpg&action=click.*

[112] M. W. Wright and E. A. Bruford: Human and orthologous gene nomenclature. *Gene*, 2006, vol. 369, 1–6.

Chapter 8

Bioinformatics of Viral Omics

Thematic details concerning bioinformatic aspects of viral omics form the scope of this chapter. Analyses based on entropy, energetic, and Fourier spectral domain approaches are indicated thereof to ascertain unique structural features in viral biosequences, and the use of associated motif details gathered on a set of query sequences pertinent to a virus and its serovars is described vis-à-vis rational vaccine design considerations.

Without effective human intervention, epidemics and pandemics typically end only when the virus or bacteria has infected every available host and all have either died or become immune to the disease.

Alan Huffman

8.1 Introduction

Episodes of epidemics and prevalence of pandemics are karmic curses levied on living systems due to countless extent of viruses and/or bacteria that are present and pervade across the stretch of entire globe over the known centuries of time. Viruses and/or bacteria are tiny morsels of living systems and perennially infect every available host. They challenge the defined parlance of life being "a terrestrial phenomenon, a characteristic of terrestrial living organisms".[1–3] Such terrestrial phenomenon is countered by epidemics and hassled by pandemics. That is, the inevitable presence of viruses/bacteria implies countless hassles of epidemics that eventually manifest as events of societal diseases placing a query on "life", and relevant influence represents a karmic dictum of hazard on the retention of life. Hence, in contexts of bioinformatics, a set of theoretical approaches, efforts of computational analytics, and methods of wet-lab experiments have emerged in the past to emphasize exclusive studies on viral and bacterial omics[4–8] (adjunct to medical and health sciences).

Commensurate with this trend, developed and presented in this chapter are pertinent bioinformatic details comprehensively addressing the methods of analyses of biosequences of viruses. Further considered are schemes to ascertain unique omic details present in *serogroup** of viruses toward envisaging vaccine designs that are compatible for inculcating immunity in human/animal species across the span of viral diversity. (*serogroup* or *serotypes**: in the context of virology, a particular virus may prevail in different forms of *serotypes,* for example, DEN1, DEN2, DEN3, and DEN4 strains in the case of dengue virus, and such strains could possess common and distinct omic features). Pertinent details on viral nomenclature and classification and reproductive forms of virus (known as *virions*[9]) can be availed from literature, such as Bândea,[9] Holmes,[10] Lwoff *et al.*,[11] International Committee on Taxonomy of Viruses,[12] and Baltimore.[13]

Described in this chapter are details presented in two parts: Part I devoted to study viral microbiological details and describe related biosequence analyses based on entropy, energetic, and spectral-domain methods. The derived results thereof are used in understanding the structures of viral sequences. In Part II, the diverse nature of viruses and serogroup is considered and corresponding analytics of viral omics are described. Also, necessary details are identified for applications in *rational vaccine designs* (RVDs),[14–17] and such RVD considerations are applied to a set of serovars.[18–28]

In summary, notwithstanding the avenues of existing *Human Genome Project* (HGP*)[29,30] enabling extensive frontiers of microbiological and bioinformatic studies (on humans and other living systems), a viably needed subject refers to elucidating omic details of primitive organisms — viruses and bacteria that are responsible for the observed epidemics and pandemics, and the underlying societal interest objectively implies expanding the concepts of bioinformatics to evolve the heuristics of "*viral bioinformatics*" — a study devoted to address the omic perspectives of viral (and bacterial) species. Hence, viral bioinformatics (as conceived in this chapter) refers to a conceivable topic on virology-specific informatic analyses.

8.2 Virus-Specific Microbiology

Virus, in general, is a tiny, living system that thrives and replicates inside living cells of other organisms.[1–5] It represents an agent that can infect all types of life forms, from animals and plants. Also, a virus can exist in different guises, and as such, it has been a concern in immunological studies to find a common vaccine for the entire set of serovars or *strains* depicting differently guised versions of the same virus. Typically, in such serovars, *common omic features** could prevail. Ascertaining such details is a focal point of interest in viral microbiology. Notwithstanding several notable wet-lab assertions are in vogue to deduce the aforesaid common features between multiple viral strains, bioinformatic pursuits are viable avenues as considered in this chapter. (*The *common omic features* in serovars of a virus can be regarded as possible *epitopal* formats** existing across the entire set of the strains), and such formats can be adopted in designing a common vaccine. (*Epitope***: in the events of realizing immunity in living systems, an antibody binds to a specific segment of a protein known as an epitope).

The underlying considerations in performing viral omic analyses refer to *single-* as well as *double-stranded viral DNA* and/or *viral RNA* structure. Considering the suite of analyzing a biosequence (established through traditional analytics and/or their variants described in earlier chapters) and applying such methods to the gamut of test viral serovars may not possibly (and comprehensively) identify the subtle, common features present among the strains. As such, a multiple set of distinct bioinformatic algorithms need to be cohesively applied to the set of query sequences of test viral strains, and the sets of independently deduced results (from each approach) can be combined and compared so as to elucidate the subtle (common format) details prevailing among the query sequences (of the serovars). Hence, considered here are three distinct methods (namely, entropic, energetic, and spectral-domain algorithms) and allying the results derived thereof from the query genomic sequences, concurrently yields robust details on common features being sought. Such three-prong approach so pursued is sparse or even absent, and hence, it surges the core motivation for the pedagogy presented in this chapter. Pertinent methods are comprehensively addressed in Chatterjee *et al.*[31]

Among the three pursued methods (namely, entropic, energetic, and spectral-domain algorithms), the entropic approach refers to entropy segmentation (information-theoretic-based approach) described in literature and reviewed in Chatterjee *et al.*,[31] Pandya,[32] Neelakanta *et al.*,[33] Arredondo *et al.*,[34] Bernaola-Galván *et al.*,[35] Román-Roldán *et al.*,[36] Baisnee *et al.*,[37] Neelakanta *et al.*,[38] and Benson.[39] The energetics-based scheme relies on so-called *nearest-neighbor* (NN) concept of energetic profiles of the underlying chemical structures[38] of the residue composition in the biosequence, and the third method of analyzing biosequences (including viral and its serovar sequences) refers to deducing characteristic spectral-domain peaks and troughs in the Fourier spectra of underlying spatial patterns of (physicochemical) parameters pertinent to genomic residues, as described in Benson.[39]

8.3 Viral Genomic Characteristics

Viral particles are the most abundant biological entities present on the earth.[1–5] The term "virus" in Latin means "poison" and viruses represent nonliving, microscopic particles consisting of either a RNA or DNA genome surrounded by a protective, virus-coded protein coat. The genomic material of the virus is packaged inside a structural *capsid* protein. Furthermore, in enveloped viruses, this structure is surrounded by a lipid bilayer with an outer layer of virus envelope *glycoproteins* as illustrated in the basic diagram of Fig. 8.1.

Viruses are considered nonliving as they do not possess the most basic characteristics of living system (such as metabolism, growth, reproduction, and reaction to stimuli). Forterre[2] poses the query, "What is life?" (originally, thanks to Schrödinger in 1944) and in pursuant to the claim by Crick and Watson on "the secret of life"—with the suggestion that "life is DNA" for any form of "life" present in the universe, and it is a characteristic terrestrial phenomenon. For replication, viruses depend on a host cell. In fact, before the viruses enter into the host, they are known as *virions* and, in his landmark paper, Bândea[9] distinguishes virions and viruses, aptly explaining the difference as follows: virions are "spores" or reproductive forms of virus, possessing "life" only as a

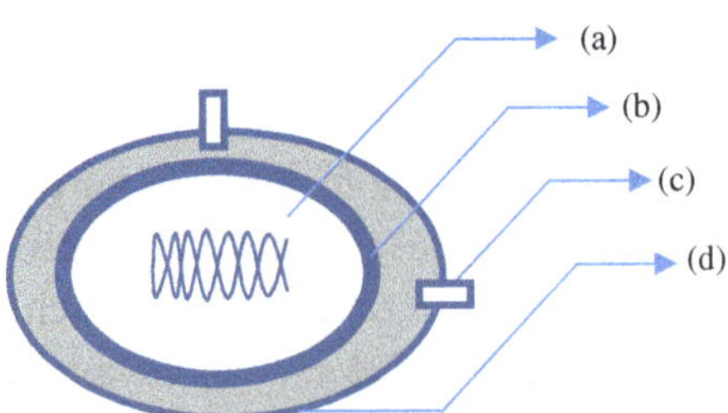

Fig. 8.1 Basic structure of a virus: (a) DNA/RNA gene; (b) capsid; (c) protein molecule; and (d) a fat-bubble envelope.

potential property, and they denote packaged genetic materials, which can be passed through direct contact or *carrier* to the host, where it replicates. Thus, a virus is simply a coat of protein wrapped around genetic material as shown in the simple diagram of Fig. 8.1. (An elaborate set of relevant illustrations on the structure of a virus is available at: Images for structure of a virus.[40])

Classification of Viruses

Inasmuch as viruses depict the most abundant biological entity, their classification and nomenclature are significant in biological contexts. The first system of classification of viruses was due to Holmes.[10] He suggested what is known as *Linnaean system of binomial nomenclature* to classify viruses into three groups under one order, *Virales.* Accordingly, viruses are classified depending on the type of living organism (fungi, bacteria, plants, or animals) that they infected. In 1962, Lwoff *et al.*[11] suggested classifying viruses using complete *Linnaean hierarchical system* for viral nomenclature (unlike binomial classification suggested by Holmes) based on their size, symmetry, nucleic acid, physicochemical properties, and presence or absence of envelope around the virus. Since a single virus can infect multiple species of organisms, Lwoff *et al.* ruled out the classification of virus based on the types of cells they infected, and the outline laid down in Holmes[10] was accepted by *International Committee on Taxonomy of Viruses* (ICTV)[12] as the standard method for nomenclature of virus with few changes and additions.

Another classification system, which is different from the one suggested by Lwoff *et al.*,[11] is due to Baltimore.[13] He argued that due to the small size of the viruses, it is difficult to identify their shapes even under electron microscope. As such, it was suggested to classify the viruses according to their genome type (namely, type of nucleic acid [DNA or RNA] and its structure [linear, circular, or segmented]) and on the method of viral mRNA synthesis. If the viruses are classified into categories, as per these characteristics, then all viruses in a given category will behave in a similar way. This suggestion was accepted by ICTV and has been included in classifying viruses along with the framework laid down by Lwoff *et al.* Relevantly, classified, three types of viruses are as follows: (i) *DNA virus,* (ii) *RNA virus,* and (iii) *reverse transcribing virus,* and their categorized versions fall into seven different versions as specified in Table 8.1.

Table 8.1 Summary of Baltimore classification of viruses.[11]

Group	*Type of genome*	*Common examples of family affecting human*
DNA virus		
I	dsDNA	*Adenovirus, Papillomaviridae* (HPV1, HPV11, HPV16, HPV18. HPV 16 and 18 can become cancerous), Herpesviridae (HHV1-8 causes diseases like Herpes, *Roseola, Epstein Barr,* Chickenpox, etc.) *Poxviridae* (smallpox), *Hepadnaviridae* (partially ds-causes *Hepatitis B*), etc.
II	ssDNA	Parvovirus B19 (causing fifth disease in children), Anelloviridae, etc.

DNA viruses have DNA as its genomic material (Fig. 8.1) and affect almost all domains of life system. The *double-stranded DNA* (dsDNA) (with two strands of DNA) replicate inside the host cell using its *DNA polymerase*. The dsDNA has linear or circular genome and follows the central dogma of molecular biology, except that it uses the machinery of the host cell for multiplying. Typically, majority of DNA viruses are dsDNA types.

There are also *single-stranded DNA* (ssDNA) versions of viruses (with a single-strand of DNA) that require the formation of an intermediate dsDNA form for genome replication. Relevant intermediate dsDNA conforms to the dsDNA assuming structures like hairpin bends or loops (as will be discussed later with reference to an example of *Parvovirus B19*).

RNA virus		
III	dsRNA	Rotavirus, etc.
IV	(+) ssRNA	*Coronaviridae* (SARS), *Picornaviridae* (Polio virus, common cold virus, Hepatitis A virus, etc.), *Hepevirus* (Hepatitis E virus), *Togaviridae* (Rubella virus, Ross River virus, *Sindbis* virus, Chikungunya virus, etc.), Flaviviridae (Yellow fever virus, West Nile virus, Hepatitis C virus, Dengue fever virus, etc.), etc.
V	(–) ssRNA	*Filoviridae* (Ebola virus, Marburg virus), *Paramyxoviridae* (Measles virus, Mumps virus), *Rhabdoviridae* (Rabies virus), *Orthomyxoviridae* — (Influenza viruses), *Bunyaviridae* (Hantavirus, Crimean-Congo hemorrhagic fever), etc.

As the name suggests, the main genomic material of RNA virus is the ribonucleic acid (RNA). Many of the deadly and widespread diseases in the world are spread by RNA virus, some of which are mentioned in Table 8.1. Most of the RNA viruses belong to the *single-stranded version* (ssRNA). Such ssRNA virus can be further classified into positive and negative senses. If the RNA base sequence is identical to the viral mRNA sequence,

Table 8.1 (*Continued*)

Group	Type of genome	Common examples of family affecting human
then it is known as positive-sense RNA virus; whereas, if the RNA base sequence is complementary to the viral mRNA sequence, it is known as negative-sense RNA virus. There are also *double-stranded RNA* (dsRNA) viruses, which represent a diverse group of viruses having varying characteristics. This version is discussed exhaustively in Patton.[18]		
The RNA viruses are genetically very unstable with a large rate of mutation. One of the principal reasons for this is the lack of the corrective mechanisms available inherently in DNA molecules. The high-rate of mutation could limit the design of vaccine and its future effectiveness. In fact, it is suggested by Holland *et al.*[19] that RNA viruses have the most important role in evolution and in the survival of different species as well as in the maintenance of diverse ecology, in general. The negative-sense RNA viruses can also be divided into two groups: (i) First group of viruses containing nonsegmented genomes for which the first step in replication is transcription arising from the negative-stranded genome by the viral RNA-dependent RNA polymerase yielding the so-called monocistronic mRNAs that code for various viral proteins. Then a positive-sense genome copy is produced that serves as the template to produce the negative-strand genome, and replication occurs within the cytoplasm. (ii) The second group of viruses possesses segmented genomes, for which, the replication occurs within the nucleus, and correspondingly the viral RNA-dependent RNA polymerase produces monocistronic mRNAs from each genome segment. The largest difference between the two groups of negative-sense RNA viruses is the location of replication site. Details on replication of RNA viruses.[8,19]		
Reverse transcribing virus		
VI	ssRNA-RT	Retroviridae (HIV)
VII	dsDNA-RT	Hepadnaviridae (Hepatitis B)
Group VI reverse transcribing virus: this category of viruses consists of some of the most-deadly as well as useful viruses. As well-known, relevant infection by HIV, a member of retrovirus is considered pandemic in modern world, and retroviruses play a role in some forms of cancer. They can also be used as *vectors* for gene therapy and in drug delivery systems.[15]		
In retrovirus, the term "retro" implies that the RNA is reverse-transcribed into DNA, which is then integrated into the cell of the host genome; RNA $\rightarrow$ DNA $\rightarrow$ RNA $\rightarrow$ protein unlike the process identified by Crick and Watson, in which the sequence changes as follows: DNA $\rightarrow$ RNA $\rightarrow$ protein.[20] Like other RNA viruses, the retrovirus mutates quickly and enormously; as such, producing effective *antiretroviral* drugs/vaccine against retroviruses viruses is difficult.		

(*Continued*)

Table 8.1 (*Continued*)

Group	*Type of genome*	*Common examples of family affecting human*
Group VII viruses: these are *hepadnaviruses* with very small genomes consisting partially of two uneven strands and partially a single-stranded circular DNA. One strand has a negative-sense orientation and the other (shorter) strand has a positive-sense orientation.[21] The mechanism of replication of *hepadnaviruses* is discussed in Holland *et al.*,[19] Kurth and Bannert,[20] and in Summers and Mason,[41] Sohn *et al.*[42] Examples of hepadnaviruses are: (i) parvovirus B19V and (ii) dengue virus.		

Structural Details of Viral Genomics

(A) *Example of B-19V ssDNA*

The *B19V* belongs to Group VII hepadnaviruses, and it is a *parvovirus*, a member of the family *parvoviridae* responsible in causing a variety of diseases in human. Furthermore, B19V is a member of the genus, *erythrovirus* in the family of parvoviridae. Structurally, B19V contains a small linear single-stranded DNA (ssDNA) genome of 5.6kb length, which harbors two identical *inverted terminal repeats (ITR)* that serve as origin of DNA replication in the host cell. The virus has a framework of overall folding of the DNA structure into hairpin formats. Such hairpins formed from a ssDNA consist of a base-paired stem structure and a loop sequence with unpaired or mismatched nucleotides as shown in Fig. 8.2. Relevant conformational studies of DNA hairpins indicate possible tri-dimensional forms (with variations), suggesting high profile of complexity of ssDNA structures. Part of this complexity can however be simplified with appropriately rationalized bioinformatic description of loop-bases and the stem part in the frame of backbone structures. The illustration in Fig. 8.3a refers to the 5′-end at the start of B19V genomic sequence. In this stretch of 5′-end, the associated 383 nucleotides tend to fold into a bulge around 365th base.

Furthermore, as shown in Fig. 8.3b, the two segment sets of bases {TCTG**a**} and {**t**GTCT} on either side of the bulge constitute the *inverted terminal repeats* (ITRs) of palindromes. The bases (**a** and **t**) shown in lower case bold fonts depict the *closing pairs* in the loop. (The bases (**a**TTTGG**t**) in the bulge can flip-flop to a form a complement set of bases, namely, (**t**AAACC**a**); that is, the bulge can format itself in two alternative "flip" or "flop" orientations.)

The details of 3′-end stem and loop (hairpin bend) are depicted in Figs. 8.3c and 8.3d. Relevant to these illustrations, the following can be

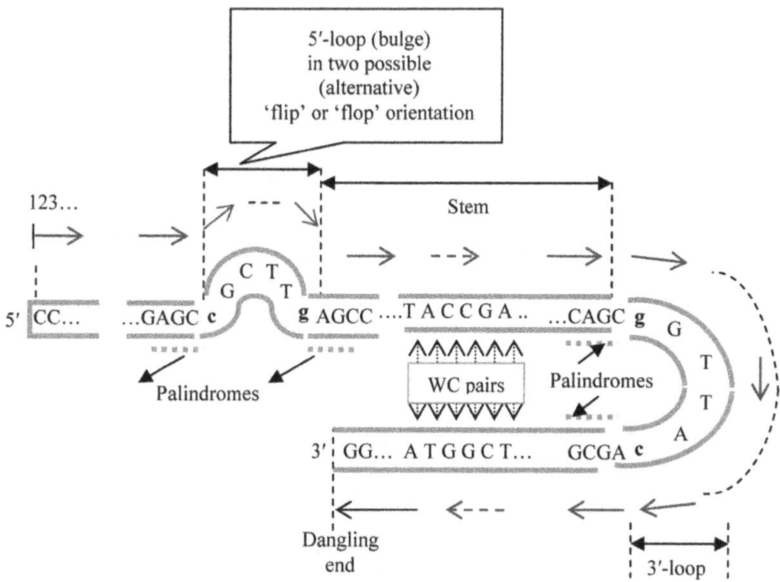

Fig. 8.2 A typical hairpin folding of a ssDNA genome (B19V parvovirus).[43–46]

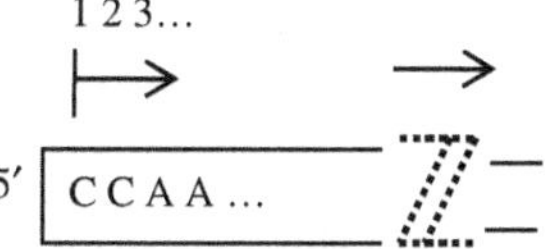

Fig. 8.3a 5′-end of the genomic sequence of B19V.

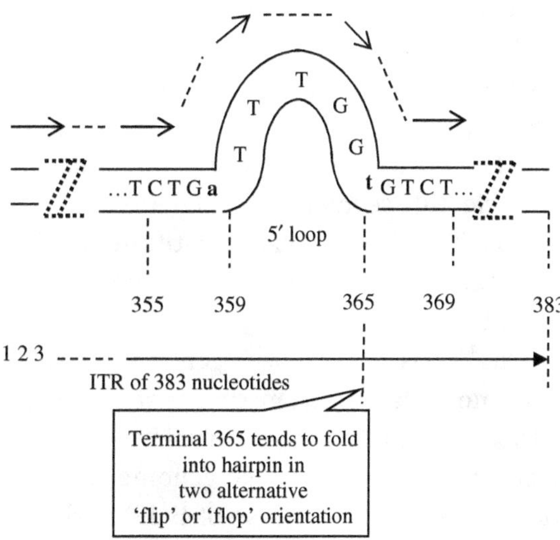

Fig. 8.3b An expanded view of B19 viral structure of bases from 1 through 5594 with details, such as a *bulge* at 5′-end.

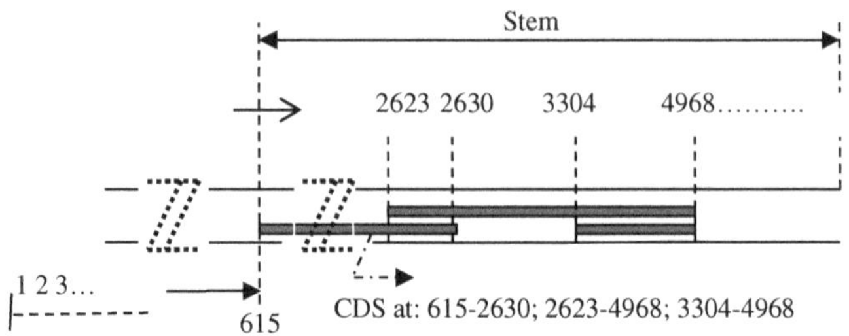

Fig. 8.3c The stem part at the 3′-end with associated locations of *coding DNA sequences* (CDS).

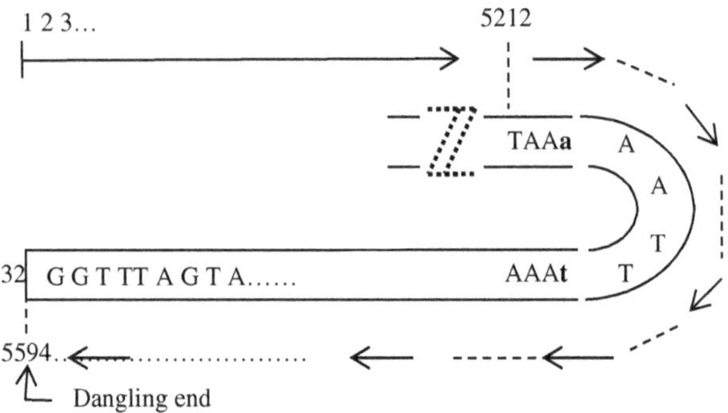

Fig. 8.3d The loop (hairpin-bend) part and the associated ITRs at the 3′-end.

observed: as shown in Fig. 8.3c, toward the 3′-end of the genomic sequence exists a stem part of nucleotides that stretches up to 5212th base site, and the stem contains CDS at: 615–2630; 2623–4968; 3304–4968. Furthermore, closer to the 3′-end, a hairpin structure (*3′-loop*) is formed with a bending and a reversed stem part, which terminates at the last nucleotide (called "the dangling end"), at the site 5594. A pair of nucleotide segment sets, {TAA**a**} and {**t**AAA} exists on either side of this 3′-end loop that constitute the palindromes of inverted repeats. Again, the set {**a**AATT**t**} in the loop can flip-flop to its complement, {**t**TTAA**a**}.

The B-19 virus[47] has a (+ sense) genome structure of ssDNA according to Baltimore classification, and as stated before, it is classified as a *parvovirus*. Normally, such a parvovirus has a nonsegmented linear ssDNA genome with an average genome size of about 5 kbps (5594 nucleotides in length in the case of B-19) with a short, double-stranded hairpin formation at the 3′-end as described and illustrated in Fig. 8.2.

The identical ITRs being present at each end of the genomes correspond to unpaired or mismatched bases in the palindromes represented by the bulges or *bubbles,* which mismatches at the stem and across noncanonical base pairs.

The hairpin folding not only enforces the stability but also solves the problem of linear DNA molecules replicating their 5′-ends due to the requirement of DNA polymerase for a primer with a free 3′-OH group.[22] The hairpin transfer mechanism solves this issue by relying on terminal palindromic sequences to self-fold, thus forming hairpin structures that prime the DNA replication. In addition to basic aspects of hairpin structure of a viral DNA, mismatched nucleotides in the 5′ terminal hairpin can be observed, and more complexity could prevail with wild type palindrome in viral DNA sequences. The complexity of hairpin structure as above is constrained by the associated molecular dynamics involving characteristic mismatches, such as tandem G-T.

Since self-complementing stability feature (*via* WC-pairing) of a double-helix does not *per se* exist in single-stranded sequence formats, the viral ssDNA sequences tend to "fold" or self-bend so as to get stabilized *via* feasible base-pair matching. More stability considerations of folded ssDNA structures are based on thermodynamic energetics as reported in Hilbers.[43] The legends and descriptions indicated in Figs. 8.3a–3d are consistent with details on human parvovirus B19V (NC_000883.1) available as GenBank data in: Human parvovirus B19[23] and at: *https://www.researchgate.net/publication/8887235_Parvovirus_B19*with pertinent information described above in Costello *et al.,*[44] Chin *et al.,*[45] and Chen *et al.,*[46] Human parvovirus B19.[47] Furthermore, with reference to B19 virus, the annotated details on proteins made by the virus are presented in Table 8.2.

Table 8.2 CDS range (from NIH website) of parvovirus B19V.[47]

Name of the protein	*CDS range of bases (NIH accession: NC_000883.2)*
Nonstructural protein (NS1)	616...2631
7.5 kDa protein	2084...2308
Minor capsid protein	2624...4969
Protein X	2874...3119
Major capsid protein	3305...4969
11 kDa protein	4890...5174

(B) *Example, Dengue viruses*

Dengue virus is a small (approximately 10.7 kb) positive-sense, single-stranded RNA virus (ss-RNA). Dengue belongs to the family *flaviviridae* and has inverted complementary sequences at the ends of the molecule that mediate long-range RNA–RNA interaction and genome cyclization. Studies have demonstrated that alternative conformations of the genome are necessary for infectivity. Dengue virus has four different serotypes or strains, namely, DEN 1, DEN 2, DEN 3, and DEN 4. The complete sequences of all the four strains are available in NCBI databank.[24–27] Relevant CDS ranges are available in the following NIH accessions: (A) Den1-(NIH accession: NC_001477); Den 2-(NIH accession: NC_001474); Den 3-(NIH accession NC_001475); and Den 4-(NIH accession NC_002640).

The genome of dengue virus encodes three structural proteins that form the coat of the virus and deliver the RNA to target cells and seven nonstructural proteins that are responsible for the production of new viruses once the virus gets access to the host cell. These four different strains have slightly varying genomic characteristics (exhibits almost 60%–80% homology between different strains),[48–51] resulting in slightly different proteins. One of the principle obstacles in developing an effective dengue vaccine is due to the need of simultaneously stimulating the immune system against all the four strains and thus generating antibodies against all the four different forms of the dengue virus. This is because when a person is infected with one serotype of the virus, and then infected later by a second serotype, the antibodies and immunity gained from the first infection appear to assist with the infection by the second subtype, instead of providing a general immunity to all serotypes. This means that an effective vaccine will have to stimulate protective antibodies against all four types at once, a feat that has not yet been achieved. Studies on nucleotide divergence characteristics among different strains of the given virus are limited. Such lack of basic information of viral diversity severely limits vaccine and antiviral therapy development efforts.

The genome of dengue virus encodes a single, long *open reading-frame* (ORF), flanked by highly structured 5′ and 3′ untranslated regions (UTRs). After entering the host cell *via* receptor-mediated endocytosis, the virus releases the genomic RNA into the cytoplasm, which serves as mRNA for translation. Like all other *Flaviviruses*, the mRNA is translated as a single polypeptide and then cleaved into constituent proteins. Dengue virus has stem loop structure at both 5′ and 3′ UTR. These

UTRs are needed for the stability and functioning of the viral RNA. Dengue virus 5′ UTRs are between 95 and 101 nucleotides long. They contain two RNA domains with distinct functions during viral RNA synthesis. The first domain consists of approximately 70 nucleotides and is predicted to fold into a *large-stem loop* (SLA), a common feature found in all viruses is included in the *Flavivirus* genus. The second domain of the dengue virus 5′ UTR is predicted to form a *short-stem loop* (SLB), which contains essential sequences for long-range RNA–RNA interaction and replication. Within these two stem loops are found some loops and bulges.

The first protein to be formed is the capsid protein. It is one of the most important proteins as it contains a hairpin bend between two AUG start codons. This structure is absolutely necessary for efficient viral replication in human and mosquito cells. The 3′ UTR is approximately 450 nucleotides long and can be divided into three domains. The domain I is located immediately after the stop codon of NS5 and is the most variable region within the viral 3′ UTR. It exhibits extensive size variation between serotypes; it can range from more than 120 nucleotides to less than 50 nucleotides. Domain II comprises of many hairpin structures. Particularly interesting is a characteristic dumbbell structure containing conserved sequences with several pseudoknot structures. Domain III of the 3′ UTR consists of a conserved sequence which is involved in a long-range RNA–RNA interaction between the ends of the viral genome, followed by a terminal stem-loop (SL) structure.[48–52]

A conserved feature of DENV and other *Flavivirus* genomes is the presence of inverted complementary sequences at the ends of the RNA that mediate long-range RNA–RNA interactions. The annotated details are available in literature[53–55] as regard to the four strains of dengue virus under discussion.

8.4 Viral Diversity: An Overview

Most infectious diseases are caused by pathogens that are genetically stable and host-specific such that a single widely administered vaccine can be used to effectively prevent widespread disease, especially epidemics.[56] However, an enormous mutating variety of genomic structures can be seen among viral species, especially ssDNA viruses and RNA viruses. Among RNA viruses, the genome is often divided into separate parts within the virion (*segmented*). Each segment often codes for one protein, and they are usually found together in one capsid. Every segment is not required to be in the same virion for the overall virus to be infectious. Antigenic diversity among ribonucleic acid (RNA) viruses occurs as a result of rapid mutation during replication, short replication times, and recombination/re-assortment between genetic materials of related strains during co-infections.[56] Hence, effective vaccination against such unstable and rapidly mutating viruses requires surveillance programs to monitor circulating serotypes and their evolution to ensure that vaccine strains match the field viruses.[57]

8.5 Viral Omic Analyses

This section (of Part I) is devoted to furnish the underlying theoretics on three methods of analyses indicated earlier toward comparison and feature extraction pursued on sequence/subsequence sets of test viruses/serovars. Relevant exercises are techniques based on: (i) Entropy-theoretics governed by the Shannon's perspectives of information theory as detailed in Fisher,[58] Shannon and Weaver,[59] Shannon,[60] Neelakanta,[61] and Kapur,[62] (ii) energetics-based approach based on thermodynamic aspects of omic physico-chemistry and energetic attributes of sequence residues as detailed in Neelakanta *et al.*,[38] and (iii) Fourier-spectrum implied methods of analyses of biosequences (by spectral domain considerations) deducing the spatial patterns of the genomic residues.[39,63–83]

Viral Genomic Analysis Based on Entropy Characterization

The entropy-based studies on biosequence analysis can be seen, for example, in Chatterjee *et al.*,[31] Pandya,[32] Neelakanta *et al.*,[33] Arredondo *et al.*,[34] Bernaola-Galván *et al.*,[35] Román-Roldán *et al.*,[36] Baisnee *et al.*,[37] and Neelakanta *et al.*[38] In Chapter 2, the concept of entropy is seen in the perspectives of Shannon's information, and how the underlying details are statistically strewn across genomic sequences are considered in Chapter 3.[61–62] Pertinent stochastic considerations and heuristics of analytics adopted in assessing comparative details of biological systems were studied as early as 1930s, for example, by Fisher (as could be evinced from the classical results in Fisher.[58])

Such studies refer to information-theoretic heuristics addressed (for example) toward: (i) Differentiating the informational profiles of coding and noncoding regions in a query DNA/RNA sequence: (ii) detecting buried signal details (such as the presence of splice-junctions within a

DNA; (iii) locating aberrant (cryptic) attributes (caused by mutational changes) that may exist in DNA sequences; (iv) observing fuzzy details across overlapping subsequences; (v) finding specific signatures of coding DNA sequence (CDS) such as CpG, TATA regions; and (vi) verifying the symmetry/asymmetry aspects of forward and reverse strands of a DNA, etc., ascertaining the secondary structural details (such as hairpin bends, loops, bulges, etc.) of ssDNA and/or in RNA sequences in terms of *NN* concept of the energetic profiles[38] of the underlying chemical structures. Stochastic considerations and pertinent analytics as regard to assessing comparative details of biological systems were studied as early as the 1930s, for example by Fisher, as could be evinced from the classical results in Fisher.[58] Information-theoretic methods and related entropy concepts can also be adopted in analyzing genomic sequences in the following exercises: (i) unique CDS characterization of viral genomic sequence patterns; (ii) elucidating fuzzy domains at codon-noncodon splice-junctions of viral genomes; (iii) deducing information redundancy in viral genomic sequences; and (iv) finding CpG motifs in viral ssDNA.

As described in Chapter 1, a DNA sequence is constituted by randomly disposed introns and exons, and relevantly as illustrated in Fig. 8.4,

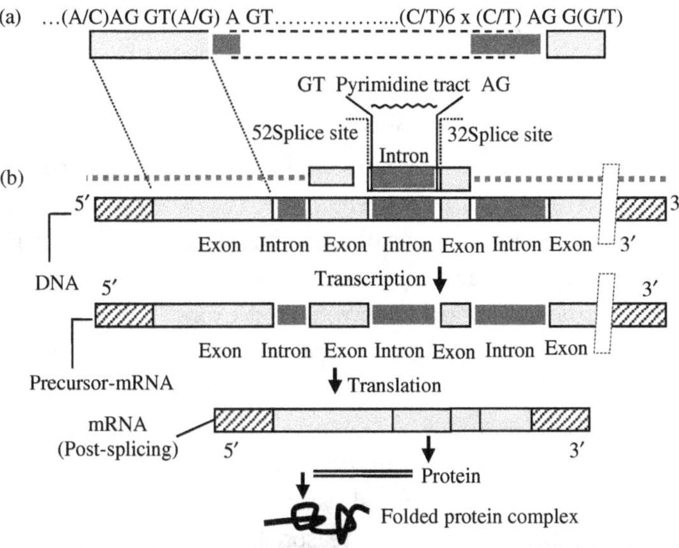

Fig. 8.4 Illustration of the formation of mRNA with: (a) splice-junction consensus and (b) splice-junctions delineating the exons and introns in the contexts of transcription through translation phases (as per central dogma) enabling the genetic information in the DNA to make the eventual protein complex (UTR: untranslated region).

the DNA sequence undergoes transcription and translational steps as per central dogma resulting in eventual protein complex. One of the bioinformatic pursuits refers to locating splice-junctions across genomic sequences (Chapters 3 and 4), and such evaluations can be extended to viral genomic sequences as required.

Splice-junctions refer to locations that delineate exons and introns as shown in Fig. 8.4; and mostly, the exact location of a splice-junction may not be reliably distinct. In a canonical sense, the splice-junction consensus follows certain rules as regard to the constitution of introns and exons. For example, introns almost always begin with the residue set {GT} and end with an {AG}. But, inasmuch as the nucleotide sequence is a statistically permutated set of the elements {A, T, C, G}, numerous putatively occurring locations (other than being in introns) may also resemble such canonical patterns. This implies that relying on such canonical details alone may not reasonably and robustly indicate the presence of splice-junctions.

As such, should a junction be recognized, statistically prevailing long-range genetic information should be analyzed so as to determine the extent to which subsequences surrounding splice junctions differ from sequence segments of adjoining spurious analogs.[84,85] Such locations could be seen distinctly in a crisp form or could be an overlap of codon-noncodon segments. In the case of such fuzzy attributes of splice-junctions, a *statistical divergence* (SD) technique can be adopted to get the fuzzy details mapped into a novel membership function. This membership function describes the fuzzy subspace of overlapping exon and intron segments and it can be defined on the basis of an "error" feature prevailing in the overlapping (noisy) segment, possibly with mutational aberrations. The underlying heuristics are as follows.

As stated earlier, the locations of splice-junction may not so reliably distinct. However, in a canonical sense, the splice-junction *consensus* may follow certain rules as regard to introns and exons.[86,87] For example, the introns almost always begin with the residue set {*gt*} at 5′-end and ends with an {*ag*} at the 3′-end. But, inasmuch as the nucleotide sequence corresponds to a set of statistically permutated elements {A, T, G, C}, numerous putatively occurring {*gt*} and {*ag*} locations (other than in the introns as indicated) may prevail and resemble such canonical patterns. This implies that relying on such canonical details alone may not reasonably and robustly show the presence of true splice-junctions. Furthermore, in the event of point mutations, stemming of aberrant splice sites is inevitable. As such, should a junction be recognized and prevailing of possible cryptic junction sites should be elucidated, it is necessary to analyze statistically prevailing long-range genetic information so as to determine the extent to which subsequences surrounding the splice junctions differ from sequence segments of adjoining spurious analogs; hence, true *versus* aberrant (*cryptic*) splice junctions can be distinguishably identified. A feasible suite of analysis is as follows: evolutionary conservation of splice junctions is invariably hampered with inevitable phylogenetic-specific mutations. If such mutations are (assumed) independent, any "noisy" change in the spatial DNA pattern of the sequence (at the splice-junctions) can be marked as a "spatial jitter" with a characteristic parameter called *spatial signal-to-noise ratio* (SSNR).

Splice junctions with a spatial jitter as above correspond to fuzzy offsets of exons and introns at their junctions. That is, the spatially jittered junction refers to an overlapping mix of codon and noncodon entities and hence constitutes a (fuzzy) universe. In other words, the splice junction information has a fuzzy structure that can only be identified and specified in norms of linguistic descriptions. Such descriptions can be characterized by a membership (function)[34,61,86] of belongingness to the attributes of exon or introns.

Described in Chatterjee *et al.*[31] and Chatterjee[83] is an appropriate *fuzzy inference engine* (FIE) designed to delineate fuzzy overlaps of codon/noncodon parts so as to elucidate the underlying cryptic (or aberrant) splice-junction. This is done on the basis of SSNR defined with reference to the spatial jitter. The SSNR mentioned earlier also represents the associated membership function of the fuzzy region.

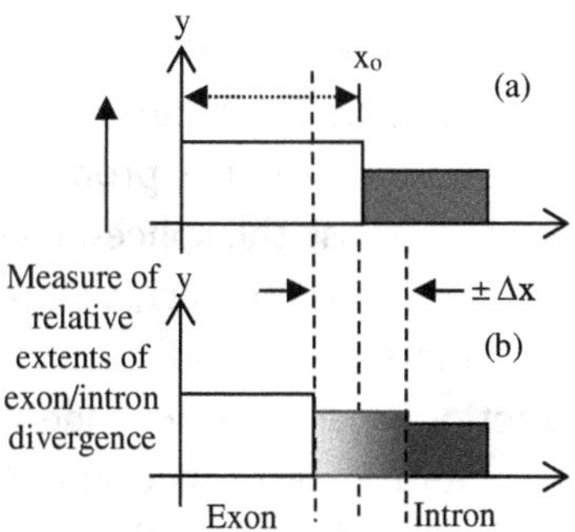

Fig. 8.5 Exon–intron (or *vice versa*) transitional residues residing along the sequence manifesting as crisp or fuzzy splice junction. (a) Unaltered (crisp) splice junction and (b) "spatially jittered" fuzzy splice-junction with graded variation of divergence (distance) of statistical features of exon/intron (or intron/exon) along the transition region. The relative extent of divergence is specified as a measure on the ordinate *y*-axis, and the abscissa (*x*-axis) depicts residue positions scale along the DNA sequence.

Now, considering an intron–exon splice junction as in Fig. 8.5, the upper figure marked as (a), it is a crisp (noise-free, uncorrupted) site with a splice junction at x_0 along the DNA sequence constituted by {A, C, T, G} residues. Should mutational corruptions have taken place, this crisp transition boundary x_0 becomes $(x_0 \pm \Delta x)$, where Δx denotes the spatial jitter, marked as in lower illustration (b) in Fig. 8.5. Furthermore, the *y*-axis in Fig. 8.5 depicts a measure of (relative) SD of exon *versus* intron (or *vice versa*) statistics (deduced in probabilistic norms) prevailing at any point, x on the sequence. (As described in Chapters 1, 3, and 4), this SD is due to the reason that an exon has a distinct stochastic distributions of {A, C, T, G} constituents *vis-à-vis* the corresponding distributions in an intron segment.

The effect of (mutation-specific) corrupts the splice junction to become unclear or fuzzy, and $(\pm \Delta x)$ indicated above is a "noisy" jitter variable superimposed on SSNR relative to crisp disposition of the splice junction at x_0. The expected *root-mean-squared* (rms) jitter J_r at any splice junction x_0 can be expressed by the "noise power", depicted as $m(x)$ imposed by the mutation errors. In terms of relevant heuristics, suppose the average length of intron-plus-exon space is $\overline{x}$. Then, the corresponding SSNR with reference to the DNA sequence space of Fig. 8.5 can be defined as follows: $\text{SSNR} = \left(x\right)^2 / J_r^2$.

Relevant to such "noisy" (mutation-infested) intron/exon (or exon/intron) transitions, the accuracy of locating the transition site, x_0 is constrained by the probability of error associated with the estimation of x_0. Hence, within the specified blurring limits of the jitter, the SSNR would implicitly predict the error probability in estimating the splice junction. As indicated in Chatterjee *et al.*[31] and Chatterjee,[83] assuming relevant mutation (error)-induced jitter parameter as a Gaussian variable (with variance, σ^2) the cumulative probability $P_c(m)$ of correct decoding of the signal, $m(x)$ across the splice-junction $(x_0 \pm \Delta x)$ can be deduced as $P_c(m; \Delta x, \sigma) = (1/2) + (1/2) \times \text{erf}\{\Delta x/[(2)^{1/2} \times \sigma]\}$, and the probability, (P_c) of correct assessment (and hence implicitly the error probability) of splice-junction location is dependent on SSNR parameter. Shown in Fig. 8.6 is the plot of P_c as a function of $(x_0 \pm \Delta x)/x_0$, where, as indicated earlier, the location of crisp splice junction (at x_0) poses a transitional error-prone width, $\pm \Delta x$.

Pertinent to the fuzzy domain, its centroid location explicitly implies elucidating the defuzzified value of the location/site of the splice-junction based on *membership of belongingness* (of the site-of-interest, x_0 in the fuzzy space). The procedure to find X_C is described in Chatterjee *et al.*[31]

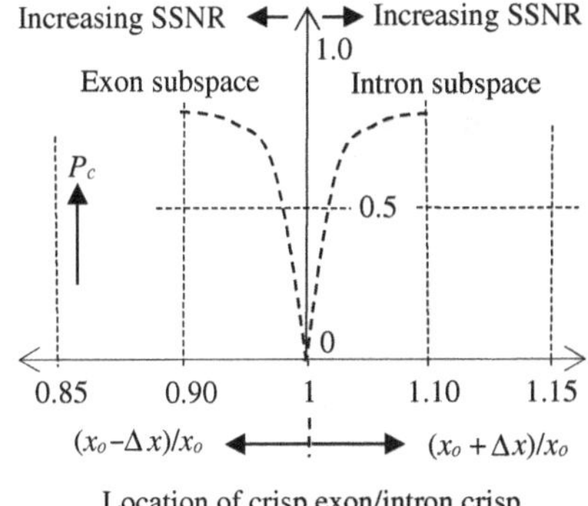

Fig. 8.6 The probability of correct estimation of a splice junction.

and Chatterjee.[83] It involves first calculating entropy/information-theoretic distances, namely, the SD values for the exon and intron regions across the fuzzy domain, and relevant SD values would show distinction in their profiles (in exon and intron regions) as illustrated in Fig. 8.5. (The SD can be any one on the divergence measure such as KL or JS mentioned in Chapter 2.)

Following the considerations presented in Neelakanta *et al.*[88] and Neelakanta *et al.*,[89] the expression for $P_c(m; \Delta x, \sigma)$ can be set as equal to: $[(1/2) + (1/2) \times \text{erf}\{\Delta x/[(2)^{1/2} \times \sigma]\}] \approx L'_q(z)/L'_q(0)$, where $L_q(z)$ denotes the *Langevin–Bernoulli function* (LBF).

The LBF with q representing a disorder entity associated with the statistics of the underlying populations as described by Neelakanta and De Groff,[88] (which is also adopted in the pertinent heuristics in other chapters). Also indicated in Neelakanta,[61] the ratio $L'_q(z)/L'_q(0)$ concurrently denotes, approximately, the *membership function* μ_q for the region (of fuzzy space or the block), $F{:}\{x_i\}$ of $(x_o \pm \Delta x)/x_o$ in Fig. 8.7, having an upper-to-lower range specified as follows: upper-bound (UB) corresponds to the isotropic disorder statistics decided with $q = ½$ and the lower-bound (LB) refers to anisotropic disorder implied by $q \to \infty$.[90]

Referring to Fig. 8.7, the mapping of computed divergence measure (SD, such as KL or JS) across intron and exon subspaces (at the slice junction) is shown; also, the corresponding membership function, μ_q(SD) mapped is shown in Fig. 8.7 along with details on UB value (when $q = ½$ in μ_q) and LB value (when $q \to \infty$ in μ_q).

Next, as illustrated in Fig. 8.7(II), the location of the splice-junction buried in the fuzzy domain is ascertained as described in Chatterjee *et al.*,[31] Pandya,[32] Neelakanta *et al.*,[33] and Arredondo *et al.*[34] This location

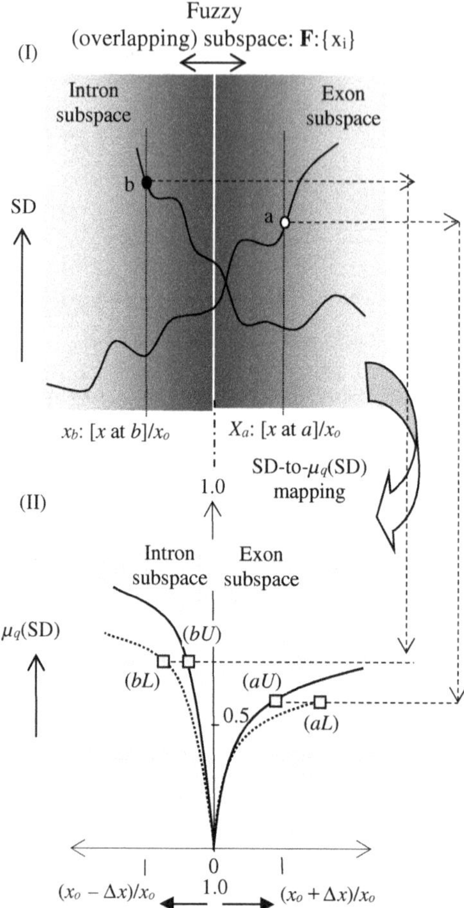

Fig. 8.7 I: Illustration of SD-to-μ_q(SD) mapping across fuzzy subspace F:{x_i}. The SD value "a" maps to upper-and lower-limits of μ_q(SD), respectively, as (*aU*) and (*aL*). Similarly, the SD value "b" maps to upper- and lower-limits of μ_q(SD), respectively, as (*bU*) and (*bL*). I: $(x_o \pm \Delta x)/x_o$ *versus* SD curves in the intron and exon subspaces. Note the SD profiles are distinct in each region. II: $(x_o \pm \Delta x)/x_o$ *versus* membership function, μ_q(SD).

corresponds to a centroid coordinate (x_C) decided by UB and LB profiles of the μ-value. In order to get the centroidal position, x_C, the translated values gathered is subjected to a defuzzification process.[88] In addition to computed details presented in Fig. 8.8 using KL-measure and the inferred splice junction (inferred at the base location 7401 *via* centroid considerations), more information/details (adjacent to the predicted splice junction) can be obtained.

For example, a splice junction also exists at the position, 7574 as per the heuristics of Krawczak[85]; furthermore, a set of cryptic splice

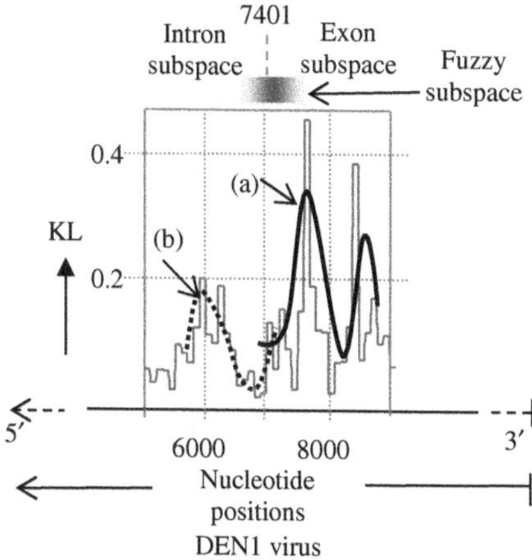

Fig. 8.8 Nucleotide position in the limited range of 5000–9000 *versus* computed KL measure of the DNA sequence of Dengue virus type 1 (*NCBI Reference Sequence: NC_001477.1*)

junctions at 7370 and 7419 also prevail (in the vicinity of the inferred centroid 7401) on the basis of heuristics of Krawczak[85] suggesting the 3′-side preferential ending in introns being, ***ag*** (i.e., the values 7370 and 7401 can be seen and picked around the centroid determined in conformance with the abutting of ***ag*** residues).

Consistently, the results on the canonical splice-junction consensus of Krawczak[85] indicate the intron subspace ending with residue set {***ag***} at 7574 as in Fig. 8.9, and notwithstanding such canonical patterns, the mutational influences could have also possibly induced aberrant splice junctions as well.

The complete list of aberrant splice junctions evaluated for the test viral DNA can be availed from the details of Chatterjee *et al.*,[31] Pandya,[32] Neelakanta *et al.*,[33] and Arredondo *et al.*[34] where relevant information are compared against the data available in NCBI GenBank showing observable overlaps of CDS domains in the query (viral) sequence. Such details eventually could facilitate identifying various protein structures. That is, the purpose of knowing correct and aberrant splice junctions in the context of viral DNA (e.g., the exemplified DEN1 virus) is pertinent in vaccine designs.[91] (Pertinent details will be discussed in Part II of this chapter.)

A systematic analysis of splice-junction sequences in eukaryotic protein-coding genes using NCBI GenBank databank has revealed a striking similarity among the rare splice junctions that may not contain ***ag*** at the 3′ splice site or ***gt*** at the 5′ splice site as could be evinced from Krishnamachari

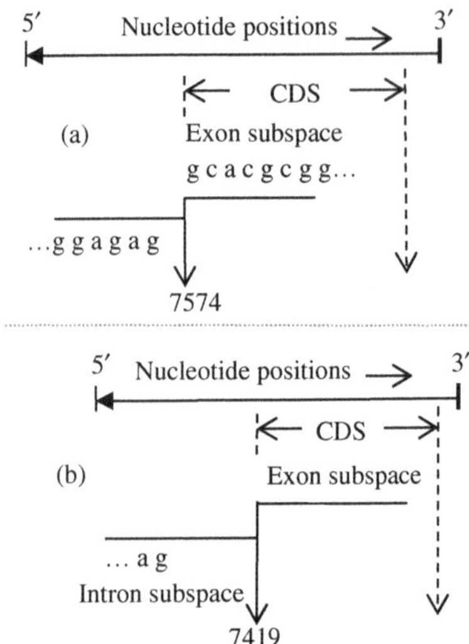

Fig. 8.9 Details on nucleotides adjacent to the predicted splice junctions: (a) As per Neelakanta *et al.*[33] and (b) as per the method described in this section. In both cases, the intron-subspace ends with a residue pair **ag** bases consistent with the canonical splice junction consensus.

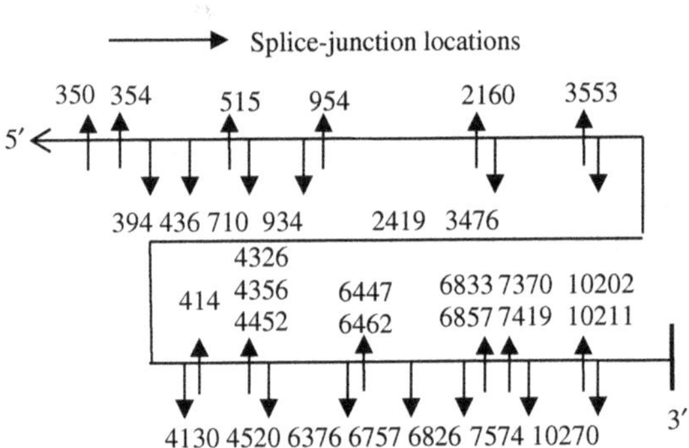

Fig. 8.10 Summary of results on the locations of splice junctions. Downward arrows indicate values of transition sites (TS) available in NCBI GenBank for DEN1 virus. Upward arrows indicated computed-estimated values that include details of cryptic sites in the fuzzy subspaces.

et al.,[92] Florea,[93] and Stephens and Schneider.[94] A summary of results on the locations of splice junctions is presented in Fig. 8.10.

Typically, the sequences between the boundaries of introns (denoting regions of DNA or precursor RNA that are not represented in m-RNA, but reside between regions) and exons (depicting regions of DNA or precursor RNA represented in mature RNA) are not random. There are several splicing events such as exon-skipping, intron-retention, cryptic splice-site usage, and alternative 3- and 5′-side splice sites could have possibly resulted in Krishnamachari *et al.*,[92] Florea,[93] and Stephens and Schneider.[94] Furthermore, specific to RNA splicing, the so-called splicing variants can be formed prior to mRNA translation due to differential inclusion or exclusion of regions in the pre-mRNA structure. In all, the descriptions in this section enable elucidating the cryptic and fuzzy aspects of splice junctions in biosequences, such as viral sequences. Pertinent analytical framework and computational aspects are augmented with the details available in Krishnamachari *et al.*,[92] Florea,[93] and Stephens and Schneider.[94]

Tutorial T-IV-8.1

This tutorial is presented to outline the details on CpG motifs present in viral ssDNA. Typically, in DNA segments, a set of CG motifs could be constituted by short stretches of guanine (G) and cytosine (C) bases. Such motifs prevail with an occurrence frequency of CG nucleotides being higher than in other regions. The CG motif section is also known as *CpG island*, where *p* implies that C and G are connected by a phosphodiester bond. The CpG islands[95–97] in essence are unmethylated regions that contain high concentration of C and G. The generally accepted definition of what constitutes a CpG island in a DNA sequence was proposed in Garden and Frommer[95] as being a 200-bp stretch of DNA with a $(C+G)$ content of 50% and an (observed *CpG*)-to-(expected *CpG*) ratio being higher than 0.6. A subsequent study[96] based on an extensive search on the complete sequence of human chromosome 21 and 22, stipulates that a DNA region with equal or greater than 500bp having a *GC* content exceeding 55% and an (observed *CpG*)-to-(expected *CpG*) ratio of 0.65 could be more likely a true *CpG* island. The *CpG* sequences are relatively rare in human DNA, but are more commonly observed in the DNA of foreign organisms such as bacteria or viruses.

There has been a general research interest in understanding the presence of *CpG* oligonucleotide in such viral and bacterial genomes. Specifically, the human immune system has been studied as regard to the way it evolved in recognizing *CpG* sequences as early signs of infection and initiating an immune response on *ad hoc* basis. The extent of *CpG* abundance and/or deficiency specific to viral genomes of a certain species has also been of research interest.[97] Thus, with reference to a ssDNA, the bioinformatic effort of interest in the present study refers to knowing the presences of *CpG* motifs along the genome sequence *via* compatible analytical and computational procedures. Identification and delineation of *CpG* islands in a test ssDNA can be done again by using the concept of entropy-based SD. Relevant details are presented in the following example.

Example EX-IV-8.1

Statement of the Exercise

This example is indicated as an exercise toward locating *CpG* islands by entropy-based approach. Location of *CpG* islands in the test sequence implies finding the fragment of bases that constitute the motifs of *CpG*. In order to determine the presence of such islands, first the entropy-dictated by the occurrence statistics of three cases, namely, *C*-alone, *G*-alone, and (*CG*) jointly are determined. For this purpose, a "junk" sequence of length *L* bases is first constructed as follows: considering a total sequence length of *L* nucleotide base locations, a uniformly distributed set of *L*/4 random locations (epoch spaces) are generated and assigned for the base *A* (in the space set *L*). Likewise, three different ensemble sets of random locations are permutatively generated within the field *L* and assigned for *C*, *G*, and *T*.

Thus, the total length *L* bases is occupied randomly by A, T, G, C each with an equal probability of $\{qA = qT = qG = qC = 1/4)$. The sequence of length so generated can be dubbed as a "junk sequence" implying that the statistics of {A, T, G, C} does not bear any information (due to certainty of equal probable occurrences of nucleotide elements {A, T, G, C} consistent with Laplacian hypothesis on equally likely occurrences of epochs).

In contrast, considering an actual genomic sequence (such as a viral sequence), the elements of {A, T, G, C} would occur randomly with unequal probabilities as dictated by the encoded genetic information. That is, corresponding probabilities of occurrence of A, T, G, and T, namely, p_A, p_T, p_G, and p_C are such that $p_A \neq p_T \neq p_G \neq p_C$, but $(p_A + p_T + p_G + p_C) = 1$.

Now, taking a window across actual and junk sequences, the occurrence probability sets, namely, $\{pC + G, pC, pG\}$ and $\{qC + G, qC, qG\}$ are determined for each window of nucleotides. Hence, the (observed *CpG*)-to-(expected *CpG*) ratio is specified by $R_{pw} = (p_{C+G})/(pC \times pG)$ and $R_{qw} = (qC + G)/(qC \times qG)$. Corresponding mutual entropy, say, in terms of Jensen–Shannon measure $JS_W(.)$ is given by: $(JS)_w = \{0.5 \times p_w \times \log_e (p/M)_w + 0.5 \times q_w \times \log_e (q/M)_w\}$ nats.

Computations pursued toward the identification of *CpG* islands in B-19 virus ssDNA lead to JS-measure *versus* nucleotide locations yield results as illustrated in Fig. 8.11. Relevant details are pertinent to mutual entropy based identification of *CpG* islands in the ssDNA of the Parvovirus B-19 available in Karlin *et al.*,[97] and Nguyen and Brooks.[98] The concept of entropy is thus used above in locating the *CpG* island segments by considering the divergence of ($C + G$, C and G) populations in the test sequence *versus* the statistics of this population set in a simulated junk sequence, as shown in Fig. 8.11. The most prevalent viruses in nature are single-stranded (DNA or RNA) viruses with the genetic material encapsulated in icosahedra-shaped capsid proteins.[98] They cannot reproduce by themselves, but infect the host cells with their genetic materials, enabling the (host) cellular machinery to produce more viruses. The viral ssDNA or ssRNA assumes invariably a hairpin format (for structural stability as has been explained in Neelakanta *et al.*,[38]) and this hairpin form consists of a base-paired stem structure plus a loop sequence having unpaired or mismatched nucleotides. Knowing the profile of hairpin structures and the loops and bulges in viral genome is largely pertinent in: (i) understanding virus replication process[44] and (ii) in drug synthesis applications, where a relevant compound being sought may act as a binding agent in a specific ssDNA/-RNA inhibiting the replication of certain viruses.[45]

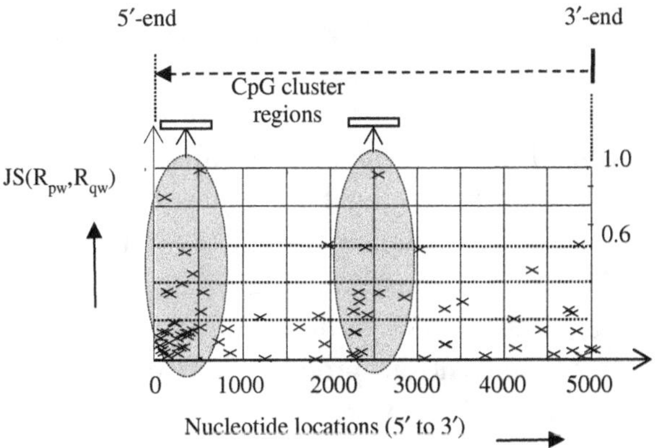

Fig. 8.11 Locating *CpG* segments in the test sequence: the computed results marked as (×) correspond to normalized JS measure evaluated as a function of R_{pw} and R_{qw} ratios. The clustered regions plus computed measure exceeding 0.6 depict plausible *CpG* islands as shown.

As regard to the framework of overall folding of DNA structure into hairpin formats, exclusive research in bioinformatic perspectives is often needed (adjunct to wet studies) so as to ascertain their energetics profiles responsible for the folding and stability of the genome backbone. While a microbiological description of DNA hairpin bends emphasizes the biological importance of such bends, a distinct pursuit of research deliberates the physics of thermodynamics on the structural aspects of the folded single-stranded viral genomes[43] yielding details on the associated energetics profile of the genome structure.

Viral Genomic Analysis Based on Energetics Considerations

Energetics aspects of viral genomic sequences: biothermodynamics and biosystem entropy are like Siamese twins.[99] One of the main steps in the viral life cycle required for replication of the virus is genome ejection into the host cell. The ejection of the viral genome from the capsid is due to very high internal pressure as a result of electrostatic forces of the nucleotides.[100,101] This pressure and electrostatic charge is partially responsible for the delivery of the viral genome into the host cell, thus making it central in the infection process. Also, as the overall folding of single-stranded DNA/RNA into hairpin structure (which is very important for its replication) and its stability depends on the energetics associated with the virus, the detailed energetics profiling is imperative in determining the particular characteristics of the virus in question.

The general prospects of relevant analyses and related computations could eventually illustrate the integration of stabilized, hairpin-folded viral DNA structure into a host genome in the perusal of replication processes. Such bioinformatic studies are focused specifically to include the following: (i) finding the site in the ssDNA/ssRNA sequences at which the hairpin bend would take place; (ii) locating possible sites in the folded structure where bulging (bubbles) may prevail; (iii) evaluating the integrity of base-pair matching within the hairpin-folded sections of the loop and the stem; (iv) ascertaining the stability of folded format of ssDNA/ssRNA sequences in terms of the associated energetics; (v) correlating energetic and entropy profiles of the folded ssDNA; (vi) using eventually, the genetic information (knowledge) in the folded structure of the ssDNA sequence, for example, understanding the mechanism of hairpin bend (folding) implications in viral ssDNA *vis-à-vis* using relevant details for rational design of vaccines; and (v) delineating codon (or CDS) and noncodon segments in a given test sequence.

In the enumerated prospects of viral genome sequence analysis as above, ascertaining the stability of ssDNA/ssRNA sequences in their folded forms *via* the associated energetics and correlating such energetic details *versus* entropy profiles form crucial exercises. Described below is an outline on relevant considerations. It first involves applying the *WC* model of base-pairing and evaluating the associated energetic profiles of ssDNA/ssRNA structures as indicated below.

The RNA (transcribed from the double-stranded DNA) and ssDNA sequences tend to become more compact by "folding" or "bending" themselves into a stabilized hairpin structure (mostly toward 3′ end) *via* nucleotide base-matching set by *Watson–Crick (WC) pairing* of $A \leftrightarrow T$ and $G \leftrightarrow C$. The RNA hairpin forms are widely addressed in the literature.[102–107] Analogous to RNA hairpin, it is hypothesized in this study that the hairpin structure of viral ssDNA is made of turn-around loop with the closing base pairs and a stem. In addition, a bulge on a strand and/or an internal loop on both strands of the hairpin can be formed due to unpaired nucleotides.[108] However, associating assertively such a bulge and/or an internal loop composition with a viral ssDNA is an open question for investigation.

The stem part plus loop constitutes the so-called *SL* structure. The stem primarily consists of *WC* base pairs (*bp*) formed between two antiparallel structures of the hairpin, 5′-through 3′-end. The stability (as indicated to prevail in RNA hairpins) is due to the so-called TNCG tetraloop composition (where *N* depicts any of the four bases A, T, G, C).

Typically, **c**TNCG**g** is an example notation that includes the loop with a closing base pair (***c*** and ***g*** denoted by lower case bold font). The closing base pair is the first base pair set next to the loop at the commencement of the stem. The hairpin-loop sequences preferentially decide on the type of closing base pairs. For example, the UNCG tetra loops in RNAs prefer a ***cgc*** losing base pair over a ***gc*** version (as decided by NN effects).

Normally, a region of unpaired nucleotides may exist at the apex of the hairpin loop, which serves as the region where the directionality of the backbone reverses making the two antiparallel strands of the stem. In a RNA hairpin, a minimum of three nucleotides can make a bend or turn (consistent with steric repulsion); however, loops of four nucleotides (mentioned above as tetra loops) are more common.

The hypothesis here conforms to the both tri- and/or tetra-loop characterizations applied to viral ssDNA genomes. Considering the viral ssDNA, it is yet to be ascertained confirmedly of stable and unstable aspects of hairpins and the existence of combinatorial base nucleotide selection in forming tetra- and/or tri-loops associated with closing base pairs. Relevant folding kinetics expressed in terms of energetics and entropy consideration could be determinants of stability/instability in the folded viral ssDNA messages as discussed in a later section.

Tutorial T-8.2

This tutorial is a note on the thermodynamic aspects of Watson–Crick (*WC*) pairs. The *WC* pairing enables stability of biosequences such as in hairpin structures. Relevant (stability) dynamics relies on free-energy minimization specified by NNs parametric attributes of base pairs present in the test sequence. That is, the stability in question conforms to the rules stipulated by each base pair depending only on the most adjacent pairs with the associated total free-energy being the sum of each contribution of the neighbors.[109] (In addition, the base-pairing enables a favored (neg)entropy state of the sequence complex involved as will be indicated in a later subsection.)

The *WC* base pairs denote significant motifs, whose thermodynamic aspects can be well represented by the *NN* model that indicates the stability of the base pairs being dependent on the identity of the adjacent pairs. Known generally as *individual nearest neighbors* (INN) model, it implies a preferential stacking of energetically conducive pairs with loop initiation leading to an eventual hairpin structure. Totally, the associated conformational free energy is constituted by paired and unpaired nucleotide stacking in the bend as well as by the nucleotides at the loop.

The free-energy increments of the base pairs in the sequence are counted as stacks of adjacent pairs. For example, the consecutive *CG* base pairs are worth about (–3.3 kcal/mol).[110–113] The loop region formed normally has unfavorable increments called *loop-initiation energy* that largely reflects an entropic cost expended in constraining the nucleotides within the loop. For example, the hairpin loop made of four nucleotides may have an initiation of energy as high as +5.6 kcal/mol. Mostly the unpaired nucleotides in the loop contribute favorable energy increments. From the literature indicated above, a set of approximate conformational free energy can be gathered on the basis of NNs considerations.

Suppose the neighboring four bases are translated with 0, 1 scoring (with 1 indicating *WC* pair and 0 denoting non-*WC* pairing). Then the following quad-bit representation *WC*- and/or non-*WC* (NWC) neighbors and the corresponding (approximate) relative levels of free energy[38] can be specified with the following abbreviations: NN-R: *neighbor on the right-side of center element*; CE: Two *Mid-(center) elements*; NN-L: *neighbor on the left-side of center element*; and EV: *free energy values* (in kcal/mol specified relative to—3.3 kcal/mol, depicting the lowest energy level and taken as zero reference level). Hence, the following set of 16 quad-bits are constructed (each depicting four adjacent matched or unmatched *WC* pairs along the hairpin stem structure):{NN-R CE NNL; EV} ≡ {1 11 1; 0};{0 11 1; 0};{1 11 0; 0}; {0 11 0; 0}; {1 10 1; 1.2}; {1 10 0; 2.2}; {0 10 1; 2.7}; {0 10 0; 6.6};{1 01 1; 1.2}; {1 01 0; 2.2}; {0 01 1; 2.7}, {0 01 0; 6.6}; {1 00 1; 4.3}; {1 00 0; 8.9}; {0 00 1; 8.9}; and {0 00 0; 10}. The data on relative energy values indicated for the *NN* composition of four adjacent bases are compiled from Tianbing *et al.*,[110] Weise *et al.*,[111] Ali and Silvey,[112] and Neelakanta.[113]

Hence, the test sequence is depicted segments of four bases, each representing {NN-R(1 base) CE(2 bases) NN-L(1 base)} and each segment is then mapped as a quad-bit plus relative free-energy (EV) indicated above. Then, using sliding-window method, the profile of free-energy variation profile across the test sequence is determined.

Example EX-8.2

Statement of the Exercise

Indicated in this example is a way to consider entropy-based segmentation method as regard to finding specific structural details of genomes of a virus. Such an effort enables characterizing subregions of genomic sequences, such as loops and bulges in ssDNA.

Suppose a hypothetical hairpin structure as shown in Fig. 8.12 is considered. It may possess along its length, stem, and/or bends/loop regions with nicked base pairs, and correspondingly, Watson–Crick pairs (*WC*) (i.e., ($A \leftrightarrow T$) and ($G \leftrightarrow C$) matching) can be seen in the stem region.

Considering the statistics of such pairs, the occurrence probability of *WC* base pairs (across the inverted stem parts) is noted and designated as p. Hence, the corresponding probability of occurrence of mismatches is designated as q (so that $q = 1 - p$). Between the event spaces of p and q defined above exists a cross-entropy (or mutual information) in Shannon's sense (Chapter 2). Such cross-entropy details (implying inherent genetic information) can be assessed as follows:

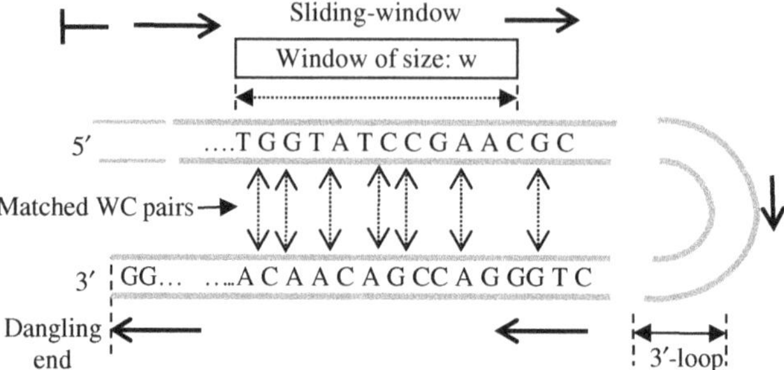

Fig. 8.12 Illustration of sliding-window procedure to compute *p* and *q* values across a hypothetical hairpin structure.

Relevant analytical and computational pursuits involve elucidating the SD of *WC versus* non-*WC* (*NWC*) pairs occurring in the sequence pattern. The SD pattern deduced along the query sequence segregates the *WC* and *NWC* entities and would enable locating the site at which loops/bulges exist and where (a hairpin) bending takes place. For example, the *WC/NWC* pairs manifesting as matched and/or unmatched inversions of the residues in the set: {A, T, G, C} and the location where hairpin bending occurs in the query ssDNA sequence are illustrated in Fig. 8.12.

A sliding-window approach can be pursued to determine the cross-entropy details along the sequence length. Suppose the *WC* base pairs (AT, CG, GC, TA) are each assigned a fitness 1, and all other (non-*WC*) pairs are treated as mismatches, each having an assigned fitness of 0. Apart from *WC*-mirrored sets, namely ($A \leftrightarrow T$) and ($G \leftrightarrow C$), the set of non-WC pairing, namely (G ↔ T) can also be regarded as stable pairs (as indicated in Tianbing *et al.*[109,110]); hence, the set {A ↔ T, G ↔ C, G ↔ T} is designated as a set of *canonical base pairs* and assigned the fitness 1.

Suppose a test ssDNA sequence constituted by *L* base residues plus a hypothetical hairpin structure is considered, and a sliding-window of size *w* (containing a total of *w* pairing and non-paring nucleotide counts) is specified along the stem and its reversed part (of the hairpin bend on the 3′ side) as shown in Fig. 8.12. Within each window, the count of 1s (meaning matching *WC* pairs) is denoted as *m*; likewise, the count of nonmatching pairs is denoted as *n*, so that $(m + n) = w$; hence, $(m/w) = p$ and $(n/w) = q$ for the window segment chosen. Thus, for each window size *w* sliding across the test ssDNA structure, the set $\{p, q\}_W$ is determined.

The SD that measures the cross-entropy (in each window segment, *w*) corresponds to an SD metric deduced in terms of p_w and q_w values (namely, the estimated p and q values in each window segment as above). For example, as defined explicitly in Chapter 2, relevant SD measure can be one of the following metrics: $(KL)_w = \{p_w \times \log_e(p/q)_w + q_w \times \log_e(q/p)_w\}$ nats; $(JS)_w = \{0.5 \times p_w \times \log_e (p/M)_w + 0.5 \times q_w \times \log_e(q/M)_w\}$ nats; and $(B)_w = -\log_e(\rho_w)$ nats. Here, $M = \{0.5 \times (pw + qw) \equiv 0.5\}$ and $\rho_W = (p_W \times q_W)^{1/2}$. (Apart from KL-, JS-, and B-measures, there are also a number of other SD metrics available as described in Chapter 2 and reported in Neelakanta,[61] and Kapur,[62] and they can also be used in the aforesaid algorithmic exercise.)

The analysis pertinent to a given test hairpin structure as above, evaluates statistically coevolving positions of *WC* pairs across the folded structure, and it decides the associated cross-entropy or mutual information. The computed divergence measure (in terms of the set $\{p, q\}$) discriminates the dispositions of *WC* and *NWC*. The complete procedure for determining the entropic profile is described in Neelakanta *et al.*[38] and Chatterjee.[83]

Example EX-8.3

Statement of the Exercise

Inspired by existing investigations on RNA folding[108] and heuristics of mapping the statistical profile of base residues in a single-stranded genomic structure (as detailed in the previous example), the present exercise is indicated to provide an exclusive entrée toward analyzing a viral ssDNA possessing hairpin bends. It refers to ascertaining various structural, entropy and/or energetic features of the test viral ssDNA, so as to understand the stability and control of the underlying gene expression; hence, the kinetics and thermodynamics of hairpin-folding, transition-state, and co-operative foldings are decided. Similar information on viral ssDNA is rather sparse in literature. Relevant study details and results are presented in a set of illustrations in Figs. 8.13a–8.13c; and the legend under each illustration explains the relevant computational suites pursued and results derived thereof.

Figs. 8.13a–8.13c illustrate details on the formation of loop/bulge and overall codon sequence CDS regimes in the sequence of a test virus: parvovirus B19. Pertinent sequence details shown on bases and base locations (at the closing ends of the bulge in the sequence) are taken from the GenBank data[23–114] on the test genome; and Figs. 8.13a–8.13c, depict, respectively, pertinent details on the following: (a) formation of a

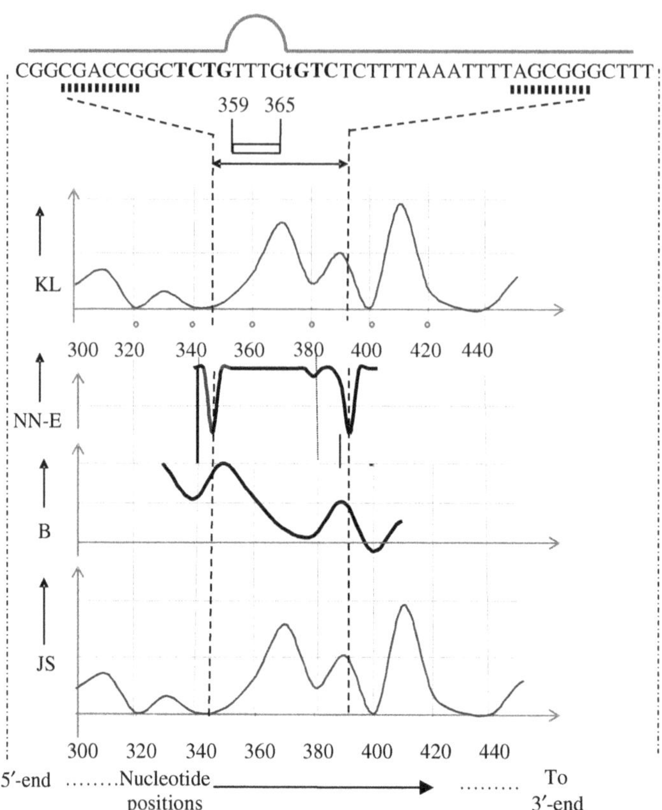

Fig. 8.13a Formation of a loop (bulge) at the 5′-end: test sequence of Parvovirus B19. Nucleotide sites from 300 to 440 *versus* computed SD-measures: KL, NN-E, B, and JS.

loop (bulge) at the 5′-end; (b) formation of a hairpin loop at the 3′-end. The normalized results plotted in Fig. 8.13b conform to SD/entropy measures of KL-, JS-, and B-metrics as well as, the NN-based energetics values (NN-E) computed in the vicinity of 3′-end; and (c) overall representation of CDS in the test sequence results and inferential details of the illustrations in Figs. 8.13a–8.13c are as follows.

Fig. 8.13a: the formation of a bulge at 5′-end on the test sequence as indicated. It refers to the bulge being at the base positions 359–365, and the analysis further shows a well-defined, almost symmetrical minimum energy (NN-E) sites on both sides of the bulge site. Also, the SD-based entropy feature (especially the B-measure) indicates symmetrical maximum entropy levels abetting the bulge bilaterally. Thus, the algorithms adopted and computations performed thus enable finding of loop/bulge formation at the 5′-end, and corresponding details on the relevant sites are in conformance with the GenBank data.

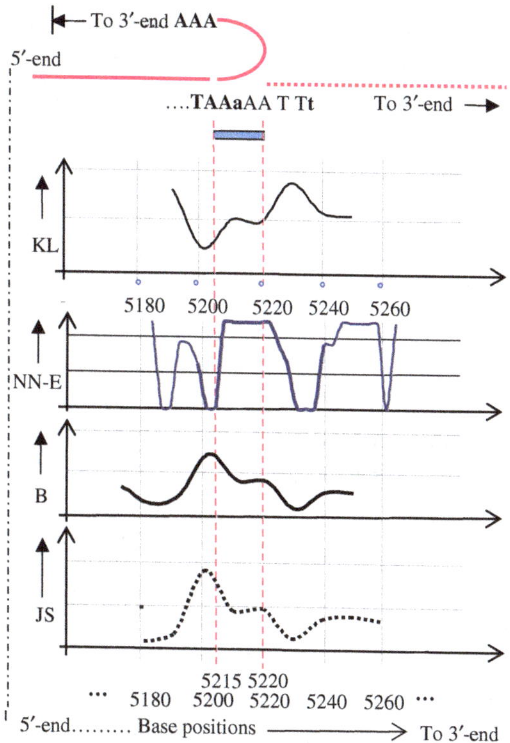

Fig. 8.13b Formation of a hairpin loop at the 3′-end: Test sequence of Parvovirus B19. Nucleotide sites from 5000 to 5300 *versus* computed SD-measures: KL, NN-E, B, and JS.

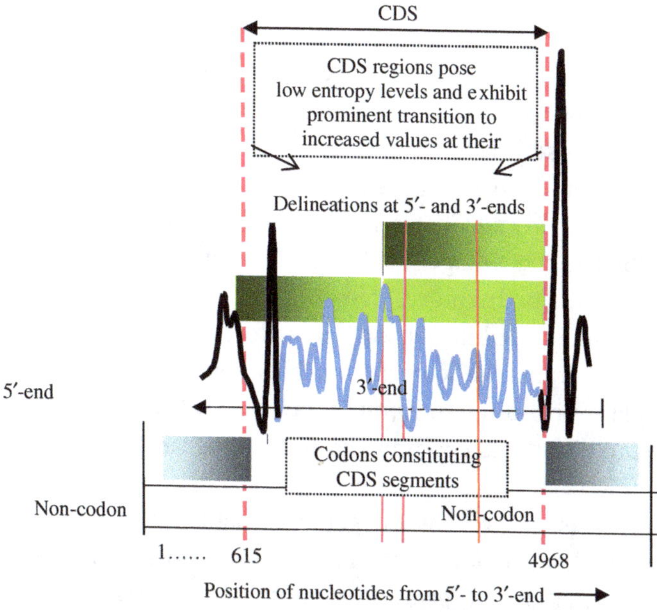

Fig. 8.13c Overall representation of CDS in the test sequence.

Fig. 8.13b: now considering 3′-end, the results project the hairpin bend at 5215–5220, and they show the loop sites with two minimum energy (NN-E) valleys on its sides as could be evinced from Fig. 8.13b. The entropy features also provide the information on the loop-site with KL-, JS-, and B-measures changing distinctly from low-to-high or high-to-low values.

Fig. 8.13c: with reference to the details in Fig. 8.15c, it can be inferred that the procedure advocated again gives confirming results on codon–noncodon transitions along the test sequence. Such details can be validated with GenBank data. However, unlike the crisp transition locations specified in GenBank data, the delineating locales are rather fuzzy. Such observations are consistent with the results in Arredondo.[34]

Similar to the methods described above, analyses of RNA sequences, (which are also single-stranded) are widely addressed in the literature[108] as regard to their structures, kinetics, thermodynamics, and biological functions. Specifically, how an RNA folds back on itself (forming a complex structure) has been studied in terms of resulting hairpin plus an extended to a stem-part posing *WC* base-pairing as well as a loop/bulge section wherein the backbone reverses its directionality. Relevant studies also include descriptions of the associated hairpin structures possessing diversity in their stem, loop, and closing base pairs.

Fourier Spectral Characterization of Viral Genomics

Use of Fourier methods for biosequence analysis is also described in Benson,[63] Cheever *et al.*,[64] Zhou,[65] Anastassiou,[66] Anastassiou *et al.*,[67] Rao and Swamy,[68] Jeng *et al.*,[69] Vieira,[70] Chen,[71] Isohata and Hayashi,[72] Herzel and Grosse,[73] Herzel,[74] Lee and Luo,[75] Fukushima *et al.*,[76] Nyeo *et al.*,[77] Hanson,[78] Silverman and Linsker,[79] Tiwari *et al.*,[80] Fickett,[81] Fickett and Tung,[82] and Chatterjee,[83] where relevant Fourier expansions are described as the "image" of the spatial sequence with relevant comparison efforts. Furthermore, detection of similarities between DNA sequences is illustrated in Benson[63] by enforcing *Fast-Fourier transform* (FFT) to ascertain the correlation between DNA sequences using complex plane encoding. In Zhou,[65] the Fourier transform (FT) method is adopted to distinguish coding and noncoding subsequences in a complete genome.

A comprehensive review on genomic signal-processing method is addressed in Anastassiou[66] and Herzel and Grosse[73] where the local texture information in genome structures is extracted by Fourier spectral mapping of test sequences. In extended contexts of using FT methods in the analysis of genomics and proteomics, *digital signal-processing* (DSP) techniques are also proposed in the literature,[68] where spectrograms are shown to be powerful tools for DNA sequence analyses providing local frequency information. Specifically, the so-called *short-time Fourier transform* (STFT) appears to provide a useful localized measure of frequency content in the spatial sequence pattern.

This STFT obtained in Rao and Swamy[68] is consistent with traditional *discrete Fourier transform* (DFT). It is applied to genome sequence analysis using sliding-window technique. For a clear Fourier spectrum to be generated, a reasonable-sized sample is required. Sampling with larger widow sizes yields relatively poor resolution, making it tedious to locate subtle features *via* FT. Sampling *via* smaller window lengths enables better resolution, but such sizes make it difficult to locate the local, distinguishing features.

In all, considering sequence-related characters depicted in terms of their numerical values (such as physicochemical parametric values) across the genomic sequence, then a signal-processing algorithm can be applied to extract the associated Fourier spectral information. For example, the Fourier spectrum of DEN1 serotype of dengue virus is indicated in Chatterjee.[83] It depicts the Fourier spectrum, *F*(.) computed using *FT* analysis of numerical values of the physicochemical parameter, *electron–ion interaction potential* (*EIIP*) pertinent to the test virus, and *F*(*EIIP*) is estimated using a (sliding) widow size traversing the residue positions of the test sequence. The $\text{EIIP}_{(A,\ T,\ G,\ C)}$ values assigned for nucleotide base set {A, T, G, C} are as follows: {$\text{EIIP}_A = 0.1260$; $\text{EIIP}_T = 0.1335$; $\text{EIIP}_G = 0.0806$; and $\text{EIIP}_C = 0.1347$}. For example, considering the following character (nucleotide) sequence: (A A A G T A G C), its translated, numerical version in terms of EIIP values can be written identically as follows: (0.1260 0.1260 0.1260 0.0806 0.1335 0.1260 0.0806 and 0.1340).

Apart from EIIP values, other parameters can be considered in representing genomic details (in terms of their translation *via* their numerical representations of parametric values). One of the earliest methods thereof was proposed by Silverman and Linsker.[79] It refers to the representation where each base is taken as the vertex of a tetrahedron in three-dimensional (3D) spaces and the genome sequence is transformed into an array composed of 3D vectors. A FT is then performed on each of the three sequences made up of a directional component from the sequence vectors. A resulting spectrum is then obtained with the sum of the three Fourier the occurrence of each base in the nucleotide sequence, summing the individual spectra to give an overall sample spectrum for the genome. Using this method, a repeating period of 3 has been found in coding regions.[80–82] Fourier analysis has also been employed as one of a number of weighting factors in the determination of introns splice sites.

In formulating the spectral domain comparison of genomic sequences of test viral strains, typically the FT-based algorithmic and computational efforts are pursued following the procedure due to Rao and Swamy,[68] wherein as mentioned above, the genomic sequences are represented by numerical sequences and the *FT* of the spatial-sequence of numerical values is determined. Implicitly, such numerical sequence-based technique of transforming a genomic sequence for decomposition *via* Fourier series depicts an *information spectrum method* (*ISM*). Relevant Fourier-transformed sequence inherently contains the physicochemical details attached to the biological functions of the genomic structure. As such, elucidating the Fourier series/transform of numerically formatted genome leads to detecting code/frequency pairs that are specific to the genomic sequence *vis-à-vis* its biological profiles. This method is insensitive to the location of motifs and therefore it warrants no prior alignment of the sequence with its counterparts.

DNA sequences exhibit *periodicity*, which can be examined by a number of methods including *autocorrelation function* (ACF) analysis, Fourier spectrum estimation, *DNA (or genome) walking**, Hurst-index estimation, de-trended fluctuation analysis, wavelet-translation and entropy, or mutual information function evaluation. (Genome walking* is a method for determining the DNA sequence of unknown genomic regions flanking a region of known DNA sequence). An important property of any genomic sequence is its "period-3 property" mentioned by Annastassiou in Anastassiou[66] and Anastassiou *et al.*[67] By virtue of the characteristic information in the gene sequence made of 64 triplets (encoded from four nucleotides, {A, T, G, C}) across exon regions, it is shown in Anastassiou[66] that the corresponding Fourier domain power spectrum of a prokaryotic DNA has a strong peak at a frequency $k = N/3$. It corresponds to a spectral component with a period 3 and N represents the discrete integer of the set denoting the sequence (in the spatial domain). Using the method described by Annastassiou,[66] the study in Chatterjee[83] shows the presence of period-3 property in all four strains of dengue virus.

Example EX-8.4

Statement of the Exercise

This example is furnished to illustrate the spatial domain analysis of genomic sequences. As mentioned above, such spatial domain representation of biological sequence analysis can be accomplished *via* FT of numerical (parametric) values substituted *in lieu* of the residues in the sequence.

As well known, given a function of a spatial variable, its FT identifies different frequency sinusoids and their amplitudes contained in that function. For example, given an analog numerical signal, $f(x)$ with x of varying amplitude (specified for example, a sequence of EIIP values assigned to each base of the genomic sequence), the STFT of $f(x)$ can be performed with: $f(x) \leftrightarrow FT(X) = F(\omega) = \Sigma_{m=1}^{R} f(x-m) \times w(m) \times \exp(-m\omega)$, where the associated entities are as follows: the whole test sequence is divided into N equally spaced intervals or frequency; for example, N can be arbitrarily taken (say, equal to 120 so as to match the window size improvised) and the frequency k is set equal to $N/3$. Furthermore, $\omega = 2\pi k/N$ and with the substitution of value of k, $\omega = 2\pi/3$. The subsequence spanning the window is denoted by $w[m]$ with the window length (size) being $R \leq N$ (= 120). The STFT, $FT(X)$ indicated above offers "a remarkable landscape of opportunity lies before next generation of biologist" toward exploring the morsels of microbiology and envisaging the intricacies of bioinformatics beneficial to the universe of living systems supported by relevant annotations[111] and descriptions as regard to a vast trove of biological species — small and big. Such studies also include viral bioinformatics, and more pertinent exercises are reviewed in the following sections.

8.6 Viral Genomics: Review Tutorials, and Exercises

In perusal of relevant topics of viral genomics, this section is designed to review the underlying considerations and formulate exemplar sets of some solved exercises and rider problems posed as projects compatible for basic and/or advanced research pursuits.

Tutorial T-8.3

This tutorial is a survey on classical pox virus studies. Pox viruses (depicting the members of the family *Poxviridae*) are the largest of all viruses, which could infect both vertebrate as well as invertebrate animals. The characteristic features of all such poxviruses are as follows: (i) replication in the cytoplasm of the host cell and (ii) the presence of viral enzymes for replication and expression of their genome.

The subfamily of pox viruses affecting the vertebrates are known as *chordopoxvirinae**. Their virions have an *ovoid* or brick-like shape with dimensions of 400 by 200 nanometers and the associated genomes consist of linear double-stranded DNA ranging from 130 to 375 kb.[115] The most distinctive characteristic of the infection due to chordopoxvirinae are the pock marks on the skin of the affected animal. Two distinct infectious virus particles exist in the subfamily of *chordopoxvirinae*: *intracellular mature virus* (IMV) and *extracellular enveloped virus* (EEV). Four *genera* of chordopoxviruses that are known to generally infect humans are *orthopox, parapox, yatapox,* and *molluscipox.* Examples of prominent members of these *genera* are as follows:

Orthopox:	smallpox virus (*variola*), vaccinia virus (VPVX), cowpox virus (CPVX), monkeypox virus (MVPX)
Parapox:	*orf* virus, *pseudocowpox, bovine papular stomatitis virus*
Yatapox:	*tanapox virus, yaba monkey tumor virus*; *Molluscipox*: *molluscum contagiosum* virus (MCV)

The genera orthopox virus (OPVX) includes 11 distinct (but closely related) species, and only three of these viruses are known to infect human.[116] The existence of gene-specific determinants[117] is responsible for the limited range of host of some poxviruses (in contrast with the broader host range of other poxviruses). Relevant to this note on pox virus, the following exercises are indicated as study problems.

Problem P-8.1

Statement of the Problem

This exercise refers to genomic-level comparison of pox viral sequences. At genomic level, the sequences of two viruses of different species in orthopoxvirus family can be compared. Existing studies inform that such sequences may commonly share at least 96% identity.

Relevantly, perform a research to determine common proteins (CPs) among such species of orthopoxvirus (with intensely shared commonness at genomic level) affecting humans. (Pertinent study can be similar to an exemplar analysis indicated in a later section of this chapter with reference to the serotypes of dengue virus).

Problem P-8.2

Statement of the Problem

Comparison of sequences of viruses infecting animals is the implied exercise here. Perform a comparison of genomic sequences of a pair of viruses isolated from pox-like infection in cattle and buffaloes[118] level to elucidate the underlying genetic characterization of such viruses. Pertinent analytical and computational studies can be done with details presented in Chapters 2 and 3.

Tutorial T-8.4

This is a tutorial note exclusively on smallpox virus. As indicated earlier, the genome of variola virus is the smallest among the species of orthopoxviruses, whereas the cow pox has the largest genome and a wider spectrum of affected host in the animal kingdom.

Over centuries, smallpox had been a highly scared and fatal type of human-specific contagious disease; now mostly eradicated thanks to worldwide vaccination efforts. Smallpox is caused by the variola virus. Controlling smallpox (*via* vaccination procedure) began "after it was noted that accidental percutaneous exposure to smallpox by a scratch on the skin reduced the severity of infection." Hence, the practice of "variolation," involving intentional administration of pustular fluids from smallpox scabs to uninfected persons. It appears that the practice of variolation had been adopted in China and India as early as in the tenth century. (However, concurrent deaths were reported due to the complications of this procedure.[119]) Later, Edward Jenner's contribution to immunization against smallpox *via* vaccination procedure ultimately led to the global eradication of smallpox.[120–126]

Problem P-8.3

Statement of the Problem

The exercise is to find the protein(s) associated with the amino acid (AA) sequence of vaccinia and variola virus AA subsequences using Protein BLAST (*blastp*).

Although the number of annotated CPs across these subsequences is seen to be higher, the one determined by divergence methods could be lesser in number. This is because different combinations of AAs could yield the same proteins. (*Note*: although the original vaccine used by Jenner was from cowpox virus, the AA sequences of variola virus[127] and vaccinia virus[128] can be exercised. The AA sequence of vaccinia virus and variola virus should be first aligned and the SD between the aligned AA sequences is determined (using, for example, *KL*, *JS*, and *B* measures). In determining such divergence measures, a window size of 100 residues can be considered, and in the regions where probability of match is higher than 40% (0.4), relevant segments of AAs are noted.

8.7 Genomic Analyses of Serotypes of a Virus

This section describes schemes of bioinformatic analyses of viral genomic sequences so as to elucidate the common and differential features of the serotypes of a test virus. The three distinct methods described earlier, namely, (a) entropy, (b) energetics, and (c) FT techniques are invoked toward the analyses involved. Descriptive outlines and implementation of relevant approaches of the three methods are presented below:

(a) *Entropy-based Analysis*

In the context of nonviral DNA sequences, their entropy details across coding and noncoding regions are comprehensively addressed in Chapter 3. Similar analytical considerations can also be applied to viral sequences, for example, to the four strains of dengue virus. Hence, the entropy segmentation method and its variations in SD sense described in Chapter 3 is adopted. Relevantly, the Kullback–Leibler (*KL*) divergence measure is used to get the information (entropy) profile of sequences pertinent to four strains of dengue virus so as to distinguish them in the entropy plane.

(b) *Energetics-based Analysis*

The energetics profile of DNA/RNA structures refers to the framework of structural stability of a sequence that conforms to the rules stipulated by the chemistry of bonding concerning each base-pair as well as it depends on the disposition of the most adjacent pairs. Relevantly, it relies on the associated total free energy (in the thermodynamic sense) depicting the sum of the contributions from the neighbors[129] That is, the INN model takes into consideration two neighboring bases (about a pair of center elements), and an *energetics value* (*EV*) for each pair is assigned depending on the neighbors on the immediate right- and left-side of the center pair. Compiled data on the energetic value for each pair of center element with reference to their adjacent neighbors are described in Neelakanta *et al.*[38] Low value of *EV* implies that the chemistry of the elements is self-selected in the sequence so as to assume a minimum potential energy profile toward thermodynamic stability.

(c) *Fourier Transform (FT)-based Analysis*

As indicated earlier, given a function of a spatial variable, its *FT* identifies different frequency sinusoids and their amplitudes contained in that function. Relevant *FT*-based algorithmic and computational efforts pursued in formulating the spectral domain comparison of genomic sequences of test viral strains, essentially follow the relevant procedure due to Rao and Swamy,[68] wherein the genomic sequences are represented by numerical sequences and hence the *FT* of this spatial-sequence of numerals is determined. The numerical values used thereof conform to the so-called *electron–ion interaction potential* (EIIP) values assigned to the nucleotides. Implicitly, such an EIIP-based numerical sequence leads to an ISM of transforming a genomic sequence for decomposition *via* Fourier series. This Fourier-transformed sequence determined from EIIP values inherently contains physicochemical details attached to the biological functions of the genomic structure. As such, elucidating the Fourier series of this numerically formatted genome (in terms of EIIP values) leads to detecting the code/frequency pairs that are specific to the genomic sequence *vis-à-vis* its biological profiles. This method is insensitive to the location of the motifs and therefore warrants no prior alignment of the sequence with its counterparts. (The EIIP values assigned for nucleotide bases were listed in an earlier section and relevant computational procedure of elucidating Fourier spectral details of a test sequence is presented in Rao and Swamy.[68])

Example Ex-8.5

Statement of the Problem

This exercise refers to identifying motif segments that indicate common features seen along 5′-end to 3′-end stretches of multiple (four) sequences.

Pertinent procedure conforms to four RNA sequences of DEN1, DEN2, DEN3, and DEN4. Outlined in Chatterjee *et al.*[31] and Chatterjee[83] are computational details toward applying entropy, energetics, and Fourier spectral methods and to ascertain the relevant set of motifs forming the fingerprint of four RNA sequences pertinent to DEN1, DEN2, DEN3, and DEN4. Also, indicated in Chatterjee *et al.*[31] and Chatterjee[83] is the procedure to determine the aligned and/or unaligned *AA* representation of a select set of seven motif sections that exhibit close (almost overlapping) degree of sequence conservation at each position of the multiple alignment. A list of translated regular expressions (obtained using PRATT) is shown in Table 8.3 with the associated fitness values of manually selected sets of motifs of the four multiple sequences of DEN1 to DEN4.

Fig. 8.14 depicts examples of two motifs constituting the fingerprint of the test multiple sequence set.

Finding CPs Among the Test Viral Strains

The method of ascertaining a set of proteins that may prevail most commonly in all the four viral strains using the selected set of motifs as above is described in Chatterjee *et al.*,[31] and Chatterjee.[83] Relevant procedure illustrates the method to find CPs across all the four viral strains using the selected set of motif sections. For each motif, the selected set of AA segments is subjected to BLAST search so as to ascertain the underlying homology and similarity.

Relevant *blastp* compares an *AA* query sequence against a protein sequence database, and protein sequences are obtained as output from this BLAST search with corresponding expected value (E-value) estimated for each *AA* sequence are listed. (A low *E*-value indicates that a score has high confidence level and all those sequences that post-high *E*-values are filtered out). Hence, the results are listed in Table 8.4 on probabilistically most common motif sections (denoted in terms of *AAs*) and possible proteins synthesized by them across all the four dengue serotypes.

In summary, the thematic objective of the narration above refers to cohesively applying three approaches namely, information-theoretic (entropy), thermodynamic-kinetics (energetics), and Fourier-spectral methods to DNA/RNA structures of a query virus (and its serovars) so as to obtain a comprehensively featured portal that identifies and classifies the distinguishable residue details buried across them. Currently, when this three-prong approach is specifically addressed to the serogroup of dengue virus, the resulting data obtained are listed in Table 8.4.

Table 8.3 Manually selected sets of motifs of the four multiple sequences of DEN1 to DEN4.

Range	*Aligned/ Unaligned*	*Fitness*	*Regular Expression*
104–1080	Aligned	131.0950	A-[AG]-[AG]-[AG]-[AC]-A-A-[AG]-C-C-[AC]-A-C-[ACG]-[CT]-T-G-G-A-[CT]-[AT]-T-[AT]-G-A-[AG]-C-T-[CGT]-[ACT]-[ACT]-[AGT]-A-A-[AG]-A
			DEN1: AAGACAAACCAACACTGGACATTGAACTCTTGAAGAcgga
			DEN2: aaAAAACAAACCAACATTGGATTTTGAACTGATAAAAAca
			DEN3: AGAACAAGCCCACGCTGGATATAGAGCTTCAGAAGAccga
			DEN4: gcccAGGGAAAACCAACCTTGGATTTTGAACTGACTAAGA
	Unaligned	79.2310	A-x(4)-A-A-x-C-C-x-A-C-x(2)-T-G-G-A-x(2)-T-x-G-A-x-C-T-x(4)-A-A-x-A
1301–1350	Aligned	136.9046	A-C-[AC]-[CG]-[CT]-[ACGT]-C-A-C-[AT]-[AC]-[AT]-G-G-[AG]-G-A-[AC]-[ACG]-[AC]-[ACG]-C-A-x(1,2)-C-A-x(0,1)-G-T-[ACGT]-G-G-A-A-A-T-G-A-[ACG]
			DEN1: tagtcACCGTACACACTGGAGACCAGCAc-CAaGTTGGAAATGAGacca
			DEN2: tgataACACCTCACTCAGGGGAAGAGCAtgCA-GTCGG AAATGACac
			DEN3: tcattACAGTGCACACAGGAGACCAACAc-CAgGTGGG AAATGAAacgc
			DEN4: ttgtaACAGTCCACAATGGAGACACCCAtgCA-GTAGG AAATGAC
	Unaligned	94.9112	A-C-x(4)-C-A-C-x(3)-G-G-x-G-A-x(4)-C-A-x(1,2)-C-A-x(0,1)-G-T-x-G-G-A-A-A-T-G-A
3001–3050	Aligned	123.1960	G-G-[ACT]-C-C-[AT]-[AGT]-T-[ACGT]-T-C [AT]-C-A-[AG]-C-A-C-A-A-[CT]-T-A-[CT]-[AC]-G-x(0,1)-C-C-[AC]-G-G-[ACG]-[CT]
			DEN1: atggaGGACCAATATCTCAGCACAACTACAGaCCAGGATatt
			DEN2: tcgctGGACCAGTGTCTCAACACAACTATAGaCCAGGCTa
			DEN3: tagctGGTCCTATCTCACAACACAACTACAGgCCCGGGTaccac
			DEN4: atgcgGGCCCTTTTTCACAGCACAATTACCG-CCAGGGC
	Unaligned	87.0711	G-G-x-C-C-x(2)-T-x-T-C-x-C-A-x-C-A-C-A-A-x-T-A-x(2)-G-x(0,1)-C-C-x-G-G

(Continued)

Table 8.3 (*Continued*)

Range	*Aligned/ Unaligned*	*Fitness*	*Regular Expression*
3601–3650	Aligned	115.7843	C-[ACT]-[ACT]-[AGT]-G-G-x(1,3)-C-x(1,3)-T-[ACG]-[AT]-C-[AC]-T-[GT]-[GT]-A-[AGT]-[AGT]-G-A-[CT]-[ACT]-T-[AG]-[ACG]-[CGT]-[ACG]-[AC]-[AG]-[ACG]-[ACG]-[CT]-[ACG]-[ACT]-[GT]-[ACGT]
			DEN1: tctcaCAATGGgaCaatTGACATGGAATGATCTGATCAGGCTATGTatca
			DEN2: ttgatCACAGGgaaCaTGTCCTTTAGAGACCTGGGAAGAGTGATGgt
			DEN3: CTCAGGg--CaaaTAACATGGAGAGACATGGCGCACACACTAataat
			DEN4: atcatCCTGGGaggCc--TCACATGGATGGACTTACTACGAGCCCTC
	Unaligned	43.8706	G-G-G-x(0,2)-A-x(2,4)-T-x(2)-C-x-T-x(2)-A-x(2)-G-A-x(2)-T
5541–5590	Aligned	148.7139	G-[ACGT]-T-C-[AG]-T-G-G-A-A-[CT]-[AT]-C-[AC]-G-G-x(2,3)-T-x(0,1)-G-A-[AC]-T-G-G-[AG]-T-[ACT]-A-C-[ACGT]-G-A-[CT]-T-[AT]-[CT]-[ACG]-[AC]-[AT]-G-G
			DEN1: tgaaaGATCATGGAACTCAGGctaT-GACTGGATCACTGATTTCCCAGGt
			DEN2: tgaacGTTCGTGGAATTCCGGacaT-GAATGGGTCACGGATTTTAAAGGga
			DEN3: GCTCATGGAATTCAGGcaaT-GAATGGATTACCGACTTCGCTGGgaaaa
			DEN4: ggaaaGGTCATGGAACACAGGgt-TcGACTGGATAACAGACTACCAAGG
	Unaligned	103.2513	G-x-T-C-x-T-G-G-A-A-x(2)-C-x-G-G-x(2,3)-T-x(0,1)-G-A-x-T-G-G-x-T-x-A-C-x-G-A-x-T-x(5)-G-G
9701–9750	Aligned	159.5583	T-T-T-C-x(0,1)-A-[CT]-[ACG]-A-[AG]-[ACT]-T-[ACG]-[AT]-T-[CT]-A-T-G-A-A-[AG]-G-A-[CT]-G-G-[ACGT]-[AC]-G-[ACG]-[AGT]-[ACT]-[AG]-[ACT]-T-[ACG]-G-T-[GT]-G-T-[GT]-C-C
			DEN1: caccaTTTCcACCAGCTGATTATGAAGGATGGGAGGGAGATAGTGGTGCC

Table 8.3 (*Continued*)

Range	*Aligned/ Unaligned*	*Fitness*	*Regular Expression*
			DEN2:
			accaTTTCcATGAGTTAATCATGAAAGACGGTCGCGTACTCGTTGT TCCa
			DEN3:
			TTTC-ATGAATTGATCATGAAAGATGGAAGAAAGTTGGTGG TTCCctgca
			DEN4:
			cTTTC-ACAAGATCTTTATGAAGGATGGCCGCTCACTAGTTGTTCC atgta
	Unaligned	103.7513	T-T-T-C-x(0,1)-A-x(2)-A-x(2)-T-x(2)-T-x-A-T-G-A-A-x-G-A-x -G-G-x(2)-G-x(5)-T-x-G-T-x-G-T-x-C-C
9751–9800	Aligned	135.4156	A-[CT]-G-A-A-C-T-[AGT]-[AG]-T-[AT]-G-G-[ACGT]-A-G-[AG]-G-C-[AC]-[AC]-G-A-[AG]-T-[AC]-T-C-[ACGT]-C-A-[AG]-G-G-[AC]-G
			DEN1: ccaagATGAACTTGTAGGTAGGGCCAGAGTATCACAAGGG
			DEN2: ccaagATGAACTGATTGGCAGAGCCCGAATCTCCCAAGGAGc
			DEN3: ccaggACGAACTAATAGGAAGAGCAAGAATCTCTCAAG GGcggga
			DEN4: ccaggATGAACTGATAGGGAGAGCCAGAATCTCGCAGG GGctgga
	Unaligned	95.9112	A-x-G-A-A-C-T-x(2)-T-x-G-G-x-A-G-x-G-C-x(2)-G-A-x-T-x-T-C-x-C-A-x-G-G-x-G

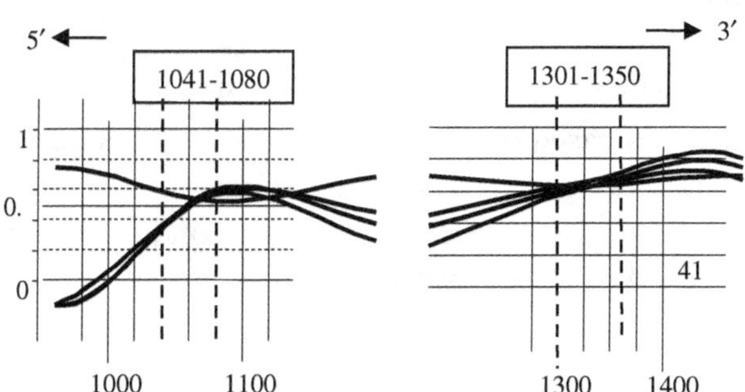

Fig. 8.14 Estimated by a cohesive combination of entropy, energetics, and Fourier spectral methods, a select set of motif segments (at base locations, 1041–1080, 1301–1350) across the multiple RNA sequences of four test serovars (in the entire 5′-to-3′) stretch exhibit almost similar feature details (measured ≥0.6 in the scale 0–1). Similar verifiable results are observed along the following segments at base sites: 3001–3050, 3601–3650, 5541–5590, and 9701–9750–9800 as indicated in Chatterjee *et al.*[31] and Chatterjee.[83]

Viable Use of the Study as Above

The gene expression in a virus morphs to different patterns at the molecular (DNA/RNA) level across its different serotypes. Such discernment features in single viral diversity indicate the need for distinct vaccine designs *vis-à-vis* differentiable pathology of the strains concerned. Diverse vaccines can be synthesized by considering the distinct underlying DNA signature features of each serotype of a given virus. For example, relevant DNA feature can be determined in terms of the expression seen in each viral DNA/RNA structure observed as: (i) CDS, CpG, TATA box features; (ii) sites of homology specified by the spatial-spectrum (in Fourier domain); (iii) long-range correlation of coding/noncoding segments; and (iv) individual nearest-neighbor energetic-interactions plus stability-seeking bends/loop formation (as in the case of single-strand DNA or in RNA sequences).

The aforesaid details refer to a motivated interest in seeking a strategy toward comprehending the inner details of genomic information *via* a framework of multiple analyses applied simultaneously on a test sequence. Hence, a three-prong approach based on entropy, energetics, and Fourier-transform methods is advocated and applied to RNA sequences of dengue viral strains. The collective details obtained thereof indicate most probable set of CP-translated motif sections among the four serovar sequences in question. These can be viably used as indicators or biomarkers in the efforts of finding a common vaccine for the serotypes in question, coping with the diversity across a single virus.

(*Disclaimer*: the list of proteins, motif sections, etc. in Table 8.4 indicated as possible useful entities in designing a common vaccine for dengue serovars is based on analytical methods, computational procedures, and various assumptions specified in making selective ensembles of data as necessary. No wet-lab studies were conducted to supplement or cross-validate the details furnished.)

8.8 Viral Omics and Vaccine Design

Preventing a disease proactively is of utmost interest in medical field. Specifically, preventing epidemics is a global concern in public health. A way to prevent the proliferation of a disease in the society is to administer the susceptible individuals with what is known as the "vaccine." Such vaccines can protect both, the people who receive them as well as those with whom they may come in contact.

Table 8.4 Probabilistically most common motif sections (denoted in terms of AAs) and possible proteins synthesized by them across all the four dengue serotypes.

Identified motif section: highly probable AA sequences translating into proteins	*Translated protein type from the AA sequence*	*Basis of selection priority*	*Typically present in*
KPTLDIELMK	Envelope protein [dengue virus]	1	Dengue virus
KPTLDFELMK	Polyprotein [dengue virus]	1	Dengue virus
TVHTGDQCK	Polyprotein [dengue virus 1 and 3]	1	Dengue virus
TVHTGDQRK	Polyprotein [dengue virus 1 and 3]	1	Dengue virus
SQHNYRPG	Polyprotein [dengue virus 1 and 2]	1	Dengue virus
QDELIGRARISQG	Polyprotein [dengue virus 2, 3 and 4]	1	Dengue virus
FHELIMKDGRELVV	Polyprotein [dengue virus 1 and 3]	1	Dengue virus
FHELIMKDGSELVV	Polyprotein [dengue virus 1, 2 and 3]	1	Dengue virus
QTNIGF*	FER-1-like protein 5 [*Homo sapiens*]	2	Human protein
QTNTGF*	Nuclear pore complex protein Nup98-Nup96 isoform 1[*Homosapiens*]	2	Human protein
SQNLTR	Dual specificity phosphatase 22 variant[*Homo sapiens*]	2	Human Protein
SQNLSR	Chain A, crystal structure of The human phosphatase (Dusp9)	2	Human Protein
SQNLAR	Unnamed protein product [*Homo sapiens*]	2	Human Protein

Note: (a) The analysis pursued also indicates existence of other 23 possible motifs in the four dengue viral sequences. But they are not observed in the BLAST search to express proteins prevalent either in dengue viral strains and/or in human. As such, they are not listed here as viable entities for vaccine design considerations on the dengue viral strains.
(b) Selection priority: the priority 1 or 2 indicated refers to the order of observations with low-end of E-values seen in BLAST tool results.
(c) The results posted are obtained by using the protein BLAST (blastp) option of the BLAST tool.

A vaccine is any preparation used as a preventive introduction of an antigen (such as microorganism or an agent pertinent to the disease) into a host organism to confer plausible immunity against a specific disease. Employed thereof is an innocuous form of a disease agent like a killed or weakened bacteria or virus that can stimulate antibody productions. Vaccine antigens by themselves are not strong enough to cause a disease, but they still are potent enough to make the immune system of the host to produce antibodies against the invading pathogens. In short, the goal of any vaccination is to induct an appropriate and effective immune response in the vaccinated person. However, it is still not so clear what precisely constitutes an effective immune response for a variety of diseases so that an appropriate vaccine design can be attempted toward the prevention of the disease.

The gene expression in a virus morphs to different patterns at the molecular (DNA/RNA) level across its different strains. These discernment features offer a viable opportunity to conceive a set of distinct vaccine designs usable to prevent the differentiable pathology likely to be caused by the strains of an invading virus. In this study, it is hypothesized such diverse vaccines can be intelligently synthesized by considering the underlying DNA signature features of various strains of a given virus. Essentially, the unique expressions seen in each viral DNA/RNA structure as regard to its CDS, CpG, TATA box, etc., and sites of homology can be candidates of interest in vaccine design applications. These entities can be ascertained by a set of bioinformatic computational efforts, such as (i) spatial-spectrum (Fourier domain) details of the genomic structure; (ii) long-range correlation of coding/noncoding segments of the DNA/RNA sequences and NN energetic-interactions of the nucleotides and related stability-seeking bend/loop formations (in the case of single-strand DNA or in RNA sequences). For example, relevant computed details on the distinguishable features pertinent to the RNA structures of dengue-virus serovars can be indicated that can be used in conceiving a smart/rational vaccine design.

Immunity: An Overview

As mentioned earlier, the immune response of a host exposed to pathogens is the result of a series of biochemical reactions in the host's body that provides a protection mechanism for the body against potentially dangerous pathogens and/or foreign substances. Two types of immunity exist, innate and acquired. Innate immunity is always present and consists of an intricate system by which the infection is recognized with the production of antimicrobial/viral activities. In addition, concurrent recruitment of neutrophils and other phagocytic cells at the site of infection would occur, so as to kill or neutralize the invading pathogens.[130–133] Such activities are triggered by the presence of cell surface and a set of internal receptors (what are known as toll-like receptors, or TLRs) that recognize certain pathogen-associated molecular patterns present inside and/or on the surface of pathogens. Acquired immunity is induced in response to the invading pathogens, but it is dependent on the innate immune system to facilitate the presentation of pathogenic antigens in order to trigger either the production of antibodies or stimulating the cellular immunity.[134,135]

Antigens are particles or protein molecules on the surface of pathogens that induce production of antibodies by the immune system. To counter the influences of antigens, the cellular part of immune system in the host has certain types of entities like white blood cells (WBCs) called *lymphocytes* (which in turn consist of B-cell and T-cell) and antibodies. The cells of *B* series (including plasma cells) are responsible in producing the antibodies, and the antibodies functionally get attached to a specific antigen and make it easier for the immune cells to destroy them.

The types of epitopes are illustrated in Fig. 8.15a. The epitope can be continuous or discontinuous as indicated. The biochemical reactions during immune response produce and select particular epitopes from the antigenic material and/or from the antigen-presenting cell (APC). The epitope essentially is a peptide that can be recognized by a T-cell and hence elicit an immune response against the foreign body. This T-cell-based immunity is essential for the induction of long-term protective immunity against infectious disease agents.[136–138]

The paratope part of the antibodies binds to the epitope by a lock-and-key mechanism.[139] The lock-and-key feature of paratope-epitope structure is illustrated in Fig. 8.15b. The epitope can be continuous or discontinuous[140] as indicated. Once *B*-cells and *T*-cells are formed, a few of these would multiply and provide some details to be preserved as a "memory" in the immune system[141] about the invading pathogen. This memorized state would allow the immune system to respond faster and act efficiently when the body is exposed subsequently to the same

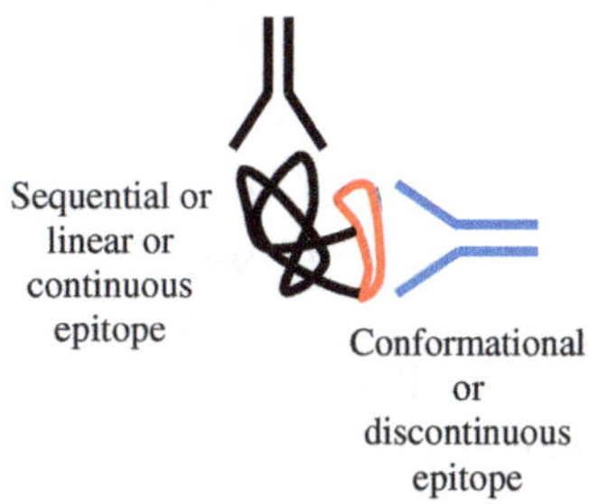

Fig. 8.15a Types of epitopes

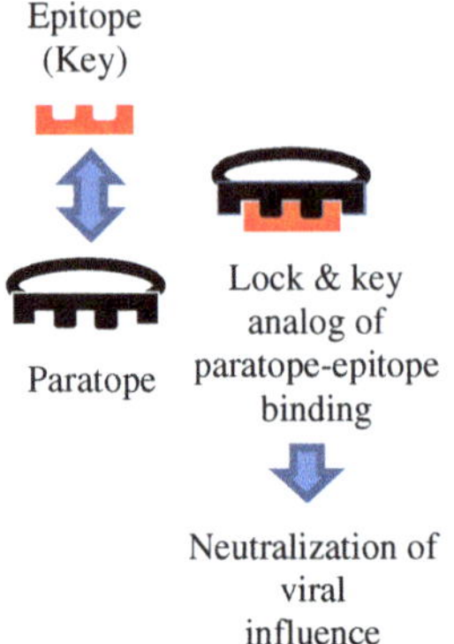

Fig. 8.15b Lock-and-key mechanism of epitope–paratope conjunction.

antigen, and such actions will prevent the host from getting infected and becoming sick due to the invading pathogen or at least the underlying mechanism would minimize the severity of the related infection. The recourse of immune response in a host who acquired immunity to a certain pathogen is shown in Fig. 8.16.

Vaccine Designs: An Overview

Consistent with the immunity response considerations as indicated above, a vaccine as defined earlier is an antigen that prepares an immune system for future protection against some pathogen without causing severe symptoms in the host. Vaccines can be of various forms, such as an organism (bacteria or virus), a protein (or peptide), or a nucleic acid sequence.[142,143] Thus, the goal of vaccine design is to create an artificial means to produce immunological memory of a particular pathogen without the risk of developing the disease in the host and offer a resistance against future possible infections.

The science of *vaccinology* started with Edward Jenner's systematic investigations into the protective effect of cowpox against smallpox (as mentioned earlier) in the late eighteenth century.[144,145] However, the momentum was provided by Louis Pasteur[146,147] for further research on vaccine development. Empirically vaccines were designed on the basis of *Pasteur's principle*: "isolate, inactivate, and inject,"[148] and it is being practiced till date. In essence, it was surmised by Pasteur that, in order to conceive a vaccine against a particular disease, the pathogen responsible for causing it has to be first "isolated" or separated from the other pathogens not causing the disease. Unlike Jenner, who used the cowpox

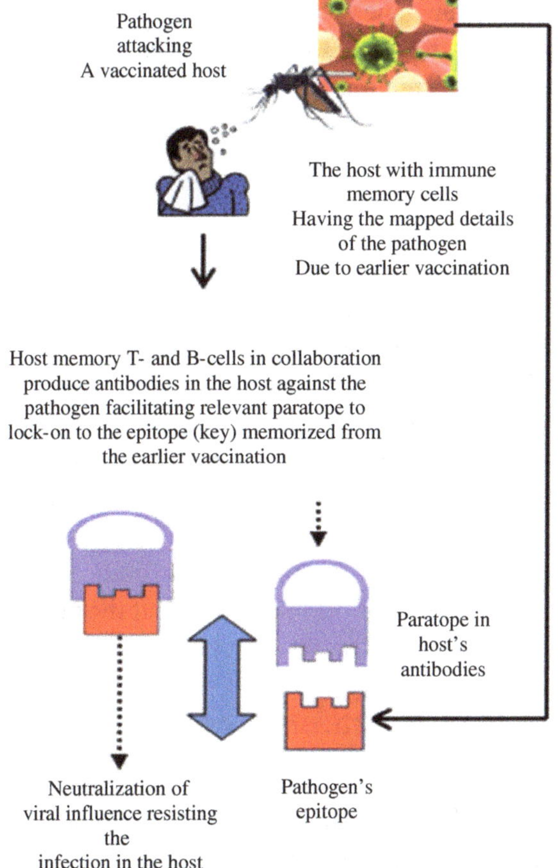

Fig. 8.16 Immune response generated in a vaccinated person's body (host) to a pathogen.

virus, which is similar to the more virulent smallpox virus (but not causing serious disease), Pasteur suggested that a weakened version of the disease causing pathogens "generated artificially" could also be effective as a vaccine; as such, a naturally weak form of the disease organism is not needed. Thus, by "inactivating" the pathogen by creating an organism (or part of it), and avoiding thereby a full-scale disease, the antigens are retained responsible for inducing the same immune response in the host as would had been caused by the unweakened form of the pathogen. This weakened form of the disease causing pathogen is then "injected" into the host for vaccinating the host.

The traditional vaccine design approach is to derive all possible vaccines from a protein sequence and test each of them for an immune response experimentally. The vaccines are designed in specialized laboratories and tested *in vitro*. However, this strategy is not only expensive, time-consuming, and often unpredictable but also it may fail to produce effective vaccine solutions against certain pathogens, especially those with very high antigen variations.[149] To reduce the high costs associated with traditional vaccine development, *in silico* based methods of vaccine design were ushered in and they are proving to be useful. The *in silico* concept of vaccine design uses computer-based algorithms to predict possible candidate for vaccine without using live cultures of pathogens in the laboratory. With the advent of modern technologies, bioinformatic tools and databases have been developed for proteomics, comparative genome analyses, and interpretation of the whole-genome sequences. Relevant computer-based techniques conceived for designing or predicting epitopes for effective vaccine design is becoming popular. They can be used to assign putative gene functions to each open-reading frame (ORF) on the basis of homology to known proteins. Systematic identification of potential antigens of a pathogen using this information without the need for cultivation of the pathogen is termed as "reverse vaccinology."[150]

Live, Attenuated, and Inactivated Vaccine

As mentioned above, the live, attenuated vaccines consist of whole, attenuated bacterial or viral particles. Relevant attenuation is achieved by growing generations of the pathogens in cells in which they do not reproduce very well. As they evolve to adapt to the new environment, they become weaker with respect to their natural host, namely, human beings. Its major advantage is in the fact that they contain the entire possible antigenic spectrum as that of the pathogen; as such, they can induce most effective and long-lasting immunity against the infection. However, on rare occasions, and especially in immune-compromised or immune-deficient host, such pathogens introduced through vaccines may prove to be fatal. Thus, this conventional method, while successful against some pathogens (especially against certain viruses like chicken pox, mumps, etc.), fail to provide a solution for many of those pathogens (especially the bacteria that may contain thousands of genes) for which a vaccine is not yet available.

Subunit and Conjugate Vaccines

Subunit vaccines include only the epitopal or antigenic part that best stimulate the immune system. As they do not contain all the other molecules that make up the pathogen, the chances of adverse reactions to the vaccine are somewhat lower. Hepatitis *B* vaccine is an example of subunit vaccine.

Conjugate vaccines versions depict a special type of subunit vaccine used for immunization against bacteria that possesses an outer coating of sugar molecules called *polysaccharides.* A polysaccharide coating on bacterial surface hides the antigens, so that the immature immune systems of infants and younger children will not be able to recognize or respond to them. Hence, while preparing conjugate vaccines, the antigenic material inside the polysaccharide-coating of the bacteria is replaced by an antigen that can be recognized by an infant's or young child's immune system. Thus, relevant immune system responds to the polysaccharide-coatings and defends against the disease-causing bacteria.

DNA and Recombinant Vector Vaccines

DNA vaccines: immunization with a "naked" DNA was introduced in late 1990s with highly reduced and/or eliminated disadvantages associated with conventional vaccines designed using live attenuated or inactivated pathogen.[151] The underlying procedure involves direct inoculation of plasmid DNA encoding sequences of viral proteins that results in the synthesis of proteins causing immune responses in the host. This version of DNA immunization is economical to produce, carries heat stability, amenable to genetic manipulation, mimics viral infection, and bears no risk of "reversion to pathogenicity." However, concerns are indicated as regard to their safety, for example, the possible integration of plasmid DNA into host chromosomes.[152]

Recombinant vector vaccines are experimental vaccines (similar to DNA vaccines). They use an attenuated virus or bacteria in introducing the antigenic DNA to cells of the host's body. The associated description "vector" implies that the virus or bacterium used is the carrier. This version of DNA vaccines does not use DNA as the immunogen,[142,143] and just a critical DNA sequence is inserted in plasmid in order to fabricate sufficient amount of protein (epitopal) copies for a given epitope. This is the crucial distinction between the two concepts. A detailed review of the history of vaccinology is provided in Bramwell and Perrie.[153]

Toxoid Vaccines

Toxoid vaccines are used in cases where the pathogen itself is harmless and toxins produced by it causes diseases. Such vaccines contain modified toxin produced by the pathogen, which initiates an immune response in the vaccinated host. When the immune system of the host receives this "detoxified vaccine," it learns to fight off the natural toxin of the pathogen. Vaccines used against diphtheria and tetanus are examples of toxoid vaccines.

Rational Design of Vaccines (RVD)

The so-called *reverse vaccinology* has become popular and gained acceptance[154] across different ways of conceiving vaccine designs in laboratories. Supplemented by technological breakthroughs in molecular biology with the generation of extensive data and research details at genomic or proteomic levels, the new art of *immune-informatics*[155] has emerged; pertinently, computational research methods (which are mostly faster than biological experiments) with reduced time-factor required in the identification of critical candidate vaccines, enable new solutions for conceiving vaccines that were considered once impossible to develop. The techniques toward this new discipline of such *rational vaccine design* (RVD) are (i) *epitope-mapping** and (ii) *reverse vaccinology***. (*Epitope-mapping refers to the process of identifying experimentally the binding site, (or *epitope*), of an antibody on its target antigen (usually, on a protein).

Finding and characterizing such antibody binding sites help in the discovery and development of new therapeutics, vaccines, and diagnostics. Experimental epitope-mapping data can be combined with robust algorithms so as to facilitate *in silico* prediction of *B*-cell epitopes based on sequence and/or structural data.

(**Reverse vaccinology is an improved scheme on vaccinology based on bioinformatic considerations. Rappuoli initially adopted it against *Serogroup B meningococcus*, and subsequently, it has been used on other bacterial vaccines.)

RVD makes use of computers for large-scale screening of potential vaccines solely on the basis of genomic and proteomic information.[156] Adopted thereof are intelligent algorithms and specific tools for setting up a relevant standardized approach. Furthermore, modern possibility of using genomic information allows vaccine development *in silico*, without the need of cultivating the pathogen. The whole concept has been summarized in Fig. 8.17. Various computational approaches and algorithms have been proposed and used for recognition/design of possible epitopal candidates and reverse vaccinology.[157–163]

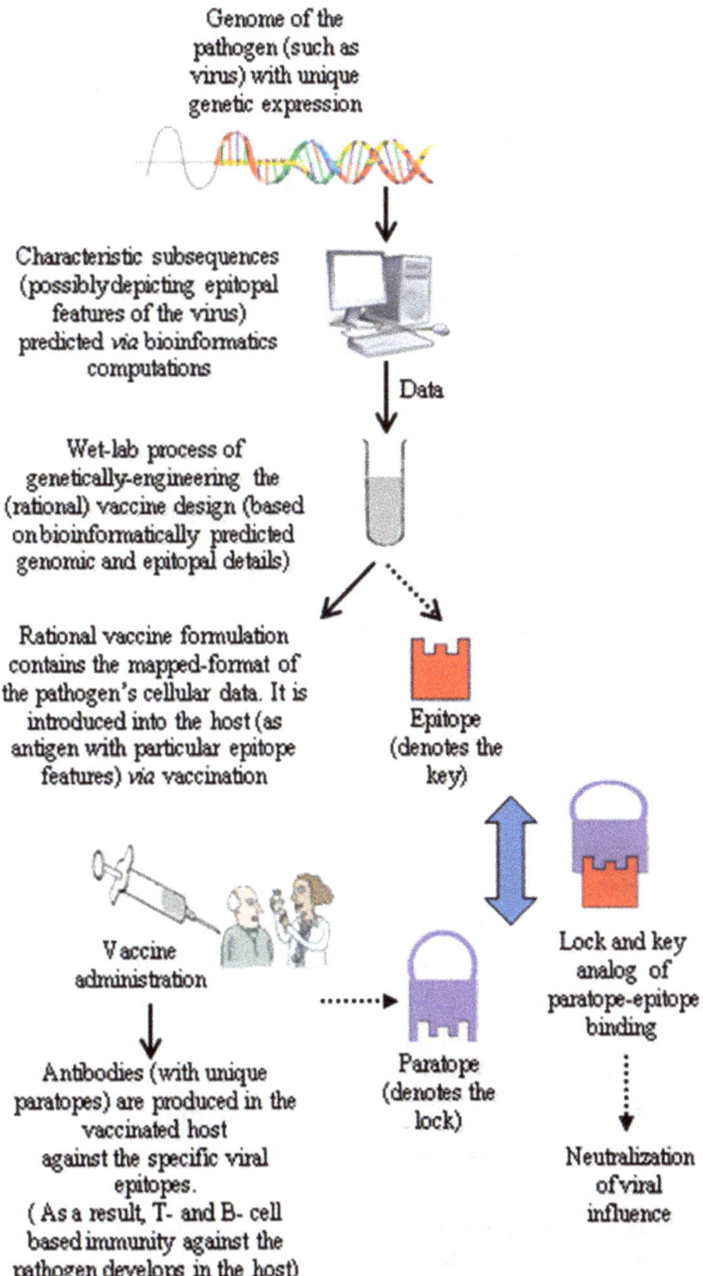
Genome of the pathogen (such as virus) with unique genetic expression
Characteristic subsequences (possibly depicting epitopal features of the virus) predicted via bioinformatics computations
Data
Wet-lab process of genetically-engineering the (rational) vaccine design (based on bioinformatically predicted genomic and epitopal details)
Rational vaccine formulation contains the mapped-format of the pathogen's cellular data. It is introduced into the host (as antigen with particular epitope features) via vaccination
Epitope (denotes the key)
Vaccine administration
Lock and key analog of paratope-epitope binding
Paratope (denotes the lock)
Antibodies (with unique paratopes) are produced in the vaccinated host against the specific viral epitopes.
(As a result, T- and B- cell based immunity against the pathogen develops in the host)
Neutralization of viral influence

Fig. 8.17 Concept of rational design of vaccine.

Tutorial 8.5

A Tutorial Note on Zika Virus: Vaccine for Zika Virus Strains

As indicated earlier, for a particular pathogen (such as virus), which may prevail in different forms of serotypes (as, for example, in the case of four dengue viral strains) each strain may be having genomic variability as well as common genomic features with reference to others. As such, finding relevant genomic information of a serogroup may be useful in knowing epitopal details for designing unique vaccines for the immunity across multiple serotypes.

Zika virus (ZIKV)[164–166]: ZIKV is a member of the virus family *Flaviviridae* and the genus *Flavivirus*; thus, it is related to the dengue, yellow fever, Japanese encephalitis, and West Nile viruses. Being similar to other *Flaviviruses*, ZIKV is enveloped and icosahedral. It has a non-segmented, single-stranded, 10-kb and positive-sense RNA genome. Studies show that six hours after cells are infected with the ZIKV, the vacuoles and mitochondria in the cells begin to swell. This swelling becomes so severe, it may cause cell death (known as *paraptosis*). This programmed cell-death requires gene expression.

Problem P-8.4

As indicated above, it is planned at NIH/NIAID to add *rZIKV/D4Δ30-713 Zika* component to the dengue vaccine candidate so as to create and evaluate a single vaccine that would protect against both Zika and all four dengue viruses. Hence, following the suite of genomic analyses on dengue strains (DEN1 through DEN4) toward vaccine design, consider the genomic sequence details of ZIKV and perform a sequence comparison study *vis-à-vis* dengue viral serotypes. The goal of such a study is to design or suggest independently a potential strategy illustrated toward the development of a vaccine against ZIKV and all the four serotypes of dengue virus. This study could be comparable and parallel with rZIKV/D4Δ30-713 related to on-going exercises. Possibly, the study can be extended to include yellow fever, Japanese encephalitis, West Nile viruses, and Covid-19 virus.

Problem P-8.5

This exercise tries to make a comparative genomic analysis on leprosy and tuberculosis bacteria. The following is a brief tutorial note on these bacterial pair and a relevant exercise is then indicated.

Leprosy and tuberculosis are diseases caused, respectively, by *Mycobacterium leprae* and *Mycobacterium tuberculosis.* With the advent of antibiotic-resistant strains of tuberculosis, research into *mycobacteria* has been focused in combating these mutants of ancient pathogens.[167–169] The genomes of *M. tuberculosis* and *M. leprae* have been sequenced in order to ascertain how to defeat these infamously successful pathogens.

The genome of *M. tuberculosis* is 4,411,522 base pairs long with 3,924 predicted protein-coding sequences and a relatively high $G + C$ content of 65.6%. At 4.4 Mbp, *M. tuberculosis* is one of the largest known bacterial genomes, coming in just short of *E. coli,* and a distant third to *Streptomyces coelicolor.*

The genome of *M. leprae* is 3,268,203 base pairs long, with only 1,604 predicted protein-coding regions, and a $G + C$ content of about 57.8%. Only 49.5% of the *M. leprae* genome contains ORFs (protein-coding regions), the rest of the genome is comprised of pseudogenes, which are inactive reading frames with recognizable and functional counterparts in *M. tuberculosis* (27%), and regions that do not appear to be coding at all and may be gene remnants mutated beyond recognition (23.5%). Of the genome of *M. tuberculosis,* 90.8% of the genome contains protein-coding sequences with only 6 pseudogenes, compared to the 1,116 pseudogenes on the *M. leprae* genome. *Mycobacteria* are rod-shaped, Gram-positive aerobes, or facultative anaerobes.

Problem statement: consistent with the details indicated in this chapter, a conceivable project exercise is to perform a comparative sequence analysis of the species: *M. tuberculosis* and *M. leprae* so as to deduce the common features of their coding segments *via* entropy, energetics, and Fourier spectral techniques outlined earlier (with reference to the viral sequences). The details obtained thereof could be useful toward realizing a common vaccine for leprosy and tuberculosis. A study in this direction is indicated in: Rivero,[170] which can be expanded in this exercise. Pertinent sequence data on the test species are to be first accessed and analyzed by perusing the methods suggested in this chapter.

Problem P-8.6

Problem Statement

A study exercise can be on an example virus of current interest (such as ZIKV.[164]) As described earlier for dengue virus, necessary genomic studies can be done on the test virus, and vaccine design considerations can be formulated.

8.9 Closure

Inferences on distinct as well as common genomic features extracted by three different techniques on viral sequences as described in this chapter are annotatable with viably possible vaccine design applications. That is, finding genomic details uniquely present in a serogroup of a test virus is useful in understanding related information required in vaccine designs that are compatible in ascertaining human/animal immunity across such viral diversity. In pertinent cases, comparison of genomes (*via* multiple strategies such as entropy, energetics, and Fourier spectral methods) of serovars of a virus and decide common and differential genomic details is pragmatically feasible and useful in relevant vaccine designs as described in this chapter.

References

[1] C. P. McKay: What is life-and how do we search for it in other worlds? *Public Library of Science Biology,* 2004, vol. 2(9), e203.

[2] P. Forterre: Defining life: The virus viewpoint, *Origins of Life and Evolution of the Biosphere,* 2010, vol. 40(2), 151–160.

[3] K. H. Nealson and P. G. Conrad: Life: Past, present and future, *Philosophical Transactions of the Royal Society of London. (Series B: Biological Sciences),* 1999, vol. 354, 1923–1939.

[4] H. R. Gelderblom: Structure and classification of viruses. In S. Baron (Ed.), *Medical microbiology,* Chapter 41, University of Texas Medical Branch at Galveston, Galveston, TX, USA: 1996.

[5] A. R. Gould: Virus evolution: Disease emergence and spread. *Australian Journal of Experimental Agriculture,* 2004, vol. 44(11), 1085–1094.

[6] B. Roizman: Multiplication. In S. Baron (Ed.), *Medical Microbiology,* Chapter 42, University of Texas Medical Branch at Galveston, Galveston, TX, USA: 1996.

[7] G. Sumbali and R. Mehrotra: *Principles of Microbiology,* Tata-McGraw Hill Education Pvt. Limited, New Delhi, India: 2009, pp. 158.

[8] D. White and F. Fenner: *Medical Virology,* Academic Press, San Diego, CA, USA: 1994.

[9] C. I. Bândea: A new theory on the origin and the nature of viruses, *Journal of Theoretical Biology,* 1983, vol. 105(4), 591–602.

[10] F. Holmes: Problems of viral nomenclature and classification. *Annals of the New York Academy of Sciences*, 1953, vol. 56, 414–421.
[11] A. Lwoff, R. Horne and P. Tournier: A system of viruses. *Proceedings of Cold Spring Harbor Symposia on Quantitative Biology,* (Spring Harbor, NY, June 7–13, 1962), vol. 27, pp. 51–55.
[12] International Committee on Taxonomy of Viruses. Available online at: *http://ictvonline.org/virusTaxonomy.asp?version=2009.*
[13] D. Baltimore: Expression of animal virus genomes. *Bacteriological Reviews,* 1971, vol. 35(3), 235–241.
[14] J. R. Stephenson: Rational design of vaccines against enveloped RNA viruses. *Vaccine,* 1985, vol. 3(1), 69–72.
[15] H. M. Blau and M. Springer: Gene therapy–A novel form of drug delivery. *New England Journal of Medicine,* 1995, vol. 333, 1204–1207.
[16] R. Curtiss, III: The impact of vaccines and vaccinations: Challenges and opportunities for modelers. *Mathematical Biosciences and Engineering.* 2011, vol. 8(1), 77–93.
[17] J. A. Mumford: Vaccines and viral antigenic diversity. *Scientific and Technical Review of the Office International des Epizooties (Paris),* 2007, vol. 26(1), 69–90
[18] J. T. Patton: *Segmented Double-stranded RNA Viruses: Structure and Molecular Biology,* Caister Academic Press, Norfolk, UK: 2008.
[19] J. Holland, K. Spindler, F. Horodyski, E. Grabau, S. Nichol and S. VandePol: Rapid evolution of RNA genomes. *Science,* 1982, vol. 215(4540), 1577–1585.
[20] R. Kurth and N. Bannert (Eds.), *Retroviruses: Molecular Biology, Genomics and Pathogenesis,* Caister Academic Press, Norfolk, UK: 2010.
[21] C. Seeger and W. S. Mason (Eds.), *Hepadnaviruses: Molecular Biology and Pathogenesis,* Springer-Verlag, Berlin, Germany: 1991.
[22] K. C. Chen, J. J. Tyson, M. Lederman, E. R. Stout, and R. C. Bates: A kinetic hairpin transfer model for parvovirus DNA Replication. *Journal of Molecular Biology,* 1989, vol. 208(2), 283–296.
[23] Human parvovirus B19: complete genome. Available online at: *http://www.ncbi.nlm.nih.gov/nuccore/356457872.*
[24] Dengue virus 1: Complete genome. Available online at: *http://www.ncbi.nlm.nih.gov/nuccore/9626685.*
[25] Dengue virus 2: Complete genome. Available online at: *http://www.ncbi.nlm.nih.gov/nuccore/158976983.*
[26] Dengue virus 3: Complete genome. Available online at: *http://www.ncbi.nlm.nih.gov/nuccore/163644368.*
[27] Dengue virus 4: Complete genome. Available online at: *http://www.ncbi.nlm.nih.gov/nuccore/12084822.*
[28] N. G. Iglesias, and A. Gamarnik: Dynamic RNA structures in the dengue virus genome. *RNA Biology,* 2011, vol. 8(2), 249–257.

[29] NIH/National Human Genome Research Institute: An overview of the human genome project. Available online at: *https://www.genome.gov/12011238/an-overview-of-the-human-genome-project/.*

[30] S. Tripp and M. Grueber: *Economic Impact of the Human Genome Project.* A Report: Battelle Memorial Institute, HQ-Columbus, Ohio, USA: 2011.

[31] S. Chatterjee, P. S. Neelakanta, and M. Pavlovic: A cohesive analysis of DNA/RNA sequences *via* entropy, energetics and spectral-domain methods to assess genomic features across single viral diversity, *International Journal of Bioinformatics Research and Applications,* vol. 11(4), 2015, 281–307.

[32] S. B. Pandya: Binary Representation of DNA Sequences towards Developing Useful Algorithms in Bioinformatics. *MSEE Thesis,* Fall 2003, Florida Atlantic University, Boca Raton, FL, USA.

[33] P. S. Neelakanta, S. Chatterjee, M. Pavlovic, A. Pandya, and D. De Groff: Fuzzy splicing in precursor-mRNA sequences: Prediction of aberrant splice-junctions in viral DNA context. *Journal of Biomedical Science and Engineering,* vol. 4, 2011, 270–279.

[34] T. V. Arredondo, P. S. Neelakanta, and D. De Groff: Fuzzy attributes of a DNA: A fuzzy inference engine for codon- "junk" codon delineation. *Artificial Intelligence in Medicine,* 2005, vol. 35, pp. 87–105.

[35] P. Bernaola-Galván, I. Grosse, P. Carpena, J. L. Oliver, R. Román-Roldán, and H. E. Stanley: Finding borders between coding and noncoding DNA regions by entropic segmentation method. *Physical Review Letters,* 2000, vol. 85, 1342–1345.

[36] R. Román-Roldán, P. Bernaola-Galván, and J. L. Oliver, "Application of information theory to DNA sequence analysis: A review," *Pattern Recognition,* vol. 29, pp. 1187–1194, 1996.

[37] P. F. Baisnee, S. Hampson, and P. Baldi: Why are complementary DNA strands symmetric? *Bioinformatics,* 2002, vol. 18(8), 1021–1033.

[38] P. S. Neelakanta, S. Chatterjee, G. A. Thengum-Pallil: Computation of entropy and energetics profiles of a single-stranded viral DNA. *International Journal Bioinformatics and Applications,* vol. 7(3), 2011, 239–261.

[39] D. C. Benson: Digital signal processing methods for biosequence comparison, *Nucleic Acids Research,* 1990, vol. 18(10), 3001–3006.

[40] Images for structure of a virus. Available online at: *www.sciencelearn.org.nz/.*

[41] J. Summers and W. S. Mason: Replication of the genome of a hepatitis B-like virus by reverse transcription of an RNA intermediate. *Cell,* 1982, vol. 29, 403–415.

[42] J. Sohn, S. Litwin, and C. Seeger: Mechanism for CCC DNA synthesis in hepadnaviruses. *PLoS ONE,* 2008, vol. 4(11), e8093 (6 pages).

[43] C. W. Hilbers, H. A. Heus, M. J. Van Dongen, and S. S. Wijmenga: The hairpin elements of nucleic acid structure: DNA and RNA folding. In F. Eckstein and D. M. J. Lilley (Eds.), *Nucleic Acids and Molecular Biology*, Springer-Verlag, Berlin, Germany, 1994, pp. 56–104.
[44] E. Costello, R. Sahli, H. Bernhard, and P. Beard: The mismatched nucleotides in the 59-terminal hairpin of minute virus of mice are required for efficient viral DNA replication. *Journal of Virology*, 1995, vol. 69(12), 7489–7496.
[45] K. Chin, F. Chen, and S. Chou: Solution structure of the ActD–5prime-CCGTT3GTGG-3prime complex: Drug interaction with tandem G•T mismatches and hairpin loop backbone. *Nucleic Acids Research*, 2003, vol. 31(10), 2622–2629.
[46] K. C. Chen, J. J. Tyson, M. Lederman, E. R. Stout and R. C. Bates: A kinetic hairpin transfer model for parvoviral DNA replication. *Journal of Molecular Biology*, 1989, vol. 208(2), 283–296.
[47] Human parvovirus B19, complete genome. Available online at: *http://www.ncbi.nlm.nih.gov/nuccore/356457872.*
[48] Taxonomy Dengue virus 1. Available online at: *https://www.uniprot.org/taxonomy/11053.* (Complete genome in: [II.3.24]
[49] Taxonomy Dengue virus 2. Available online at: *https://www.uniprot.org/taxonomy/11060.* (Complete genome in: [II.3.25]
[50] Taxonomy Dengue virus 3. Available online at: *https://www.uniprot.org/taxonomy/11069.* (Complete genome in: [II.3.26]
[51] Taxonomy Dengue virus 4. Available online at: *https://www.uniprot.org/taxonomy/11070.* (Complete genome in: [II.3.27]
[52] J. A. Mumford: Vaccines and viral antigenic diversity. *Revue scientifique ettechnique (International Office of Epizootics)*, 2007, vol. 26(1), 69–90.
[53] W. Resch, L. Zaslavsky, B. Kiryutin, M. Rozanov, Y. Bao, and T. A. Tatusova: Virus variation resources at the national center for biotechnology information: dengue virus *BMC Microbiology*, 2009, vol. 9, 65.
[54] Dengue. Available online at: *https://www.cdc.gov/dengue/index.html.*
[55] D. E. Alvarez, M. F. Lodeiro, S. J. Luduena, L. I. Pietrasanta and A. V. Gamarnik: Long-range RNA-RNA interactions circularize the dengue virus genome. *Journal of Virology*, 2005, vol. 79(11), 6631–6643.
[56] R. Curtiss, III: The impact of vaccines and vaccinations: Challenges and opportunities for modelers. *Mathematical Biosciences and Engineering.* 2011, vol. 8(1), 77–93.
[57] A. R. Gould: Virus evolution: Disease emergence and spread. *Australian Journal of Experimental Agriculture*, 2004, vol. 44(11), 1085–1094.
[58] R. A. Fisher: The use of multiple measurements in taxonomic problems: *Annals of Eugenics*, 1936, vol. 7, 179–188.

[59] C. E. Shannon and W. W. Weaver. The Mathematical Theory of Communication, University of Illinois Press, Urbana, USA: 1949.

[60] C. E. Shannon: Transmission of information, *The Bell System Technical Journal*, 1948, vol. 27, 379–423.

[61] P. S. Neelakanta. *Information-Theoretic Aspects of Neural Networks*, CRC-Press, Boca Raton, FL, USA: 1999.

[62] J. N. Kapur: Measures of uncertainty in mathematical programming and physics. *Journal of the Indian Society of Agricultural Statistics, 1972*, vol. 24, 47–66.

[63] D. C. Benson: Fourier methods for biosequence analysis. *Nucleic Acids Research*, 1990. vol. 18(21), 6305–6310.

[64] E. A. Cheever, G. C. Overton, and D. B. Searls: Fast Fourier transform-based correlation of DNA sequences using complex plane encoding. *Computer Applications in the Biosciences*, 1991, vol. 7(2), 143–154.

[65] Y. Zhou, L. Zhou, Z. Yu, and V. Anh: Distinguish coding and noncoding sequences in a complete genome using fourier transform. *Proceedings of the Third International Conference on Natural Computation (*August 24–27 2007), Haikou, Hainan, China (*ICNC 2007)*, pp. 295–299.

[66] D. Anastassiou: Genetic data processing, *IEEE Signal Processing*, 2001, vol. 18(4), 8–20.

[67] D. Anastassiou, H. Liu, and V. Varadan: Variable window binding for mutually exclusive alternative splicing. *Genome Biology*, 2006, vol. 7(1), Article, 7:R2.

[68] K. D. Rao and M. N. S. Swamy: Analysis of genomics and proteomics using DSP techniques, *IEEE Transactions on Circuits and Systems I (Regular papers)*, 2008, vol. 55(1), 370–378.

[69] C. Jeng, I. Yang, and K. Hsieh: Bacteria classification on power spectrums of complete DNA sequences by self-organizing map: *Neural Information Processing*, (*Letters and Reviews*), 2005, vol. 9(3), 53–58.

[70] M. D-S. Vieira: Statistics of DNA sequences: A low-frequency analysis. *Physical Review E*, 1999, vol. 60(5), 5932–5937.

[71] Y. H. Chen, S. L. Nyeo, and J. P. Yu: Power-laws in the complete sequences of human genome. *Journal of Biological Systems*, 2005, vol. 13(2), 105–115.

[72] Y. Isohata and M. Hayashi: Analyses of DNA base sequences for eukaryotes in terms of power spectrum method. *Japanese Journal of Applied Physics*, 2005, vol. 44(2), 1143–1146.

[73] H. Herzel and I. Grosse: Measuring correlations in symbolic sequences. *Physica A*, 1995, vol. 216, 518–542.

[74] H. Herzel, O. Weiss, and E. N. Trifonov: 10–11 bp periodicities in complete genome reflect protein structure and DNA folding. *Bioinformatics*, 1999, vol. 15, 3187–3193.

[75] W. J. Lee and L. F. Luo: Periodicity of base correlation in nucleotide sequence, *Physical Review E,* 1997, vol. 56(1), 848–851.

[76] A. Fukushima, T. Ikemura, M. Kinouchi, T. Oshima, Y. Kudo, H. Mori, and S. Kanaya: Periodicity in prokaryotic and eukaryotic genomes identified by power spectrum analysis. 2002, *Gene,* vol. 300(1–2), 203–211.

[77] L. Nyeo, I. C. Yang, and C. H. Wu: Spectral classification of archaeal and bacterial genomes. *Journal of Biological Systems,* 2002, vol. 10(3), 233–241.

[78] R. W. Hanson: Fast fourier transform analysis of DNA sequences. *BA Thesis,* The Division of Mathematics and Natural Sciences, Reed College, Portland, OR, USA, May 2003.

[79] B. D. Silverman and R. Linsker: A measure of DNA periodicity. *Journal of Theoretical Biology.* 1986, vol. 118, 295–300.

[80] S. Tiwari, S. Ramachandran, S. Bhattacharya and R. Ramaswamy: Prediction of probable genes by fourier analysis of genomic sequences. *CABIOS,* 1997, vol. 113, 263–270.

[81] J. W. Fickett: Recognition of protein coding regions in DNA sequences. *Nucleic Acid Research,* 1982, vol. 10, 5303–5318.

[82] J. W. Fickett and C. S. Tung: Assessment of protein coding measures. *Nucleic Acid Research,* 1992, vol. 20, 6441–6450.

[83] S. P. Chatterjee: *Bioinformatic analysis of viral genomic sequences and concepts of genome-specific rational vaccine design. Ph.D. Dissertation.* College of Engineering and Computer Science, Florida Atlantic University, Boca Raton, FL, USA: 2013.

[84] M. Farach, M. Noordewier, S. Savari, L. Shepp, A. Wyner, and J. Ziv: On the entropy of DNA: Algorithms and measurements based on memory and rapid convergence. *Proceedings of the Sixth Annual ACM-SIAM Symposium on Discrete Algorithms* (SODA' 95), (22–24 January 1995, San Francisco, CA, USA), pp. 48–57.

[85] M. Krawczak, J. Reiss, and D. N. Cooper: The mutational spectrum of single base-pair substitutions in mRNA splice junctions of human genes: Causes and consequences. *Human Genetics,* 1992, vol. 90(1–2), 41–54.

[86] J. S. R. Jang, C. T. Sun, and E. Mizutani: *Neuro-fuzzy and Soft Computing,* Springer Verlag., New Jersey, USA: 1997.

[87] Y. Jinand L. Wang: *Fuzzy Systems in Bioinformatics and Computational Biology (Studies in Fuzziness and Soft Computing vol. 242),* Springer-Verlag, Heidelberg, Germany: 2009.

[88] P. S. Neelakanta, S. T. Abusalah, D. De Groff, and J. C. Park: Fuzzy nonlinear activity and dynamics of fuzzy uncertainty in the neural complex. *Neurocomputing,* 1998, vol. 20, 123–153.

[89] P. S. Neelakanta, J. C. Park, and D. De Groff: Complexity parameter *vis-à-vis* interaction systems: Application to neurocybernetics, *Cybernetica,* 1997, vol. XL(4), 243–253.

[90] P. S. Neelakanta and D. De Groff: *Neural Network Modeling: Statistical Mechanics and Cybernetic Perspectives*, CRC Press, Boca Raton, FL, USA: 1994.

[91] M. Pavlovic, M. Cavallo, A. Kats, A. Kotlarchyk, H. Zhuang, and Y. Shoenfels: From Pauling's Abzyme concept to the new era of hydrolytic anti-DNA autoantobodies: A link to rational vaccine design? A review. *International Journal of Bioinformatics Research and Applications*, 2011, vol. 7(3), 220–238.

[92] A. Krishnamachari, V. M. Mandal, and Karmeshu: Study of binding sites using Rényi parametric entropy measure. *Journal of Theoretical Biology*, 2004, vol. 227, 429–436.

[93] L. Florea: Bioinformatics of alternative splicing and its regulation. *Briefing in Bioinformatics*, 2006, vol. 7(1), 55–69.

[94] R. M. Stephens and T. D. Schneider: Features of spliceosome evolution and function inferred from an analysis of the information at human splice sites. *Journal of Molecular Biology*, 1992, vol. 228(4),1124–1136.

[95] G. Garden and M. Frommer: CpG islands in vertebrate genomes, *Journal of Molecular Biology*, 1987, vol. 196(2), 261–282.

[96] D. Takai and P. A. Jones: Comprehensive analysis of CpG islands in human chromosomes 21 and 22, *Proceedings of the National Academy of Science (USA)*, 2002, vol. 99(6), 3740–3745.

[97] S. Karlin, W. Doerfler, and L. R. Cardon: Why is CpG suppressed in the genomes of virtually all small eukaryotic viruses but not in those of large eukaryotic viruses? *Journal of Virology*, 1994, vol. 68(5), 2889–2897.

[98] H. D. Nguyen and C. L. Brooks, III: Generalized structural polymorphism in self-assembled viral particles, *Nano Letters*, 2008, vol. 8(12), 4574–4581.

[99] G. N. Alekseev: *Energy and Entropy*. Moscow, Mir Publishers, USSR: 1986.

[100] V. A. Belyi and M. Muthukuma: Electrostatic origin of the genome packing in viruses: *PNAS*, 2006, vol. 103(46), 17174–17178.

[101] C. L. Ting, J. Wu, and Z. G. Wang: Thermodynamic basis for the genome to capsid charge relationship in viral encapsidation. *PNAS*, 2011, vol. 108(41), 16986–16991.

[102] V. P. Antao and I. Tinoco, Jr: Thermodynamic parameters for loop formation in RNA and DNA hairpin tetraloops: *Nucleic Acids Research*, 1992, vol. 20(4), 819–824.

[103] P. Guillaume, H. Santini, C. Pakleza, and J. A. H. Cognet: DNA tri and tetra-loops and RNA tetra-loops hairpins fold as elastic biopolymer chains in agreement with PDB coordinates. *Nucleic Acids Research*, 2003, vol. 31(3) 1086–1096.

[104] E. M. Moody, J. C. Ferrer, and P. C. Bevilacqua: Evidence that folding of an RNA tetraloop hairpin is less cooperative than Its DNA counterpart. *Biochemistry*, 2004, vol. 43(25), 7992–7998.

[105] P. C. Bevilacqua and J. M. Blose: Structures, kinetics, thermodynamics, and biological functions of RNA hairpins. *Annual Review of Physical Chemistry*, 2008, vol. 59, 79–103.

[106] T. Xia, J. SantaLucia, Jr., M. E. Burkard, R. Kierzek, S. J. Schroeder, X. Jiao, C. Cox, and D. H. Turner: Thermodynamic parameters for an expanded nearest-neighbor model for formation of RNA duplexes with Watson-Crick base pairs. *Biochemistry*, 1998, vol. 37(42.1), 4719–4735.

[107] A. D. Baxevanis and B. F. Oullette: *Bioinformatics: A Practical Guide to the Analyses of Genes and Proteins*. John Wiley and Sons, Hoboken, NJ, USA: 2005, pp. 146–147.

[108] D. H. Mathews, J. Sabina, M. Zuker, and D. H. Turner: Expanded sequence dependence of thermodynamic parameters improves prediction of RNA secondary structure. *Journal of Molecular Biology*, 1999, vol. 288, 911–940.

[109] X. Tianbing, J. A. McDowell, and D. H. Turner: Thermodynamics of nonsymmetric tandem mismatches adjacent to G & C base pairs in RNA. *Biochemistry*, 1997, vol. 36, 12486–12497.

[110] X. Tianbing, J. SantaLucia, Jr., M. E. Burkard, R. Kierzek, S. J. Schroeder, J. Xiaoqi, C. Cox, and D. H. Turner: Thermodynamic parameters for an expanded nearest-neighbor model for formation of RNA duplexes with Watson-Crick base pairs. *Biochemistry*, 1998, vol. 37, 14719–14735.

[111] A. Weise, M. Gross, K. Mrasek, H. Mkrtchyan, B. Horsthemke *et al.*: Parental-origin-determination fluorescence *in situ* hybridization distinguishes homologous human chromosomes on a single-cell level. *International Journal of Molecular Medicine*. 2008, vol. 21, 189–200.

[112] S. M. Ali and S. M. Silvey: A general class of coefficients of divergence of one distribution from another. *Journal of the Royal Statistical Society*, Series B (Methodological), 1966, vol. 28(1), 131–142.

[113] P. S. Neelakanta, T. V. Arredondo, and D. De Groff, D., Redundancy attributes of a complex system: Application to bioinformatics. *Complex Systems*, 2003, vol. 14, 215–233.

[114] Website on: Human parvovirus B19, complete genome (NCBI Reference Sequence: NC_000883.1). Available on-line at: *http://www.ncbi.nlm.nih.gov/nuccore/9632996.*

[115] Chordopoxvirinae: Available on-line at: *http://viralzone.expasy.org/all_by_species/172.html.*

[116] Variola virus and other orthopoxviruses. Chapter 2. In R. Briere (Ed.), *Assessment of Future Scientific Needs for Live Variola Virus*, The National Academy Press, Washington D. C., USA: 1999, pp. 15–18

[117] S. J. Werden, M. M. Rahman, G. McFadden: Poxvirus host range genes. *Advances in Virus Research*, 2008, vol. 71, 135–171.

[118] S. Yadav, M. Hosamani, V. Balamurugan, V. Bhanuprakash, and R. K. Singh: Partial genetic characterization of viruses isolated from pox-like

infection in cattle and buffaloes: Evidence of buffalo pox virus circulation in Indian cows. *Archives of Virology*, 2010, vol. 155(2), 255–261.

[119] F. Fenner, D. A. Henderson, I. Arita, Z. JeZek, and I. D. Ladnyi: Chapter 6 Early efforts at control: Variolation, vaccination, and isolation and quarantine. In *Smallpox and its Eradication*, pp. 245–276, World Health Organization, Switzerland, 1988.

[120] S. Riedel: Edward Jenner and the history of smallpox and vaccination. *Proceedings-Baylor University Medical Center.* 2005, vol. 18(1), 21–25.

[121] F. Fenner, D. A. Henderson, I. Arita, Z. JeZek, and I. D. Ladnyi: Chapter 7: Developments in vaccination and control between 1900 and 1966. In *Smallpox and Its Eradication*, World Health Organization, Switzerland, 1988, pp. 278.

[122] CDC (Centers for Disease Control and Prevention). Vaccinia (Smallpox) vaccine recommendations of the immunization practices advisory committee (ACIP). *Morbidity and Mortality Weekly Report*, 2001, 50(RR10), 1–25.

[123] World Health Organization: Declaration of global eradication of smallpox. *The Weekly Epidemiological Record*, 1980, vol. 55, 145–152.

[124] R. C. Condit, N. Moussatche, and P. Traktman: In a nutshell: Structure and assembly of the vaccinia virion. *Advances in Virus Research*, 2006, vol. 66, 31–124.

[125] R. C. Hendrickson, C. Wang, E. L. Hatcher, and E. J. Lefkowitz: Orthopox virus genome evolution: The role of gene loss. *Viruses*, 2010, vol. 2(9), 1933–1967.

[126] I. K. Damon: Poxviruses. In D. M. Knipe and P. M. Howley (Eds.), *Fields Virology*. Lippincott Williams & Wilkins; Philadelphia, PA, USA: 2007. pp. 353–370.

[127] Amino acid sequence of variola virus. Available online at: *http://www.ncbi.nlm.nih.gov/nuccore/NC_001611.1.*

[128] Amino acid sequence of vaccinia virus. Available online at: *http://www.ncbi.nlm.nih.gov/nuccore/NC_006998.1.*

[129] X. Tianbing, J. A McDowell, and D. H. Turner: Thermodynamics of nonsymmetric tandem mismatches adjacent to G & C base pairs in RNA. *Biochemistry*, 1997, vol. 36, 12486–12497.

[130] S. Wright: *Evolution and the Genetics of Populations: Genetics and Biometric Foundations vol. 4 Variability within and Among Natural Populations.* University of Chicago Press, Chicago, IL, USA: 1984.

[131] S. Wright: *Evolution and the Genetics of Populations: Genetics and Biometric Foundations Vol. 3 (Experimental Results and Evolutionary Deductions),* University of Chicago Press, Chicago, IL, USA: 1984.

[132] K. P. Murphy, P. Travers, M. Walport, and C. Janeway, *Janeway's Immunobiology.* Garland Science, New York, NY: 2008.

[133] P. Parham and C. Janeway, *The Immune System* (3rd ed.). Garland Science, New York, NY, USA: 2009.

[134] S. Akira, S. Uematsu and O. Takeuchi: Pathogen recognition and innate immunity. *Cell*, 2006, vol. 124, 783–801.

[135] D. Masopust, V. Vezys, E. J. Wherry, and R. Ahmed: A brief history of CD8 T cells. *European Journal of Immunology*, 2007, vol. 37(Suppl. 1S), 103–110.

[136] J. R. Lees and D. L. Farber: Generation, persistence and plasticity of CD4 T-cell memories. *Immunology*, 2010, vol. 130, 463–470.

[137] K. K. McKinstry, T. M. Strutt, and S. L. Swain: The potential of CD4 T-cell memory. *Immunology*, 2010, vol. 130, 1–9.

[138] M. Zanetti, P. Castiglioni, and E. Ingulli: Principles of memory CD8 T-cells generation inrelation to protective immunity. *Advances in Experimental Medicine and Biology*, 2010, vol. 684, 108–125.

[139] I. Kumagai and K. Tsumoto: Antigen–antibody binding. *Encyclopedia of Life Sciences*, 2010, 1–7.

[140] R. Goldsby, T. J. Kindt, B. A. Osborne, and J. Kuby: Antigens. Chapter 3 (pp. 57–75). In J. Kuby and W. H. Freeman: *Immunology*, W. H. Freeman and Company, New York, NY, USA: 2003.

[141] P. Delves, S. Martin, D. Burton, and I. Roitt: *Essential Immunology*, Wiley-Blackwell, Oxford, UK: 2006.

[142] M. Pavlovic, A. Kats, M. Cavallo, R. Chen, J. X. Hartmann, and Y. Shoenfeld: Pathogenic and epiphenomenal anti-DNA antibodies in SLE. *Autoimmune Disease*, 2010, vol. 2010.

[143] M. Pavlovic, M. Cavallo, A. Kats, A. Kotlarchyk, H. Zhuang, and Y. Shoenfeld: From Pauling's abzyme concept to the new era of hydrolytic anti-DNA autoantibodies: A link to rational vaccine design? A review. *International Journal of Bioinformatic Research and Application*, 2011, vol. 7(3), 220–238.

[144] E. Jenner: An inquiry into the causes and effects of the variolae vaccine: A disease discovered in some of the western counties of England, particularly Gloucestershire, and known by the name cow pox, *Classics of Medicine Library*, Birmingham, AL, USA: 1978.

[145] S. Y. Tan: Edward Jenner (1749–1823): Conqueror of smallpox. *Singapore Medical Journal*, 2004, vol. 45(11) 507–508.

[146] P. Debré: *Louis Pasteur*. The Johns Hopkins University Press, Baltimore, MD, USA: 1994.

[147] A. W. Artenstein and G. A. Poland: Vaccine history: The past as prelude to the future: *Vaccine*, 2012, vol. 30(36), 5299–5301.

[148] R. Rappuoli: From Pasteur to genomics: Progress and challenges in infectious diseases: *Nature Medicine*, 2004, vol. 10(11), 1177–1185.

[149] R. Rappuoli and F. Bagnoli: *Vaccine Design: Innovative Approaches and Novel Strategies*, Caister Academic Press, Norfolk, UK: 2011.

[150] M. Mora, D. Veggi, L. Santini, M. Pizza, and R. Rappuoli: Reverse vaccinology. *Drug Discovery Today*, 2003, vol. 8(10), 459–464.

[151] A. Henke: DNA immunization–A new chance in vaccine research? *Medical Microbiology and Immunology*, 2002, vol. 191(3–4), 187–190.

[152] S. A. Abdulhaqq and D. B. Weiner: DNA vaccines: Developing new strategies to enhance immune responses. *Immunologic Research*, 2008, vol. 42(1–3), 219–232.

[153] V. W. Bramwell and Y. Perrie: The rational design of vaccines. *Drug Discovery Today*, 2005, vol. 10(22). 1527–1534.

[154] R. Rappuoli: Reverse vaccinology, a genome-based approach to vaccine development: *Vaccine*, 2001, vol. 19(17–19), 2688–2691.

[155] A. S. DeGroot, H. Sbai, C. Saint Aubin, J. McMurry, and W. Martin: Immuno-informatics: Mining genomes for vaccine components. *Immunology and Cell Biology*, 2002, vol. 80, 255–269.

[156] M. Hagmann: Computers aid vaccine design. *Science*, 2000, vol. 290(5489), 80–82.

[157] M. Ardito, J. Fueyo, R. Tassone, F. Terry, K. DaSilva, S. Zhang, W. Martin, A. De Groot, S. Moss, and L. Moise: An integrated genomic and immunoinformatic approach to H. Pylori vaccine design. *Immunome Research*, 2011, vol. 7(2), Article 1(12 pages), (Open-access online).

[158] F. R. Burden and D. A. Winkler: Predictive bayesian neural network models of MHC class II peptide binding. *Journal of Molecular Graphics and Modelling*, 2005, vol. 23, 481–489.

[159] P. Donnes and A. Elofsson: Prediction of MHC class I binding peptides, using SVMHC. *BMC Bioinformatics*. 2002, vol. 3, 25 (Open-access online).

[160] H. Noguchi, R. Kato, T. Hanai, Y. Matsubara, H. Honda, V. Brusic, and T. Kobayashi: Hidden Markov model-based prediction of antigenic peptides that interact with MHC class II molecules. *Journal of Bioscience and Bioengineering*, 2002, vol. 94(3), 264–270.

[161] M. N. Davies and D. R. Flower: Harnessing bioinformatics to discover new vaccines. *Drug Discovery Today*. 2007, vol. 12(9–10), 389–395.

[162] R. M. Welsh and R. S. Fujinami: Pathogenic epitopes, heterologous immunity and vaccine design. *Nature Reviews Microbiology*, 2007, vol. 5, 555–563.

[163] B. J. Yoon: Hidden Markov models and their applications in biological sequence analysis. *Current Genomics*. 2009, vol. 10(6), 402–415.

[164] I. Rabe: Zika virus. BPAC-CDC Presentation, 111716, Centers for disease control and prevention, November 18, 2016, 1–39. Available on-line at: *https://www.fda.gov/downloads/AdvisoryCommittees/Committees-MeetingMaterials/BloodVaccinesandOtherBiologics/BloodProducts-AdvisoryCommittee/UCM533373.pdf*.

[165] A. D. Haddow, A. J. Schuh, C. Y. Yasuda, M. R. Kasper, V. Heang, R. Huy, H. Guzman, R. B. Tesh, and S. C. Weaver: Genetic characterization of Zika virus strains: Geographic expansion of the Asian lineage. *PLOS Neglected Tropical Diseases.* February 28, 2012. Available on-line at: *https://doi.org/10.1371/journal.pntd.0001477.*

[166] NIH-NIAID: Zika virus vaccines. Available online at: *https://www.niaid.nih.gov/diseases-conditions/zika-vaccines.*

[167] S. T. Cole, K Eiglmeier, J. ParkhillandK. D. James: 2001. Massive gene decay in the leprosy bacillus. *Nature,* 2001, vol. 409, 1007–1011.

[168] S. T. Cole, R. Brosch, J. Parkhill, T. Garnier, C. Churcher, *et al.*: Deciphering the biology of *Mycobacterium tuberculosis* from the complete genome sequence. *Nature,* 1998, vol. 393, 537–543.

[169] T. Parish and A. Brown (Eds.). *Mycobacterium: Genomics and Molecular Biology.* Caister Academic Press, Poole, UK: 1998.

[170] E. Rivero: New vaccine shows promise as stronger weapon against both tuberculosis and leprosy. Available online at: *http://newsroom.ucla.edu/releases/new-vaccine-shows-promise-as-stronger-weapon-against-both-tuberculosis-and-leprosy.*

Section V
Phylogenetic and Species Informatics

Commensurate with classical and modern views on evolution theory pertinent to the origin and diversity of species, details on phylogenetic concepts are presented in this section in the context of bioinformatics. The two chapters presented thereof are:

Chapter 9

Evolution-Theoretics and Phylogeny: An Overview

Evolutionary history of organisms concerning the origin and diversity of living systems viewed in the perspectives of morphological, paleontological, and biogeographical phylogeny is the pedagogy outlined in this chapter.

What has happened in the course of the evolution of the life?
— George G. Simpson: *The Meaning of Evolution*

9.1 Evolution Theory

Pursuant to his historical visit to Galapagos Islands in 1837, Charles Darwin presented his observations on living systems with a bold proposition that "all the organic beings, which have ever lived on this earth have descended from some one primordial form." His celebrated book, *The Origin of Species by Means of Natural Selection, or the Preservation of Favoured Races in the Struggle for Life*[1] evolved thereof elaborated his hypothesis exemplifying the observed details. Thus emerged, Darwin's evolution-theoretics proposed to infer the relationships among species in terms of morphological, paleontological, and biogeographical information of living systems, and it systematically grew into the art of understanding the phylogenetic relationships among species as they exist today with a wide scale of diversity.

9.2 Natural Selection

The mechanism of evolution relies on *natural selection*, which refers to a process that acts on populations of individual species in order to produce diversity of life as seen now. It hypothesizes that any population in nature produces a gamut of offspring more than the environment can support. As such, there is always an inevitable struggle for survival and existence within the population, and those individuals that are "best adapted to the environment" bear a better chance of survival and produce offspring.

Furthermore, in the biological ecosystem having favorable ambient conditions, the *biotic* species (such as plants, animals, fungi, bacteria) successfully proliferate (manifesting as epochs of survival, existence, and reproduction) and colonize the area (typically barren and unoccupied). Such environmental conditions are dictated by *abiotic factors* such as habitat, (pond, lake, ocean, desert, mountain) and weather features (such as temperature, rain, snow). The species in an ecosystem also pose influencing factors (*biotic factors*) on fellow species, and the biotic and abiotic factors cohesively combine to create a *complex biological ecosystem* that represents a community or unit of living organisms extensively comingled among nonliving entities.

Most often, the template of natural selection implies how fortuitous individuals get constantly morphed due to external (exogenous) factors or abiotic influences. However, the extent of a population adapting to its environment does not guarantee the future survival of species. With climatic changes, reduction in food supply and increased dynamics of prey–predator jointly make the species to suffer from a degenerative state (that possibly could eventually lead to their extinction, as well).

Theory of evolution thus points out the plausibility of a unique, unifying and cohesive force that explains the origin and existence of all forms of life, and "...nothing in biology makes sense except in the light of evolution."[2] As such, evolution-theoretics "...is to the life sciences what the long sought holy grail of the unified field theory is to astrophysics...."[3] Darwin's description of evolving life conforms to an extensive variation sorted out through drift and selection of lineages that diverge, and it implies a "descent with modification" making the organisms to bear a 'history', and the probabilistic epochs of such (largely mutation-implied) changes depict a stochastic framework of informatics on ethereal of history of living systems.

9.3 Modern Views on Evolution: Synthetic Theory

Through Darwin's enunciation of evolution-theoretics, the underlying process and operational aspects of natural selection were not fully explainable. This is mainly due to the absence of genetic considerations *vis-à-vis* inheritance that was proposed by Mendel (in 1866) revealing the secrets of heredity and variability that may prevail in populations, and only in the twentieth century, neo-Darwinism came into being with the *synthetic theory of evolution* that blends the fields of genetics and evolution.[4] This neoteric concept of *genetic mutation* and closely observed details on protein sequences (in the early 1960s) showed that the level of mutation among populations could be much higher than that expected by advantageous mutations. Hence, the theory such as *Kimura's neutral theory of evolution* specifies that most mutations at the genetic level are neutral meaning that they are neither advantageous nor disadvantageous toward natural selection, and they are determined largely by the mutation rate and effective population size.[5]

Though the implicit knowledge of genetic science[6] has been in vogue even in the prehistoric times (through selective breeding and domestication of animals). *Genetics* (with *geno* meaning "to give birth" in Greek) emerged gracefully in the twentieth century (with the proposal by William Bateson to describe the study of inheritance, variation, and heredity). As described in earlier chapters, subsequent contributions by Watson and Crick (in 1953 relevant to the physical and chemical structure of DNA that resides in each and every cell of living systems) provided more details on the exact format of genetic instructions/coding in terms of information associated with the chain of genes formed by simple molecules of DNA/RNA sequences. In the post-omic era, rapid sequencing methods,[7] invention of the polymerase chain reaction, PCR,[8] and automation of DNA sequencing[9] came into being leading to the so-called *phylogenetic analysis* (described in the next chapter).

9.4 Basics of Phylogeny

Considering the diversity of life, relevant details gathered (from morphological, biochemical, and gene sequence data) "suggests that all organisms on Earth are genetically related"; and the genealogical relationships of living things can be represented by a vast (topology of) evolutionary tree...", and this *tree-of-life* depicts the *phylogeny** of organisms. (*Phylogeny** or *phylogenesis* implies a race history of an animal or a plant type.) In essence, phylogeny is a collection of information about the origin and diversity of species. In all, "different species arise from previous forms *via* descent, and that all organisms, from the smallest microbe to the largest plants and vertebrates are connected by the passage of genes along the branches of the phylogenetic tree that links all of life."[10,11] Darwin recognized and suggested thereof that all life, that has ever existed, is related through the process of natural selection and by a "great Tree-of-Life,"[1] and his recognition today reverberates in reconstructing such trees-of-life with the discovery of the last universal common ancestor of almost all life.[12–14]

9.5 Convergent and Divergent Evolutions

Acquisition of the same biological trait in unrelated lineages defines the *convergent evolution* process in species. As a classical example, the last common ancestor of birds and bats did not have wings, but presently these species are capable of flying (with wings of almost similar shapes across all birds). The wings depict a morphology seen with limbs modified (as evidenced by their bone structure), and convergent evolution implies specific traits with respect to such structures (termed as *analogous structures*). Specific to species having a common origin, the convergent evolution refers to *homologous structures*. That is, the wings of bat and *pterodactyl* (a kind of "winged lizard") depict analogous structures. But the bat wing is homologous to human and mammal forearms, in the sense that an ancestral state is shared despite serving different functions. (The similarity in species of different ancestry, which is the result of convergent evolution is known as *Homoplasy*).

Furthermore, convergent evolution is similar to, but distinct from what are known as, the phenomena of *evolutionary relay* and *parallel evolution*. Evolutionary relay portrays how independent species acquire similar characteristics through their evolution in similar ecosystems at different times seen, for example, as the dorsal fins of extinct ichthyosaurs and sharks, and parallel evolution takes place whenever two independent species evolve simultaneously in the habitat of same ecology. Also, they acquire similar characteristics, (as, for instance, seen in extinct browsing-horses and *paleotheres*).

The opposite of convergent evolution is *divergent evolution*. Here, related species evolve different traits. This can happen at molecular level as a result of random mutation unrelated to adaptive changes. Similar to convergence in evolution, a rationale for divergent evolution can be seen *vis-à-vis* the accumulation of differences between groups leading to the formation of new species. That is, a "divergent evolution results from diffusion of the same species adapting to different environments, leading to natural selection defining the success of specific mutations." For example, the vertebrate limb is a result of divergent evolution. Though the limb in several species has a common origin, it has diverged somewhat in overall structure and function. Divergent evolution could prevail both in orthologous and paralogous genes defined in Chapter 7.

9.6 Phylogenetic Trees

Studies on *phenotype* (depicting the set of observable characteristics of an individual species arising from the interaction of its genotype with the environment) precede sequencing efforts exercised on omics commenced in 1960–1970s with evolutionary positionings of species on the basis of anatomical features and by the taxonomy of plant and animal classification. Subsequently, phylogeny is indicated toward deciphering the history of descent of a group of organisms described by a branching tree-like diagram (known as *phylogenetic tree*) as illustrated in Fig. 9.1. The phylogenetic approach offers rigorous mathematical background and computational methods to infer the trees or, better known as *dentograms.*

In constructing the tree-of-life, each branch is called a *clade* and living organisms are placed as *leaf-ends* at the tips of the branches, and their evolutionary history can be represented by a series of ancestors shared hierarchically *via* different subsets of organisms (that are seen surviving today). Such organisms seen alive are depicted as the leaf-ends on the dentograms, leading to down-tracing their history back to the

Fig. 9.1 An evolutionary tree-of-life: concept of phylogeny implies that existing species (depicted as end leaves) can be linked by an underlying structure in a tree-like manner.

branches encountering their various ancestors (and, eventually culminating in an *ancestral root* of origin). On the time scale, these ancestral existence denote thousands or millions or even possibly hundreds of millions of years ago. In short, a notion exists that all of life is genetically connected through a mammoth phylogenetic tree.

Does it mean that there could be a common ancestor for humans and beetles? Plausibly, *yes*! Relevant common ancestral organism could have been some sort of a worm, and somewhere along the evolutionary timeframe, this species of ancestral worm divided into two separate (worm) species, which thereafter divided again (and again), each division (or *speciation*) resulting in new, independently evolving lineages along the branches of the tree with the end result being two possible leaves: human and beetles!

Across various ancestral forms derived, the new lineages retain mostly their ancestral features. However, these would gradually get modified and supplemented with necessary traits that make them congenially adjust to (and survive in) the environment of their habitation. The pedagogy of phylogeny of organisms explains relevant similarities and differences among plants, animals, and microorganisms that have evolved through the formalism of the tree of life. Thus, the phylogenetic tree as shown in Fig. 9.1 offers a systematic framework for various subdisciplines of biology in understanding the organizational aspects of biological knowledge.

As is evident from its name, a phylogenetic tree as in Fig. 9.1 is a conceptual representation of the tree of life. It depicts a tree structure identifiable with a root-of-origin and species located as leaf nodes. In essence, it is a diagrammatic representation of the evolutionary relationships among a set of species. Branches in the tree depict that a particular species has evolved from another through internal nodes (which denote the *epochs of speciation*) resulting in a new species (Chapter 7). Quantitatively, the *length of a branch* can be regarded as some measure of *evolutionary distance* between two nodes. Implicitly, it indicates the *degree of dissimilarity* of two (DNA) sequences in the branched parts of the tree at speciation node. Constructing a phylogenetic tree is based on the following heuristics: (i) species exhibit variations or *diversity;* (ii) a set of similar species can be identified and grouped together; and (iii) grouped species can be connected (or linked) to a common ancestor.

A dentograms can be constructed to represent a tree with all the leaf-nodes being of equidistant from the root, implicitly representing the passage of time of sequences up to the leaf-nodes from the root.

Descriptively, the components of a phylogenetic tree and the set of terminology associated in constructing such structures are as follows:

- *Tree*: this is a line diagram that provides a visual means of representation for a group of sequences or species and indicates their time series of origin. A tree consists of nodes, *branches,* and *leaves.*
- *Nodes*: these depict ends of the tree and are represented by a small circle. A node can be internal or at an end (in which case, it is called a "leaf").
- *Leaf*: as indicated above, it is the loose-end (terminal) node of a branch in the tree.
- *Bifurcating node*: this node explicitly carries two distinct lineages arising from it.
- *Nodes*: a tree consists of nodes (each represented by a small circle) connected by branches. A node can be internal or at an end (in which case, it is called a "leaf"; that is, the leaf is the loose-end [terminal] node of a branch in the tree). *Internal nodes* denote hypothetical nodes, and a unique (internal) node can be identified as the root of the tree depicting the ancestor of all the sequences. The terminal nodes represent a set of organisms (for which the sequences data are compiled to reconstruct the phylogenetic tree). Each terminal node (leaf) is designated as an *operational taxonomical unit* (OTU). That is, an OTU depicts a terminal node in phylogenetic analysis and represents an organism, and a group of such OTUs constitute a *clade* representing a set of several evolving organisms and their common ancestral nodes. A *bifurcating node* explicitly carries two distinct lineages arising from it. The defined nodes as above are identified in the illustration of Fig. 9.2.

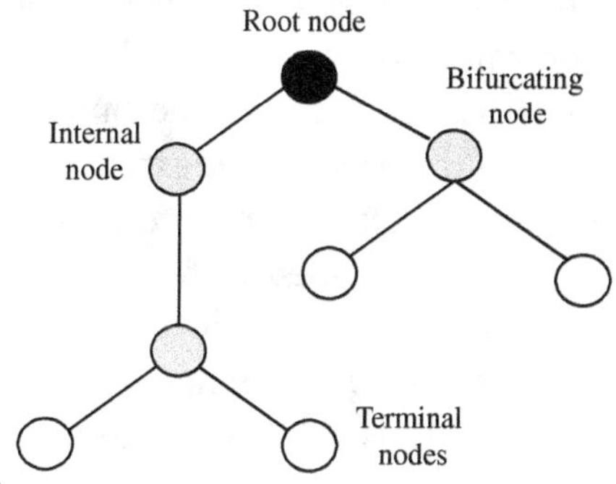

Fig. 9.2 Types of nodes in a phylogenetic tree.

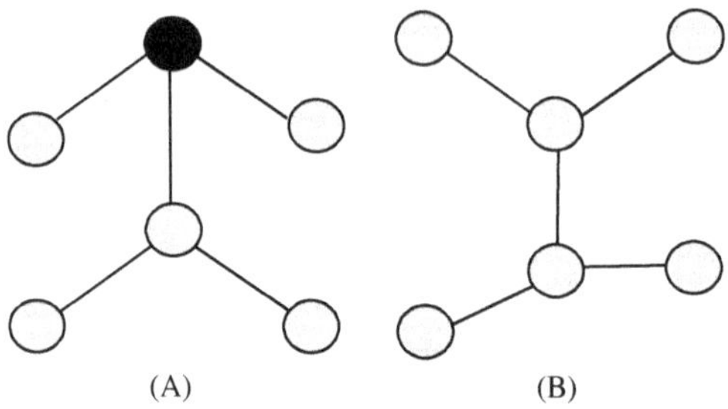

Fig. 9.3 Phylogenetic tree configurations: (A) rooted tree and (B) unrooted tree.

Table 9.1 The number of nodes and branches in rooted and unrooted trees.

M: Number of OTUs			
Type of tree	**Type of node**	**Number of branches**	**Number of nodes**
Rooted tree	Interior	M−2	M−1
	Total	2M−2	2M−1
Unrooted tree	Interior	M−3	M−2
	Total	2M−3	2M−2

The types of trees exist in two versions, namely, the *rooted* and *unrooted trees* as illustrated in Fig. 9.3. A rooted tree implies a structure in which the direction of evolution is specified with respect to a "root" or branch-off site of a single node. That is, the tree structure is shown with a designated "root" (depicting an ancestral origin), and *via* adequate divergence encountered across the phases of evolution, multitudes of species (denoted as different OTUs of the tree) have evolved.

An unrooted tree simply displays the underlying connections or links between the species. That is, no specified root node of ancestral implication is seen on this structure. The nodes seen explicitly denote simply the mutual relativeness (such as the branch lengths) between them, and this branch length measures the extent of divergence between the nodes. As regard to two versions of tree as above, the number of branches and nodes can be obtained in terms of the number (M) of OTUs. Relevant details are shown in Table 9.1.

Scaling the tree (or a *scaled tree*) implies elucidating the differences between adjoining nodes. It is done by determining the length of branches involved. Another definition of interest is the *gene tree,* which

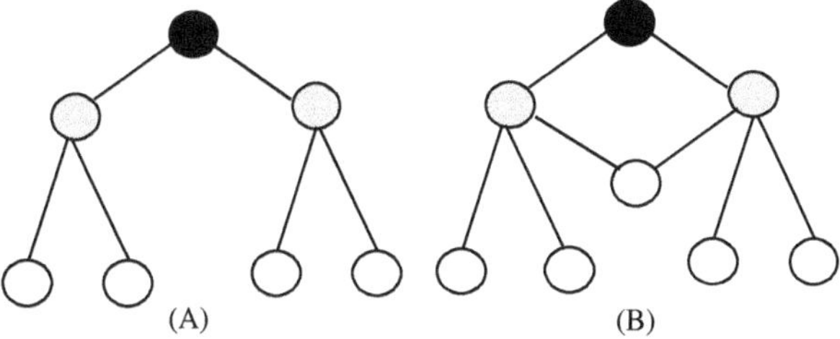

Fig. 9.4 (A) A simple tree and (B) a tree network mesh.

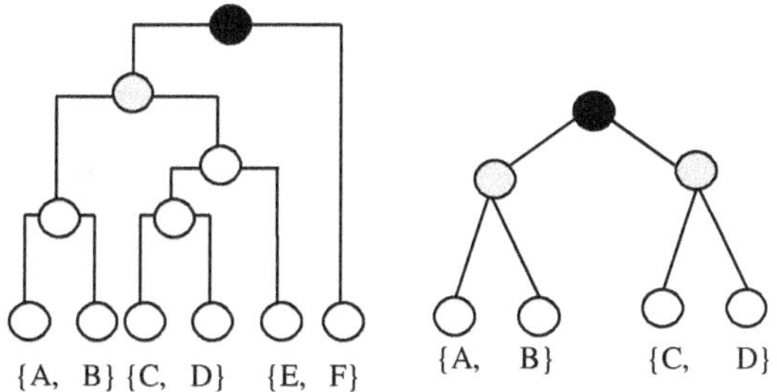

Fig. 9.5 Newick tree format.

denotes a tree structure that results from analyzing homologous genes. While trees signify only one path between any pair of nodes, a *tree network* is a mesh that has more than one path between any pair of nodes as shown in Fig. 9.4.

A mathematical way to represent graph-theoretic trees in terms of edge-lengths and using parentheses and commas results in what is known as the *Newick formatted phylogenetic tree* (alternatively known as *Newick notation* or *New Hampshire tree* format). The Newick format of a tree is a shorthand notation and is illustrated in Fig. 9.5. When an unrooted tree is represented in Newick format, an arbitrary node is chosen as its root. Whether rooted or unrooted, typically the representation of the tree is rooted on an internal node and rarely (but permissibly) rooting a tree on a leaf node is done.

A tree-like network that expresses ancestor–descendant relationships is called a *cladogram*. It describes the topology of a rooted phylogenetic tree *via* relative ancestral origins of sequences, but without any considerations of branch length. In essence, cladistic classifications of trees illustrate cladograms shown in Fig. 9.6 with the intention to reflect the relative recency of common ancestry or the sharing of homologous features.

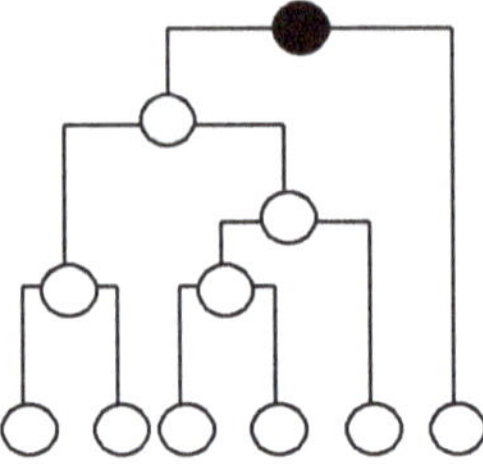

Fig. 9.6 A cladogram showing the relative order of common ancestry.

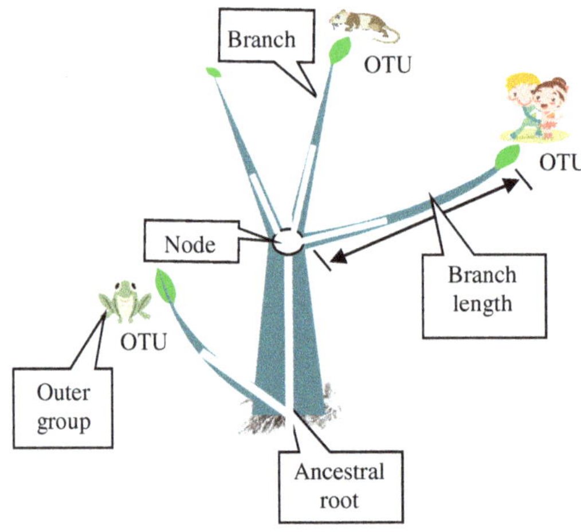

Fig. 9.7 Components of a phylogenetic tree. (OTU: Operational taxonomic units; leaves and root are the common ancestor of all OTUs).

9.7 Phylogenetic Trees Based on Morphological Features

Phylogenetic trees based on numerical taxonomy do not indicate the buried subjectivity of evolution. For example, in a particular phylogenetic analysis, the numerical approach does not specify the relative importance of, say for example, the skin color and tail-length.

In all, a phylogenetic tree that displays morphological aspects across the diverging species of evolution can be indicated. An example of such morphological features is that a chimp has furs and a bird does not. A morphological tree can be depicted as shown in Fig. 9.7. The morphological aspect across diverging species of evolution as shown in Fig. 9.8, for example, is that a chimp has furs and a bird does not and so on.

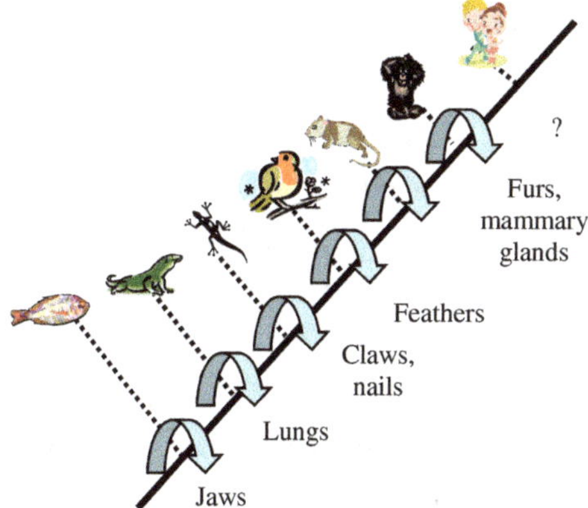

Fig. 9.8 A tree of organisms based on overall similarity and morphology or other observable traits across their phylogeny or evolutionary relation.

9.8 Salient Terms Associated with A Cladogram

In expressing the ancestor–descendant relationships *via* cladogram, the associated terminology can be defined as follows:

- *Cladistics*: It literally means the "branch" and it forms biological systematics in classifying species of organisms into hierarchical monophyletic groups. More specifically, it is defined as the study of the pathways of evolution. How many branches there are among a group of organisms, which branch connects to which other branch, and what is the branching sequence are queries of interest to *cladists.* Typically, cladistics strives to identify monophyletic clades — a group that represents a species and all its descendants. Closely related clades are called *sister groups* and other groupings are known as *paraphyletic* depicting common ancestor and some of its descendants and *polyphyletic* (which signifies sister groups but not the common ancestor).
- *Taxon (and taxa)*: A taxon (with *taxa* being its plural) represents a group of (one or more) organisms (which a taxonomist adjudges to be a unit). A clade is a special form of taxon in phylogenetic sense.
- *Additive tree*: It is a cladogram specified with branch lengths. It is also known as *phylograms* or *metric trees.* An example of additive tree is shown in Fig. 9.9.

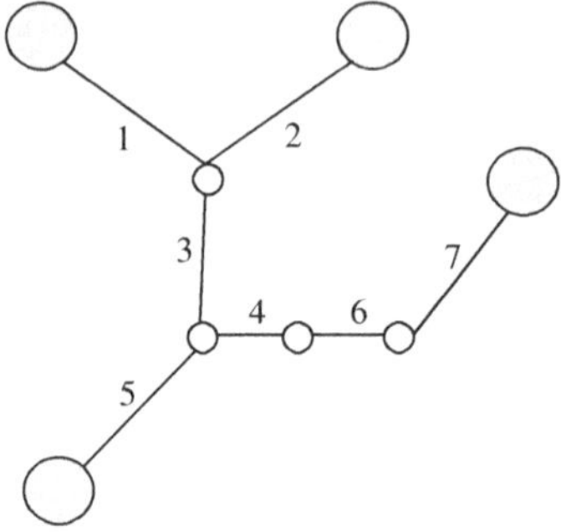

Fig. 9.9 Example of an additive tree. The numbers shown are some hypothetical branch lengths.

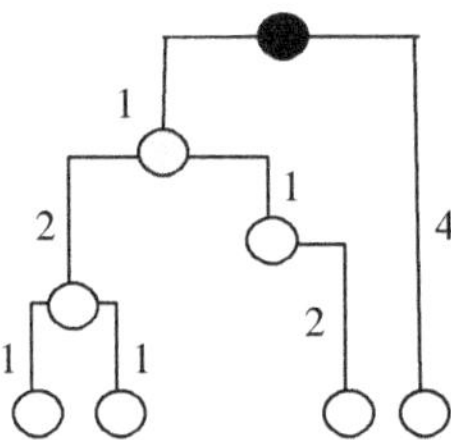

Fig. 9.10 An example of ultrametric tree. (All the tips of the tree are of equidistance from the root. In this case, the equidistance is equal to four).

- *Ultrametric tree*

An ultrametric tree refers to a dendogram denoting a specific type of an additive tree in which the tips of the tree are all equidistant from the root as illustrated in Fig. 9.10 where the relative order of common ancestry can be observed.

- *Cladistic apomorphy, synapomorphy,* and *pleisomorphy*
    - *Apomorphy* refers to a derived character shared by a species and its descendants, but not in the ancestral species
    - *Synapomorphy* is used to define an ancestor and its descendants
    - *Pleisomorphy* defines the ancestral characteristics
- *Phenetics* is the study of relationships among a group of organisms on the basis of degree of similarity between them, and this similarity may refer to molecular, phenotypic, or anatomical features
- *Phenogram* is a tree-like network expressing phenetic relationships

- *Pleisomorphy*: This defines some characteristics pertinent to the ancestor, which are sequel in all further sequences. And it is a character state present in both *outgroups*, as well as in the ancestors
- *Homoplasy*: This refers to *similarity* that has evolved independently without being indicative of common phylogenetic origin. Similarity seen in species of different ancestry is the result of convergent evolution, and it denotes the homoplasy
- *Polytomies*: *Soft polytomy* implies a lack of information about the order of divergence, and *hard polytomy* hypotheses that multiple divergences occurred simultaneously
- *Autapomorphy* is a derived trait seen uniquely in a particular taxon
- *Synapomorphy*: In contrast with autapomorphy, synapomorphy denotes the characteristics shared with the ancestor in a specific phylogeny and then derived from the ancestor. In cladistics defined earlier, a synapomorphy or synapomorphic character denotes a trait, which is shared ("symmorphy") by two or more taxa and their last common ancestor, whose ancestor in turn does not possess the trait. A synapomorphy is thus an apomorphy (or a derived characteristic of a clade)

The homologous genes derived from a common ancestor can manifest in three types of relations as homologs, orthologs, and xenologs. Formation of orthologs and paralogs occurs as a result of duplication and speciation, and *xenologs* result from a *lateral transfer* between two organisms (where a lateral transfer implies a direct DNA transfer between two species).

9.9 Review Section: Examples and Exercises

Example Ex-9.1

Statement of the Exercise

The concept of evolution is the key to reconstructing phylogenetic trees. It means tracing the genealogy from the root (ancestor) culminating into a gamut of species (namely, "leaves" or "taxa" as will be elaborated in the next chapter).

> *With reference to the three options of hypothetical phylogenies indicated in Fig. 9.11 for a set of species {lizard, bird, rat and gorilla}, what is the scope of finding a "tree-of-evolution", which is simple and parsimoniously adequate in portraying the observed evolution?*

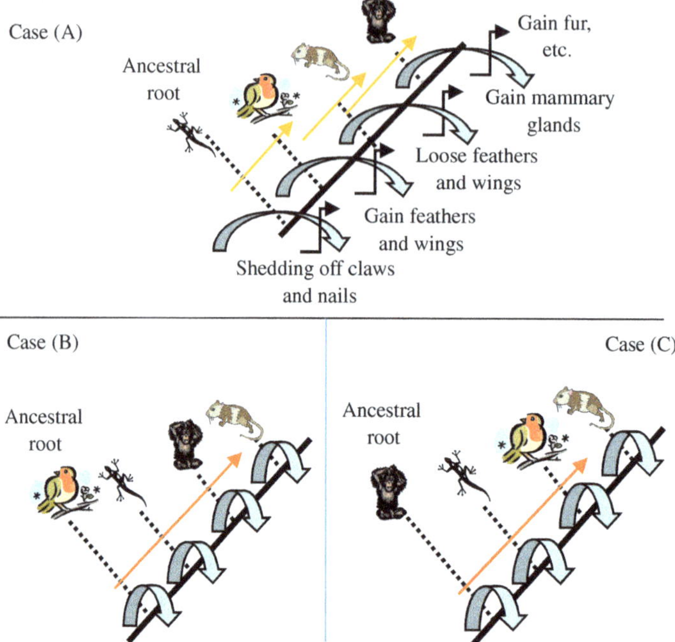

Fig. 9.11 Three cases of hypothetical lineages: *which one is the most plausible descendancy (assuming mutational changes as in Case (A) lineage)*?

Case (A) — Lineage: lizard (assumed as ancestral root) → bird → rat → gorilla.
Case (B) — Lineage: bird (assumed as ancestral root) → lizard → gorilla → rat.
Case (C) — Lineage: gorilla (assumed as ancestral root) → lizard → bird → rat.

Solution

The evolutionary lineage assumed in each case is dictated by the mutational changes explicitly illustrated in Case (a) option in Fig. 9.11. Hence, a matrix as shown in Table 9.2 can be stipulated explaining qualitative and conceptual details as needed.

Problem P-9.1

Statement of the Problem

This is a rider exercise on EXAMPLE EX-9.1 on parsimonious phylogenetic reconstruction conceivable across a lineage of species depicting the following set: {fish, crocodile, bird, gorilla, and human} as shown in Fig. 9.12.

Justify the evolutionary lineage illustrated in Fig. 9.12, introducing appropriate mutational changes as needed. Hence, develop a matrix

Table 9.2 Descriptions of three lineage options of descendancy shown in Fig. 9.11.

Plausible mutational changes (MC-I, II, III)

MC-I Shedding off claws and nails and gain feathers and wings
MC-II Lose feathers and wings
MC-III Developing mammary glands
MC-III+ Gain fur, etc.

Plausible opposite trends of mutational changes (MC-I, II, III)*

(MC-I)* Gaining claws and nails and gain feathers and wings
(MC-II)* Acquiring feathers and wings
(MC-III)* Loosing mammary glands
(MC-III+)* Shedding off the fur, etc.

Lineage options

Case	**Ancestral species**		**Species seen after**					
			First transformation		**Second transformation**		**Third transformation**	
(a)	Lizard		Bird		Rat		Gorilla	
	→	MC-I	→	MC-II	→	MC-III	→	MC-III+

Comment: Sequences of changes in Case (a) appear to be parsimoniously explainable and logical stages of mutations involving first the (assumed ancestral) lizard shedding off its claws and nails and then acquire feathers and wings; next, with the advent of developing mammary glands, emergence of the species of rat is feasible and its further culmination into the taxon, namely, the gorilla.

(b)	Bird		Lizard		Gorilla		Rat	
	→	MC-II	→	MC-II and (MC-I)*	→	MC-III	→	(MC-III+)*

Comment: in Case (b), the sequences of changes appear to be questionable as regard to the stages of mutations: first, the (assumed ancestral) bird has to lose its feathers and wings and then acquire claws and nails. Subsequently, the gaining of mammary glands should occur, leading to the species of gorilla, then a lesser version of animal kingdom, namely, the rat has to be seen as the taxon. Such a phylogeny cannot be parsimoniously explainable and logically feasible.

(Continued)

Table 9.2 (*Continued*)

(c)	Gorilla		Lizard		Bird		Rat
	→	(MC-III)* and (MC-III+)*	→	(MC-I)*	→	MC-I, (MC-II)*	→	MC-II and MC-III

Comment: in Case (c) also, the sequences of mutational changes are hardly feasible: First, the (assumed ancestral) gorilla has to lose its mammary glands and furs, and then, it should acquire claws and nails. Subsequently, the gaining of feathers and wings should take place, to be a bird, then a reversed version into a mammal of animal kingdom, namely, the rat has to be realized as the taxon. This phylogeny is highly illogical and cannot be parsimoniously explainable.

Conclusion

Case (a) is the correct answer, i.e., depicting possible mutational changes to be parsimoniously explainable and logically feasible. In all, it represents the simplest solution in Occam razor sense.

Fig. 9.12 A hypothetical lineage of: fish → crocodile → bird → gorilla → human constituting an evolutionary tree.

similar to Table 9.2 to stipulate that any other option of phylogenetic tree of the test species would prove to be futile as regard to its implied conjectural parsimony.

Problem P-9.2

Problem Statement

This problem refers to an exercise on deciding the parsimonious phylogeny across the descendancy conceivable across a lineage of species generally regarded as: *Model organisms**. (A *model organism** refers to a set of widely studied species and are considered to maintain and breed in a laboratory setting posing specific experimental advantages. That is, model organisms denote nonhuman species that are used in laboratory studies to understand biological processes.)

Typically, the following are examples of model organisms that are used to study genetics: yeast (*Saccharomyces cerevisiae*), fruit fly (*Drosophila melanogaster*), nematode worm (*Caenorhabditis elegans*), western clawed frog (*Xenopus tropicalis*), mouse (*Mus musculus*), and zebra fish (*Danio rerio*).

Construct a table (such as Table 9.2) to study the descendancy line of model organisms listed above. In each case, collect relevant genomic and proteomic details from resources and expand the scope of your study for further research.

9.10 Closure

Evolutionary heuristics presented in Darwin[1] on the origin and diversity of organisms as decided by natural selection with the preservation of favored species that continuously struggle for their existence in the natural ambient of biotic and abiotic constituents, and his proposition implies that all living systems have descended from some "primordial form" and evolved over an extensive period of time and seen today as the remnants of a common ancestor.

Darwin's theory could also be seen in the philosophical heuristics of classical religions (like Hinduism and Buddhism), wherein the evolutionary hierarchy (known as *Dashavatara*) of life is divided into 10 strata across four phases of *era* of phenomenal size (or *Yuga*, each *Yuga* depicting a divine scale of 4,320,000 years!).[15] Relevant ancient wisdom indicates discrete stages of evolution commencing from the cradle of a primitive *stratum of life* and laddering up into upper-levels and modeling the concomitant disambiguation that frames the crux of phylogeny. It is stated that "... the origin of life commenced from primitive grass-roots and plants, and it swept through worm-like living things evolving further and expanding sequentially as reptiles, birds and animals; eventually, it culminated into higher forms of life — depicting the coarse and superior versions of human ...":

> ...Verse 30 of *Thiruvachagam,* (a Tamil poetic literature of *circa* ninth century of common era, written by the Saint *Manickavasagar).*

References

[1] C. Darwin: *The Origin of Species by Means of Natural Selection, or the Preservation of Favoured Races in the Struggle for Life.* John Murray, London, UK: 1859.

[2] T. Dobzhansky: Nothing in biology makes sense except in the light of evolution. *The American Biology Teacher,* 1973, vol. 35, 125–129.

[3] S. M. Thompson: A very basic introduction to molecular evolutionary phylogenetic inference software. Available online at: *http://bio.fsu.edu/~stevet/IntroBioInfo/Lab8.pdf.*

[4] J. Huxley: *Evolution: The Modern Synthesis.* Allen and Unwin, Sydney, Australia: 1974.

[5] M. Kimura: Evolutionary rate at the molecular level. *Nature,* 1968, vol. 217, 624–626.

[6] W. S. Klug and M. R. Cummings: *Concepts of Genetics.* Macmillan Publishing Co., New York, NY: 1991.

[7] F. Sanger and A. R. Coulson: A rapid method of determining sequences in DNA primed synthesis with DNA polymerase. *Journal of Molecular Biology,* 1975, vol. 94(3), 441–446.

[8] R. K. Saiki, S. Scharf, F. Faloona, K. B. Mullis, G. T. Horn, H. A. Erlich and N. Arnheim: Enzymatic amplification of beta-globin genomic sequences and restriction site analysis for diagnosis of sickle cell anemia. *Science.* 1985, vol. 230(4732), 1350–1354.

[9] T. Hunkapiller, R. J. Kaiser, B. F. Koop and L. Hood: Large-scale and automated DNA sequence determination. *Science,* 1991, vol. 254(5028), 59–67.

[10] W. Hennig: Phylogenetic systematics. *Annual Review of Entomology,* 1965, vol. 10, 97–116.

[11] E. Zuckerkandl and L. Pauling: Molecules as documents of evolutionary history. *Journal of Theoretical Biology,* 1965, vol. 8, 357–366.

[12] C. R. Woese, O. Kandler and M. L. Wheelis: Towards a natural system of organisms: Proposal for the domains archaea, bacteria, and eucarya. *Proceedings of the National Academy of Science (USA),* 1990, vol. 87(12), 4576–4579.

[13] B. Korber, M. Muldoon, J. Theiler, F. Gao, R. Gupta, A. Lapedes, B. H. Hahn, S. Wolinsky and T. Bhattacharya: Timing the ancestor of the HIV-1 pandemic strains. *Science,* 2000, vol. 288(5472), 1789–1796.

[14] S. Mahanty, A. Saul and L. H. Miller: Progress in the development of recombinant and synthetic blood-stage malaria vaccines. *Journal of Experimental Biology,* 2003, vol. 206, 3781–3788.

[15] C. M. Brown: *Hindu Perspectives on Evolution: Darwin, Dharma, and Design.* Routledge (Taylor & Francis Group LLC), Florence, KY, USA: 2012, pp. 163–164.

Chapter 10

Bioinformatic Aspects of Phylogeny

In terms of classical and modern views on evolution theory, phylogenetic concepts and related analyses are described; hence, the heuristics of (re)constructing phylogenetic trees are detailed across the hierarchical span of . . . gene-to-species.

"... the evolution of a group of populations of living organisms can be handled by *exact methods...*"

— Cavalli-Sforza and Edwards (1965)

10.1 Phylogenetic Concepts and Analyses

Phylogeny is a latter-day perception on evolutionary theoretics. It emerged as a result of historically observed descriptive notions of organisms and causal-analytic considerations in the developmental studies on omics in living systems. Consistent with evolution-theoretic concepts, phylogenetics leads to conceiving a *phylogenetic tree* with basic constitutive components as illustrated in Fig. 10.1.

In making of a phylogenetic tree, the following heuristics are presumed as outlined in the previous chapter: (i) each species represented in the tree (denotes a *taxon*), which bears a relation to a common ancestor; (ii) the phylogenetic tree, in essence, is a branching structure that bifurcates at designated nodes; and (iii) inevitable mutations (that presumably have taken place exist randomly along the time scale of evolution) decide the eventual structure of a complete phylogenetic tree.[1]

10.2 Making a Phylogenetic Tree

Further details of a set of query multiple sequences of the species are required to initiate the (re)construction efforts of a phylogenetic tree. Supportive considerations on such efforts are listed as follows:

(i) Ensuring that the query multiple sequences (MS) are optimally aligned and the associated data are efficacious in leading to a

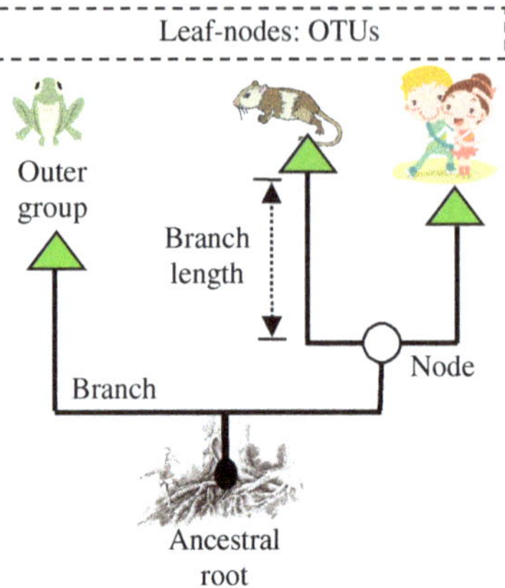

Fig. 10.1 Illustration of a simple life-tree model and phylogenetic terms.

"right tree": the set of MS should be devoid of gap-rich regions and the poorly conserved *N-* and *C-termini* segments retaining mostly conserved parts (motif) that depict the trace of family history

(ii) Preparing the MS data for phylogenetic analyses: this implies avoiding incomplete sequences and/or sequence fragments, and the genes that result from any lateral transfer (i.e., as xenologs), should be excluded (except in cases where such xenologs are to be studied).

(iii) Rendering the MS data compatible for analytical models and computational programs toward *ad hoc* formatting of test multiple sequences

Preparing MS dataset for constructing a phylogenetic tree relies on two considerations: (a) an approach based on similarities and (b) another based on dissimilarities observed in the data. Similarity aspect is concerned with relatedness across compared entities. It is divided into the following two types: *phenetic version*–depicting a character-based *phenogram*, for example, "comparing a set of plants of related characteristics (*taxa*) in terms of their associated characters (such as petals, sepals, anthers, ovary, size, or style, etc.)" and *cladistic version*, which refers to a method of hypothesizing relationships among organisms in reconstructing evolutionary trees. Relevant analysis is based on the data pertinent to the characters and/or *traits* of organisms.

10.3 Process of Reconstructing a Tree

A class of phylogeny is known as *molecular phylogenetics* or *molecular systematics.* It uses the structure of molecules in order to gain information on the evolutionary relationships for an organism. The result of such molecular phylogenetic analysis can be expressed in a phylogenetic tree.

The process of building or (re)constructing a phylogenetic tree structure using a set of aligned and prepared multiple query sequences is as follows: suppose a set of N multiple query sequences are considered, then it can be shown that the corresponding number of evolvable (possible) trees, M, will be extremely large values, if the value of N is excessive.[2] For example, when $N = 10$, $M = 1{,}027{,}025$, and so on. Next, building the tree of evolution implies assessing phylogeny, which can be done by two approaches, namely, (i) *distance-based approach* and (ii) *clustering-based approach.*

Considering rooted and unrooted trees indicated in Chapter 9, the rooted version has a node that corresponds to ancestral species as shown in Fig. 10.2(A), and the unrooted tree bears no assumption as regard to the position of an evolutionary ancestor (root) in the tree as depicted in Fig. 10.2(B). Relevant to the example of an unrooted tree in Fig. 10.2(B), there are six vertices of leaf-nodes (or *taxa*), indicated as, I, II, III, IV, V, and VI, and a positive weight (or length) is assigned between any two consecutive nodes as shown. This length, for example, may depict the number of mutations on the evolutionary path. Quantitatively, the

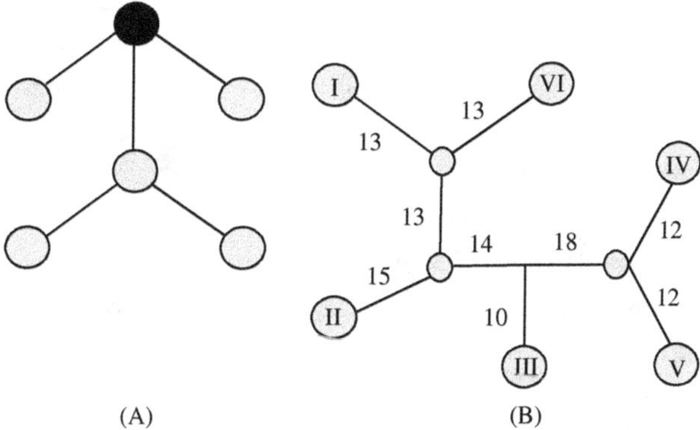

Fig. 10.2 Phylogenetic tree configuration depicting: (A) rooted tree and (B) weighted unrooted (star) topology with taxonic vertices marked as I–VI.

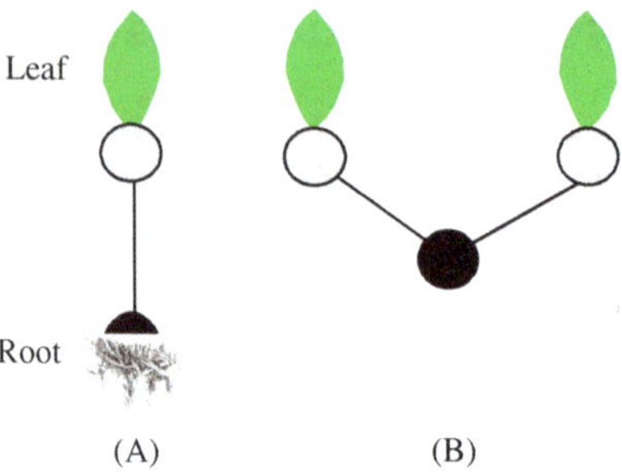

Fig. 10.3 Given the number of nodes (ν), realization of ε_R and ε_{UR}: (A) with $\nu = 2$ in a rooted tree and (B) with $\nu = 3$ in an unrooted (star topology) tree.

length (d) of the path between any two vertices can be specified as the sum of the weights in the path between them. For example, the length d between nodes I and V is given by: $\boldsymbol{d_{I-V}} = (13 + 13 + 14 + 18 + 12) = 70$. In general, given a weighted tree (T) with n taxa (end-nodes), computation of the path, $\boldsymbol{d_{i,j}(T)}$ between any two leaves $(\boldsymbol{i, j})$ can be done as indicated in the above example as regard to $\boldsymbol{d_{I-V}}$.

Now, considering an inverse problem, suppose a distance matrix $[\boldsymbol{n \times n}]$ with $\Delta_{i,j} = [\boldsymbol{d_{i,j}(T)}]$, (for every two leaf-nodes $(\boldsymbol{i, j})$, is available (as mostly gathered *via* biological experiments). A method is then required to search for a tree T that has n leaf nodes and consistent with the data in hand. Whenever the pertinent matrix size is small say, $[\boldsymbol{3 \times 3}]$, symmetric as well as non-negative, then the construction of the tree could be trivial. But for larger matrix sizes, the number of trees to be constructed becomes unwieldy. This could be evinced from the following algorithmic relations[2]: given ν nodes, the number of rooted trees that can be designed is $\boldsymbol{\varepsilon_R} = (2\boldsymbol{\nu} - 3)!! = (2\boldsymbol{\nu} - 3)(2\boldsymbol{\nu} - 5)(2\nu - 7) \ldots \infty$, and number of conceivable unrooted trees is $\varepsilon_{UR} = (2\boldsymbol{\nu} - 5)!! = (2\boldsymbol{\nu} - 5)(2\boldsymbol{\nu} - 7)(2\boldsymbol{\nu} - 9) \ldots \infty$. Illustrated in Fig. 10.3 is a couple of simple examples.

The assessment of evolutionary distance (ED) relies on the so-called "distance measures" specified in terms of the number of nucleotide or amino acid substitutions. Examples of this distance-based method (as will be explained later) are (A) The *unweighted pair group method with arithmetic mean* (UPGMA), (B) the *neighbors relation method,*[3] and (C) the *transformed distance method.*

The steps along the pipeline of framing a distance method-based phylogenetic tree are as follows: (i) first, the test genome sequences of the species are considered; (ii) second, for all pairs of species considered, the homologous genes are ascertained; (iii) third, the homologs being represented are permuted; (iv) hence, the rearranged distance for all pairs of species are computed. (v) then the phylogenetic tree is built using the distances obtained; and (vi) the associated distance matrix depicts the distances for the *taxa* (leaves) of the test phylogenetic tree, which specifies the distances between the internal nodes.

In the context of evolutionary notions, the distance mentioned above (and specified as d_{ij}) denotes a *metric* subject to satisfying certain conditions, namely: (i) positivity ($\mathbf{d_{ij} \geq 0}$); (ii) identity ($\mathbf{d_{ij} = 0}$, ***iff i = j***); (iii) symmetry ($d_{ij} = d_{ji}$ for each (i, j)); and (iv) the associated triangle inequality is: ($\mathbf{d_{ij} \leq d_{ik} + d_{kj}}$) for each i, j, k. Furthermore, the constraints involved are based on evolutionary mechanism and are consistent with the following properties: *additivity.* Suppose every edge in a tree is labeled with a distance, d_{ij}, then the tree is additive, if for every pair of *taxa*, the distance between the *taxa* is the sum of the edge distances on the path between the *taxa*; for example, considering a distance matrix shown in Fig. 10.4, it corresponds to a tree topology shown in Fig. 10.5. An additive tree with a root corresponds to an *ultrametric tree*, if the distances between any two taxa (i and j) and their common ancestor root (k) are equal, that is, $\mathbf{d_{ik} = d_{jk}}$. The time elapsed since divergence of i and j from the common root is equal to $d_{ij}/2$. This implies that the edge lengths are measured by a molecular clock with a constant rate. In the case of an ultrametric tree, suppose i, j, k are any three species, $\mathbf{d_{ik} = d_{jk} \geq d_{ij}}$.

Raw distance values *vis-à-vis* phylogenetic character data can be decided by simple counts on the number of pairwise differences in character states. Specifically, as described in Chapter 2, the raw distance values in question refer to the *Manhattan distance* that conforms to

	α	β	χ	σ
α	0	6	12	12
β		0	12	12
χ			0	6
σ				0

Fig. 10.4 An example of distance matrix.

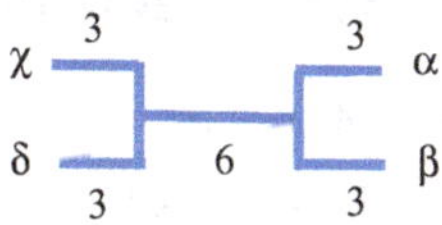

Fig. 10.5 Tree topology of the distance matrix of Fig. 10.4.

what is known as *taxicab geometry* (proposed by Hermann Minkowski in the nineteenth century). Inasmuch as the distance-matrix approach requires "genetic distance" evaluation between the sequences being classified, they need as an input, the multiple sequence alignment (MSA) described in earlier chapters. Relevant genetic distance is often defined as the fraction of mismatches at aligned positions with gaps either ignored or counted as mismatches.[4] Distance methods lay foundation for progressive and iterative types of MSA. But such methods may not use efficiently the information about any local high-variation regions that may appear across multiple subtrees.[5]

The so-called methods of *character state approach* rely on the state of the character, namely (i) the nucleotide or amino acid at a particular site and (ii) the presence or absence of an indel at a certain DNA location. An example of character state methods is the *maximum parsimony* (MP) *method.* Yet another statistical strategy of phylogenetic tree reconstruction using molecular data refers to *maximum likelihood* (ML) *method,* which uses all the information available in the sequence.

The genetic distance concept is a *data clustering* strategy that leads to *neighbor-joining* (NJ) *approach*[3] enabling reconstruction of unrooted trees. In its approach, the NJ exercise does not assume a constant rate of evolution across lineages. As such, the time of evolutionary divergence cannot be ascertained from mutations incurred. The main virtue of NJ (relative to other methods) is its computational efficiency. That is, NJ is a polynomial time algorithm, and it can be used on very large datasets for which other means of phylogenetic analyses (such as minimum evolution (ME), MP, and ML, which accommodate computationally intense of large data).

The NJ approach is statistically consistent with several models of evolution. Hence, given data of sufficient length, NJ will reconstruct the true tree with high probability. Atteson[6] proved that if each entry in the distance matrix differs from the true distance by less than half of the shortest branch length in the tree, then NJ would enable constructing the correct tree.

10.4 Descriptions of Tree Reconstruction

The distance-based approach of tree (re)construction as indicated earlier is based on the heuristics of specifying certain *distance measures,* such as the number of nucleotide or amino-acid substitutions that had occurred along the evolutionary process. The underlying methods of distance-based approach are outlined below.

A. UPGMA Method

The technique of UPGMA refers to a simple agglomerative (bottom-up) hierarchical clustering method. It is generally attributed to Sokal and Michener,[7,8] and the tree construction strategy was developed originally in the context of taxonomic phenograms that correspond to trees reflecting phenotypical similarities between *operational taxonomic units* (OTUs). The underlying consideration in the UPGMA is based on the relationship existing between organisms viewed in terms of similarity (instead of probing their genealogy). That is, similar organisms are grouped (clustered) together.

The method of such grouping exercised *via* the phenetic approach of UPGMA involves first identifying the groups to make clusters of varying degrees of similarity (on the basis of least-distant to most-distant relationships). It makes the tree clock-like, specific to the so-called *molecular clock hypothesis* (MCH) conceived by Zuckerkandl and Pauling.[9] The molecular data used in relevant calculations are typically nucleotide sequences for DNA or amino acid sequences for proteins.

Corresponding to the *gene clock* (or *evolutionary clock*), the construction of a phylogenetic tree can be viewed as being *ultrametrically* consistent. That is, an *ultrametric tree* as defined earlier has all the path lengths from the root up to the leaf nodes being equal; The ultrametric trees conceived in terms of gene clock imply that the time of divergence and number of changes due to accepted mutations (in amino acids) is proportional to time. Pertinent construction implies that the resulting tree has real numbers at its nodes with decreasing order as illustrated with an example in Fig. 10.6(A). Suppose the distances between the set of *taxa*, $\{\alpha, \beta, \chi, \delta, \varepsilon\}$ in Fig. 10.6(A) are considered, then an *ultrametric matrix* can be constructed as shown in Fig. 10.6(B). It is a symmetric matrix with diagonal values being zero and off-diagonal values are positive.

The UPGMA thus refers to constructing taxonomic phenograms that depict trees reflecting the phenotypic similarities between OTUs. The underlying method of phenetic approach involves identifying groups or clusters of species with varying degrees of similarity (i.e., on the basis of least distant to most distant relationships). Hence, UPGMA can be adopted arbitrarily with a large dataset. But a single tree can be obtained

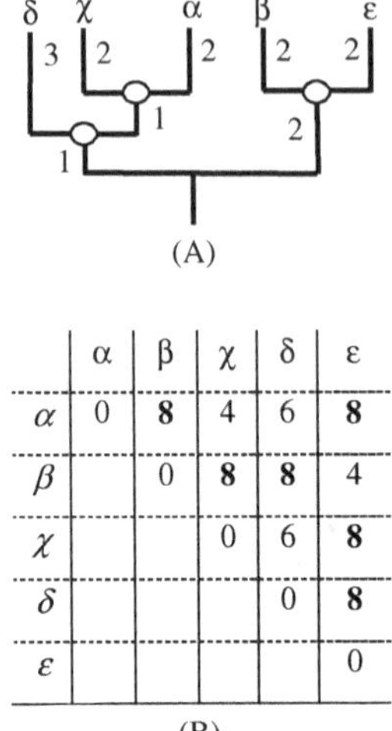

	α	β	χ	δ	ε
α	0	**8**	4	6	**8**
β		0	**8**	**8**	4
χ			0	6	**8**
δ				0	**8**
ε					0

Fig. 10.6 (A) An example of ultrametric tree and (B) corresponding ultrametric matrix of the tree shown.

in a broad sense of similarity and does not look if the relationships considered are historically significant.

In summary, UPGMA employs a sequential clustering algorithm, in which local topological relationships are identified in order of similarity, and the phylogenetic tree is built in a ladder format. First, identified from among all the OTUs are those two OTUs that are most similar to each other and then these two are combined as a new single OTU. This combined OTU is called a *composite OTU*. Subsequently, a distance matrix is constructed with the composite OTU plus the rest of the OTUs. Again, the pair with the highest similarity is identified to make the new composite and so on, until only two OTUs are left out. Thus, the UPGMA employs a sequential clustering algorithm, in which the local topological relationships are identified in their order of similarity so as to build the phylogenetic tree in a ladder format as illustrated *via* a pseudocode in Table 10.1.

B. NJ Method

As mentioned before, the NJ method refers to an algorithmic procedure to find a tree with the shortest, root-to-*taxa* distance. This method essentially involves in reconstructing the phylogenetic trees computation of the lengths of the branches of the tree. In each stage of reconstruction, the two nearest nodes of the tree are chosen and

Table 10.1 Pseudocode illustrating the computation on reconstructing a phylogenetic tree *via* UPGMA method.

%% Reconstructing a phylogenetic tree *via* UPGMA method can be done with a set of *taxa* and a corresponding distance matrix of OTU set

Input

→ Suppose the pairwise evolutionary distance matrix (EDM) of the set of six OTUs set: {α, β, χ, δ, ε, ϕ} is as in Fig. 10.7:

	α	β	χ	δ	ε
β	2				
χ	4	4			
δ	6	6	6		
ε	6	6	6	4	
ϕ	8	8	8	8	8

Fig. 10.7 Input data: given pairwise test distance matrix for the test tree construction using UPGMA procedure.

Perform

Step 1:

→ In the given input phylogeny dataset, the smallest distance can be identified between α and β, ($\alpha \leftrightarrow \beta$).

→ It is equal to 2 and the branching point is positioned at a distance of 2/2 = 1 substitution.

→ Therefore, the resulting subtree is as shown in Fig. 10.8

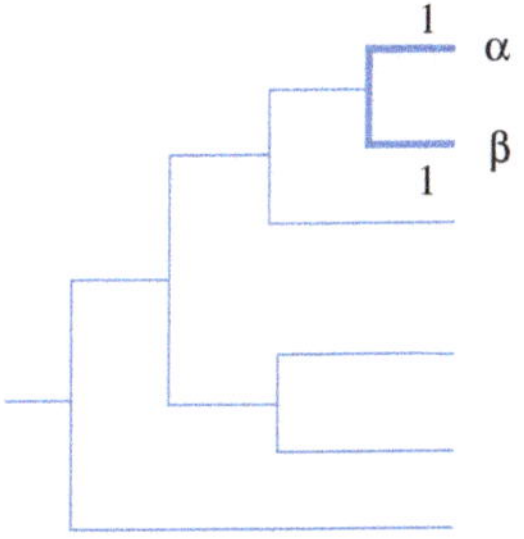

Fig. 10.8 Step-1: subtree construction.

→ Using the first clustering of α and β *via* Step 1, the resulting single composite Out(α, β) is specified; and hence, a new subtree and distance matrix can be elucidated as in Step 2 indicated in Figs. 10.9 and 10.10.

(Continued)

Table 10.1 (*Continued*)

Next

Perform

→ Step 2: ($\alpha \leftrightarrow \beta$)

Denoting the "distance" by the notation—DIST(.), the follow-up set of clustering with respect to the composite OTU(α, β) can be written as follows and the reconstructed subtree is shown in Fig. 10.9:

DIST(α, β), χ = {DIST($\alpha\chi$) + DIST($\beta\chi$)}/2 = 4
DIST(α, β), δ = {DIST($\alpha\delta$) + DIST($\beta\delta$)}/2 = 6
DIST(α, β), ε = {DIST($\alpha\varepsilon$) + DIST($\beta\varepsilon$)}/2 = 6
DIST(α, β), ϕ = {DIST($\alpha\phi$) + DIST($\beta\phi$)}/2 = 8

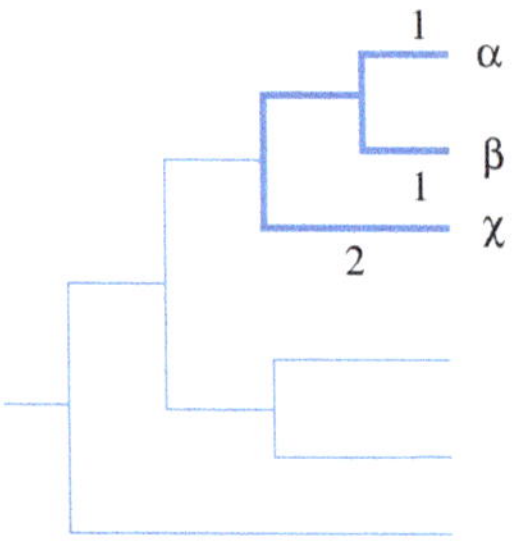

Fig. 10.9 Step-2: subtree construction.

%% In essence, the distance between a single OTU and a composite OTU is the average of the distances between the single OTU and the constituent single OTUs of the composite OTU. Hence, a fresh distance matrix can be ascertained using the newly determined distances, and this procedure is iterated leading to the following steps and results thereof.

Next

Perform

→ Step 3

← It refers to clustering of the composite OUT-($\alpha\beta$) with respect to the rest of other OTUs, and the distance matrix as in Fig. 10.10 can be written:

	$\alpha\beta$	χ	δ	ε
χ	4			
δ	6	6		
ε	6	6	4	
ϕ	8	8	8	8

Fig. 10.10 New distance matrix after Step 3.

Table 10.1 (*Continued*)

Next

Perform

→ Step 4

← It refers to clustering of the composite OTU-($\alpha\beta$), OTU-χ, and composite OTU-$\delta\varepsilon$, and the resulting distance matrix can be written as in Fig. 10.11

	$\alpha\beta$	χ	$\delta\varepsilon$
χ	4		
$\delta\varepsilon$	6	6	
ϕ	8	8	8

Fig. 10.11 New distance matrix after Step 4.

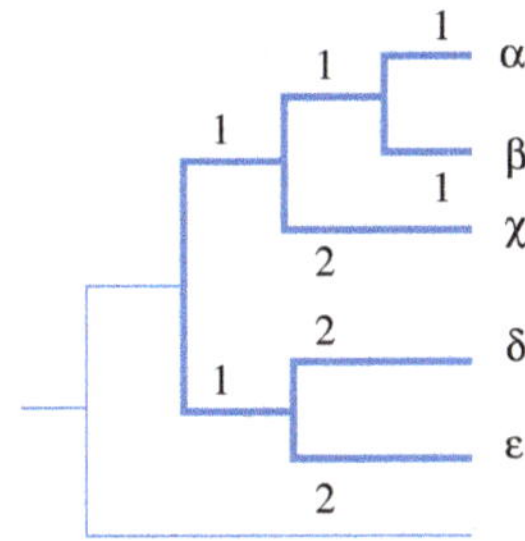

Fig. 10.12 Steps 3 and 4: subtree construction.

Next

Perform

→ Step 5

← It refers to clustering of the composite OTUs-{($\alpha\beta$), χ} and composite OTU-$\delta\varepsilon$; hence, the resulting distance matrix can be specified as in Fig. 10.13

	$\alpha\beta, \chi$	$\delta\varepsilon$
$\delta\varepsilon$	6	
ϕ	8	8

Fig. 10.13 New distance matrix after Step 5.

Next

Perform

→ Step 6

← It refers to clustering of the composite OTUs-($\alpha\beta\chi$) and composite OTU-$\delta\varepsilon$; hence, the resulting distance matrix is shown in Fig. 10.14.

(*Continued*)

Table 10.1 (*Continued*)

	$\alpha\beta\chi$, $\delta\varepsilon$
ϕ	8

Fig. 10.14 New distance matrix after Step 6.

Next

Perform

→ Step 7

← This procedure frames an unrooted tree with the presumed UPGMA implying equal rates of mutation along all branches across the evolution model considered.

→ Theoretically, the root is then at equidistant from all OTUs. As such, pursuing the method of mid-point rooting, the root of the entire tree is then located at DIST{($\alpha\beta\chi\delta\varepsilon$), ϕ}/2 = 4

→ The inferred (reconstructed) phylogenetic tree *via* the UPGMA method is as illustrated in Fig. 10.15.

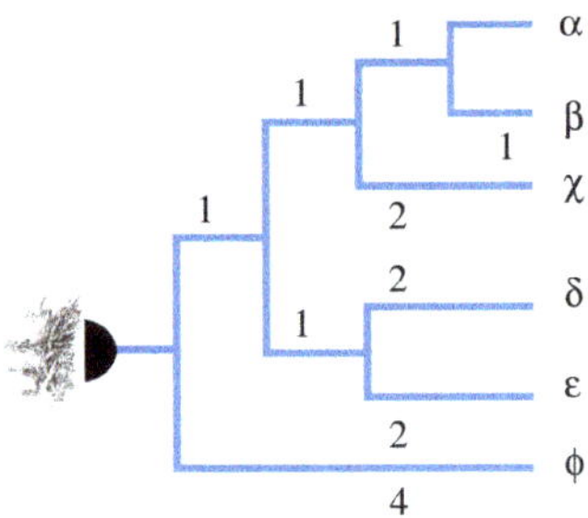

Fig. 10.15 Steps 5–7: resulting final tree construction.

%% The UPGMA method has however suffered from the following demerits: (i) The clustering method of UPGMA is highly sensitive "to unequal evolutionary rates." That is, whenever one of the OTUs has suffered more mutations over the evolutionary time scale (compared to other OTU), the resulting/reconstructed tree could be of incorrect wrong topology, (ii) The clustering procedure is valid only when the data are *ultrametric,* and (iii) such ultrametric distances are defined in terms of the so-called *three-point condition.*

This condition can be specified as follows: given a set of any three *taxa,* {α, β, χ}, then relevantly: DIST ($\alpha\chi$) corresponds to: max [DIST ($\alpha\beta$), DIST ($\beta\chi$)]. That is, "the two greatest distances are equal" implying that the UPGMA implicitly presumes that the evolutionary rate is invariant

Table 10.1 (*Continued*)

```
for all branches. However, should the rate constancy among
lineages not hold, then the UPGMA may yield an invalid
topology. This will be illustrated by an example in the
review section along with more examples and problems on
UPGMA procedure.

   End
```

defined as *neighbors.* This is recursively followed until all of the nodes are paired together. The algorithm was originally developed by Saitou and Nei[3] with subsequent corrections on the proof of the algorithm plus some minor changes (to the algorithm) by Studier and Kepler.[10]

The NJ method of tree reconstruction can be summarized as follows: first, the phylogenetic neighbors are defined as a pair of OTUs connected at a single node between them. For instance, consider a tree illustrated in Fig. 10.16 with a (non-internal) node set $\{\alpha, \beta, \chi, \delta\}$, and the set $\{X, Y\}$ depicts an internal node pair. The associated branch lengths correspond to $\{\ell_a, \ell_b, \ell_c, \ell_d\}$ as illustrated. Furthermore, α and β are neighbors since they are connected through a common internal node.

Likewise, χ and δ are also neighbors. (However, as is evident, $\{\alpha, \delta\}$ and $\{\beta, \chi\}$ are not neighbors.) The underlying algorithm in constructing the NJ method based tree assumes that the trees are *additive trees.* An additive tree means a tree where, for example, in Fig. 10.16, the distance between the nodes α and β is equal to the distance between nodes, α and X plus the distance between nodes, β and X. Furthermore, a *four-point condition* is stipulated to specify the general observation that neighbors are closer than non-neighbors. Referring to Fig. 10.16, with the observed details,

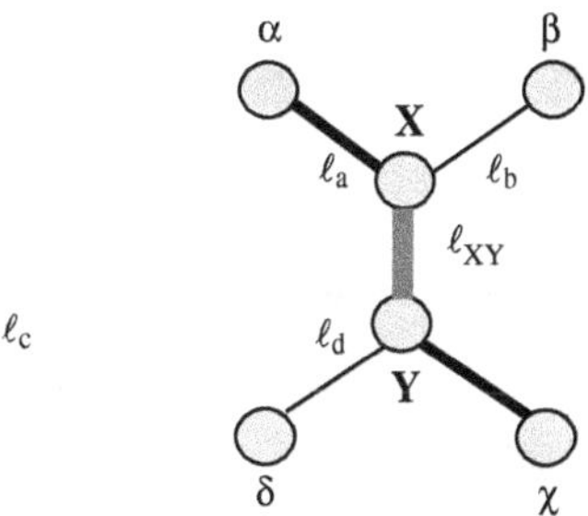

Fig. 10.16 Illustration of neighbors in the phylogenetic trees.

namely, $(d_{\alpha\chi} + d_{\beta\delta}) = (d_{\alpha\delta} + d_{\beta\chi}) = (\ell_a + \ell_b + \ell_c + \ell_d + 2\ell_{XY}) = (d_{\alpha\beta} + d_{\chi\delta} + 2\ell_{XY})$, the corresponding *four-point condition* can be stated as follows:

[Sum of all distances between neighboring nodes: for example, $(d_{\alpha\beta} + d_{\chi\delta})$ in the sets $\{\alpha, \beta\}$ and $\{\chi, \delta\}$] < [Sum of all distances between non-neighbor nodes: for example, $(d_{\alpha\chi} + d_{\beta\delta})$ in the sets $\{\alpha, \chi\}$ and $\{\beta, \delta\}$] Eq. (10.1a)

[Sum of all distances between neighboring nodes: for example, $(d_{\alpha\beta} + d_{\chi\delta})$ in the sets $\{\alpha, \beta\}$ and $\{\chi, \delta\}$] < [Sum of all distances between non-neighbor nodes: for example, $(d_{\alpha\chi} + d_{\beta\delta})$ in the sets $\{\alpha, \delta\}$ and $\{\beta, \chi\}$] Eq. (10.1b)

Suppose, there are N number of OTUs with a matrix in which the EDs between every pair of OTUs are known. Then, the process of building the tree starts with no initial clustering. That is, every OTU is considered as being in an equal relationship with every other OTU. This leads to initiating a nonhierarchical (star-like) structure shown in Fig. 10.17(A).

Let the distance between OTU pair i and j be d_{ij} and ℓ_{ab} be the distance between any nodes, a and b. The sum of the branch lengths for the nonhierarchical star-like tree centered at X in Fig. 10.17(A) is then given by:

$$S_o = \left[\sum_{i=1}^{N} \ell_{iX}\right] = \left[\frac{1}{N-1}\sum_{(i<j)} d_{ij}\right] \qquad \text{Eq. (10.1c)}$$

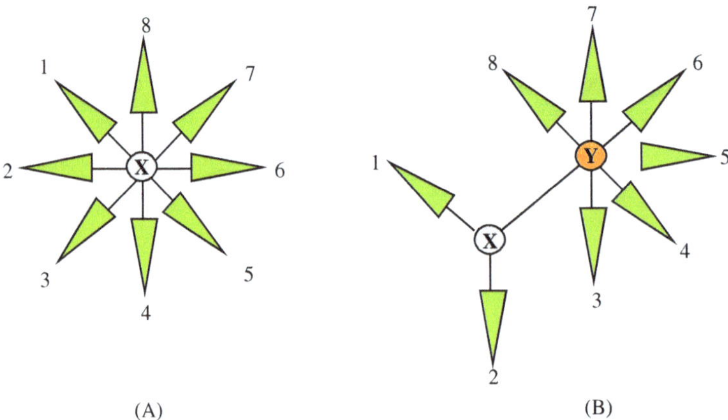

Fig. 10.17 (A) Initialized star-like, nonhierarchical tree structure and (B) the tree realized with the given set of OTUs.

The above relation is due to the fact that each branch is counted $(N-1)$ times when all distances are added.

The next step of tree construction (*via* NJ method) involves identifying the first pair of OTUs to be clustered, implying that they diverge from the same node in the tree. Considering the example (with eight OTUs, as above) forming the star is as shown in Fig. 10.17(A). Suppose the nodes 1 and 2 constitute the first pair being clustered at X, and the rest of OTUs are clustered at another node Y as illustrated in Fig. 10.17(B). That is, for this assumed tree, the OTU pair {1, 2} diverges from a single node (X) and the other OTU set {3, 4, 5, 6, 7, 8} stems from another node (Y). Correspondingly, the "length" (S_{12}) of the tree can be determined using a relevant algorithm is as follows: the length of the tree (S_{ij}) for a given node pair $\{i, j\}$ can be deduced in terms of number of OTUs (N) and pairwise distances d_{ij} for every two leaf nodes (*taxa*/OTUs) (i, j). Hence, with respect to Fig. 10.17(B), the branch length between X and Y (ℓ_{XY}) is given by

$$\ell_{XY} = \left[\frac{1}{\{2(N-2)\}}\sum_{(k\,=\,3,\,N)}(d_{1k}+d_{2k})\right] - [(N-2)(\ell_{1X}+\ell_{2X})] - \left[2\sum_{i=3}^{N}\ell_{iY}\right] \qquad \text{Eq. (10.2)}$$

where the term $[\frac{1}{\{2(N-2)\}}\sum_{(k\,=\,3,\,N)}(d_{1k}+d_{2k})]$ specifies the sum of all distances that include ℓ_{XY}.

Furthermore, $[(N-2)(\ell_{1X}+\ell_{2X}]$ and $[2\sum_{i=3}^{N}\ell_{iY}]$ are terms that depict irrelevant entities being excluded in computing ℓ_{XY}. Furthermore, if the interior branch (X to Y) in Fig. 10.17(B) is removed, there will be two independent star-like trees, one for the OTU pair {1 and 2} and the other for the remaining set of $(N-2)$ OTUs. Then, the corresponding branch lengths $(\ell_{1X}+\ell_{2X})$ and $[\sum_{i\,=\,3}^{N}\ell_{iY}]$ can be deduced by applying Eq. (10.2). That is,

$$(\ell_{1X}+\ell_{2X}) = d_{12} \qquad \text{Eq. (10.3a)}$$

and

$$\left[\sum_{i=3}^{N}\ell_{iY}\right] = \left[\frac{1}{N-3}\sum_{(3\,\le\,i\,<\,j)} d_{ij}\right]. \qquad \text{Eq. (10.3b)}$$

Adding these branch lengths, the sum of all branch lengths of the tree configuration of Fig. 10.17(B), namely, S_{12} is then obtained as follows:

$$S_{12} = [\ell_{XY} + (\ell_{1X} + \ell_{2X}] + \left[\sum_{i=3}^{N} \ell_{iY}\right] + \left[\frac{1}{\{2(N-2)\}} \sum_{(k=3,\, N)} (d_{1k} + d_{2k})\right]$$

$$= \left[\frac{1}{\{2(N-2)\}} \sum_{(k=3,\, N)} (d_{1k} + d_{2k})\right] + \left[\frac{1}{2} d_{12}\right] + \left[\frac{1}{(N-2)} \sum_{(3 \le i < j)} d_{ij}\right] \qquad \text{Eq. (10.4)}$$

where the three terms on the right-hand side denote explicitly the following: $[\frac{1}{\{2(N-2)\}} \sum_{(k=3,\, N)} (d_{1k} + d_{2k})] =$ Mean distance from OTU1 (with $k = 1$) and OTU2 (with $k = 2$) to the rest of the OTUs with $k = 3, 4, \ldots,$ and N $[\frac{1}{2} d_{12}] =$ Semi-pairwise distance from OTU1 to OTU2 and $[\frac{1}{N-2} \sum_{(3 \le i < j)} d_{ij}] =$ Average of all pairwise distances between the OTUs with $k = 3$ to N. Eq (10.4) refers to the sum of least-squares estimate of branch lengths.

Thus, S_{12} denotes the total tree length when OTU1 and OTU2 are considered to form the first cluster. Likewise tree lengths $S_{1j} \rightarrow S_{13}, S_{14}, \ldots, S_{1N}$ can be calculated, and the lowest of these would indicate the "true" pair of OTUs to be first clustered (implying the first neighbors to be joined). These first two OTU neighbors are then merged to depict a single (composite) OTU. For example, suppose S_{12} is found to be smallest among all values of the set $\{S_{ij}\}$. Then OTUs 1 and 2 are designated as a pair of (nearest) neighbors and these are joined to form a combined OTU (1–2). Relevant distance between this combined OTU with respect to any other OTU j is given by $D_{(1-2)j} = (D_{1j} + D_{2j})/2$ for $(3 \le j \le N)$. Now, the number of OTUs involved is reduced by one, and for the resulting new distance matrix, the previous procedure is applied so as to find the next pair of (nearest) neighbors. This cycle of steps is iterated until the number of OTUs becomes three upon a single unrooted tree.

In all, the NJ topology gives the least total branch length at each step of the algorithm. Hence, it specifies the minimum evolution criterion for phylogenetic trees. But, as stated before, the NJ solution may not lead to the true tree topology with this least total branch length specification inasmuch as it is based on a step-wise approach and conforms to the so-called *greedy algorithm*. Furthermore, the NJ method is computationally efficient and conforms to polynomial time algorithm. Very large datasets can be handled by NJ approach. (Other phylogenetic analyses/ inferences like *ME**, *MP***, and *ML**** are, however, computationally intensive.)

(*Note:* ME method*: based on the assumption that the tree with the smallest sum of branch length estimates is most likely to be the true one; MP method**: the phylogenetic tree that minimizes the total number of character state changes is a preferred result of interest, and ML method*** determines the tree topology, branch lengths, and parameters of the evolutionary model that maximize the probability of observing the sequences at hand).

The following details are salient characteristics of the NJ method described above:

(a) The NJ strategy is statistically consistent in yielding a reconstructed "true tree" with high probability under many models of evolution.[6,11]
(b) NJ topology gives the least total branch length is preferred at each step of the algorithm. Hence, it specifies minimum evolution criterion for phylogenetic trees.
(c) NJ method is computationally efficient and conforms to polynomial time algorithm. Very large datasets can be handled by NJ approach.
(d) NJ is a form of star decomposition, and, heuristically, it is a standalone approach yielding reasonably acceptable trees.

In essence, the NJ method is based on an iterative algorithm, which assumes an additive tree or the summing procedure in its each iteration as could be evinced from the pseudocode presented in Table 10.2.

Table 10.2 Pseudocode on NJ method of phylogenetic tree construction.

```
%% NJ method of phylogenetic tree construction
Input
  ← A set of test multiple sequences data
Call
  → A 'subroutine' for sequence alignment
    → It enables global multiple alignment and selecting reli-
    able data of aligned sequences devoid of any gaps; that is,
    any gap present is skipped or ignored—Working alignment
Next
← Build distance matrix between sequences
Define
← A distance parameter (d_i,j) between the nodes 1 through N
   → The distance parameter can be defined by different
   notions, such as:
   (a) Jukes-Cantor method[12,13] (b) Kimura method[14]
   (c) p-distance method, etc.

%% Descriptive notes on distance methods estimation
    Distance estimates refer to estimating the mean number
    of changes per site, since two species (seen via their
    nucleotide sequences) split from each other. If the number
    of differences is simply counted (yielding the so-called
    p-distance*), it may underestimate the extent/amount of
    changes taken place, especially whenever the sequences
    are very dissimilar as a result of multiple hits. (Note:
    *the p-distance method defines the distance between two
    sequences as equal to the ratio of count of different
    positions [of a nucleotide, for example] to total number
    of positions. More appropriate estimates in vogue are
    based on a model that includes parameters reflecting the
    way the sequences might have evolved. The distance models
    in general are based on common strategies such as ML-
    Jukes Cantor approach, which assumes that all changes are
    equally likely.
    That is, in the Jukes and Cantor model,[12,13] the rate of
    nucleotide substitution remains unaltered for all pairs of
    the four nucleotides A, T, C, and G. It corresponds to an
    ML estimate of the number of nucleotide substitutions
    between two sequences. The underlying assumptions are that
    the equality of substitution rates among sites and equal
    nucleotide frequencies and the higher rate of transitional
    substitutions as compared to transversional substitutions
    are not corrected.
```

Table 10.2 (*Continued*)

	A	T	C	G
A	–	α	α	α
T	α	–	α	α
C	α	α	–	α
G	α	α	α	–

Fig. 10.18 An example of $[N \times N]$ distance matrix.

Note that the distance matrix shown in Fig. 10.18 is symmetric, and as such, only the values above/below the diagonal need to be computed; that is, the matrix is symmetric regardless of the diagonal and it is suffice to show normally only the top half (or lower half) filled.

The Jukes-Cantor distance is computed as follows: denoting the *distance* depicting the number of nucleotide substitutions per site as *d*, it is given by: $\mathbf{d = [-(3/4) \times \log_e\{1-(4/3) \times p\}]}$, where *p* is the proportion of sites with different nucleotides inclusive of transitions plus transversions, and corresponding variance of *d* is given by Nei and Kumar **(p. 36)**[13]: $\mathbf{Var(d) = \{p \times (1 - p)\}/\{L \times [1-(4/3) \times p]^2\}}$ where *L* denotes the number of valid common sites relevant to number of sites being compared.

Kimura distance models[14]: there are two versions of this model:(i)the *Kimura two-parameter model*, which corrects for multiple hits by taking into account both transitional as well as transversional substitution rates, and at the same time, it assumes that the frequencies (of occurrence) of four nucleotides {*A*, *C*, *T*, *G*} are equal; also, the rates of substitution remain unaltered among sites. Second,(ii)the so-called *Kimura two-parameter gamma model* corrects for multiple hits, while taking into account the transitional and transversional substitution rates; furthermore, the differences in substitution rates among sites are also duly considered. In this context, the evolutionary rates among sites are modeled in terms of gamma distribution statistics, and relevantly, a gamma parameter is prescribed while computing this distance.

	A	T	C	G
A	–	β	β	α
T	β	–	α	β
C	β	α	–	β
G	α	β	β	–

Fig. 10.19 An example of $[N \times N]$ distance matrix.

(*Continued*)

Table 10.2 (*Continued*)

Kimura two parameter model corrects for multiple hits, taking into account transitional and transversional substitution rates, while assuming that the four nucleotide frequencies are the same and that rates of substitution do not vary among sites. Correspondingly, the Kimura two-parameter model is specified by the matrix show in Fig. 10.19 (where α and β denote transitional and transversional substitution rates, respectively).

Relevantly, the following parameters can be computed using the algorithmic details given by Day[15]:

Number of transitions plus transitions = Number of nucleotide substitutions per site: d
Number of transitions only = Number of transitional substitutions per site: s
Number of transversions only = Number of transversional substitutions per site: v
Ratio: (Number of transitions/Number of transversions) = $R = s/v$
Number of valid common sites = Number of sites compared = L

The algorithms toward the required computations of the entities as above are as follows: suppose P and Q denote the frequencies of occurrence of the sites with transitional and transversional differences, respectively, suppose $w_1 = (1 - 2P - Q)$ and $w_2 = (1 - 2Q$, then $d = [\log_e(1/w_1)^{1/2} + \log_e(1/w_2)^{1/4}]$; $s = [\log_e(1/w_1)^{1/2} + \log_e(w_2)^{1/4}]$; $v = [\log_e(1/w_2)^{1/2}]$; and $R = (s/v)$. The entities $\{d, s, v, R\}$ conform to a set of random variates.

Next
Enter

← Distance parameter as elements in a matrix.
→ Resulting distance matrix (DM) is used to identify the first branch of the tree as consisting of the two OTUs that have the shortest distance between them

%% An example of an $[N \times N]$ distance matrix is illustrated in Fig. 10.20, where the columns and rows of this matrix denote nodes, and the values $(\mathbf{d}_{i,j})$ of the matrix elements represent the distance between nodes i and j.

Table 10.2 (Continued)

	$i \rightarrow$					
	d_{11}	d_{12}	-	-	-	d_{1N}
	d_{21}	d_{22}				d_{2N}
	-					
j	-			d_{ij}		
$\downarrow$	-					
	d_{N1}	d_{N2}				d_{NN}

Fig. 10.20 An example of $[N \times N]$ distance matrix.

Node		1	2	3	4	5
	Taxa*	BSu	Bst	Lvi	Amo	Mlu
1	Bsu	0	0.172	0.215	0.309	0.233
2	Bst		0	0.299	0.340	0.206
3	Lvi			0	0.280	0.394
4	Amo				0	0.429
5	Mlu					0

Taxa*

Bsu: *Bacillus subtilis*

Bst: *Baclillus stearothermophilus*

Lvi: *Lactobacillus viridescens*

Amo: *Acholeplasma modicum*

Mlu: *Micrococcus luteus*

Fig. 10.21 An example of [5 × 5] distance matrix with real (numerical) data matrix for five ribosomal RNA sequences of a taxa set

A practical [5 × 5] distance matrix with numerical values is shown in Fig. 10.21. It corresponds to real data matrix for five *ribosomal RNA* (rRNA) sequences. Each value denotes the estimated number of nucleotide residue substitutions per position separating the corresponding pair of the presently existing sequences.

Implement

← Computation toward realizing the tree structure

Initialize

→ Step 1

← Start off with a star tree: Fig. 10.22

(Continued)

Table 10.2 (*Continued*)

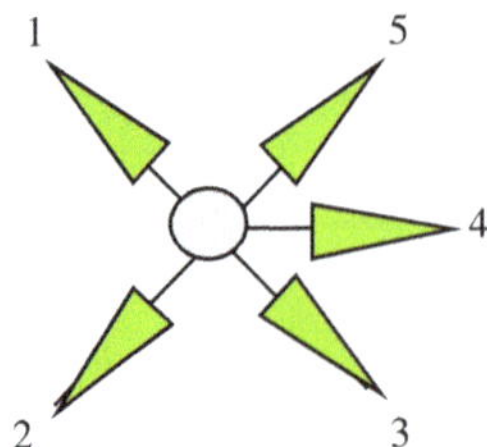

Fig. 10.22 Star-like tree initialized.

→ Step 2

← Define/identify the neighbors

← Two nodes with the lowest values in the matrix are chosen and they are defined as *neighbors*. In the example shown above, the nodes 1(Bsu) – 2 (Bst) show the lowest value (0.172) in the distance matrix. So, they are the "nearest" and identified as "neighbors."

← Merge these neighbors to form a single composite (new) OTU and label it as (1-2)

→ Step 3

← Compute the branch lengths for:
(i) Between this composite node (1-2) and (ii) the rest of the other nodes

← This computation involves taking the unweighted arithmetic mean of all the pairwise distances of the new OTU(1-2) with respect to all other OTUs (namely, 3, 4, 5). Hence, the results are:

- Distance between (1-2) and [3 = $d_{(1-2)3}$ = $(d_{13} + d_{23})/2$]
- Distance between (1-2) and [4 = $d_{(1-2)4}$ = $(d_{14} + d_{24})/2$]
- Distance between (1-2) and [5 = $d_{(1-2)5}$ = $(d_{15} + d_{25})/2$]

→ Step 4

← Construct the resulting table as shown in Fig. 10.23

Table 10.2 (*Continued*)

Node		1	2	3	4	5
		Bsu	Bst	Lvi	Amo	Mlu
1	Bsu	0	**0.172**	0.215	0.309	0.233
2	Bst		0	0.299	0.340	**0.206**
3	Lvi			0	0.280	0.394
4	Amo				0	0.429
5	Mlu					0

Node		1-2	3	4	5
		Bsu-Bst	Lvi	Amo	Mlu
1-2	Bsu-Bst	0	0.257	0.325	0.220
3	Lvi		0	0.280	0.394
4	Amo			0	0.429
5	Mlu				0

Fig. 10.23 Numerical example of [5 × 5] distance matrix: Step 4.

→ Step 5

← Repeat the process from Step 2 again looking for the two nearest nodes and constructing the resultant matrix of Fig. 10.24

→ OTU 5 (Mlu) and the composite OTU (1–2)(Bsu/Bst) constitute the nearest neighbors

→ Construct the new matrix with:

$d_{(1\text{-}2)5} = (d_{15} + d_{25})/2 = (d_{1x} + d_{2x})$

$d_{[(1\text{-}2)5]3} = (d_{1x} + d_{13})/2 + (d_{2x} + d_{23})/2$

$d_{[(1\text{-}2)5]4} = (d_{1x} + d_{14})/2 + (d_{2x} + d_{24})/2$

Node		(1-2)5	3	4
		(Bsu/Bst)Mlu	Lvi	Amo
(1-2)5	(Bsu/Bst)Mlu	0	0.282	0.435
3	Lvi		0	0.280
4	Amo			0

Fig. 10.24 Numerical example of [5 × 5] distance matrix: Step 5.

(*Continued*)

Table 10.2 (Continued)

→ Step 6

← Repeat the process from Step 2 again looking for the two nearest nodes and constructing the resultant matrix in Fig. 10.25

→ Composite OTU(1-2)5: (Bsu-Bst)Mlu and the composite OTU(3-4): (Lvi-Amo)form nearest-neighbors

→ Construct the new matrix with:

$d_{(1-2)5} = (d_{1x} + d_{2x})$

$d_{[(1-2)5](3-4)} = (d_{1x} + d_{1x-3} + d_{2x} + d_{2x-3})/4$

Node		(1-2)5 (BSu-BSt)Mlu	(3-4) (Lvi-Amo)
(1-2)5	(Bsu-Bst)Mlu	0	0.289
(3-4)	(Lvi-Amo)		0

Fig. 10.25 Numerical example of [5 × 5] distance matrix: Step 6.

%% The solution described above is regarded as the required final tree inasmuch as no prior information is available on more possible reductions.

End

Fitch–Margoliash Method: Determining Branch Lengths

Toward phylogenetic tree construction, another computation indicated refers to determining the branch lengths. This can be done *via Fitch–Margoliash (FM) algorithm*[11] as outlined as follows:

Suppose, for example, the OTU pair {1, 2} is the first pair being joined in the tree of Fig. 10.26, then the corresponding λ_{1X} and λ_{2X} are estimated by:

$$\lambda_{1X} = (d_{12} + d_{1Z} - d_{2Z})/2 \quad \text{Eq. (10.5a)}$$
$$\lambda_{2X} = (d_{12} + d_{2Z} - d_{1Z})/2 \quad \text{Eq. (10.5b)}$$

where Z denotes a group of all OTUs (leaf nodes) excluding the OTU pair {1, 2}. It represents intuitively the whole cluster of all included OTUs as shown in Fig. 10.26, and d_{1Z} and d_{2Z} are distances between (1 and Z) and (2 and Z), respectively, which are determined explicitly by the following relations [8.3]: $d_{1Z} = [\frac{1}{N-2}\sum_{i=3}^{N} d_{1i}]$ and $d_{2Z} = [\frac{1}{N-2}\sum_{i=3}^{N} d_{2i}]$.

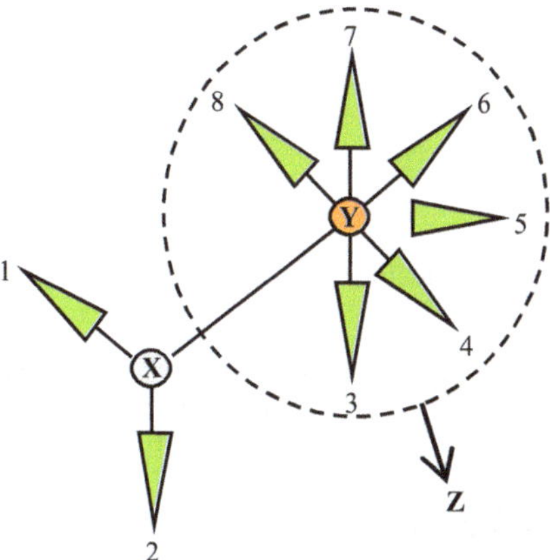

Fig. 10.26 Cluster (Z) of all the leaf nodes excluding {1 and 2}.

And, λ_{1X} and λ_{2X} denote least-square estimates for the tree of the illustration in Fig. 10.26.

The FM method uses the genetic distance and follows a weighted least-squares method for clustering. That is, more weight in the tree construction process is given to closely related sequences. This approach would reduce the inaccuracy associated in the distance measurement on distantly related sequences. In the relevant computation, the distances used as input to the algorithm are first normalized so as to prevent large deviational artifacts in (computing) relationships between closely related *versus* distantly related groups. Furthermore, the distances adopted for this method must be linear. This linearity criterion for distances requires that the expected values of the branch lengths for two individual branches must equal the expected value of the sum of the two branch distances. This property in the context of biological sequences applies only when such sequences are corrected for the possibility of back mutations at individual sites. (Relevant correction can be done *via* substitution matrix such as those derived from the Jukes–Cantor model of DNA evolution). The distance correction is, however, necessary in practice only when the evolution rates differ among branches.

Though the least-squares criterion advocated enables the distances in FM approach to be more accurate it, burdens the computational efforts more heavily than the NJ methods. Finding the optimal least-squares tree with any correction factor is *NP-complete** warranting maximum parsimony analysis in the search through tree space. (Note: *NP complete**: suppose given a solution to a problem and it is verifiable

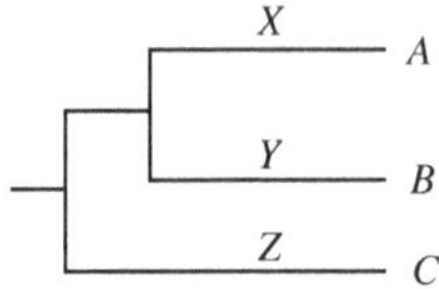

Fig. 10.27 Illustrative example of the FM method.

quickly (in polynomial time) and the associated class of complexity[15] is *NP complete* (abbreviated NP-C or NPC).

Illustratively, the FM method can be explained with a simple example as follows: given a simple tree with branch lengths, X, Y, and Z and the taxonic set $\{A, B, C\}$ as shown in Fig. 10.27, the estimation of X, Y, and Z can be done *via* FM approach as follows: $X = (d_{AB} + d_{AC} - d_{BC})/2$; $Y = (d_{AB} + d_{BC} - d_{AC})/2$ and $Z = (d_{AC} + d_{BC} - d_{AB})/2$.

10.5 Statistical Aspects of Tree Construction

Consistent with the introductory details indicated earlier, the three versions of tree reconstruction, namely, (i) parsimony method, (ii) distance-based or distance-matrix method, and (iii) the *ML method* conform to the following considerations: in inferring phylogenetic trees from a set of paralogous sequences, the internal nodes in the tree represent duplication events. Relevantly, the phylogenetic trees are viewed as unrooted trees inasmuch as the choice of the two nodes between which the root has to be placed is not obvious; and, typically, there are two techniques usually adopted to infer where the root of the tree is located. One approach is called *outgroup* rooting*.[5,16] (*Note*: The term *outgroup** implies the following: suppose three monophyletic groups of organisms are compared and two of them are more closely related to each other than either to the third, then the third group would depict the outgroup). The outgroup rooting technique is used to determine the locations of the root of the tree with the *a priori* assumption that the outgroup has diverged earlier from a common ancestor than the other groups in the tree.

Another method of constructing the tree refers to assuming a *molecular clock*, which implies that the underlying evolution on all lineages of the tree had occurred at the same rate (i.e., a homogeneous rate of evolution is implicitly assumed). This assumption makes the distance from the root to tip being equal for all paths emanating from the root. Thus, the root of the tree can be found by locating its place in the tree where the distance along all lineages to all tips is approximately equal.[5,16] In formulating a phylogenetic tree, the following concepts defined in the previous chapter are useful: (i) *plesiomorphy* and *symplesiomorphy* or *symplesiomorphic*; (ii) *synapomorphy*; and (iii) *homoplasy.*

Distribution of evolutionary characters (such as phenotypic traits or *alleles*) follows directly the branching pattern of evolution. Thus, it can be stated that if two organisms possessing a shared character should be more closely related to each other than to a third organism that lacks this character. "For example, bats and monkeys are more closely related to each other than either is to an elephant because, the male bats and monkeys possess external testicles, which elephants lack. (However, it is fallacious to state that the bats and monkeys are more closely related to one another than they are to whales, though the two have external testicles absent in whales."[17])

In all, the parsimonious approach in phylogenetics is a class of character-based tree estimation, which uses a matrix of discrete (phylogenetic) characters in order to infer one or more optimal phylogenetic trees for a given set of *taxa*. The term MP refers to an optimality criterion set forth in constructing a phylogenetic tree. By stipulating MP criterion, the resulting shortest tree is possibly the "best" consistent with the data being considered. Basic concept behind MP is due to James Farris[18] and Walter Fitch.[19]

The "best" tree obtained *via* maximum parsimony criterion could underestimate actual evolutionary change that might have occurred. Moreso, MP may not be statistically consistent guarantying to yield the "true tree" with high probability, given sufficient data.[20] In general, parsimony approach falsely assumes that the convergence is rare. Though convergently derived characters have some value in maximum-parsimony-based phylogenetic analyses, the presence of convergence may "not systematically affect the outcome of parsimony-based methods." Furthermore, computationally the time required for parsimony or phylogenetic analysis is proportionately decided by the number of *taxa* (and characters) considered in the analysis.

MP involves trees being scored *vis-à-vis* the degree to which the trees accommodate a parsimonious distribution of the character data. "The most parsimonious tree for the dataset represents the preferred hypothesis of relationships among the taxa in the analysis." Furthermore, the parsimony analysis often yields a number of equally *most-parsimonious trees* (MPTs) due to the following: (i) the number of missing entries ("?") in the dataset; (ii) the existence of characters showing too much homoplasy; and (iii) the presence of topologically labile "wildcard" taxa with innumerable extent of missing entries. As such, several methods have been proposed to reduce the number of MPTs by removing characters or *taxa* with large amounts of missing data prior to analysis, removing or down-emphasizing highly homoplastic characters or by exercising *a posteriori* removal of the wildcard taxa and then reanalyzing the data. Even with the removal of multiple MPTs, the parsimony analysis would still give a point estimate with no-confidence intervals.[21] That is, as regard to reconstruction of phylogenetic trees, the confident level on the evolved phylogenetic is important. Felsenstein suggested that bootstrapping can be applied across characters of a taxon-by-character data matrix so as to produce a set of replicated datasets (*via* bootstrapping), and a consensus tree can be constructed to summarize the results of all pseudo-replicates. The proportion of trees/replicates in which a grouping is recovered is presented as a measure of support for that group. Studies exist on the application of bootstrapping to large datasets and the relative performance of *bootstrapping* and *jackknifing*.[22–25] Bootstrapping refers to a test procedure or a metric that involves random sampling with replacement.[24–26] In phylogenetic platform, bootstrapping method is used only on characters, because adding duplicate *taxa* does not change the result of a parsimony analysis. The bootstrap is much more commonly employed in phylogenetics. Parallel to bootstrapping method also exists another type of *statistical resampling* procedure adopted to construct pseudo-replicate of samples. It is known as jackknifing and it involves resampling a given set of data without replacement (i.e., it depicts a "leave-one-out" approach).[27,28] In phylogeny, jackknifing method can be employed on a sample set of characters or *taxa*. In general, interpretation of validly using such artificially generated pseudo-replicates is difficult, since the eventual variable of interest is the tree, and comparison of trees with different *taxa* is not, in general straightforward. Both jackknifing and bootstrap methods involve, in general, an arbitrary, but a large number of repeated iterations involving perturbation of the original data.

Summarizing, the MP method described above is popularly used in sequence-based tree reconstructions, and it is based on *Occam Razor* considerations so as to rely on the simplest method that would be the most parsimonious explanation of an observation (unless new observations dictate enforcing a more complex theory). In phylogenetic analysis, this MP problem implies finding a phylogenetic tree that explains a given set of aligned sequences using a minimum number of "evolutionary events"; and, relevantly, the tree denotes a MPT.

As mentioned earlier, UPGMA and NJ are other two algorithms prescribed to reconstruct a phylogenetic tree in addition to the parsimony method described above. However, UPGMA and NJ methods may not necessarily produce the "best tree" for the data. In practice, pairwise distance data availed may depict underestimates of the path-distance between *taxa* on a phylogram. In a given pairs of taxa, some character changes that have occurred in ancestral lineages may be undetectable, inasmuch as later changes could have erased the underlying evidence. (Such erasures refer to *multiple hits* and *back mutations* in sequence data.) Though such erasure problem could be seen in all phylogenetic estimations, it is more of concern in distance methods, because only two samples are used for each distance calculation.

To obviate the complications noted above and decide on the best tree for a given dataset, the distance-based method can include a tree-search protocol that requires satisfying explicitly certain optimality criteria, such as ME and *least-squares inference* (LSI) stipulations. The LSI approach, in general, refers to a regression algorithm that minimizes the residual differences between path distances along the tree, and pairwise distances in the data matrix effectively fit the tree so as to correspond empirical distances. In contrast, the ME criterion allows the tree to assume the shortest sum of branch lengths (and thus, it minimizes the total amount of evolution assumed). The ME approach closely resembles the parsimony method; that is, the ME analysis of distances-based method of tree construction depends on a discrete character dataset and it may "favor the same tree as conventional parsimony analysis of the same data."

Distance methods are quite useful since tree realization could be rendered fast and the results are reasonable estimate of phylogeny. Compared to methods that directly use characters, distance methods permit using data, which may not yield itself for conversion into character data, such as in "DNA–DNA hybridization assays." Furthermore, distance methods accommodate analyses with the possibility that "the rate at which particular nucleotides are incorporated into sequences may vary over the tree."

ML method: the phylogeny inference *via* ML approach involves finding the tree topology, branch lengths, and related parameters of evolutionary model that maximize the probability of observing the given sequences at hand *vis-à-vis* the set of taxa being evolved. This approach is based on standard statistical techniques adopted while inferring probability distributions toward assigning probabilities to possible phylogenetic trees being reconstructed. Relevant method warrants thereof a *substitution model* in order to assess the probability of associated mutations. When a tree has more mutations at interior nodes in explaining the observed phylogeny, it is assessed and identified with a lower probability.[29]

The ML method is akin to the maximum parsimony method, but it allows more statistical flexibility "by permitting varying rates of evolution across both lineages and sites." As such, this method stipulates that evolution at different sites and along different lineages must be statistically independent; therefore, it is well-suited in the analysis of distantly related sequences. However, this method is regarded as being computationally intractable to compute as a result of its NP hardness defined earlier.

Likelihood in the above context implies realizing specified probabilities of the sequences toward modeling evolution on a tree. The more probable the sequences *versus* the tree being reconstructed, it is supposedly, the preferred tree. To realize this preferred tree, all possible trees must be first considered; so, this makes the ML method computationally intense. However, inasmuch as this approach allows choosing a model of evolution, it is widely adopted for divergent groups and complex phylogeny evaluations.

In the contexts of phylogeny, *Bayesian hypothesis* (based on *Bayes' theorem**) is envisaged toward deducing an inference using a likelihood function so as to evolve the associated *a posteriori* probability of trees *via* models of evolution. The underlying conditional probability produces the most likely phylogenetic tree for the given data specified with some *a priori* probabilities. With the advent of high-speed computations and integrated *Markov Chain Monte Carlo* (MCMC) algorithms**, the Bayesian inference technique indicated above has proliferated into various applications of molecular phylogenetics and related systemics.

(**Bayesian hypothesis* or *Bayes' theorem* is based on notions of inverse probability, and it provides the basis of the so-called *Bayesian inference.* Relevant Bayesian approach combines the *a priori* probability, $P(\alpha)$ of realizing an entity (say, a tree in the context of phylogeny) with the likelihood of a certain data (β) in producing the associated *a posteriori* probability distribution, $P(\alpha|\beta)$. This posterior probability implies that, should the probability of a predicted/chosen entity (for example, a phylogenetic tree) be correct to an extent of the highest posterior probability), the chosen entity would represent the "best-entity" or the best-tree in phylogenetic sense.

*MCMC methods*** are used to approximate the posterior distribution of a parameter of interest by random sampling in a probabilistic space. With the advent of MCMC methods, the Bayesian inference method has been widely adopted in phylogeny.[30] Common algorithms adopted in MCMC methods are as follows: (i) the *Metropolis–Hastings algorithms,*[31] (ii) the Metropolis-Coupling MCMC (MC^3),[32] and (iii) the LOCAL algorithm of Larget and Simon.[33]

10.6 Review Section: Examples, Exercises, and Tutorials

This section reviews the salient topics of the chapter and present example exercises and rider problems to enable the readers to comprehend the scope of phylogeny.

Example Ex-10.1

Statement of the Exercise

This example is concerned with the NJ method of phylogenetic tree reconstruction. Suppose there are three distances specified as in Fig. 10.28 with reference to a set: {α, β, γ}.

In reconstructing an unrooted tree configured as shown in Fig. 10.29, the problem is to determine the values of branch lengths *a, b,* and *c.*

Solution

	β	γ
α	25	30
β		35

Fig. 10.28 A hypothetical distance matrix of the test unrooted tree.

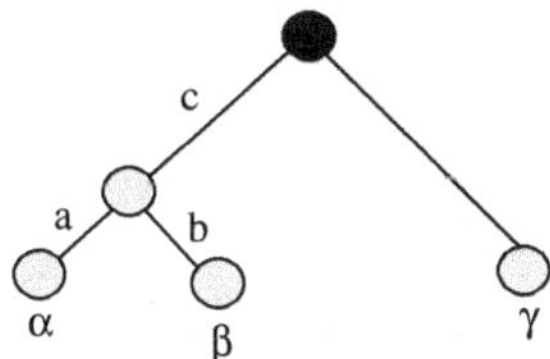

Fig. 10.29 Reconstruction of the test unrooted tree for the distance matrix of Fig. 10.28.

From the distance matrix, the following can be ascertained:

Distance from α to β $\rightarrow$ $(a + b) = 25$
Distance from α to γ $\rightarrow$ $(a + c) = 30$
Distance from β to γ $\rightarrow$ $(b + c) = 35$

Solving the three independent, linear equations as above, the following results are obtained: ($a = 10$, $b = 15$, and $c = 20$).

Example Ex-10.2

Exercise Statement

By deducing the distance matrix between a given set of hypothetical set of multiple sequences indicated below, construct a unrooted tree for the distance matrix derived.
Given set of hypothetical multiple query sequences:

α CAG CGT TGG GCG ATG GCA ACC...
β CAG CGT TGG GCG ACG GTA ATC...
δ CAG CAT TGA ATG ATG ATA ATC...
ε CAG CAT TGA GTG ATA ATA ATC...

Solution

The distances depict number of changes seen in the nucleotides (or amino acids) between any two sequences under comparison. Hence, the required matrix is shown in Fig. 10.30, and the unrooted tree evolved for the given set of sequences is illustrated in Fig. 10.31.

The distance scoring used can be empirical. However, only a single tree is required to be realized. In building this single tree, the raw pairwise distances may not always be perfectly additive. That is, with reference to the above example, the sum of the three

	α	β	δ	ε
α		3	7	8
β			6	7
δ				3
ε				

Fig. 10.30 Test distance matrix.

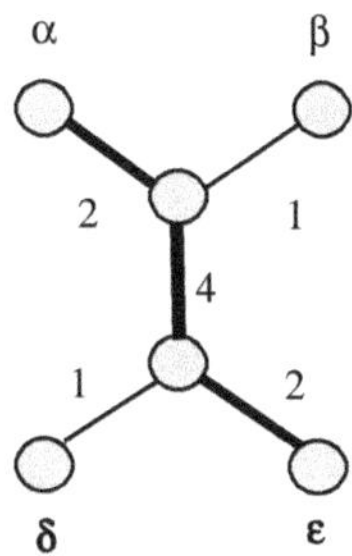

Fig. 10.31 Unrooted tree evolved.

simultaneous equations (=sum of the branch lengths) can be seen exactly equal to the sum of pairwise distances. Such exactness may not prevail in raw data. Hence, a situation may arise when there is undetected homoplasy.

NJ approach can be regarded as a star decomposition technique of heuristic nature. It is computationally not very intensive and may produce reasonable trees. But its tree search does not pose an optimality criterion, and as such, no guarantee is given to say that the "recovered tree is the one that best fits the data." More attempts to produce a starting tree are necessary, and then a tree search exercising an optimality criterion should be done so as to ensure that the best tree is obtained.

Tutorial T- 10.1

Q- and distance-matrices in the context of phylogeny: Q matrices are useful in NJ method of tree reconstruction. Relevant considerations are as follows: suppose a set of four taxa $\{\alpha, \beta, \delta, \varepsilon\}$ is specified, and the exercise is to deduce the associated Q matrix, which is a representation of a stationary, homogeneous, continuous-time, finite-state-space Markov chain parameterized in terms of a *rate matrix*. An example of Q matrix is shown below:

$$Q = [q_{ij}] = \begin{bmatrix} -q_{11} & q_{12} & q_{13} & q_{14} \\ q_{21} & -q_{22} & q_{23} & q_{24} \\ q_{31} & q_{32} & -q_{33} & q_{34} \\ q_{41} & q_{42} & q_{43} & -q_{44} \end{bmatrix} \quad \text{Eq. (10.6)}$$

In the above Q matrix, the off-diagonal rates as shown are non-negative, and diagonal terms are negative. Furthermore, the diagonal element in each row has a magnitude equal to the sums of the other (positive) off-diagonal elements in that row. As such, sum of the entities in each row is equal to zero. That is, in all: (q_{11}) is negative and $|q_{11}| \equiv \sum_{j=2,3,4} q_{1j}$; ($q_{22}$) is negative and $|q_{22}| \equiv \sum_{j=1,3,4} q_{2j}$; ($q_{33}$) is negative and $|q_{33}| \equiv \sum_{j=1,2,4} q_{3j}$; lastly, ($q_{44}$) is negative and $|q_{44}| \equiv \sum_{j=1,2,3} q_{4j}$.

The NJ method considers a distance-matrix specifying the distance between each pair of taxa as the input. It is a bottom-up (agglomerative) clustering method developed in the contexts of phylogenetic trees by Saitou and Nei in 1987.[3] Usually, it is used for trees based on DNA or protein sequence data, and its algorithm requires knowledge of the distance between each pair of taxa (e.g., species or sequences) to form the tree.

The Q matrix is calculated using a distance matrix relating n taxa; relevantly, NJ on a set of n taxa would require (n–3) iterations. At each step, one has to build and search a Q matrix. Initially, the Q matrix has a size $[n \times n]$, then in the next step, it is $[(n-1) \times (n-1)]$, and so on. The algorithm of NJ approach commences with a completely unresolved tree, whose topology corresponds to that of a star network; then, it iterates over a set of steps until the tree is completely resolved and all branch lengths are determined:

$$Q(i,j) = (n-2)d(i,j) - \sum_{k=1}^{n} d(i,k) - \sum_{k=1}^{n} d(j,k) \qquad \text{Eq. (10.7)}$$

where $d(i, j)$ is the distance between taxa i and j. More details on Q matrix is available in Neighbor Joining Method[34] as regard to finding the distance of the pair members to the new node and the distance of the other taxa to the new node.

Distance of the pair members to the new node: suppose for each neighbor in the pair just joined, calculation of the distance to the new node with reference to the taxa set: {θ and ϕ assumed as the paired taxa, and ω is the newly generated node} is attempted. It can be done by using the following algorithm:

$$d(\theta,\omega) = \frac{d(\theta,\phi)}{2} + \frac{1}{2(r-1)}\left(\sum_{k=1}^{k=r} d(\theta,k) - \sum_{k=1}^{k=r} d(\phi,k)\right) \qquad \text{Eq. (10.8)}$$

Distance of the other taxa to the new node: for each taxon not considered above, the distance to the new node is obtained from:

$$d(\omega,k)=\frac{1}{2}\left[d(\theta,k)-d(\theta,u)\right]+\frac{1}{2}\left[d(\phi,k)-d(\phi,u)\right] \qquad \text{Eq. (10.9)}$$

where, as stated earlier, ω is the new node; and (θ and ϕ) are the members of the pair just joined. Furthermore, k is the node for which it is required to calculate the distance.

Problem P-10.1

Statement of the Problem

For a given set of four taxa {α, β, δ, ε} and the associated distance matrix as shown in Fig. 10.32, construct the relevant Q matrix:

With reference to this problem, the resulting Q matrix will indicate two pairs of taxa that have the lowest value, and one can select either of the two pairs for the next step of the algorithm. Suppose the taxa α and β are joined together, then adding more branches is iterative. If more sequences are included, corresponding branches can be introduced in the tree. This method scales linearly with the number of sequences, and it is pertinent to adding a new node to the tree. Given a set of four taxa {α, β, δ, ε} as shown in Fig. 10.32, the exercise is to deduce the corresponding Q matrix.

In the example above, two pairs of taxa have the lowest value, namely, –40. One can select either of them for the second step of the algorithm. Next following the example by assuming that the taxa α and β are joined together. Adding more branches is then iterative. If more sequences are included, corresponding branches can be introduced in the tree. This method scales linearly with the number of sequences, and it involves adding a new node to the tree.

	α	β	δ	ε
α	0			
β	7	0		
δ	11	6	0	
ε	14	9	7	0

Fig. 10.32 Test distance matrix: a set of four taxa {α, β, δ, ε} and their genetic differentials.

Solution Hint

Table 10.3 **Q matrix of the test set of four taxa {α, β, δ, ε} Q matrix.**

	α	β	δ	ε
α	0			
β	−40	0		
δ	−34	−34	0	
ε	−34	−34	−40	0

Distance of the pair members to the new node: suppose for each neighbor in the pair is just joined, calculation of the distance to the new node with reference to the taxa set, {θ and ϕ: the paired taxa; ω: the newly generated node} can be done by using the algorithm of Eq. (10.8).

Distance of the Other Taxa to the New Node

For each taxon not considered above, the distance to the new node is obtained from Eq. (10.9), where as said before, ω is the new node, and θ and ϕ are the members of the pair just joined, and k is the node for which it is required to calculate the distance, and f and g are the members of the pair just joined.

Problem P-10.2

With reference to a given set of four taxa {α, β, δ, ε} with the associated distance matrix indicated in Fig. 10.33, the following exercises are indicated:

Statement of the Problems

With reference to Fig. 10.33 and Eqs. (10.8) and (10.9) determine the following: (1) the distance between α and the new node, (2) the distance

	α	β	δ	ε
α				
β	8			
δ	12	7		
ε	15	10	8	

Fig. 10.33 Test distance matrix.

Table 10.4 Some real-world taxa set and their aligned codon sequence segments useful in phylogenetic analyses.[35]

Taxa set	Aligned codon sequences
Alligator	GTG AAC TTC CAC- - -CGT TGA CTC ...
Emu	GTG ACA TTC ATT ACT CGA TGA TTT ...
Kiwi	GTG ACC TTT ACT ACT CGA TGA CTC ...
Ostrich	GTG ACC TTC ATT ACT CGA TGA CTT ...
Swan	GTG ACC TTC ATC AAC CGA TGA CTA ...
Goose	GTG ACC TTC ATC AAC CGA TGA CTA ...
Chicken	GTG ACC TTC ATC AAC CGA TGA TTA ...
Woodpecker	GTG ACC TTC ATC AAC CGA TGA TTA ...
Finch	ATG ACA TAC ATT AAC CGA TGA TTA ...
Ibis	GTG ACC TTC ATC AAC CGA TGA CTA ...
Stork	GTG ACC TTC ATT ACC CGA TGA CTA ...
Osprey	ATG ACA TTC ATC AAC CGA TGA CTA ...
Falcon	GTG ACC TTC ATC AAC CGA TGA CTA ...
Vulture	ATG ACA TTC ATC AAT CGA TGA CTA ...
Penguin	GTG ACC TTC ATT AAC CGA TGA CTA ...

between β and the new node, (3) the distance between δ and the new node, and (4) the distance between ε and the new node.

Tutorial T-10.2

In Larget,[35] a multiple set of taxa and their aligned segments of codon sequences as shown in Table 10.4 are presented.

Relevant details in Table 10.4 can be adopted in deliberating solutions for any exercises and examples toward establishing the associated phylogenetic trees.

10.7 Phylogenetic Tree Reconstruction: A Descriptive Outline

Reconstructing a phylogenetic tree involves a systematic set of steps as outlined and illustrated *via* pseudocode format in Table 10.5:

Table 10.5 A pseudocode formatted description of reconstructing a phylogenetic tree *via* phylogenetic analysis.

```
Initialize
%% Step I: choosing the test sequences
%% Subroutines I,II, and II (outline below) can be availed for
   use
Call Subroutine I*
  → Subroutine I: outlines the method on:
     → Choosing the query sequences
        ← Write SeQ
           ← Database search is made, and the test sequences
              are listed
Perform MSeQA
     ← It refers to MS alignment (MSA)
     ← Perform MSA (as per a designated/chosen method)
Goto
    Next
Call Subroutine II*
  → Subroutine II: Outlines the method on:
     ← Multiple alignment preparation
Goto
    Next
Check similaity of query sequences
%% This procedure enables choosing the model of evolution by
   checking similarity
Define
  SIM: similarity as strong, moderate, and low/none
Check SIM:
  If SIM defines "strong similarity", then
Goto PM
  ← PM corresponds to performing "parsimony method" (as in
    Subroutine III) or else,
Check SIM:
  If SIM defines "moderate similarity", then
     Goto DM
        ← DM corresponds to "distance method"
          (as in Subroutine III) or else,
Check SIM:
  If SIM defines "low or no similarity", then
     Goto ML
        ← ML corresponds to "maximum-Likelihood method" (as
          in Subroutine III)
```

Table 10.5 (*Continued*)

Next
Perform tree-building/reconstruction strategy
← This refers to constructing the required phylogenetic tree using the set of multiple sequences aligned and prepared (as above)
Goto
Subroutine III*
Next
← Perform tree-construction *via* algorithm compatible for PM, ML, or DM using the set of multiple sequences aligned and prepared (as above)
Evaluate the quality of the end result
%% This is done by applying *consensus methods* to the set of trees to figure out what is the reliable tree that depicts the (true) evolution history of the sequences addressed
Goto
← **Subroutine Q**
→ This is written to evaluate the quality of the end result
End

%% Subroutines (I, II, and III)*:
%% Details
Subroutine I
← **Choosing the query sequence**
← This refers to choosing homologous query sequences
→ Relevant choice is based on checking the following:

Check: whether the selection is not a sequence fragment
- Incomplete sequences are not friendly toward MSA nor tree reconstruction. At least same fragment is used for all multiple sequences

Check: whether the sequences chosen are not xenologs
- Unless the purpose is to study xenologs, avoid including genes that result from lateral transfer

Check: whether the selection is not a recombinant Sequence
- Some proteins could result from a combination of multiple proteins (as is common in viruses). Such proteins carry two ancestors (instead of one)not being compatible for regular treereconstruction.

Check: whether the sequences are pertinent to large and complex families containing various domains and repeats

(*Continued*)

Table 10.5 (*Continued*)

```
        - Working on smaller and more uniform subsets is
          preferable
Check: whether the sequences are pertinent to nucleotides or
proteins
Check: whether they are from closely related species
        - In conformance to an extent of being at least 70% iden-
        tical with or without the underlying mutations being high
If
  … the sequences do not conform to an extent of being at
    least 70% identical as above (meaning, that they are more
    divergent, then
Goto
        The method using protein sequence or conserved
        nucleotides(such as ribosomal RNA) or else
          Use the chosen nucleotide sequence
Return
  Selected sequence
Next
  Go to Subroutine II:
    ← Perform MSeQA:
  → Building MSA and preparation for tree
    construction
%% Typical multiple alignment programs are Clustal, T-Coffee,
   MAFFT, Muscle, etc.
  → Subroutine II involves two steps:
      Step (i)  Retrieving homologs
      Step (ii) Arranging multiple sequences and preparing the
                multiple sequences for tree construction,
                which refers to "cleaning up" the chosen mul-
                tiple sequences for alignment by a set of
                procedures as indicated below:
Retrieve: Step (i)
        → Given a sequence, its homolog can be retrieved as
        follows: Access NCBI BLAST by typing the URL cited as[36]
        into the address line of the web browser and pursue the
        options available.
        With the BLAST search pursued the FASTA format sequence
        of six ORF can be retrieved from the BLAST output.
Arrange/Prepare: Step (ii)
        ← This refers to placing the chosen multiple sequences
        one below the other forming a column of sequences as an
        alignment procedure and preparing the multiple sequences
        for tree construction. The preparation involves "cleaning
        up" the chosen multiple sequences for alignment by a set
        of procedures.
```

Table 10.5 (*Continued*)

%% Example-1: Multiple DNA sequences of hypothetical homologous species: Set-I

```
AAGCA-AGGTAAATGCATGC ATGGA- - AGTCCTGGAATGGTA
AGAT - - AGGTAAATGCAGCTAGCAT-AAGTCCTGGACCGGAT
GCAATTAGGTAAAACCAAGGTAC CT- -AGTCCTGGAGAGATA
GTGATTAGGTAAAACCAACGCAACGCAGTCCTGG ACGTAGG
```

%% Example-2: Multiple protein sequences of hypothetical homologous species: Set-II

```
AS L IFR- SDAYS KNRTV IPVWNEGFQ- - DQSSLHVVVKMQEY
AGVA- - SDAYS KKRTV IKNSVNPVWNEDQGSLHVVVKKEN
CSAVFASDAYS KKRTVVIKNSVNNVWN D QS SLHVV ELL- - - -
CVVVF- SDAY S KKRTVI IKNSVNPVWNEDQSSLHVVVKTKEE
```

Prepare

← This refers to removing certain sets of columns in the arranged multiple sequences shown in the above examples as shaded sections.

→ *Criteria for such removals*?

- o Sections of gap-free columns are mostly retained. Gaps invariably cause phylogenetic tree-forming.(In tree-construction, programs like ClustalW follows complete-deletion policy ignoring every column that contains the gap)
- o Extremities of the multiple sequences are removed inasmuch as *N*-terminus and *C*-terminus tend to be poorly conserved, and as such, they are not well aligned
- o Gap-rich columns can be removed. They often spond to loops. As such, even when a program returns an alignment with gap-rich columns, it may not be a meaningful
- o Most informative blocks should be retained. Ideally, in building a tree, high alignment of sequences possessing low level of identity is preferable since it would contain a trace of the family history
- o "Good blocks" (typically with 20–30 amino acids long) with a few conserved positions are useful in realizing a correct tree

%% Programs for column removal:

- Removal of columns that are unlikely to be correctly aligned can be done *via* T-Coffee server. T-Coffee is a progressive method for sequence alignment.[37]
- Editing multiple alignments can be done with Jalview. It is a multiple alignment editor written in Java. It is used widely in a variety of web pages (e.g., the EBI Clustalw server and the Pfam protein domain database and is available as a general purpose alignment editor.

(*Continued*)

Table 10.5 (*Continued*)

```
%% Programs for multiple alignment: Clustal, T-Coffee, MAFFT,
Muscle, etc.

Return
  → Aligned multiple sequence
      End
```

Example Ex-10.3

Statement of the Exercise

With reference to a set of six OTUs {α, β, χ, δ, ε, ϕ}, whose sequences are listed below, the exercise is to establish an EDM and construct a UPGMA-based ultrametric phylogenetic tree for the test EDM.

The given set of six OTUs {α, β, χ, δ, ε, ϕ} is:

α G A A C G C T G C G T G G T G T A G T C G T C T G C G A G A T A T G G C T G G

β G A A C G C T G C G T G G T G T A G T C G T C T G C G A G A T A T G G C T C T

χ G A A C G C T G C G T C G T G T T G T C G T C T G T G A G A T A T G G C T C G

δ G A A G C C T G C G T G T G G T T G T C G T C T G C G A G A T A T G G C T C G

ε G A A G G T T G C G T G G T G T T G T C T G C T G C G A G A T A T G G C T C G

ϕ T C A G G C C G C G T G G T G T T G T C G T C T G C G A G A A T T G G C T C G

And the corresponding common ancestral root (R) is:
R T A A G G C T G C G T G G T G T T G T C G T C T G C G A G A T A T G G CT C G

Solution Outline

Assuming that the given set of OTUs had the common ancestral root (R) indicated, the corresponding EDM can be constructed as illustrated in Fig. 10.34.

The ED values indicated in the matrix of Fig. 10.34 correspond to the extent (number) of changes in the nucleotide residue set of R in going to each taxonic leaf (OTU). That is, ED corresponds to number of dissimilarities observed between the sequences being compared. This is illustrated in the following set of comparisons in Table 10.6.

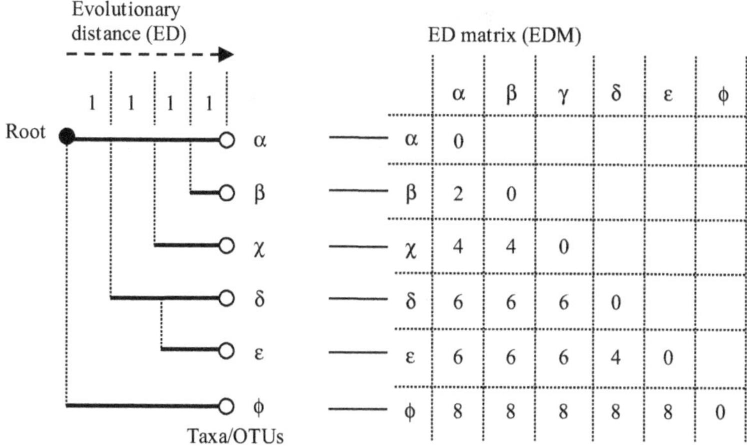

Fig. 10.34 Construction of the evolutionary distance matrix for the OTU set of Example Ex-10.3.

Table 10.6 Steps of comparison of each query sequence in the multiple set against the root sequence as well as with other OTU sequences.

Check I

Distance of each OTU of the set six given sequences {α, β, χ, δ, ε, ϕ} from the root *R*. (Dissimilar residues are shown in bold)

R	**T**AA**G**GCTGCGTGGTGT**T**GTCGTCTGCGAGATATGGCT**C**G
α	**G**AA**C**GCTGCGTGGTGT**A**GTCGTCTGCGAGATATGGCT**G**G
R	**T**AA**G**GCTGCGTGGTGT**T**GTCGTCTGCGAGATATGGCTC**G**
β	**G**AA**C**GCTGCGTGGTGT**A**GTCGTCTGCGAGATATGGCTC**T**
R	**T**AA**G**GCTGCGT**G**GTGTTGTCGTCTG**C**GAGATATGGCTCG
χ	**G**AA**C**GCTGCGT**C**GTGTTGTCGTCTG**T**GAGATATGGCTCG
R	**T**AAG**G**CTGCGTG**GT**GTTGTCGTCTGCGAGATATGGCTCG
δ	**G**AAG**C**CTGCGTG**TG**GTTGTCGTCTGCGAGATATGGCTCG
R	**T**AAGG**C**TGCGTGGTGTTGTC**GT**CTGCGAGATATGGCTCG
ε	**G**AAGG**T**TGCGTGGTGTTGTC**TG**CTGCGAGATATGGCTCG
R	T**A**AGGCTGCGTGGTGTTGTCGTCTGCGAGA**TA**TGGCTCG
ϕ	T**C**AGGCCGCGTGGTGTTGTCGTCTGCGAGA**AT**TGGCTCG

(Continued)

Table 10.6 (*Continued*)

Check II

Distance of OTUs {β, χ, δ, ε, ϕ} from α. (Dissimilar residues are shown in bold)

α	GAACGCTGCGTGGTGTAGTCGTCTGCGAGATATGGCT**GG**
β	GAACGCTGCGTGGTGTAGTCGTCTGCGAGATATGGCT**CT**
α	GAACGCTGCGT**G**GTGT**A**GTCGTCTG**C**GAGATATGGCT**G**G
χ	GAACGCTGCGT**C**GTGT**T**GTCGTCTG**T**GAGATATGGCT**C**G
α	GAA**CG**CTGCGTG**GT**GT**A**GTCGTCTGCGAGATATGGCT**G**G
δ	GAA**GC**CTGCGTG**TG**GT**T**GTCGTCTGCGAGATATGGCT**C**G
α	GAA**C**G**C**TGCGTGGTGT**A**GTC**GT**CTGCGAGATATGGCT**G**G
ε	GAA**G**G**T**TGCGTGGTGT**T**GTC**TG**CTGCGAGATATGGCT**C**G
α	**GA**A**C**GC**T**GCGTGGTGT**A**GTCGTCTGCGAGA**TA**TGGCT**G**G
Φ	**TC**A**G**GC**C**GCGTGGTGT**T**GTCGTCTGCGAGA**AT**TGGCT**C**G

Check III

Distance of OTUs {χ, δ, ε, ϕ} from β. (Dissimilar residues are shown in bold)

β	GAACGCTGCGT**G**GTGT**A**GTCGTCTG**C**GAGATATGGCTC**T**
χ	GAACGCTGCGT**C**GTGT**T**GTCGTCTG**T**GAGATATGGCTC**G**
β	GAA**CG**CTGCGTG**GT**GT**A**GTCGTCTGCGAGATATGGCTC**T**
δ	GAA**GC**CTGCGTG**TG**GT**T**GTCGTCTGCGAGATATGGCTC**G**
β	GAA**C**G**C**TGCGTGGTGT**A**GTC**GT**CTGCGAGATATGGCTC**T**
ε	GAA**G**G**T**TGCGTGGTGT**T**GTC**TG**CTGCGAGATATGGCTC**G**
β	**GA**A**C**GC**T**GCGTGGTGT**A**GTCGTCTGCGAGA**TA**TGGCTC**T**
ϕ	**TC**A**G**GC**C**GCGTGGTGT**T**GTCGTCTGCGAGA**AT**TGGCTC**G**

Check IV

Distance of OTUs {δ, ε, ϕ} from χ. (Dissimilar residues are shown in bold)

χ	**G**AA**C**GCT**G**CGT**C**GTGTTGTCGTCTG**T**GA**G**ATATGGCTCG
δ	**C**AA**T**GCT**C**CGT**G**GTGTTGTCGTCTG**C**GA**T**ATATGGCTCG

Table 10.6 (*Continued*)

χ	GAA**C**G**C**TGCGT**C**GTGTTGTC**GT**CTG**T**GAGATATGGCTCG
ε	GAA**G**G**T**TGCGT**G**GTGTTGTC**TG**CTG**C**GAGATATGGCTCG
χ	**GA**A**C**GC**T**GCGT**C**GTGTTGTCGTCTG**T**GAGA**TA**TGGCTCG
ϕ	**TC**A**G**GC**C**GCGT**G**GTGTTGTCGTCTG**C**GAGA**AT**TGGCTCG

Check V

Distance of OTUs {ε, ϕ} from δ. (Dissimilar residues are shown in bold)

δ	GAAG**CC**TGCGTG**TG**GTTGTC**GT**CTGCGAGATATGGCTCG
ε	GAAG**GT**TGCGTG**GT**GTTGTC**TG**CTGCGAGATATGGCTCG
δ	**GA**AG**C**C**T**GCGTG**TG**GTTGTCGTCTGCGAGA**TA**TGGCTCG
ϕ	**TC**AG**G**C**C**GCGTG**GT**GTTGTCGTCTGCGAGA**AT**TGGCTCG

Check VI

Distance of OTU, ϕ from ε. (Dissimilar residues are shown inbold)

ε	**GA**AGG**TT**GCGTGGTGTTGTC**TG**CTGCGAGA**TA**TGGCTCG
ϕ	**TC**AGG**CC**GCGTGGTGTTGTC**GT**CTGCGAGA**AT**TGGCTCG

With reference to the EDs indicated in the ED matrix (EDM) of Fig. 10.34, the problem in hand is to construct an additive tree using clustering algorithm of UPGMA. For this purpose, as mentioned earlier, the number of observed nucleotide or amino acid substitutions can be used in the UPGMA sequential clustering algorithm, where local topological relationships are identified in the order of similarity. Hence, building of the phylogenetic tree is done *via* the following steps:

Step A: given an EDM, the two OTUs that are most similar (bearing the closest ED) are identified from among all the OTUs, and these two

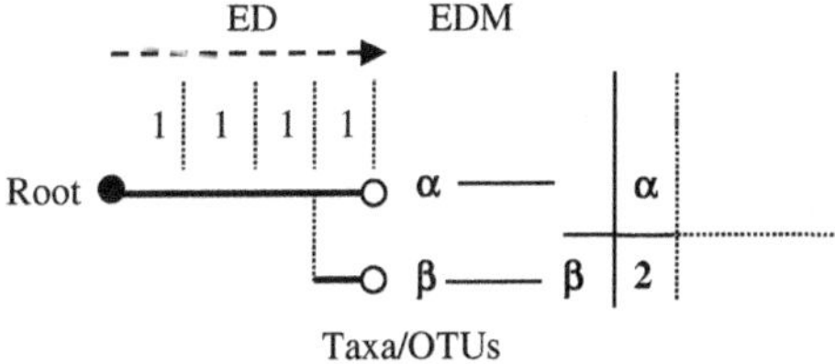

Fig. 10.35 Selection of two OTUs α and β that are most similar in the given EDM (bearing the smallest ED equal to 2).

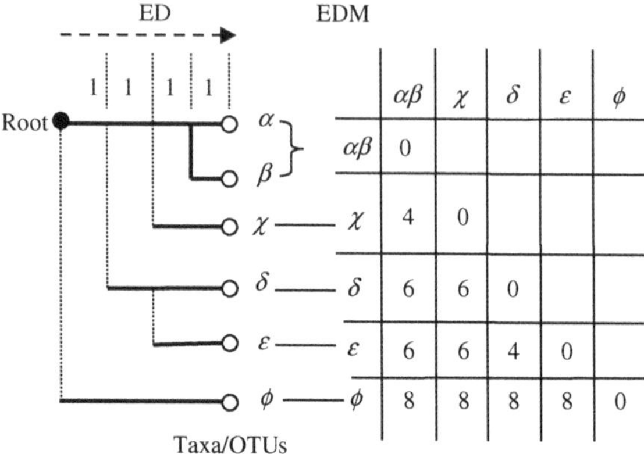

Fig. 10.36 Construction of the new EDM using composite OTU, {α, β}.

OTUs are treated as a new single composite OTU. As shown in Fig. 10.35, the OTU-pair α and β is the chosen pair by virtue of their smallest ED of 2 in the EDM, and the composite OTU is {α, β}.

Step B: with the composite OTU {α, β} introduced in the EDM, the new matrix is constructed by choosing again the most similar pair and clustering them together as illustrated in Fig. 10.36. Relevant set of EDs is elucidated as follows:

ED between {α, β} and χ = ½ × [ED(α and χ) + ED (β and χ)]
ED between {α, β} and δ = ½ × [ED(α and δ) + ED (β and δ)]
ED between {α, β} and ε = ½ × [ED(α and ε) + ED (β and ε)]
ED between {α, β} and ϕ = ½ × [ED(α and ϕ) + ED (β and ϕ)]

Step C: as in the previous steps, among the new group of OTUs, the pair with the highest similarity is identified and the corresponding composite OTU is specified. This procedure is repeated until only two OTUs are left out. Corresponding EDM constructions are shown in the set of Figs. 10.37–10.40. The resulting ultrametric tree for the test EDM is illustrated in Fig. 10.40.

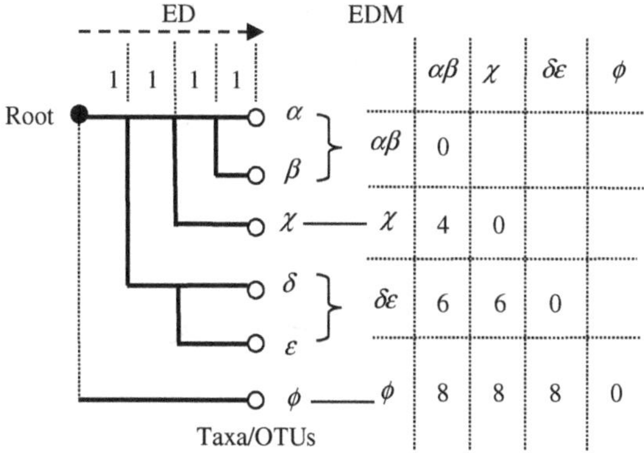

Fig. 10.37 Construction of the new EDM with the composite OTUs, {δε} and {δε}.

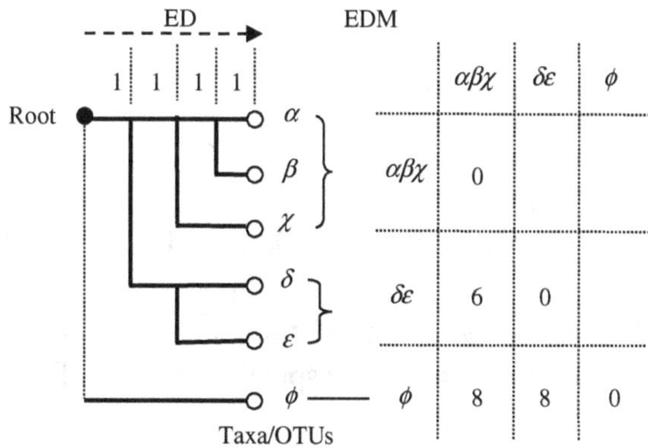

Fig. 10.38 Construction of the new EDM with the composite OTUs, {αβχ} and {δε}.

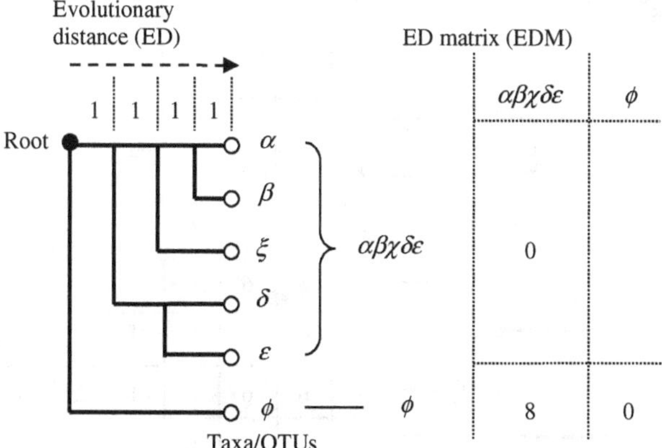

Fig. 10.39 Construction of the new EDM with the composite OTUs, {αβχ δε}.

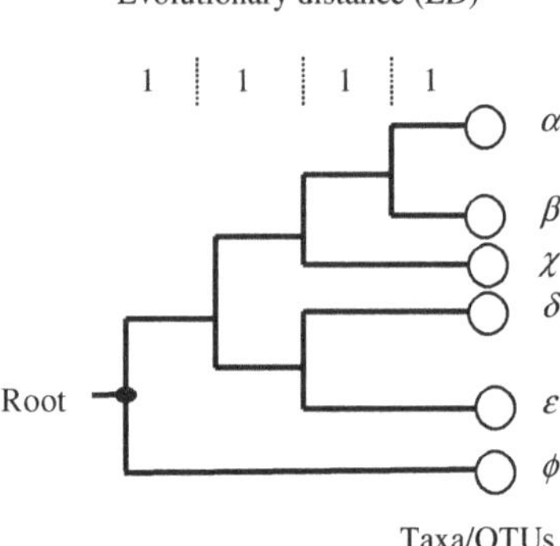

Fig. 10.40 Final UPGMA-based ultrametric phylogenetic tree for the test EDM.

Problem P-10.3

Statement of the Problem

R: G A A T G T T G C G T G G T G T T G T G G T C T G C
A T A T A A C T C G AATGCCT

The EDM with metrics of distances depicting a hypothetical ultrametric phylogenetic tree is shown in Fig. 10.41. Suppose the sequence, *R* shown corresponds to a common ancestral root of given EDM in Fig. 10.41, determine the sequences that can be specified as the OTU set, {*a, b, c, d, e, f, g*} of the tree.

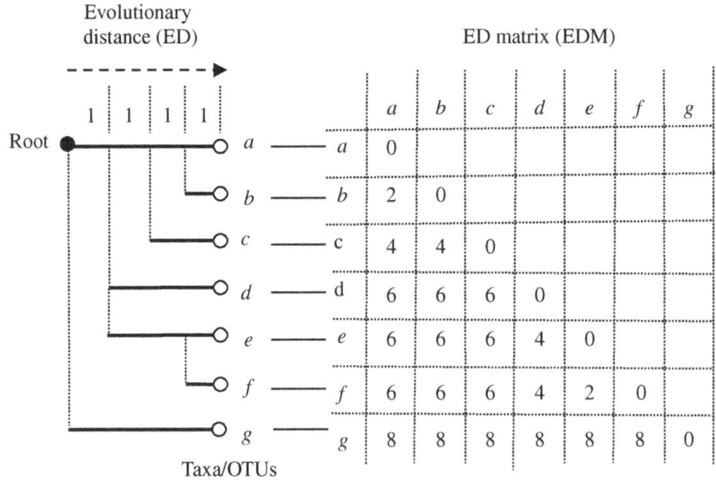

	a	b	c	d	e	f	g
a	0						
b	2	0					
c	4	4	0				
d	6	6	6	0			
e	6	6	6	4	0		
f	6	6	6	4	2	0	
g	8	8	8	8	8	8	0

Fig. 10.41 A hypothetical EDM supplied for Problem P-10.3.

Problem P-10.4

Consider an EDM with metrics of distances as shown in Fig. 10.42. For the ED values of the matrix indicated, relevant topology is as illustrated.

Statement of the Problem

Determine the three-point condition and show that the tree reconstructed by considering the evolutionary history *via* UPGMA leads to a wrong topology of Fig. 10.43, implying that UPGMA on unequal rates of mutation will show a completely different topology from the original EDM-specific tree.

(*Solution hint*: Considering divergence phases of α and β, the taxon β has faced mutations at a much higher rate than the taxon α. Check, therefore, the three-point criterion is violated and the possible UPGMA-based tree is erroneous. In such cases, the neighborhood-joining procedure (described below in Example Ex-10.4) will yield the correct topology).

(Ultrametric distances are constrained by the so-called "three-point conditions," which stipulates that for any given set of three taxa {*A, B, C*}, the two largest distances are equal, that is, *ED*(*A*–*C*) ≤ maximum of [*ED*(*A*–*B*), *ED*(*B*–*C*)]. The two largest distances being equal signifies that the evolutionary rate is the same for all branches. Should this condition of rate constancy fails among lineages, an erroneous topology would result in).

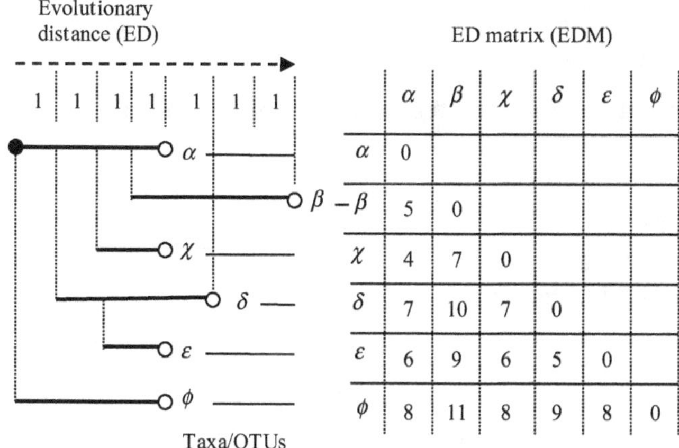

	α	β	χ	δ	ε	φ
α	0					
β	5	0				
χ	4	7	0			
δ	7	10	7	0		
ε	6	9	6	5	0	
φ	8	11	8	9	8	0

Fig. 10.42 EDM of an phylogenetic tree having no evolution rate constancy.

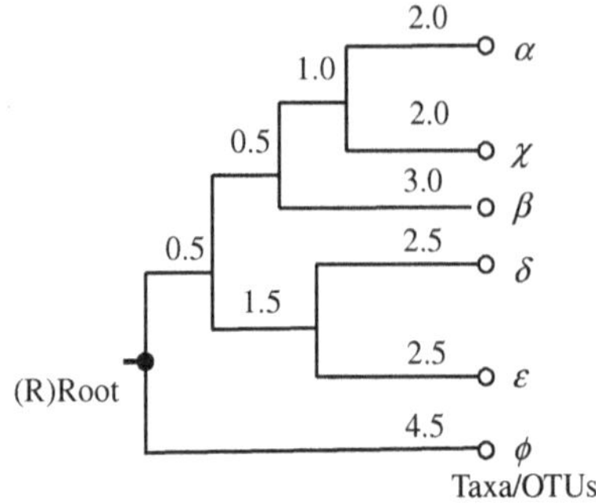

Fig. 10.43 UPGMA-based topology for the EDM of Fig. 10.43.

Example Ex-10.4

An EDM indicated below corresponds to real data matrix for five rRNA sequences. Each value denotes the estimated number of nucleotide residue substitutions per position separating the corresponding pair of the presently existing sequences.

		1	2	3	4	5
Node	Taxa	BSu	Bst	Lvi	Amo	Mlu
1	Bsu	0	0.172	0.215	0.309	0.233
2	Bst		0	0.299	0.340	0.206
3	Lvi			0	0.280	0.394
4	Amo				0	0.429
5	Mlu					0

Statement of the exercise

Construct an ultrametric tree. Show the steps and ultrametric details involved.

Solution

The tree topology is shown in Fig. 10.44. The steps involved is left as a do-it-yourswelf exercise.

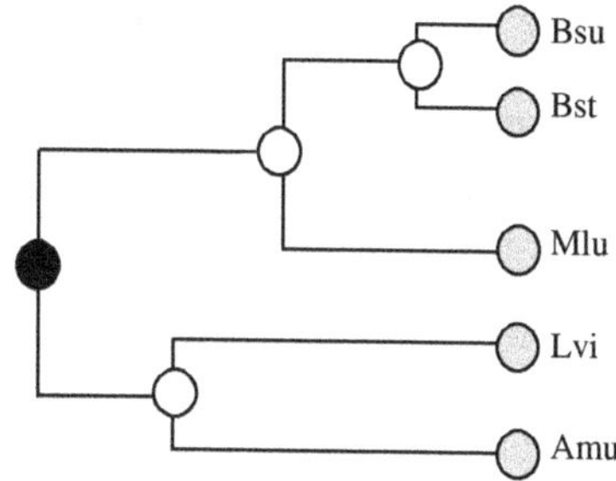

Fig. 10.44 Test tree topology.

Tutorial T-10.5

Genetic Differentiation: Genetic Identity and Genetic Distance

Genetic differentiation is measured with two parameters: *genetic identity* (I), which estimates the proportion of genes that are identical in two populations and *genetic distance* (GD), which estimates the proportion of gene changes that have occurred in the separate evolution of two populations. Genetic identity (I) is calculated according to the method due to Nei.[38] Table 10.7 gives the genetic identity, I, and genetic distance, GD, for all pairwise comparisons between the taxa in Tables 10.7a and 10.7b.[39]

Distance matrices adopted in phylogeny are nonparametric distance methods; and, they were originally conceived with phenetic data using a matrix of pairwise distances. These distances are then reconciled to produce a "tree" (i.e., a phylogram, with informative branch lengths). The distance matrix can be assessed from a number of different resources, such as measured distance (e.g., from immunological studies) or morphometric analysis, various pairwise distance formulas (like Euclidean distance indicated in Chapter 2).

Relevant algorithms are applied to discrete morphological characters, or genetic distance from sequence, restriction fragment, or allozyme data. Furthermore, for phylogenetic character data, raw distance values can be determined by enumeration of the number of pairwise differences in character states (expressed *via* Hamming distance). The distance matrices of phylogenetic analysis as above explicitly depend on GD between the sequences being classified; hence, they require a MSA as an input. The GD is mostly defined as the fraction of mismatches at aligned positions (with the gaps either ignored or counted as mismatches). Relevant distance methods attempt to reconstruct an "all-to-all matrix from the

Table 10.7a Genetic identity (I) between pairs of taxa representing the members of great apes: *Homo sapiens* (Humans); *Pan troglodytes* (Chimps); *Gorilla gorilla* (Gorilla); *Pongo abelii* (Orangutan); and *Hylobatidae lar* (Gibbons).

OTU	*Homo*	*Pan t*	*Gorilla*	*Pongo a*	*H. lar*
Homo	0	0.680	0.689	0.707	0.489
Pan t		0	0.688	0.738	0.510
Gorilla			0	0.671	0.618
Pongo a				0	0.583
H. lar					0

Table 10.7b Genetic distance (GD) between pairs of taxa representing the members of great apes: *Homo sapiens* (Humans); *Pan troglodytes* (Chimps); *G. gorilla* (Gorilla); *P. abelii* (Orangutan); and *Hylobatidae lar* (Gibbons).

OTU	*Homo*	*Pan t*	*Gorilla*	*Pongo a*	*H. lar*
Homo	0				
Pan t	0.386	0			
Gorilla	0.373	0.373	0		
Pongo a	0.347	0.304	0.238	0	
H. lar	0.716	0.673	0.482	0.598	0

sequence query set" describing the distance between each sequence pair. From this, a phylogenetic tree is evolved, which places closely related sequences under the same interior node, and the branch lengths closely denote the observed distances between sequences. Distance matrix methods may produce either rooted or unrooted trees, depending on the algorithm as detailed in the following examples and exercises.

Evolutionary distance (ED expressed in % of normalzed values) in contrast refers to the estimated number of nucleotide substitutions per site between two homologous DNA sequences or the number of amino acid substitutions per site between two homologous protein sequences. The ED between sequences can be estimated by computing the proportion of nucleotide differences between each pair of sequences using MEGA, for example, of pertinent models of *Drosophila* species.[40]

Example Ex-10.5

Aligned nucleotide or amino acid sequences typically lead to the computation of the matrix of genetic distances (or EDs) between all pairs of sequences. The compiled approximate values ED that exists between a pairwise species in a set of OTUs as indicated in Table 10.8. This data refer to the following taxa denoting human–ape lineage: Humans: *Homo sapiens,* abbreviated as *Homo*/Hu; Chimps: *Pan troglodytes,* abbreviated as *Pan t*/Ch; Gorilla: *Gorilla gorilla,* abbreviated as *Gorilla*/Go; Orangutan: *Pongo abelii,* abbreviated as *Pongo a*/Or; and, Gibbons: *Hylobatidae lar,* abbreviated as *H. lar*/Gi.[41,42] In 2005, millions of bases of human–ape lineage species were sequenced, assembled, and compared to the human genome. The sequence divergence has generally the following pattern: (Hu-Ch) < (Hu-Go) < (Hu-Or) < (Hu-Gi).

Table 10.8a Given initial data on ED (in %) across pairwise OTUs.

OTU	Human	Chimp	Gorilla	Orangutan	Gibbon
Human	0	94 ± 1.16	110 ± 1.47	180 ± 1.96	205 ± 2.00
Chimp		0	115 ± 1.40	195 ± 1.97	220 ± 2.47
Gorilla			0	190 ± 1.97	220 ± 2.50
Orangutan				0	220 ± 2.60
Gibbon					0

Table 10.8b Initial data on mean values of ED in % of pairwise OTUs.

OTU	*Homo*	*Pan t*	*Go*	*Pongo a*	*H. lar*
Homo	0	94	111	180	206
Pan t		0	115	194	218
Go			0	188	218
Pongo a				0	217
H. lar					0

Statement of the Exercise

Using the UPGMA concept, construct an evolutionary tree for the dataset in Table 10.8a on the OTUs of *Homo*/Hu; *Pan t*/Ch; *Gorilla*/Go; *Pongo a*/Or; and *H. lar*/Gi.

Solution on Tree-building

The data on OTUs indicated in Table 10.8a represent approximate error-barred statistical values attributed to EDs (in %) estimated across pairwise test OTUs as listed. Relevant mean-values are as shown in Table 10.8b.

Tree reconstruction

Step-1
From the dataset of Table 10.8b, it can be identified that humans and chimps are the closest neighbors having *ED* = 94% being the smallest entry.

Step-2
After combining humans-and-chimps marked as Hu-Ch, (Table 10.8c) is obtained, and the next lowest entry on *ED* = 113% pertinent to Go is noted.

Step-3
Merging humans–chimps (Hu–Ch) with gorillas (Go) having an ancestral distance: *ED* = 113%, the details as in Table 10.8d are obtained.

Table 10.8c Humans and chimps (Hu–Ch) merge with ancestral distance 94% of *Pan t.*

OTU	Hu–Ch	Go	Or	Gi
Hu–Ch	0	113	187	212
Go		0	188	218
Or			0	217
Gi				0

Hu
Hu–Ch
ED = 94%
Ch

Table 10.8d Combined data on Hu-Ch with Go and the next lowest entry on *ED* = 187.5% pertinent to Or is noted.

OTU	Hu-Ch-Go	Or	Gi
Hu-Ch-Go	0	187.5	215
Or		0	217
Gi			0

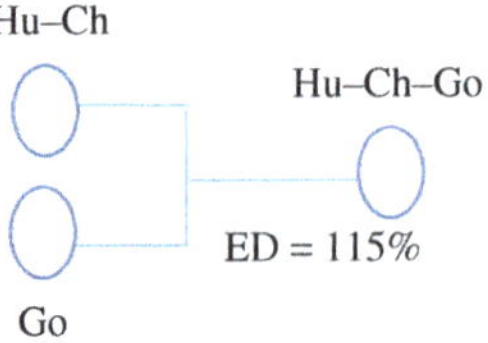

Fig. 10.45 Merged results of Step-3: Humans–chimps (Hu–Ch) and gorillas (Go) with combined ancestral distance 113%.

Step-4
Next, the Hu–Ch–Go merged with Or (having ancestral distance: *ED* = 187.5%) as detailed in Table 10.8e, and the next lowest entry on *ED* = 216% pertinent to Gi is noted.

Table 10.8e Result of Step-3 on humans–chimps–gorillas (Hu–Ch–Go) and orangutans (Or) merging with ancestral distance 187.5%.

OTU	Hu–Ch–Go–Or	Gi
Hu–Ch–Go–Or	0	215
Gi		0

The resulting phylogenic tree obtained is as illustrated in Fig. 10.46.

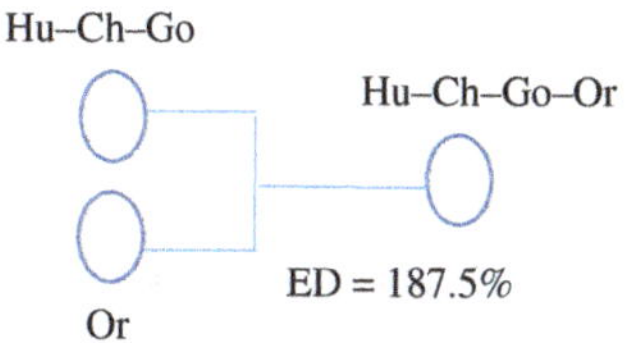

Fig. 10.46 Merged results of Step-4: Humans–chimps–gorillas (Hu–Ch–Go) with combined ancestral distance, *ED* = 187.5%.

Step-5
The Hu–Ch–Go–Or is next merged with the species having lowest entry on *ED* = 216% pertinent to Gi yielding results as in Fig. 10.47.

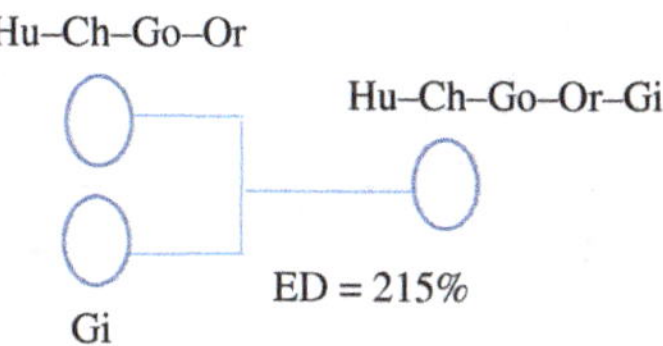

Fig. 10.47 Humans–chimps–gorillas–orangutans (Hu–Ch–Go–Or) and gibbons (Gi) merge with ancestral distance 215%.

Step-7
Overall state of the mergers of Hu-Ch-Go-Or-Gi is illustrated in Fig. 10.48. The final stage of phylogenic tree construction is shown in Fig. 10.48.

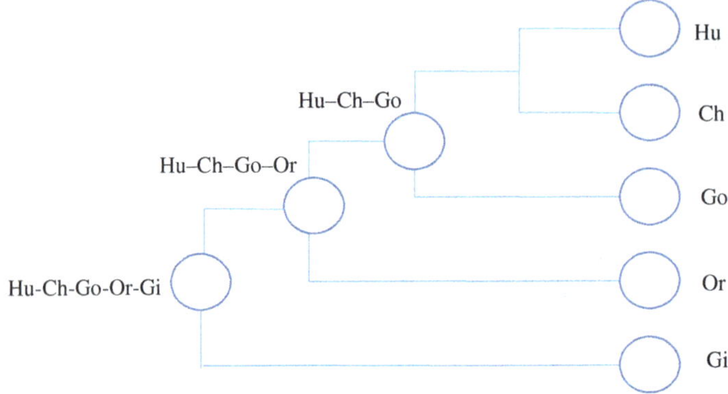

Fig. 10.48 Overall state of the mergers indicated: Hu-Ch-Go-Or-Gi.

Problem P-10.5

Statement of the Problem

Using the UPGMA concept, construct an evolutionary tree for the data on pairwise species differences indicated in Table 10.9.

Table 10.9 Pairwise species differences of Problem P-10.5.

OTU	*A*	*B*	*C*	*D*	*E*
A	0	5	30	45	35
B		0	28	42	32
C			0	10	15
D				0	20
E					0

Problem P-10.6

Statement of the Problem

Table 10.10 Pairwise species differences of Problem P-10.6.

OTU	*A*	*B*	*C*
B	4		
C	4	2	
D	8	8	6

Using the UPGMA concept, construct an evolutionary tree for the data on pairwise species differences indicated in Table 10.10: (*Solution hint*: useful reference ExPASy[43])

Example Ex-10.6

Statement of the Exercise

Using the FM concept, construct an evolutionary tree for the data on pairwise species differences indicated in Table 10.11a.

Alternatively, the above data can be represented as follows:

Table 10.11a Pairwise species genetic differences of the tutorial exercise: EXAMPLE Ex-10.6.

OTU	*A*	*B*	*C*	*D*
A	0			
B	20	0		
C	25	30	0	
D	40	45	50	0

Table 10.11b Pairwise species differences of the tutorial exercise/ EXAMPLE Ex 10.6, an alternative representation of the data.

OTU	*A*	*B*	*C*	*D*
A	0	20	25	40
B		0	30	45
C			0	50
D				0

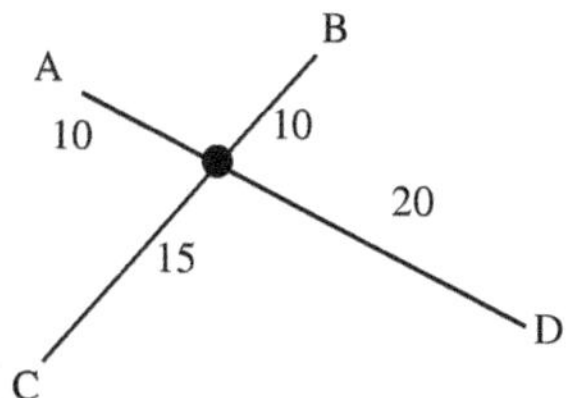

Fig. 10.49 An unrooted tree drawn with branches stemming from a common ancestral node consistent with the data in Table 10.11b.

An unrooted tree can be drawn with branches stemming from a common ancestral node (with the give values in the matrix in Table 10.11a) as illustrated in Fig. 10.49.

Step 1 — Calculation of the sum of branches algebraically as follows:

$$\begin{aligned}\text{Distance from } A \text{ to } B &= a+b = 20\\ \text{Distance from } A \text{ to } C &= a+c = 25\\ \text{Distance from } A \text{ to } D &= a+d = 40\\ \text{Distance from } B \text{ to } C &= b+c = 30\\ \text{Distance from } B \text{ to } D &= b+d = 45\\ \text{Distance from } C \text{ to } D &= c+d = 50\end{aligned} \qquad \text{Eq. (10.10)}$$

These calculations as above are made to find out the *ED* of each individual (*A, B, C, D*) from the ancestral node. (This distance can be either equal or nonequal for different individual taxon.)

Step 2 — Finding the values of *a, b, c,* and *d* by solving the linear simultaneous equation results of Eq. (10.10). The results are as follows: a =10; b = *10*; c = 15; and d = 20. These denote the lengths of branches.

Problem P-10.7

Statement of the Problem

OTU	*H*	*C*	*G*	*O*	*A*
H	0	95	110	185	205
C		0	118	195	220
G			0	190	215
O				0	215
A					0

Fig. 10.50 Genetic differences across a set of five taxa, {*H, C, G, O, A*}.

Using the data on genetic differences across a set of five taxa, {*H, C, G, O, A*} as shown in the table of Fig. 10.50, construct an unrooted tree formulating the lengths of branches from the common ancestral node. The answer hint is indicated in Fig. 10.51.

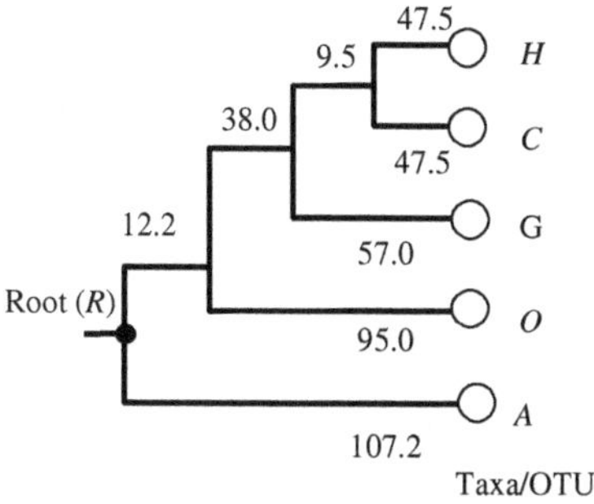

Fig. 10.51 Solution hint for the Problem P-10.7.

Problem P-10.8

Statement of the Problem

Given the state of EDs as in Table 10.12, create a distance matrix *via* NJ method on the resulting taxa.

Table 10.12 State matrix of evolutionary distances between the taxa set {*A, B, C, D, E, F*}.

OTU	*A*	*B*	*C*	*D*	*E*
B	5				
C	10	20			
D	15	25	35		
E	45	55	60	65	
F	70	75	80	85	90

Solution Hint:

The new distance matrix (m) for each pair of nodes is calculated as follows: $m\ (i, j) = d(j) - [r(i)] + r(j)/(N - 2)$, where N is the number of taxa

Example: net divergence r for each OTU from all other nodes is determined as follows:

$$
\begin{aligned}
r(A) &= d(AB) + d(AC) + d(AD) + d(AE) + d(AF) \\
&= (5 + 10 + 15 + 45 + 70) \quad = 145 \\
r(B) &= (5 + 20 + 25 + 55 + 75) \quad = 180 \\
r(C) &= (10 + 20 + 35 + 60 + 80) = 205 \\
r(D) &= (65 + 85 + 15 + 25 + 35) = 225 \\
r(E) &= (45 + 55 + 60 + 65 + 90) = 315 \\
r(F) &= (70 + 75 + 80 + 85 + 90) = 400
\end{aligned}
$$

Furthermore:

$$m(AB) = d(AB) - [r(A)+ r(B)]/(6 - 2)] = 5- [145 + 180]/4$$
$$= -76.25$$

Similarly, $m(AC)$, $m(AD)$, ... $m(EF)$ can be determined. Hence the new distance matrix is as in Table 10.13.

Table 10.13 Resulting new distance matrix of Problem P-10.8.

OTU	*A*	*B*	*C*	*D*	*E*
B	−76.25				
C	−77.50	−76.25			
D	−77.50	−76.25	−72.50		
E	−70.00	−68.75	−70.00	−70.00	
F	−66.25	−70.00	−71.25	−71.25	−88.75

Common values cluster together leading to the above modified matrix, so that the tree can be constructed by linking the least distant pair of nodes. The neighbors are chosen such that m_{ij} is the smallest (implying the least value) in the matrix. In this problem, it is at [*B*, *C*]. With *B* and *C* as neighbors, a new node *U* is found, and from this internal node *U*, the branch lengths $d(BU)$ and $d(CU)$ are deduced as follows: $d(BU) = d(BC)/2 + r(A) - r(B)/[2 \times (N - 2)]$ and $d(CU) = d(BC) - d(BU)$. Similarly, other distances are determined as follows: $[d(UA) = d(BA) + d(CA) - d(BC)/2]$; $[d(UC) = d(BC) + d(BD) - d(BC)/2]$; $[d(UD) = d(BD) + d(CD) - d(BC)/2]$; and, $[d(UF) = d(BF) + d(CF) - d(BC)/2]$. These calculations are continued to find the resulting final NJ tree.

NJ tree construction: the principle of this method is to find pairs of OTUs (=neighbors) that would minimize the total branch length at each stage of clustering of OTUs starting with a star-like tree. NJ is a bottom-up (agglomerative) clustering method for the creation of phylogenetic trees, created by Saitou and Nei in 1987.[3] The pseudocode in Table 10.14 illustrates the procedure on reconstruction of a phylogenetic tree *via* iterative algorithm of NJ method.

Table 10.14 Reconstruction of a phylogenetic tree: iterative algorithm of neighbor-joining (NJ) method.

Initialize

Input

→ $A[n \times n[$ distance matrix $\Delta_{i,j} = [d_{i,j}\ (T)]$ for every two leaves (taxa) (i,j)

Do

Iteration (I) over $k = 1, 2, \ldots, r$ over r taxa

← Each I involves the following calculation

Calculate

Q matrix based on the current distance matrix

← For a given distance $d(i, j)$ between taxa i and j and using the corresponding distance matrix relating r taxa, Q matrix is specified as follows:

$$\leftarrow \quad Q_{i,j} = (r-2)d(i,j) - \sum_{k=1}^{r} d(i,k) - \sum_{k=1}^{r} d(j,k) \qquad \text{Eq. (10.11)}$$

Determine

→ The pair of taxa with the lowest value in Q

Create

→ A node on the tree that joins this pair of taxa.

← The closest neighbors are joined

Calculate

→ The distance of each of the taxa in the pair to this new node

Determine

→ The distance of all taxa outside of this pair to the new node

GoTo

→ I $(k = 1, 2, \ldots, r)$, considering the pair of joined neighbors as a single taxon and using the distances calculated in the previous step.

Stop ← Termination of r**th** iteration

End

10.8 Concluding Remarks

Phylogeny stands out as a primary bioinformatic tools in elaborating the systematics of genetics and in the evolution of life-tree reconstruction. Pertinent efforts are enabled by the concepts of phylogenies estimated from the aligned sets of multiple DNA sequence data toward reconstruction of phylogenetic trees as outlined in this chapter.

In all, classification of life conforms to a tree-of-life that gets pruned with the passage of time, and it becomes bushier through the expanding knowledge bases of biology, and as it creeps into intricacies of microbiology, concepts of genetics, heuristics of bioinformatics, and thematics of phylogeny, relevant convergence themed as the gist of narrations outlined in this chapter; a tree-of-life is always a tree-of-evolution, no matter how it is looked at!

References

[1] W. J. Ewens and G. R. Grant: *Statistical methods in Bioinformatics — An Introduction.* Springer Science + Business Media Inc., Berlin/Heidelberg, Germany: 2005.

[2] G. Li: Generation of Rooted Trees and Free trees. *MS. Thesis,* Department of Computer Science, University of Victoria, 1996. Available on-line at: *http://webhome.cs.uvic.ca/~ruskey/Theses/GangLiMScThesis.pdf.*

[3] N. Saitou and M. Nei: The neighbor-joining method: A new method for reconstruction of phylogenetic trees. *Molecular Biology and Evolution,* 1987, vol. 4, 406–425.

[4] D. M. Mount: *Bioinformatics: Sequence and Genome Analysis.* Cold Spring Harbor Laboratory Press: Cold Spring Harbor, New York, NY, USA: 2004.

[5] J. Felsenstein: *Inferring Phylogenies.* Sinauer Associates, Sunderland, MA, USA: 2004.

[6] K. Atteson: The performance of neighbor-joining methods of phylogenetic reconstruction. *Algorithmica,* 1999, vol. 25, 251–278.

[7] R. Sokal and C. Michener: A statistical method for evaluating systematic relationships. *University of Kansas Science Bulletin,* 1958, vol. 38, 1409–1438.

[8] C. D. Michener and R. R. Sokal: A quantitative approach to a problem of classification. *Evolution,* 1957, vol. 11, 490–499.

[9] E. Zuckerkandl and L. B. Pauling: Evolutionary divergence and convergence in proteins. In V. Bryson and H. J. Vogel (Eds.), *Evolving Genes and Proteins.* Academic Press, New York, NY, USA: 1965, pp. 97–166.

[10] J. A. Studier and K. J. Kepler: A note on the neighbour-joining method of Saitou and Nei. *Molecular Biology and Evolution,* 1988, vol. 5, 729–731.

[11] W. M. Fitch and E. Margoliash. Construction of phylogenetic trees. vol. 55, *Science,* 1967, 279–284.

[12] T. H. Jukes and C. R. Cantor: Evolution of protein molecules. In H. N. Munro (Ed.), *Mammalian Protein Metabolism.* Academic Press, New York, NY, USA: 1969. pp. 21–132.

[13] M. Nei and S. Kumar: *Molecular Evolution and Phylogenetics.* Oxford University Press, New York, NY, USA: 2000.

[14] M. Kimura: A simple method forestimating evolutionary rates of base substitutions through comparative studies of nucleotide sequences. *Journal of Molecular Evolution,* 1980, vol. 16(2), 111–120.

[15] W. H. E. Day: Computational complexity of inferring phylogenies from dissimilarity matrices. *Bulletin of Mathematical Biology,* 1986, vol. 49, 461–467.

[16] J. Felsenstein: Phylogenies from molecular sequences: Inference and reliability. *Annual Review of Genetics,* 1988, vol. 22, 521–565.

[17] *Mammal* — Wikipedia, the free encyclopedia. Available on-line at: *https://en.wikipedia.org/wiki/Mammal*

[18] J. S. Farris: Methods for computing Wagner trees. *Systematic Zoology,* 1970, vol. 19, 83–92.

[19] F. M. Fitch: Toward defining the course of evolution: Minimum change for a specified tree topology. *Systematic Zoology,* 1971, vol. 20(4), 406–416.

[20] J. Felsenstein: Cases in which parsimony or compatibility methods will be positively misleading. *Systematic Zoology,* 1978, vol. 27(4), 401–410.

[21] S. L. Baldauf: Phylogeny for the faint of heart: A tutorial. *Trends in Genetics,* 2003, vol. 19(6), 345–361.

[22] P. S. Soltis and D. E. Soltis: Applying the bootstrap in phylogeny reconstruction. *Statistical Science,* 2003, vol. 18(2), 256–267.

[23] A. C. Davison: *Bootstrap Methods and Their Application.* Cambridge Series in Statistical and Probabilistic Mathematics. Cambridge University Press, Cambrige, UK: 1997.

[24] B. R. Efron and J. Tibshirani: *An Introduction to the Bootstrap.* Chapman & Hall/CRC Press: Boca Raton, FL, USA: 1998.

[25] R. R. Wilcox: *Fundamentals of Modern Statistical Methods: Substantially Improving Power and Accuracy.* Springer Science + Business Media LLC, New York, NY, USA: 2010.

[26] A. C. Davison: *Bootstrap Methods and Their Application.* Cambridge Series in Statistical and Probabilistic Mathematics. Cambridge University Press, Cambridge, UK: 1990.

[27] C. F. J. Wu: Jackknife, bootstrap and other resampling methods n regression analysis (with discussions). *Annals of Statistics,* 1986, vol. 14, 1261–1350.

[28] B. Efron: *The jackknife, the bootstrap, and other resampling plans.* Society of Industrial and Applied Mathematics CBMS-NSF Monographs, 1982, vol.38.

[29] C. Venditti, A Meade and M. Pagel: Detecting the node density artifact in phylogeny reconstruction. *Systematic Biology,* 2006, vol. 55(4), 637–643.

[30] Z. Yang and B. Rannala: Bayesian phylogenetic inference using DNA sequences: A Markov Chain Monte Carlo Method. *Molecular Biology Evolution*, 1997, vol. 14(7), 717–724.

[31] W. K. Hastings: Monte Carlo sampling methods using Markov chains and their applications. *Biometrika*, 1970, vol. 57, 97–109.

[32] C. J. Geyer: Markov Chain Monte Carlo maximum likelihood. In E. M. Keramidas and S. M. Kaufman (Eds.), *Computing Science and Statistics: Proceedings of the 23rd Symposium on the Interface*, Seattle Wa. USA, April, 21–24, 1991, Interface Foundation of North America, Fairfax Station, VA, USA: 1991, pp. 156–163.

[33] B. Larget and D. L. Simon: Markov Chain Monte Carlo algorithms for the Bayesian analysis of phylogenetic trees. *Molecular Biology Evolution*, 1991, vol. 16, 750–759.

[34] Neighbor joining Method. In Wikipedia, (the free encyclopedia) — Available on-line at: *https://en.wikipedia.org/wiki/Neighbor_joining*

[35] B. Larget: An Introduction to Phylogenetics — Available on-lie at: *https://www.stat.wisc.edu/~larget/phylogeny.pdf*

[36] NCBI Blast — On-line URL: *http://www.ncbi.nlm.nih.gov/blast*

[37] C. Notredame, D. Higgins and J. Heringa: T-Coffee: A novel method for multiple sequence alignments. *Journal of Molecular Biology*, 2000, vol. 302, 205–217.

[38] M. Nei: Genetic distance between populations. *American Naturalist*, 1972, vol. 106, 238–292.

[39] E. J. Bruce and F. J. Ayala: Phylogenetic relationships between the man and the apes: Electrophoretic evidence. *Evolution*, 1979, vol. 33(4), 1040–1056.

[40] MEGA — Available on-line at: *https://www.megasoftware.net/webhelp/walk_through_mega/estimating_evolutionary_distances_from_nucleotide_sequences.htm*

[41] Z. Cheng, M. Ventura, X. She, P. Khaitovich, T. Graves, K. Osoegawa, D. Church, D. P. DeJong, R. K. Wilson, S. Paabo, M. Rocchi, E. E. Eichler: A genome-wide comparison of recent chimpanzee and human segmental duplications. *Nature*, 2005, vol. 437 88–93.

[42] G.V. Glazko and M. Nei: Estimation of divergence times for major lineages of primate species. *Molecular Biology and Evolution*, 2003, vol. 20(3), 424–434.

[43] ExPASy. Available on-line at: *www. Expasy.org/phylogeny evolution ExPASy*

Bibliography

A

E. Aarts and J. Korst: *Simulated Annealing and Boltzmann Machines.* A John Wiley & Sons, Ltd., New York, NY, USA: 1989.

E. Ackerman: *Biophysical Science.* Prentice-Hall, Ltd., Englewood Cliffs, NJ, USA: 1962.

G. Alexander: *General Zoology.* Barnes and Nobel, Inc., New York, NY, USA: 1951.

E. S. Allman and J. A. Rhodes: *Mathematical Models in Biology: An Introduction.* Cambridge University Press, Cambridge, UK: 2004.

T. K. Attwood, D. J. Parry-Smith and S. Phukan: *Introduction to Bioinformatics.* Pearson Education Ltd., Chennai, TN, India: 1999.

M. Auxiliya: *Aritificial Intelligence.* A. R. S. Publications, Chennai, TN, India: 2009.

B

H. P. Bal: *Bioinformatics Principles and Applications.* Tata McGraw-Hill Ltd., New Delhi, India: 2005.

H. P. Bal: *Programming for Bioinformatics.* Tata McGraw-Hill Publishing Company Ltd., New Delhi, India: 2003.

P. Baldi and S. Brunak: *Bioinformatics the Machine Learning Approach.* The MIT Press, Cambridge, MA, USA: 2001.

A. Barani and Hari: *Projects in Bioinformatics.* Scitech Publications Pvt. Ltd., Chennai, TN, India: 2005.

A. D. Baxevanis and B. F. Francis Ouellette: *Bioinformatics A Practical Guide to the Analysis of Genes and Proteins.* A John Wiley & Sons, Inc., Publication, Hoboken, NJ, USA: 2005.

G. Benson and R. Page: *Algorithms in Bioinformatics.* Springer-Verlag, Heidelberg, Germany: 2004.

S. A. Berger, W. Goldsmith and E. R. Lewis: *Introduction to Bioengineering.* Oxford University Press, Inc., New York, NY, USA: 1996.

B. Bergeron: *Bioinformatics Computing.* Pearson Education, Inc., Upper Saddle River, NJ, USA: 2003.

R. Bhar and S. Hamori: *Hidden Markov Models: Applications to Financial Economics.* Kluwer Academic Publications, Norwell, MA, USA: 2004.

C. M. Bishop: *Neural Networks for Pattern Recognition.* Oxford University Press, Inc., New York, NY, USA: 1995.

A. Borem, F. R. Santos and D. E. Bowen: *Understanding Biotechnology.* Pearson Education, Inc., Upper Saddle River, NJ, USA: 2003.

P. E. Bourne and H. Weissig: *Structural Bioinformatics.* John Wiley & Sons, Inc., Somerset, NJ, USA: 2003.

B. P. Buckles and F. E. Petry: *Genetic Algorithms.* The Institute of Electrical and Electronics Engineers, Inc., CA, USA: 1992.

J. A. Bucklew: *Large Deviation Techniques in Decision, Simulation and Estimation.* A John Wiley & Sons, Inc., New York, NY, USA: 1990.

J. Burdge: *Chemistry.* The McGraw-Hill Companies, Inc., New York, NY, USA: 2009.

C

N. A. Campbell: *Biology.* The Benjamin/Cummings Publishing Company, Inc., San Francisco, CA, USA: 1987.

A. M. Campbell and L. J. Heyer: *Discovering Genomics, Proteomics and Bioinformatics.* Pearson Education, Inc., New Delhi, India: 2007.

N. A. Campbell, J. B. Reece and L. G. Mitchell: *Biology.* An imprint of Addison Wesley Longman, Inc., CA, USA: 1987.

C. R. Cantor and C. L. Smith: *Genomics, The Science and Technology Behind the Human Genome Project.* John Wiley and Sons, New York, NY, USA: 1999.

A. Chattopadhyay: *Mathematics and Statistics in Engineering and Technology.* Narosa Publishing House, New Delhi, India: 1999.

S. Choudhuri: *Bioinformatics for Beginners: Genes, Genomes, Molecular Evolution, Databases and Analytical Tools.* Elsevier, Inc., San Diego, CA, USA: 2014.

W. E. L. Clark: *History of Primates.* The University of Chicago Press, Chicago, IL, USA: 1949.

J. Claverie and C. Notredame: *Bioinformatics for Dummies.* Wiley Publishing, Inc., New York, NY, USA: 2003.

D

S. Datta: *Operation Gene.* Publications and Information Directorate, New Delhi, India: 1994.

A. Datta and S. Bhattacharya: *Gene Power.* Publications & Information Directorate, New Delhi, India: 1995.

G. Deco and D. Obradovic: *An Information-Theoretic Approach to Neural Computing.* Springer-Verlag, Inc., New York, NY, USA: 1996.

M. M. Domach: *Introduction to Biomedical Engineering.* Pearson Education, Inc., Upper Saddle River, NJ, USA: 2010.

R. C. Dubey: *A Textbook of Biotechnology.* S. Chand and Company Ltd., New Delhi, India: 1993.

E

P. R. Ehrlich, R. W. Holm and I. L. Brown: *Biology and Society.* McGraw-Hill, Inc., USA: 1976.

D. A. Evans and V. L. Patel: *Cognitive Science in Medicine.* The MIT Press, Boston, MA, USA: 1989.

W. J. Ewens and G. R. Grant: *Statistical Methods in Bioinformatics: An Introduction.* Springer Science + Business Media, Inc., New York, NY, USA: 2005.

F

A. Feinstein: *Foundations of Information Theory.* McGraw-Hill Book Company, Inc., New York, NY, USA: 1958.

G

N. Gautham: *Bioinformatics Databases and Algorithms.* Narosa Publishing House Pvt. Ltd., New Delhi, India: 2006.

Z. Ghosh and B. Mallick: *Bioinformatics Principles and Applications.* Oxford University Press, New Delhi, India: 2008.

C. Gibas and P. Jambeck: *Bioinformatics Computer Skills.* O'Reilly and Associates, Inc., Sebastopol, CA: 2001.

S. A. Glantz: *Primer of Biostatistics.* The McGraw-Hill Companies, Inc., New York, NY, USA: 2005.

S. Gopal, A. Haake, R. P. Jones and P. Tymann: *Bioinformatics A Computing Perspective.* The McGraw-Hill, Inc., New York, NY, USA: 2009.

V. A. Greulach: *Botany Made Simple.* Doubleday and Company, Inc., New York, NY, USA: 1968.

J. S. Griffith: *Mathematical Neurobiology: An Introduction to the Mathematics of the Nervous System.* Academic Press Ltd., London, UK: 1971.

M. M. Gromiha: *Protein Bioinformatics: From Sequence to Function.* Elsevier, India Pvt. Ltd., New Delhi, India: 2010.

H

E. R. Hanauer: *Biology: Made Simple.* W. H. Allen & Company, Inc., London, UK: 1967.

M. He and S. Petoukhov: *Mathematics of Bioinformatics Theory, Practice and Applications.* John Wiley and Sons, INC, Hoboken, NJ, USA: 2011.

P. G. Higgs and T. K. Attwood: *Bioinformatics and Molecular Evolution.* Blackwell Science Ltd., Oxford, UK: 2005.

M. Ho, F. Popp and U. Warnke: *Bioelectrodynamics and Bio communication.* World Scientific Publishing Co., Pte. Ltd., Singapore: 1994.

X. Hu and Y. Pan (Ed.): *Knowledge Discovery in Bioinformatics: Techniques, Methods, and Applications Wiley Interscience.* John Wiley and Sons, Hoboken, NJ, USA: 2007.

S. E. Huether and K. L. McCane: *Understanding Pathophysiology.* Mosby, Inc., St. Louis, MO, USA: 2004.

S. P. Hunt and F. J. Livesey: *Functional Genomics.* Oxford University Press, Inc., New York, NY, USA: 2000.

D. Husmeier and R. Dybowski: *Probabilistic Modeling in Bioinformatics and Medical Informatics.* Springer Publishing, New York, NY: USA 2005.

I

S. Ignacimuthu: *Basic Bioinformatics.* Narosa Publishing House Pvt. Ltd., New Delhi, India: 2005.

A. Isaev: *Introduction to Mathematical Methods in Bioinformatics.* Springer-Verlag, Heidelberg, Germany: 2004.

J

D. C. Jamison: *Perl Programming for Bioinformatics and Biologists.* John Wiley and Sons Ltd., Chichester, UK: 2003.

G. B. Johnson: *The Living World.* The McGraw-Hill Companies, Inc., New York, NY, USA: 2000.

N. C. Jones and P. A. Pavzner: *An Introduction to Bioinformatics Algorithms.* Ane Books Pvt. Ltd., New Delhi, India: 2005.

P. Joshi: *Genetic Engineering and its Applications.* Agrobios, India: 2003.

K

K. Kakollu: *Bioinformatics.* Pragati Prakashan Educational Publishers, Meerut, UP, India: 2006.

J. N. Kapur and H. K. Kesavan: *Entropy Optimization Principles with Applications.* Academic Press, Inc., San Diego, CA, USA: 1992.

J. C. Kendrew: *The Thread of Life: An Introduction to Molecular Biology.* Harvard University Press, Boston, MA, USA: 1966.

I. A. Khan: *Elementary Bioinformatics.* Pharma Book Syndicate, Hyderabad, AP, India: 2005.

I. A. Khan and A. Khanum: *Recent Advances in Bioinformatics.* Ukazz Publications, Hyderabad, AP, India: 2002–03.

R. C. King: *Genetics.* Oxford University Press, Inc., New York, NY, USA: 1962.

M. E. Kinn and M. A. Woods: *The Medical Assistant, Administrative and Clinical.* W. B. Sauders Company, USA: 1999.

G. J. Klir and T. A. Folger: *Fuzzy Set, Uncertainty and Information.* Prentice Hall, Englewood Cliffs, NJ, USA: 1988.

W. S. Klug and M. R. Cummings: *Concepts of Genetics.* Macmillan Publishing Company, New York, NY, USA: 1991.

S. Knudsen: *A Biologist's Guide to Analysis of DNA Microarray Data.* A John Wiley & Sons, Inc., New York, NY, USA: 2002.

N. Kolchanov and R. Hofestaedt: *Bioinformatics of Genome Regulation and Structure.* Kluwer Academic Publishers, New Delhi, India: 2004.

R. R. Kouser and B. Gunasundari: *Data Warehousing and Data Mining.* Lakshmi Publications, Chennai, TN, India: 2013.

D. E. Krane and M. L. Raymer: *Fundamental Concepts of Bioinformatics.* Pearson Education, Inc., Upper Saddle River, NJ, USA: 2003.

H. D. Kumar: *Molecular Biology.* Vikas Publishing House Pvt. Ltd., Noida, UP, India: 1998.

V. Kumaresan: *Biotechnology.* Saras Publication, Nagercoil, TN, India: 1994.

L

R. W. Larsen: *Engineering with Excel.* Pearson Education, Inc., Upper Saddle River, NJ, USA: 2013.

K. Layon: *Mobilizing Web Sites: Develop and Design.* Peachpit Press, San Francisco, CA, USA: 2012.

A. M. Lesk: *Introduction to Bioinformatics.* Oxford University Press, New Delhi, India: 2002.

C. C. Lin and L. A. Segal: *Mathematics Applied to Deterministic Problems in the Natural Sciences.* Macmillan Publishing Co., Inc., New York, NY, USA: 1974.

P. S. Lohar: *Bioinformatics.* MJP Publishers, Chennai, TN, India: 2009.

M

K. Mani and N. Vijayaraj: *Bioinformatics A Practical Approach.* Aparnaa Publication, TN, India: 2004.

K. Mani and N. Vijayaraj: *Bioinformatics: For Beginners.* Kalaikathir Achchagam, Coimbatore, TN, India: 2002.

K. Manikandakumar: *Dictionary of Bioinformatics.* MJP Publishers, Chennai, TN, India: 2009.

S. Markel and D. Leon: *Sequence Analysis in a Nutshell.* O'Reilly & Associates, Inc., Cambridge, MA, USA: 2003.

D. W. Mount: *Bioinformatics Sequence and Genome Analysis.* Cold Spring Harbor Laboratory Press, New York, NY, USA: 2004.

N

S. Nanavati, M. Thieme and R. Nanavati: *Biometrics: Identify Verification in a Networked World.* John Wiley and Sons, Inc., Toronto, ON, Canada: 2002.

A. J. Nair: *Introduction to Biotechnology and Genetic Engineering.* Infinity Science Press LLC, Sudbury, MA, USA: 2008.

P. Narayanan: *Bioinformatics A Primer.* New Age International (P) Ltd., New Delhi, India: 2005.

P. S. Neelakanta: *Information Theoretic Aspects of Neural Networks.* CRC Press, Inc., Boca Raton, FL, USA: 1999.

P. S. Neelakanta and D. De Groff: *Neural Network Modeling: Statistical Mechanics and Cybernetic Perspectives.* CRC Press, Inc., Boca Raton, FL, USA: 1994.

O

J. H. Otto and A. Towle: *Modern Biology.* Holt, Rinehart and Winston Publishers, New York, NY, USA: 1977.

P

S. Palanichamy and M. Shunmugavelu: *Outlines of Biological Chemistry and Biotechniques.* Palani Paramount Publications, Palani, TN, India: 1933.

Y. H. Pao: *Adaptive Pattern Recognition and Neural Networks.* Addision-Wesley Publishing Company, Inc., Boston, MA, USA: 1989.

G. Paun, G. Rozenberg and A. Salomaa: *DNA Computing New Computing Paradigms.* Springer-Verlag Berlin-Heidelberg, Germany: 1998.

M. Pavlovic and B. Balint: *Bioengineering and Cancer Stem Cell Concept.* Springer, New York, NY, USA: 2015.

J. K. Percus: *Mathematics of Genome Analysis.* Cambridge University Press, Cambridge, UK: 2002.

P. Peters: *Biotechnology: A Guide to Genetic Engineering.* WCB/McGraw-Hill Companies, New York, NY, USA: 1993.

P. A. Pevzner: *Computational Molecular Biology: An Algorithmic Approach.* The MIT Press, Cambridge, MA, USA: 2000.

J. Pfeiffer: *The Cell.* Time Incorporated, New York, NY, USA: 1964.

Q

R

A. Ragland and N. Arumugam: *Biochemistry and Biophysics.* Saras Publication, Nagercoil, TN, India: 2000.

J. J. Ramsden: *Bioinformatics: An Introduction.* Springer Science + Business Media, Inc., New York, NY, USA: 2004.

M. M. Ranga: *Bioinformatics.* Agrobios, Jodhpur, RN, India: 2003.

H. H. Rashidi and L. K. Buehler: *Bioinformatics Basics: Applications in Biological Science and Medicine.* CRC Press LLC, Boca Raton, FL, USA: 2000.

S. C. Rastogi, N. Mendiratta and P. Rastogi: *Bioinformatics Concepts, Skills & Applications.* CBS Publishers & Distributers Pvt. Ltd., New Delhi, India: 2003.

S. C. Rastogi, N. Mendiratta and P. Rastogi: *Bioinformatics: Methods and Applications Genomics, Proteomics and Drug Discovery.* Prentice-Hall of India Private Ltd., New Delhi, India: 2006.

J. Rifkin: *Entropy A New World View.* The Viking Press, New York, NY, USA: 1980.

D. Roy: *Bioinformatics.* Narosa Publishing House Pvt. Ltd., New Delhi, India: 2009.

S

D. R. Sanadi: *Current Topics in Bioenergetics.* Academic Press, Inc., New York, NY, USA: 1979.

V. C. Scanlon and T. Sanders: *Essentials of Anatomy and Physiology.* F. A. Davis Company, Philadelphia, PA, USA: 1999.

R. Schalkoff: *Pattern Recognition: Statistical, Structural and Neural Approaches.* John Wiley & Sons, Inc., New York, NY, USA: 1992.

M. Schena: *Microarray Analysis.* A John-Wiley and Sons, Inc., Edison, New Jersey: 2003.

C. W. Sensen: *Essentials of Genomics and Bioinformatics.* Wiley-Vch Verlag, GmbH, Weinheim, Germany: 2002.

H. Shimizu: *Biological Complexity and Information.* World Scientific Publishing Co. Pte. Ltd., Singapore: 1990.

G. G. Simpson: *The Meaning of Evolution.* The new American Library of World Literature, Inc., New York, NY, USA: 1951.

R. Sing and R. Sharma: *Bioinformatics Basics, Algorithms and Applications.* Universities Press Pvt. Ltd., India: 2010.

S. Singhal and S. Singhal: *Bioinformatics.* Pragati Prakashan Educational Publishers, Meerut, UP, India: 2006.

J. E. Smith: *Biotechnology.* Cambridge University Press, Cambridge, UK: 1996.

V. R. Srinivas: *Bioinformatics A Modern Approach.* Prentice-Hall of India Pvt. Ltd., New Delhi: 2005.

W. D. Stansfield: *Theory and Problems of Genetics,* 3rd ed. The McGraw-Hill Companies, Inc., USA: 1991.

M. L. Steinberg and S. D. Cosloy: *Dictionary of Biotechnology and Genetic Engineering.* Jaico Publishing House, Mumbai, India: 2003.

D. Stekel: *Microarray Bioinformatics.* The Press Syndicate of the University of Cambridge, Cambridge, UK: 2003.

P. Strong: *Biophysical Measurements.* Tektronix, Inc., Oregon, USA: 1970.

L. Stryer: *Biochemistry.* W. H. Freeman and Company, Westminster, MD, USA: 1975.

Y. M. Svirezhev and D. O. Logofet: *Stability of Biological Communities.* MIR Publishers, Moscow, Russia: 1983.

T

T. Thiel, S. T. Bissen and E. M. Lyons: *Biotechnology: DNA to Protein.* Tata McGraw-Hill Publishing Company Ltd., New Delhi, India: 2002.

G. J. Tortora, B. R. Funke and C. L. Case: *Microbiology: An Introduction.* An Imprint of Addison Wesley Longman, Inc., Chicago, IL, USA: 1988.

A. Tozeren and S. W. Byers: *New Biology for Engineers and Computer Scientists.* Pearson Education, Inc., Hoboken, NJ, USA: 2004.

H. C. Tuckwell: *Stochastic Process in the Neurosciences.* Society for Industrial and Mathematics, Philadelphia, PA, USA: 1989.

U

V

K. M. Van De Graaff and J. L. Crawley: *A Photographic Atlas for the Biology Laboratory.* Morton Publishing Company, Englewood, CO, USA: 1994.

E. O. Voit: *A First Course in Systems Biology.* Garland Science, Taylor & Francis Group, LLC, New York, NY, USA: 2013.

W

J. M. Walker and R. Rapley: *Molecular Biology and Biotechnology.* Panima Publishing Corporation, New Delhi, India: 2002.

J. G. Webster: *Medical Instrumentation: Application and Design.* A John Wiley & Sons, Inc., New York, NY, USA: 1998.

D. R. Westhead, J. H. Parish and R. M. Twyman: *Bioinformatics.* Viva Books Private Ltd., India: 2003.

E. Wong: *Stochastic Process in Information and Dynamical Systems.* Robert E. Krieger Publishing Company, New York, NY, USA: 1971.

P. M. Woodward: *Probability and Information Theory, with Applications to Radar.* Pergamon Press Ltd., London, UK: 1953.

X

D. Xu, J. M. Keller, M. Popescu and R. Bondugula: *Application of Fuzzy Logic in Bioinformatics.* Imperial College Press, London, UK: 2008.

Y

Z

S. Zodgekar: *Bioinformatics the Future Science.* The Icfai University Press, Hyderabad, India: 2008.

W. Zucchini and L. MacDonald: *Hidden Markov Models for Time Series: An Introduction Using R.* CRC Press, Boca Raton, FL, USA: 2018.

Index

www.ingramcontent.com/pod-product-compliance
Lightning Source LLC
Chambersburg PA
CBHW052115230326
41598CB00079B/3682